Beginning Statistics: A to Z

Beginning Statistics: A to Z

William Mendenhall
University of Florida, Emeritus

An Alexander Kugushev book

Duxbury Press
An Imprint of Wadsworth Publishing Company
Belmont, California

Print Buyer: Karen Hunt
Designer: Nancy Benedict Graphic Design
Technical Illustrator: Lori Heckelman
Cover: Vargas/Williams/Design
Cover Background: West Light
Compositor: Polyglot Compositors, Pte., Ltd.; GTS Graphics, Inc.
Printer: R. R. Donnelley

This book is printed on acid-free recycled paper.

Duxbury Press
An Imprint of Wadsworth Publishing Company
A division of Wadsworth, Inc.

Printed in the United States of America

1 2 3 4 5 6 7 8 9 10—97 96 95 94 93

Library of Congress Cataloging-in-Publication Data

Mendenhall, William.
Beginning statistics—A to Z / William Mendenhall.
p. cm.
Includes index.
ISBN 0-534-19122-3
1. Statistics. I. Title. II. Title: Beginning statistics.
QA276.12.M45 1993
51925—dc20 92-31635
CIP

preface

Most of us would like our students in an introductory course to acquire a usable understanding of statistics that stays with them one, two, or five years (or even longer) after they have completed the course. This may be an impossible dream but a change in emphasis may bring us closer to this goal.

Do we need to spend so much time on the basic theory of probability? Does the student need to grind and grind through numerous calculations when, in this day of the computer, few, if any, calculations are performed on a hand calculator?

A student should understand the role that probability plays in statistical inference. But how much probability theory does a student really need to know in order to understand these concepts? Calculating our way through a simple example to better understand why and how a statistic works is useful, but maybe one is enough. The time that we currently spend on probability and calculating will be better spent on statistical concepts, learning about some of the most useful statistical methods, how and why they work, and, most important, what they mean.

This text is designed to focus on **concepts, statistical analyses,** and their **interpretation.** Downplaying probability and calculating increases the time available for these topics. It also allows us to increase course coverage. The book presents a complete introduction to statistics and statistical methods in one term. We would expect courses to include the standard topics contained in Chapters 2–7 along with all or most of the methods chapters that follow.

A Comment on Probability

We use the concept of the "rare" (i.e., improbable) outcome throughout the text to explain the reasoning employed in statistical inference. The student learns in Chapter 3 that it is very improbable that an observation will fall more than three standard deviations from its mean and we use this concept to explain how control charts are used in process quality control. Similarly, sampling distributions are used to identify rare or improbable values of statistics and enable us to establish upper limits on errors of estimation (what the news media call sampling error). We apply the same reasoning when we base a decision in a statistical test on the p-value of the observed test statistic. These difficult inferential concepts are introduced in Chapter 5 and reinforced repeatedly in the methods chapters that follow.

A Comment on the Use of Data

Most examples and exercises are based on data selected from newspapers or research journals. These data add realism to the presentation and, more important, provide evidence that a knowledge of statistics is a very helpful tool for most professions. They also provide realistic data sets for instructor-generated examples and exercises.

A computer printout containing the relevant statistics for each confidence interval or statistical test is shown and interpreted. The student is also shown how to calculate pertinent statistics because many journal articles include only sample descriptive statistics. These examples and exercises can be omitted to further reduce calculations if the instructor wishes to do so.

A Comment on Flash Cards

We think with words and concepts! True, we can learn the words and concepts in the process of "doing"—i.e., solving problems, etc.—but it won't be as fast or as efficient as memory reinforcement *along with* the "doing." To this end, the student is supplied with "flash (memory) cards" that are shrink-wrapped with the book. They are accompanied by an explanation on their use.

Supporting Materials

The following supporting material is available for the instructor:

1. An instructor's manual that contains suggestions by the author on the use of the text and solutions to the exercises
2. A floppy disk that contains the three data sets listed in Appendix 1 of the text
3. A teaching disk that can be used to demonstrate statistical concepts (e.g., the concept of a sampling distribution, the Central Limit Theorem, etc.)
4. An examination bank

The author acknowledges with thanks the assistance of many talented people. Ken Brown (College of San Mateo), Carolyn Apperson Hansen (University of Virginia Health Sciences Center), Laurence Johnson (Cuesta College), James Lang (Valencia Community College), Austin Meek (Canada College), Kirk Steinhorst (University of Idaho), and Bruce Trumbo (California State University, Hayward) provided early reviews of the manuscript and helpful suggestions for its improvement. Dennis Wackerly of the University of Florida provided many suggestions over lunch and also reviewed Chapter 11. One of his most useful suggestions was directing me to James Lang of Valencia Community College to obtain the excellent instructional disk that accompanies the text. I particularly thank Carolyn Hansen for producing the computer printouts, for finding the cholesterol data set, and for suggestions throughout the writing. I am grateful to Terry Sincich, J. J. Cerda, M.D., journal editors, and others for permission to use data sets and tables, and to Teresa Bittner of Laurel Tutoring for preparing the solutions manual and answers to exercises.

I thank Alex Kugushev, long-time editor and friend, for reading and critiquing early chapters of the manuscript and helping to improve my writing style. And, particularly, I acknowledge the outstanding and meticulous copyediting and comments of Susan Reiland. Finally, I thank my wife for her support and for tolerating my writing "just one more book."

table of contents

Preface v

1 To the Reader 1

1.1 The Role of Statistics in the Modern World 2

1.2 The Language of Statistics 3

1.3 The Role of the Computer in Statistics 4

1.4 How to Use This Book 4

2 Language, Purpose, and Uses of Statistics 6

2.1 Elements, Observations, and Variables 7
Qualitative variables—Quantitative variables

2.2 Universe, Population, and Sample 11

2.3 Bivariate and Multivariate Data 13
Fish contaminated with DDT in the Tennessee River—A University of Florida study of human cholesterol levels

2.4 The Objective of Statistics 16
How we make inferences about a population—Of what value is statistical inference?

2.5 The Design of Experiments and Sample Surveys 21

2.6 The Misuses of Statistics 22

2.7 Key Words and Concepts 23

3 Graphical Descriptions of Data 26

3.1 Describing Qualitative Data 27
Notation used in graphical description of data—Bar graphs and pie charts

3.2 Relative Frequency Distributions for Quantitative Data 34
Interpreting a relative frequency distribution—Probabilistic interpretations of areas under a relative frequency distribution—The sample and population relative frequency distributions—Constructing a relative frequency distribution—Computer-generated relative frequency distributions

TABLE OF CONTENTS

3.3 Graphical Descriptions of the Cholesterol Data Set, Case 2.3 43
Background information on body cholesterol—The cholesterol data set

3.4 Stem-and-Leaf Displays 64
How to construct them—Modified stem-and-leaf displays—Computer-generated stem-and-leaf displays

3.5 Key Words and Concepts 67

4 Numerical Descriptions of Quantitative Data 71

4.1 Why We Need a Numerical Method for Describing a Data Set 72

4.2 Measures of Central Tendency 74
The Mean—Some essential notation—The Median

4.3 Measures of Variation or Spread 78
The range—The variance—The standard deviation—The Empirical Rule—Calculating the variance and standard deviation for a data set—Computer printouts of the mean and measures of variation

4.4 Applying the Empirical Rule: Approximating a Standard Deviation 96

4.5 Applying the Empirical Rule: Processes and Statistical Control 98
A case—Statistical control—Control charts

4.6 Measures of Relative Standing 103
The terminology—The location of the median and skewness

4.7 Looking for Outliers 107
What is an outlier?—Detecting outliers: A box plot for the 30 TCHOL measurements—Outliers: What to do with them—Computer-generated box plots

4.8 Numerical Descriptions of TCHOL and RATIO, Case 2.3 110

4.9 Statistics and Parameters 116
Measuring error and goodness

4.10 Key Words and Concepts 118

5 How Probability Helps Us Make Inferences 122

5.1 Shifting Gears 123

5.2 Probability and Inference 123
What is probability?—How probability is used to make statistical inferences

5.3 Some Concepts of Probability 125
The relative frequency concept of probability—Mutually exclusive outcomes and the additive property of probabilities—Independent outcomes and the multiplicative property of probabilities—The complement of an outcome

5.4 Random Samples 129
A probability sample—A random sample—How to draw a random sample—Difficulties in drawing random samples

5.5 Probability Distributions 132
Discrete random variables and their probability distributions—Continuous random variables and their probability distributions

5.6 The Binomial Probability Distribution 136
Binomial experiments—The binomial probability distribution—The mean and standard deviation of a binomial probability distribution—An example of statistical inference using the binomial probability distribution—Binomial probability tables—Other discrete probability distributions

5.7 The Normal Probability Distribution 148
Finding areas under the normal curve—The standard normal z distribution—The normal approximation to the binomial and other probability distributions

5.8 Sampling Distributions 156
The sampling distribution of the sample mean—When will the sample size be large enough for the sampling distribution of $\bar{x}$ to be approximately normal?

5.9 Making Statistical Inferences: Estimation 163
Point estimators—Confidence intervals—How does probability play a role in statistical estimation?

5.10 Making Statistical Inferences: Tests of Hypotheses 167
Measuring risk: The probability of making the wrong decision—Choosing the value of α—An analogy—Reading computer printouts

5.11 Key Words and Concepts 172

6 Inferences About a Population Mean 175

6.1 The Problem 176

6.2 The z and the t Statistics 176
The sampling distribution of $\bar{x}$—The standard normal z statistic—Finding the probability that $\bar{x}$ deviates from μ by a specified amount—Limitations on the use of the z statistic—Student's t statistic—Properties of a Student's t distribution—A comparison of the z and the t distributions

6.3 Confidence Intervals for a Population Mean 189
How a confidence interval is constructed and why it works—A confidence interval for μ when σ is known—A confidence interval for μ when σ is unknown—An example of a small-sample confidence interval—Which method should we use?

6.4 Another Way to Report an Estimate 200

6.5 Why Test Hypotheses About a Population Mean? 205

6.6 Tests of Hypotheses About a Population Mean 206
One- and two-tailed tests for μ: What are they?—Large- and small-sample tests for a population mean μ

6.7 Choosing the Sample Size 217
Factors that affect the width of a confidence interval

6.8 Why Assumptions Are Important 222

6.9 Key Words and Concepts 222

7 Comparing Two Population Means 229

7.1 The Problem 230

7.2 Independent Random Sampling and Matched-Pairs Designs 230
The design of an experiment: What it is and why we do it—An independent random samples design—A matched-pairs design

7.3 Comparing Population Means: Independent Random Samples 234
The sampling distribution of $(\bar{x}_1 - \bar{x}_2)$—A large-sample confidence interval for $(\mu_1 - \mu_2)$—A small-sample confidence interval for $(\mu_1 - \mu_2)$—Tests of the difference between population means based on independent random samples—Comments on the assumptions—Which method should we use?

7.4 Comparing Population Means: A Matched-Pairs Design 259
Confidence intervals for the difference in population means based on matched pairs—Matched-pairs tests for a difference in population means—How much money can we save by using a matched-pairs design?

7.5 Choosing the Sample Size 270

7.6 Key Words and Concepts 273

8 Correlation and Linear Regression Analysis 279

8.1 The Problem 280

8.2 Scattergrams 281

8.3 A Linear Relation Between x and y 283
Describing the relationship between two variables—The equation of a line

8.4 The Method of Least Squares 288
Finding the best-fitting straight line for a set of data—The method of least squares—The least squares line

8.5 The Pearson Coefficient of Correlation 290
A coefficient of correlation: What it is—A practical interpretation of r^2

8.6 A Simple Linear Regression Analysis: Questions It Will Answer 299
What is a prediction interval?

8.7 A Simple Linear Regression Analysis: Assumptions 300
Comments on the assumptions

8.8 A Simple Linear Regression Analysis: Typical Computer Outputs 302
Finding the equation of the least squares line—An estimator of σ^2—A test and confidence interval for β_1—Testing the adequacy of the prediction equation—Calculating the p-value when testing the slope of a line—Estimating the slope of the least squares line—Estimating the mean value of y and predicting a particular value of y for a value of x—A word of caution

8.9 A Multiple Linear Regression Analysis 320
The general linear model—The computer output for a multiple regression analysis: What it contains—A multiple regression analysis: An example

8.10 Key Words and Concepts 329

9 Inferences from Qualitative Data 334

9.1 The Problem 335
The types of inferences we want to make

9.2 Assumptions 336
The characteristics of a multinomial experiment—The relationship between a binomial and a multinomial experiment

9.3 Inferences About a Population Proportion 338
Sampling procedure, symbols, and objectives—The sampling distribution of the sample proportion—A confidence interval and test for a population proportion *p*

9.4 Comparing Two Population Proportions 347
Sampling procedure, symbols, and objectives—The sampling distribution of the difference between two sample proportions—A confidence interval and test for the difference between two population proportions

9.5 Choosing the Sample Size 354

9.6 A Contingency Table Analysis 356
Definitions—The purposes of a contingency table—Step-by-step: How a chi-square test statistic is calculated—The chi-square test for contingency (dependence)—Contingency table analysis when either row or column totals are fixed—Finding the approximate *p*-value for a chi-square test

9.7 Key Words and Concepts 378

10 An Analysis of Variance for Designed Experiments 385

10.1 The Problem 386

10.2 How an Analysis of Variance Works 387

10.3 An ANOVA Table for Comparing Two or More Population Means 389

10.4 An ANOVA *F* Test for Comparing Two or More Population Means 393
Assumptions for an ANOVA—An analysis of variance *F* test—The tabulated values of *F*

10.5 Tests and Confidence Intervals for Individual Means 398

10.6 Multiple Comparisons: Tukey's HSD Procedure 402
The HSD procedure—Applying the HSD procedure to the drug potency data

10.7 Experimental Design: Factorial Experiments 406
A single-factor experiment—Multifactor experiments: The one-at-a-time approach—Factor interactions—Factorial experiments

10.8 The Analysis of Variance for a Two-Factor Factorial Experiment 410
The ANOVA table for *r* replications of a two-factor factorial experiment—A test for factor interaction—Tests for factor main effects—Comparing treatment means

10.9 Other Analyses of Variance 422

10.10 Key Words and Concepts 422

11 Nonparametric Statistical Methods 427

11.1 The Problem 428
Another look at a Student's t test—How a nonparametric test works

11.2 The Mann–Whitney U Test for Independent Random Samples 430
Comparing results for the Mann–Whitney U and the Student's t tests—A computer printout for a Mann–Whitney U test—The Mann–Whitney U test for large samples

11.3 The Wilcoxon Signed Ranks Test for a Matched-Pairs Design 441
The logic behind the Wilcoxon signed ranks test—A computer printout for a Wilcoxon signed ranks test

11.4 The Kruskal–Wallis H Test 449
A computer printout for a Kruskal–Wallis H test

11.5 The Spearman rank correlation coefficient 455
Spearman's rank correlation coefficient: What is it?—Interpreting the value of r_s—Computer printouts for a Spearman's rank correlation analysis—Consumer preference experiments

11.6 Key Words and Concepts 465

APPENDIX 1: Data Sets 471
Table 1 Iron Content (Percentage) of 390 1.5-Kilogram Specimens of Iron Ore from a 20,000-Ton Consignment of Canadian Ore 472
Table 2 DDT Analyses on Fish in the Tennessee River, 1980 474
Table 3 University of Florida Study of Body Cholesterol for 107 People 477

APPENDIX 2: Statistical Tables 480
Table 1 Random Numbers 481
Table 2 Binomial Probabilities (n = 5, 10, 15, 20, 25) 483
Table 3 Areas Under the Normal Curve 485
Table 4 Upper-Tail Values for the Student's t Distribution 486
Table 5 Upper-Tail Values of the Chi-Square Distribution 487
Table 6 Upper-Tail Values for an F Distribution 488
Table 7 Upper-Tail Values for the Studentized Range 498
Table 8 Lower-Tail Values for the Mann–Whitney U Statistic 504
Table 9 Lower-Tail Values for the Wilcoxon Signed Ranks Test 509
Table 10 Upper-Tail Values of Spearman's Rank Correlation Coefficient 510

APPENDIX 3: Answers to Selected Exercises 511

Index 522

Beginning Statistics: A to Z

one

To the Reader

1.1 The Role of Statistics in the Modern World
1.2 The Language of Statistics
1.3 The Role of the Computer in Statistics
1.4 How to Use This Book

1.1 The Role of Statistics in the Modern World

We live in a world described by numbers—numbers that monitor every aspect of our daily lives and of the world in which we live. The data to support this contention? See your daily newspaper, news magazine, or one of the many professional or scientific journals. The varied applications of statistics can be seen in the following examples.

Manufacturing

The Japanese did not achieve superiority over some segments of American industry solely on the basis of labor costs. They surpassed us on quality. To do this, they collected data on the quality characteristics of a product, say a Toyota automobile. The number of breakdowns per 100,000 miles, the causes of breakdowns, and many other data on the current quality of a model were recorded. Japanese engineers had to find out what the quality was today. Then, from an analysis of the data, they were able to make a better product for tomorrow. The tools used to deduce the necessary changes? The techniques of W. Edwards Deming, the world recognized statistician and expert on statistical quality control—and statistics.

The Social Sciences

An article in *The Gainesville Sun* (October 26, 1991) is headlined "Study Indicates Bias in Bar Exam." It goes on to say that Hari Swaminathan, a statistician at the University of Massachusetts, examined the test results of 3,777 law school graduates who took the bar exams that year. Seventy-five percent of the 3,554 white graduates passed the exams given in February and July. Only 39% of the black graduates passed the February exam; 46% passed the July tests. Is this difference in passing percentages due to random variation or is it due to a real difference in passing rates of whites and blacks? If a real difference exists in the passing rates, can we say that it is due to "bias"? What can statistics tell us about this complex question?

Business

In the fall of 1991, the economic picture looked bad to some economists. Why? The statistics that characterize the state of the economy—the unemployment rate, the measure of consumer confidence, new housing construction starts, consumer spending, etc.—were all headed in the wrong direction. How do we monitor the state of the economy? Statistics.

Politics

The New York Times (July 14, 1991) reports that, "Poll Finds G.O.P. Growth Erodes Dominant Role of the Democrats." The article indicates that of the 14,685 persons polled, 34% called themselves Democrats and 31% called themselves Republicans. These statistics, numbers that summarize and describe the political pulse of the nation, tell us the party affiliation of voters at the time the poll was taken. But the pulse changes from day to day. As an election draws near, polls are taken more and more frequently to monitor changes. The tool used to evaluate the changes and to deduce their implications? Statistics.

Research

A headline in *The Wall Street Journal* (November 5, 1991) states, "Pfizer Gets Clearance from FDA to Market Zithromax Antibiotic." This announcement of approval by the U.S. Food and Drug Administration (FDA) to market a new and powerful antibiotic carries an unspoken message. Behind the approval, as for all new food and pharmaceutical products, lie years of testing, by both the manufacturer and the FDA. The object is to prove that the new product works and to establish safe dosages for the consumer. The method used to draw these conclusions from experimental data? Statistics.

These examples illustrate just a few of the many applications of statistics. Most important, they show that statistics enters into almost all aspects of human endeavor. Why?

Statistics is about describing and predicting. It is a response to the limitations of our ability to know. Any complex aspect of human knowledge is either difficult for us to understand, or lends itself to misunderstanding—unless we possess the right tool with which to examine and analyze its complexity. Statistics is that tool. You will need it as you enter the 21st century.

1.2 The Language of Statistics

How does statistics differ from many other courses that you have taken, or will take, in college? Obviously, the subject matter is different, but it is more than that. The material in most courses in the social and biological sciences, business, etc., is presented in everyday language. You encounter new words and concepts, of course, but you have had some prior exposure to these subjects in reading news or magazine articles.

Statistics, in contrast, will seem like a new language. True, you have read articles similar to those described in Section 1.1. But it is likely that you have had minimal exposure to the language and concepts of statistics. Therefore, you need to develop a new attitude and be prepared to learn a new language.

We think with words and concepts, and one topic builds on another. **In order to follow your instructor's lectures and to read this book, you need to commit new words, symbols, formulas, and concepts to memory. You will also need to stay with the course daily.** If you do, you will find statistics interesting—in fact, fascinating and useful.

1.3 The Role of the Computer in Statistics

Statistical methods often involve extensive, tedious numerical calculations. Prior to the invention of the computer, these calculations were done on a desk calculator. All statistical courses taught students how to use the appropriate formulas and how to simplify the calculations, if possible. Some calculations were so lengthy that it was impossible to perform them on a desk calculator.

The advent of the computer solved this problem. It is still necessary to know what a statistical quantity is and to understand why and how it works. But, in this day and age, statistical calculations are done on a computer. Consequently, we will downplay calculating. Instead, we will learn how to read and interpret statistical computer printouts and published statistical statements. If you have access to a computer and a statistical software package, you may want to try producing your own output for a statistical analysis of some data.

There are a number of good statistical software packages available. The printouts for these packages differ in minor respects. Once you are familiar with one, you can easily learn how to read another. We will show primarily printouts for the Minitab, SAS, and Data Desk statistical software packages with a few others used for exercises.*

Some statistical software packages are better than others and some are easier to use. If you want to perform some statistical analyses on your computer, ask your instructor to recommend a suitable software package.

1.4 How to Use This Book

This book explains the basic concepts of statistics and shows how they are applied. The material in the text will be presented in the following format:

* Minitab is a statistical software package produced by Minitab, Inc., 215 Pond Laboratory, University Park, Pennsylvania 16802.

SAS is a statistical software package produced by the SAS Institute, Box 8000, Cary, North Carolina 27512.

Data Desk is a computer software package developed by Paul Velleman, Data Description, Inc., Box 4555, Ithaca, New York 14852.

1. Each chapter begins with a short paragraph, entitled *In a Nutshell*, that explains how the chapter material fits into the overall objective of the text.

2. Definitions and concepts are printed in bold type. Key words are shown in bold italics. Formulas and symbols are shown in boxes. This is the material that you might highlight with a soft pen for review. We have done it for you. Most of the explanatory material is shown in plain text.

3. To help you commit new words and concepts to memory, use the accompanying memory cards. The more times you try to recall a fact, the easier it is to recall. Look at the question on the face of a memory card. Try to recall the answer. Then flip the card to the opposite side and read the answer. Use the cards for 5 or 10 minutes several times a day.

4. After we have presented new words and concepts, we will show you how to use them. Examples will show you how to apply them to real-world situations. Then we will present exercises and you can see what you have learned.

The exercise sets start with basic questions and short problems to reinforce your understanding of the language of statistics. Then we will ask you to apply the new concepts to the interpretation of computer outputs and the statistical analyses of real data sets taken from newspapers, magazines, and scientific journal articles.

Language, Purpose, and Uses of Statistics

In a Nutshell

The words that we know limit our understanding of the complex world around us. This chapter will introduce you to the new language of statistics. It will tell you what statistics is, what it is supposed to accomplish, and why it is an invaluable tool in interpreting the vast amounts of data used to describe the social and physical phenomena with which we live.

2.1 Elements, Observations, and Variables
2.2 Universe, Population, and Sample
2.3 Bivariate and Multivariate Data
2.4 The Objective of Statistics
2.5 The Design of Experiments and Sample Surveys
2.6 The Misuses of Statistics
2.7 Key Words and Concepts

2.1 Elements, Observations, and Variables

Sampling a Consignment of Iron Ore

Case 2.1 and Data Set Consider the following problem. We want to buy a shipload of iron ore but the amount that we are willing to pay depends on the percentage of iron in the shipment. That problem should be easy to solve. We take a 1.5-kilogram specimen of ore from the shipment, submit it to chemical tests, and find that it contains 66.08% iron. Problem solved? Not quite!

As a check, we take another 1.5-kilogram specimen from the shipment and test it. It contains 65.92% iron. The next specimen contains 65.83%. In fact, if we were to select many different specimens of ore throughout the shipment, we would find that the percentage of iron in the specimens varies from one specimen to another.

Why do these percentage measurements differ? Part of the reason is that our chemical analyses are subject to measurement error; the other reason is that the percentage of iron in the ore varies from one spot on the earth's crust to another.

Table 1, Appendix 1, provides some real data on a particular 20,000-ton consignment of Canadian iron ore. It gives the percentage of iron for 390 1.5-kilogram specimens of ore selected from the shipment (Takahashi, U. and Imaizami, M., "Sampling Experiment of Fine Iron Ore," *Reports of Statistical Application Research, Union of Japanese Scientists and Engineers*, Vol. 18, No. 1, 1971). A portion of this data set is shown in Table 2.1 (page 8). Look at the data and consider our discussion. What can we learn from this case study?

Elements and Observations

We learn about the world we live in by observation. For example, a biologist observes and records the sex, weight, length, width, and other characteristics of a species of beetle in order to identify it and to distinguish it from other species. An engineer records the breaking strength of a type of steel cable in order to characterize its strength. A political scientist records the attitudes of voters concerning some local issues in order to predict upcoming voting results. And we want to measure the amount of iron in a 1.5-kilogram specimen of iron ore in order to describe the amount of ore in the specimen and, ultimately, in the 20,000-ton consignment.

An *element* is the object that we observe. An *observation* is the information recorded for each element. For example, the biologist observes each of a number of beetles and records the sex, weight, length, and width for each. A single beetle is an element. The information recorded for each beetle is the observation.

The engineer tests 10 pieces of steel cable and observes the breaking strength for each. A single test specimen is an element and its breaking strength is the observation. The political scientist interviews a sample of 1,000 people from a local electorate and records the opinion of each person concerning an issue on the ballot. A single person is an element and the person's opinion is the observation.

TABLE 2.1 Percentage Iron in 1.5-Kilogram Specimens of Iron Ore: A Portion of Table 1, Appendix 1

Number	% Iron	Number	% Iron	Number	% Iron	Number	% Iron	Number	% Iron
1	66.08	79	64.66	157	65.81	235	65.57	313	65.33
2	65.92	80	65.21	158	66.30	236	65.79	314	65.00
3	65.83	81	64.50	159	65.68	237	65.51	315	65.03
4	65.81	82	64.18	160	65.40	238	65.97	316	64.95
5	65.77	83	64.25	161	65.83	239	66.05	317	65.16
6	65.83	84	64.24	162	65.57	240	66.27	318	65.39
7	65.82	85	64.41	163	65.39	241	65.23	319	65.55
8	66.06	86	64.01	164	65.57	242	66.00	320	65.91
9	65.85	87	64.39	165	65.57	243	65.14	321	65.50
10	65.75	88	64.89	166	65.89	244	65.85	322	65.72
11	66.02	89	64.89	167	65.79	245	65.66	323	65.76
12	65.81	90	65.09	168	65.60	246	66.50	324	65.50
13	65.85	91	64.02	169	65.62	247	65.69	325	65.38

We measure 390 1.5-kilogram specimens of iron ore. The object observed, a single 1.5-kilogram specimen, is an element. The measurement of the percentage of iron in the element is the observation.

The point that we want to make is that there is a difference between an element and an observation made on it. An *element* is the object observed, and the *observation* is the information recorded on the element. The object on which an observation is made is often called an "experimental unit" in the biological and physical sciences and a "sampling unit" in the social sciences.

Variables

A characteristic of the elements that varies from one element to another is called a *variable*. For example, the weight of a beetle of a particular species will vary from one beetle to another. Similarly, the sex of a beetle will vary from beetle to beetle. Therefore, both weight and sex are variables. The breaking strength of a test piece of steel cable will vary from one test piece to another. Therefore, breaking strength is a variable. Public opinion, for or against, an upcoming issue on a local ballot varies from one prospective voter to another. Therefore, prospective voter opinion on a particular issue is a variable. Percentage of iron in a 1.5-kilogram specimen of iron ore varies from one 1.5-kilogram specimen to another. Therefore, percentage of iron is a variable.

A collection of observations on one or more variables is called *data*. A particular collection of data is called a ***data set*** (or, a set of data). For example, the weight measurements on 10 beetles are data and the collection is called a data set. The weight measurements on a different group of 15 beetles is another data set.

Variables and their associated observations (*data*) can be either *qualitative* or *quantitative*.

Qualitative Variables

A *qualitative variable* is one that assumes values that are not numerical but can be categorized. Observations made on a qualitative variable are called *categorical data*. For example, if we observe a beetle and record the value of the variable "sex," the observed value can only be categorized as either male or female. Therefore, a beetle's sex is a qualitative variable. Similarly, a person's political affiliation is a qualitative variable, because an observation on a person will fall into one of a number of categories chosen by the political scientist, e.g., Democrat, Republican, Socialist, Independent, or Other. Note, however, that the variable "political affiliation" cannot be measured.

Quantitative Variables

A *quantitative variable* is one that measures the quantity or the amount of something. Observations made on a quantitative variable are called *quantitative data*. In contrast to categorical data, quantitative data will be measurements. For example, the weight, length, and width of a beetle are examples of quantitative variables. The breaking strength of a test specimen of steel cable is a quantitative variable. The percentage of iron in a 1.5-kilogram specimen of iron ore is also a quantitative variable because it, like weight, length, width, and breaking strength, measures the quantity of something. The values that a quantitative variable may assume correspond to points on a line. Figure 2.1 shows five percentage iron measurements plotted on a line.

FIGURE 2.1 A plot of five percentage iron measurements: 66.08, 65.92, 65.83, 65.81, 65.77

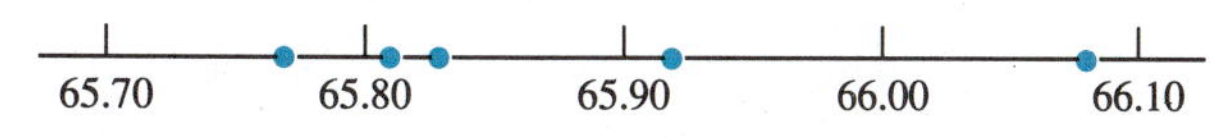

Note that quantitative data can be categorized but categorical data cannot be measured. For example, we can categorize beetle length measurements into three length categories—less than .5 inch, at least .5 but less than 1 inch, and 1 inch or more—but we cannot measure the sex of a beetle.

Example 2.1

Examine each of the six variables listed below. State (1) whether each is quantitative or qualitative and (2) identify the elements with which each is associated.

Variable

a. The price of a house
b. The length of a seatrout
c. Marital status of a person
d. A team's score in a basketball game
e. Your cholesterol level
f. The occupation of a person

Solution

(1) Type of Variable		**(2) Element**	
a.	Quantitative	a.	A house
b.	Quantitative	b.	A seatrout
c.	Qualitative	c.	A person
d.	Quantitative	d.	A single game
e.	Quantitative	e.	You, at a specified time
f.	Qualitative	f.	A person

Example 2.2

Describe the data associated with each variable in Example 2.1.

Solution

a. The price of a single house would be a number of dollars.

b. The length of a seatrout would be a number of inches.

c. The marital status of a person can only be categorized. **Choose the categories so that each observation falls into one and only one category.** Three suitable categories would be: (1) married; (2) unmarried or divorced; (3) never married. Categorical data on 100 people would be a listing, one item for each person, indicating the category into which that person's marital status falls.

d. A game score could be 0 or any positive whole number.

e. Your cholesterol level would be a positive number.

f. The occupation of a person is a categorical observation. You choose the categories. Remember that each observation must fall into one and only one category. One possibility would be the following four categories: (1) government employee; (2) employed by private business but not self-employed; (3) self-employed;

(4) unemployed. Categorical data on 50 people would be a listing, one for each person, indicating the category into which each person falls.

2.2 Universe, Population, and Sample

In Case 2.1, we are interested in the percentage iron measurements as they vary throughout the more than 12 million 1.5-kg specimens of iron ore that would be contained in the 20,000-ton shipment. This set of elements, the elements of interest to us, is called a *universe*. The corresponding collection of percentage iron measurements, one for each element in the universe, is called a *population.*

Universe and Population

Expressing these ideas in more general terms, **a *universe* is the set of all elements of interest to a researcher. The corresponding set of observations on some variable is called a *population.*** Note that in statistics, the word *population* does not refer to the number of people living in a particular area or region. A population is a collection of data of interest to us.

How do we know which elements belong in a particular universe? The universe is the collection of elements in which we are interested. The researcher must specify the universe of elements of interest to her or to him. For example, the set of all beetles of a particular species might represent a universe of interest to a biologist. Or the biologist might define the universe to be the set of all beetles of a given species that are now alive as well as all that might be born in the future, an infinitely large number of beetles. If the biologist wishes to learn something about a particular variable, say the weight of the beetle, the set of weight measurements corresponding to all of the beetles in the universe would be the population of interest. The relationships between a universe, an element, an observation, and a population are illustrated graphically in Figure 2.2 (page 12).

Several populations of measurements may be associated with the same universe. If the biologist wants to know something about the proportions of males and females in the universe of beetles, the population of interest to her would be the set of sex observations (male or female) for all beetles in the universe. The set of length measurements, one for each member of the universe, is another population. The set of all weight measurements is another.

A Sample

A *sample* is a subset of observations selected from a population (see Figure 2.3). Most often, we cannot observe all of the observations in the population. For example,

FIGURE 2.2

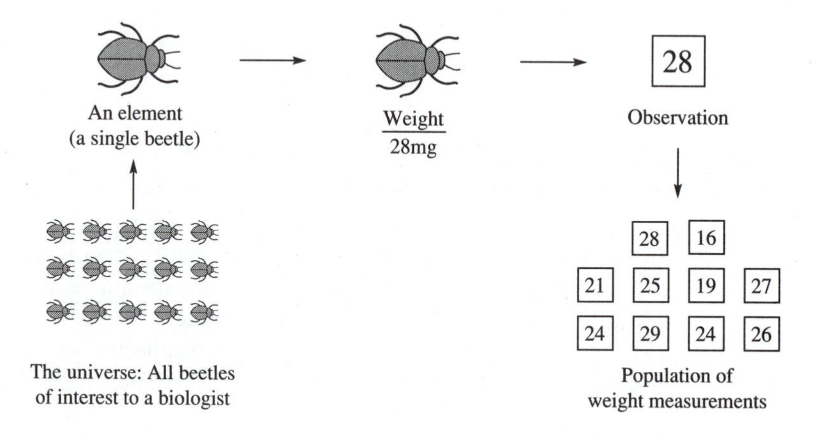

FIGURE 2.3 Sample as a subset of a population

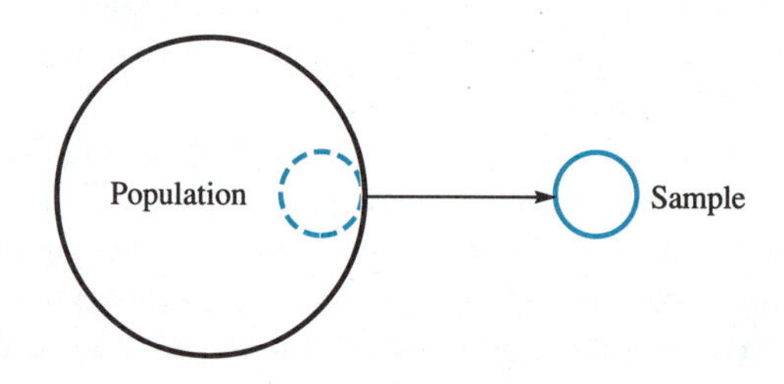

it would be impossible for a biologist to locate and observe every single member of a species of beetle. Since we cannot observe the whole population, we take a sample from it. We then attempt to use what we learn about the sample to deduce (we call it *infer*) the nature of the population based on the characteristics of the sample.

The percentage iron measurements of Table 1, Appendix 1, represent a sample of 390 measurements selected from the population of all percentage iron measurements, i.e., the population of measurements associated with the universe of more than 12 million 1.5-kilogram specimens of ore. Why do we sample? There is no way that we could (or would want to) measure the percentage of iron in every 1.5-kg specimen in the 20,000-ton consignment. It is easier to sample 390 measurements from this population and use the sample data (Table 1, Appendix 1) to *infer* the nature of the entire population.

The word *sample* is often used in the literature to refer to both the subset of elements selected from the universe and to the sample of observations made on these elements. It is also used as a verb. Instead of saying that we want "to select a sample of 10 people from . . . . ," we say that we want "to sample 10 people from ....” Our intended meaning—the sample of objects (elements) that we observe, the sample of observations on them, or whether we are using *sample* as a verb—will be made clear by the context in which the word is used.

Example 2.3

Suppose that you want to know whether eligible voters in Alachua County, Florida, favor (yes, no, or undecided) a new school bond issue. Identify each of the following:

a. The elements

b. The type of observation (qualitative or quantitative)

c. The universe

d. The population

Solution

a. An experimental unit (element) is an eligible voter.

b. The observation is qualitative because it can only be categorized as a “yes,” a “no,” or “undecided.”

c. The universe is the set of all eligible voters in Alachua County.

d. The population is the set of responses (yes, no, undecided) for each of the voters in the universe.

Example 2.4

Suppose that 2,000 eligible voters were selected from among all of the eligible voters in Alachua County and each was asked whether he or she favored the school bond issue. Describe the sample.

Solution

The sample is the set of observations (the 2,000 responses of “yes,” “no,” and “undecided”) obtained from the 2,000 eligible voters.

2.3 Bivariate and Multivariate Data

Fish Contaminated with DDT in the Tennessee River

Case 2.2 and Data Set Table 2, Appendix 1, provides a second example of a statistical study. The table gives the length, weight, and DDT concentration for 144 fish sampled from the Tennessee River, Alabama (*Source*: Sincich, T., *Statistics by*

Example, 4th edition, Dellen Publishing Co., San Francisco, 1990). One objective of the study was to establish the approximate range of values for DDT concentration in fish of different species selected from the Tennessee River. Another was to determine the proportion of fish containing DDT in excess of 5 parts per million (ppm). A portion of this table is shown in Table 2.2.

TABLE 2.2 Results of DDT Analyses on Fish Samples

OBS	Location	Species	Length	Weight	DDT
1	FCM5	Channel Catfish	42.5	732	10.00
2	FCM5	Channel Catfish	44.0	795	16.00
3	FCM5	Channel Catfish	41.5	547	23.00
4	FCM5	Channel Catfish	39.0	465	21.00
5	FCM5	Channel Catfish	50.5	1252	50.00
6	FCM5	Channel Catfish	52.0	1255	150.00
7	LCM3	Channel Catfish	40.5	741	28.00
8	LCM3	Channel Catfish	48.0	1151	7.70
9	LCM3	Channel Catfish	48.0	1186	2.00
10	LCM3	Channel Catfish	43.5	754	19.00
11	LCM3	Channel Catfish	40.5	679	16.00
12	LCM3	Channel Catfish	47.5	985	5.40
13	SCM1	Channel Catfish	44.5	1133	2.60
14	SCM1	Channel Catfish	46.0	1139	3.10
15	SCM1	Channel Catfish	48.0	1186	3.50
16	SCM1	Channel Catfish	45.0	984	9.10
17	SCM1	Channel Catfish	43.0	965	7.80
18	SCM1	Channel Catfish	45.0	1084	4.10
19	TRM275	Channel Catfish	48.0	986	8.40
20	TRM275	Channel Catfish	45.0	1023	15.00

Examining Table 2.2, you can see that each observation (OBS) gives five bits of information concerning a single fish. It gives (in code) the location where the fish was caught, the species of the fish, the length in centimeters (cm), the weight in grams (gm), and its DDT content in parts per million (ppm). Therefore, for each fish, we record data on four (ignoring the location information) variables: species, which is qualitative, and height, weight, and DDT concentration, all of which are quantitative variables. Each fish is an element of the universe and the universe is, presumably, the collection of all adult channel catfish, small-mouth buffalo, and large-mouth bass in the Tennessee River.

We envision a population of data associated with each of these four variables. For example, the population of DDT measurements would be the set of all DDT measurements, one for each fish in the universe.

When we observe the value of only a single variable on each element, the data are said to be **univariate**. If we observe two variables on each element, the data are said to be **bivariate**. In general, if two or more variables are observed on each element, the data are said to be **multivariate**. Therefore, the data in Table 2.2 are multivariate.

When a set of data is multivariate, we can treat each variable separately. For example, we could concern ourselves with only the sample of 144 DDT measurements; i.e., we could treat the data set as univariate and use it to infer the nature of the population of all DDT measurements. We could treat the length and weight measurements in a similar manner. However, a major reason for collecting multivariate data is to see whether any pairs of the variables are related. For example, we might like to know whether the DDT concentration (ppm) per fish depends on the species of the fish. Or, we might like to know whether the DDT content depends on the fish's weight. We will study problems of this type in Chapter 8.

A University of Florida Study of Human Cholesterol Levels

Case 2.3 and Data Set The news media have made us very aware in recent years of the strong relationship between high blood cholesterol levels and heart disease. Table 3, Appendix 1, gives multivariate data on the cholesterol levels of 107 healthy people, aged 17 to 62 (*Source*: J. J. Cerda, University of Florida College of Medicine). A partial reproduction of the data set is shown in Table 2.3 (page 16).

Cholesterol in the human body can be one of three types: LDL (low-density lipids) and VLDL (very low-density lipids), both of which are considered harmful, and HDL (high-density lipids), which is desirable. Total cholesterol, a weighted sum of LDL and HDL, is the more widely used measure of cholesterol level in the body but the ratio LDL/HDL is considered a better measure. Although the exact relationship between these measures and heart disease is unknown, it is currently recommended that a person's total cholesterol level be less than 200 milligrams per deciliter and the LDL/HDL ratio be less than 3.

An element in this study, the object observed, is a single person. We do not know the universe that the researchers envisioned but it was probably the collection of all healthy people in Florida (or, perhaps, in the United States) between the ages of 17 and 62. Eight variables were observed for each person.

Examining Table 2.3, you can see that the identification number of each person is recorded in column 1. The remaining columns record a person's (2) age, (3) sex, (4) TCHOL (the total cholesterol), (5) TRI (triglyceride level), (6) HDL (high-density lipids), (7) LDL (low-density lipids), (8) VLDL, and (9) RATIO = LDL/HDL.

A study of the sample data associated with the individual variables will provide information on the range of values that we might expect for healthy people. The TCHOL and RATIO data will enable the researcher to estimate the proportion

TABLE 2.3 Cholesterol Data

ID	Age	Sex	TCHOL	TRI	HDL	LDL	VLDL	Ratio
1	42	M	178.0	87.0	42.0	118.60	17.40	2.82381
2	38	M	195.9	74.4	48.3	132.72	14.88	2.74783
3	50	M	236.0	410.0	23.0	131.00	82.00	5.69565
4	22	M	158.0	138.0	35.0	95.40	27.60	2.72571
5	44	F	277.0	117.0	82.0	171.60	23.40	2.09268
6	45	F	267.0	225.0	35.0	187.00	45.00	5.34286
7	62	M	237.0	189.0	39.0	160.20	37.80	4.10769
8	42	F	248.0	211.0	31.4	174.40	42.20	5.55414
9	38	M	230.0	158.0	47.0	151.40	31.60	3.22128
10	53	M	181.0	123.0	40.0	116.40	24.60	2.91000
11	50	F	172.0	80.0	77.0	79.00	16.00	1.02597
12	46	F	161.0	102.0	65.0	75.60	20.40	1.16308
13	42	F	223.0	124.0	54.0	144.20	24.80	2.67037
14	46	M	126.0	134.0	29.0	70.20	26.80	2.42069
15	44	M	202.0	234.0	32.0	123.20	46.80	3.85000

of people in the universe whose cholesterol levels exceed the recommended values. We will also be able to examine the relationship between pairs of variables, such as the relationship between the age of a person and the person's cholesterol RATIO.

2.4 The Objective of Statistics

What Is the Objective of Statistics?

The goal of most statistical studies is to obtain information about one or more characteristics of a universe. For example, a scientist might want to obtain information on the weight or the height characteristics of all beetles of a particular species. Since it is not feasible (i.e., it is impossible, too costly, or whatever) to observe every beetle in the universe of all beetles of that species, we select a sample of observations from the universe. We observe the characteristics of the sample and use these sample characteristics to **infer**—i.e., deduce—the nature of the population. This process is shown symbolically in Figure 2.4. Phrasing our goal another way, **the objective of statistics is to make inferences about a population based on information contained in a sample.**

FIGURE 2.4 The relation between the sample and the inference

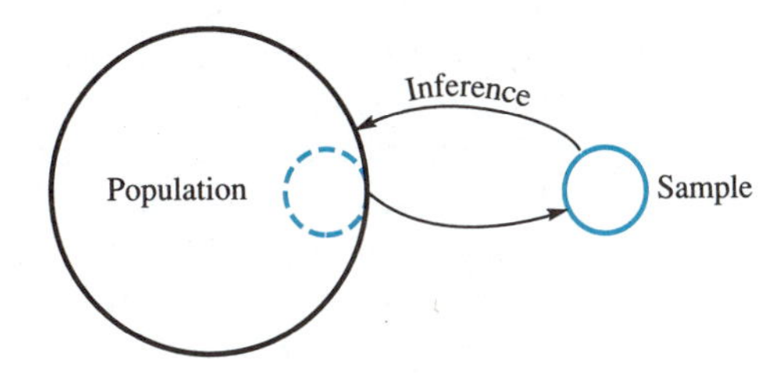

What Do We Mean by the Word *Inference*?

Drawing conclusions about something based on partial information is called *inference*. For example, suppose that you saw someone climbing into a second-story window in a residential neighborhood at night. Based on what you have seen (which is limited information) you might infer that the person is a burglar or you might draw the conclusion that the person is a homeowner who has lost his key. In this course, we will learn how to draw conclusions (make inferences) about the nature of a population of data based on the limited information in a sample. The process that we will use is called ***statistical inference***.

How We Make Inferences About a Population

You will learn in Chapters 3 and 4 that we can describe a population of measurements graphically or we can describe it using a few descriptive numbers. For example, one numerical characteristic of the population of TCHOL measurements is the *proportion* of measurements in the population (and, therefore, people in the universe) that have TCHOL measurements that are in the desirable range—i.e., less than 200. This proportion is a number and it conveys some useful information about the population.

We use the sample data to make inferences about a population of data by either **estimating** the value of or **making a decision** about a numerical descriptive measure. For example, we could use the proportion of the 107 sample TCHOL measurements in Case 2.3 to *estimate* the proportion of all measurements in the population that are less than 200. Or you might use the sample data in Case 2.1 to *decide* whether the percentage of iron in the complete 20,000-ton consignment is less than 60%. We will explain the reasoning employed in making inferences and decisions in Chapter 5. Then in Chapters 6–11, we will show you how to make some very practical inferences from sample data.

Of What Value Is Statistical Inference?

What is the difference between statistical inference and any other kind of inference? Can't anyone examine sample data and use them to make inferences about the sampled population? What advantage do statistical methods have over other methods of inference?

Every method of inference about a population is subject to error because the inference is based on the limited information contained in the sample. Theoretical statisticians study the different methods for making inferences and they evaluate the magnitude of the errors associated with each. For example, if we wanted to use the sample of 390 percentage iron measurements, Case 2.1, to estimate the amount of iron in the 20,000-ton shipload of ore, theoretical statistics can tell us how to obtain a good estimate of the total. More important, it can tell us by how much our answer could be in error. Summarizing, **the advantages of statistical methods are**:

1. **They tend to make inferences with smaller errors than other methods.**
2. **More important, statistical methods enable us to make an inference about a population with a specified measure of error; i.e., we can make a statement about how large the error is likely to be.**

Anybody can make an inference. You and I can look at a set of sample data and each of us can make an inference about the sampled population. Which of us made the better inference, a statement that more clearly describes the true nature of the population? We will never find out because we rarely ever get to see the entire population.

In contrast to our personal, intuitive inference making ability, statistical methods give us something that other inference making procedures do not. Not only do they make a good inference about the population, but they also give a measure of "how good" the inference actually is. That is, they tell us how close an estimate is likely to be to a population numerical descriptive measure that we are attempting to estimate. They also tell us the chance of being wrong when we make a decision.

Example 2.5

An article in *The New York Times* (July 26, 1988) describes the results of a Media General–AP public opinion poll of adult attitudes toward the nation's space program. The national survey of 1,223 adults found, among other things, that approximately half of those questioned lost confidence in the National Aeronautics and Space Administration (NASA) after the space shuttle *Challenger* exploded in January 1986. The article states that the poll's margin of sampling error was plus or minus 3 percentage points. Describe each of the following:

a. The universe of interest to the pollsters
b. The population
c. The sample
d. The type of data
e. The inference
f. The measure of goodness associated with the inference

Solution

a. The universe is the set of all adults in the United States.

b. There is a population of responses for each question asked in the survey. The population for this example would be the set of "yes" and "no" answers to the question about "loss of confidence."

c. The sample would be the set of 1,223 responses for the 1,223 adults included in the survey.

d. The data are qualitative.

e. Since 50% of those in the sample lost confidence in NASA, the pollsters estimate that 50% of all adults in the United Stated would state that they had lost confidence in NASA (if there were any way to contact all of them).

f. Do we really believe, based on the responses of only 1,223 adults, that 50% of *all* (millions and millions) adults in the United States lost confidence in NASA? I'd bet it's close! The article states that the margin of sampling error was plus or minus 3 percentage points. This means that, with a high probability, the pollster's estimate is within 3 percentage points of the true population percentage of adults who lost confidence in NASA after the *Challenger* explosion. This sampling error gives us a measure of the accuracy of the pollsters' estimate and enables us to decide how much faith we wish to place in it.

How did the pollsters in Example 2.5 arrive at such an accurate estimate of the opinion of so many millions of Americans based on the opinions of only 1,223 adults? The answer is that they used statistical inference, the topic of Chapters 5–11.

EXERCISES

2.1 What is the difference between a universe and a population? Give two examples.

DEM **2.2** An article in *The New York Times* (April 7, 1990) states that "The United States ranks in the middle of 33 developed nations in terms of life expectancy and death rate...." The life expectancies for the 33 countries, obtained from the Federal Centers for Disease Control, are shown in Table 2.4 (page 20).

a. On what elements were the observations made in order to collect the data of Table 2.4?

b. Why, do you suppose, did the Federal Centers for Disease Control collect these data?

c. Do you think that the data in Table 2.4 represent a sample or a population? Justify your answer.

2.3 State whether each of the following variables is quantitative or qualitative.

a. The number of people in a supermarket waiting line

b. The daily income for a particular taxi driver

c. The blood type of a human

d. The acidity of a chemical solution

e. Weight of a box of a particular type of cereal

f. Type of tennis court surface

2.4 Identify an element for each of the variables listed in Exercise 2.3.

TABLE 2.4 Life Expectancies

Country	Life Expectancy	Country	Life Expectancy
Australia	76.3	Italy	75.5
Austria	75.1	Japan	79.1
Belgium	74.3	Luxembourg	74.1
Britain	75.3	Malta	74.8
Bulgaria	71.5	The Netherlands	76.5
Canada	76.5	New Zealand	74.2
Czechoslovakia	71.0	Norway	76.3
Denmark	74.9	Poland	71.0
East Germany	73.2	Portugal	74.1
West Germany	75.8	Rumania	69.9
Finland	74.8	Soviet Union	69.8
France	75.9	Spain	76.6
Greece	76.5	Sweden	77.1
Hungary	69.7	Switzerland	77.6
Iceland	77.4	United States	75.0
Ireland	73.5	Yugoslavia	71.0
Israel	75.2		

MED **2.5** Repeated analyses of human blood may give readings that vary from one analysis to another. To reduce the possibility of getting an extreme or misleading reading, an analyst will divide a patient's blood into a specified number of portions, say three, and obtain an analysis for each portion.

a. Describe an element associated with a reading.
b. Describe the universe.
c. Describe the population of interest to the analyst.
d. Is the analyst capable of observing the whole population?
e. What do the three readings on a patient's blood represent?

MFG **2.6** A manufacturer conducted an investigation of the productivity (the number of items produced per hour) per worker in a particular assembly operation. Each of 20 workers performed the assembly operation for 1 hour and the number of items produced was recorded for each.

a. What is the variable observed in the study?
b. What was the element upon which an observation was made?
c. What universe, do you think, did the manufacturer have in mind?
d. Will the 20 recorded productivity measurements be equal in value?
e. What do the 20 productivity measurements represent, the population or the sample?

SOC **2.7** In 1988, the New York City police conducted an undercover investigation of the city's 11,787 cab drivers (*The New York Times*, September 23, 1988). Fifty cab drivers were hired at Kennedy Airport by undercover detectives to drive them to the World Trade Center. Twenty-two of the drivers overcharged their passengers from \$10 to \$75 for the \$30–\$35 ride and 20 were arrested (two drove off before apprehension).

a. Describe an element of the universe.
b. Describe the universe.

c. Identify the observed variable and state whether the variable is quantitative or qualitative.

d. Describe the population.

e. Describe the sample.

f. What was the objective of the sampling?

EDUC **2.8** A survey of 382 college and university presidents (*The Wall Street Journal*, April 30, 1990), conducted by the Carnegie Foundation for the Advancement of Teaching, concluded that, "College social problems, such as excessive drinking and racial tension, reflect murky standards of conduct and administrative ambivalence about enforcing the rules that do exist." For example, 67% of the presidents responded that alcohol abuse was a problem on their campuses, "yet many colleges don't take a firm stand against alcohol abuse, even by minors."

a. Describe the element of interest in this study.

b. Describe the universe.

c. Is the observation on alcohol abuse quantitative or categorical?

d. Describe the population associated with the variable of part **c**.

e. Describe the sample.

f. What does the 67% represent?

2.9 The General Accounting Office claims that the U.S. Office of Education "may be paying banks millions of dollars each year in excess subsidies on student loans" (*The New York Times*, September 1, 1988). An audit of 2,038 loan accounts found that 18% were in error or lacked documentation to support the amount billed.

a. Describe the element of interest to the General Accounting Office.

b. Suppose that the observed variable is whether a loan account was found to be in error or lacked documentation. State whether this variable is quantitative or qualitative.

c. Describe the universe.

d. Describe the population.

e. Describe the sample.

f. Describe the inference that was made.

2.5 The Design of Experiments and Sample Surveys

We stated in Section 2.4 that theoretical statistics not only tells us how to make good inferences about a population, but it also tells us "how good" a particular inference, an estimate or a decision, is likely to be. A third contribution of statistics to inference making is in the planning of the sampling procedure.

The plan that we use to collect sample data—deciding on the variables that we intend to measure, the number of elements to include in the sample, and how to collect them—is called the **design of an experiment** in the experimental sciences (biology, physics, psychology, engineering, etc.). It is called a **sample survey** or **poll** in the social sciences (e.g., opinion and consumer preference polls).

Experimental designs and survey designs are important because they address two problems:

1. It is often very difficult to obtain a sample that is representative of the population. Ideally, we number all of the elements in a universe and the design tells us, by their numbers, which elements to select. The problem is that we often cannot number the elements in the universe or, if we can, we do not have access to each of them. For example, see Case 2.1. How could we number, by exact location, each 1.5-kg specimen of iron ore in a shipload? And, if we could, how could we reach a particular specimen so that we could measure it? If we sample only the elements that we can reach (e.g., those on the surface), will the elements in the sample be representative of the population?

The universes associated with most laboratory experiments are often infinitely large because they exist solely in the mind of the experimenter. For example, an engineer who produces 10 synthetic diamonds using a new method of synthesis views the 10 as a sample from a universe of diamonds, perhaps infinitely large, that could be generated if the process were repeated over and over again indefinitely. This universe and its associated population of measurements, like many others, is conceptual. It exists solely in the mind of the experimenter and therefore the elements of the universe cannot be numbered Even if they were, there would be no way of including element #4,519 in a sample of only 10 elements.

2. The second problem encountered is cost. If the element that we are observing is a beetle, the cost of collecting the beetle and observing its sex, weight length, and width would probably not be too expensive. If the element is a synthetic diamond, it could be more expensive. If the element is a large rocket used to launch commercial payloads into space, the cost of a single observation could be quite expensive. It is sometimes possible to obtain spectacular cost savings by using an appropriate design. For example, running an experiment one way might cost $100,000, whereas running the same experiment using a good design might provide the same amount of information for $5,000.

A data collector (i.e., an experimenter) tries to select a design for a particular situation that gives promise of producing a representative sample and that minimizes the cost of acquiring a specific amount of information.

We will wait to introduce the concepts of design until you are able to understand why they are needed and how they accomplish their intended objective. In Chapter 5 we will introduce random sampling. This is a method that will allow us to make inferences from a sample to a population and to do so with a measured degree of error. Chapter 7 will introduce the concept of matching, a method that can greatly increase the information in an experiment. Chapter 10 will show you how to increase the information in multivariable experiments—experiments where you change the levels of one or more variables and then observe the effect of the changes on the elements of a universe.

2.6 The Misuses of Statistics

Statistics (the methodology), like a paring knife, can be a very valuable tool when it is properly used. When it is misused, it can lead to some very bad results.

You have probably heard it said that you can prove anything with statistics. Properly stated, you can distort the truth in any language but it is easier to lie with statistics. The reason? Most people are unfamiliar with the concepts and language of statistics. Presented with an argument backed by data and summarizing statistics, it is not surprising that many people accept the word of a person who appears knowledgeable. You have undoubtedly heard it before: In the land of the blind, a one-eyed man is king!

Many people, accidentally or on purpose, use data sets and their summarizing statistics as evidence of the truth of some position that they support. Like any argument, the reality of a situation can be distorted by withholding data, analyzing only part of the data, or presenting only certain statistical facts that support a particular point of view.

There are good and bad physicians, and good and bad lawyers, despite the fact that both professions require a license to practice. There are both good and bad statisticians. But the practice of statistics is not regulated. Any person can call himself or herself a statistician.

How can we decide whom to believe? As with any profession, you believe a person who has established a record of authoritative and honest reporting, someone who has proved himself or herself professionally. I would have greater faith in the diagnosis by a physician who has a long record of correct diagnoses. I would have greater faith in the statistical pronouncements of one of the major polling firms or of a statistical consultant who has a demonstrated ability to practice statistics. Then I would test what I see or hear against the basic concepts and methods of statistics to which I have been exposed!

Statistics do not lie. People do! Some of the ways to intentionally or inadvertently distort the truth with statistics are described in the book *How to Lie with Statistics* by D. Huff (New York: Norton, 1954).

2.7 Key Words and Concepts

- We learn about the world we live in by ***observation***.
- The object or person upon which an observation is made is called an ***element*** (or an *experimental unit* or *sampling unit*).
- The set of all such elements of interest to the observer is called the ***universe***.
- A characteristic of the elements that varies from one element to another is called a ***variable***.
- Variables and their associated observations can be either *quantitative* or *qualitative*.
- A ***quantitative variable*** measures the quantity of something and is measured on a numerical scale.
- A ***qualitative variable*** can only be categorized.
- If a variable is observed for each element of a universe, the resulting collection of observations is called a ***population***.

- A ***sample*** is a subset of observations selected from the population. The word *sample* is also used to represent a subset of elements selected from a universe.
- A ***statistical inference*** is an estimate or decision about some characteristic of a population based on the limited information contained in a sample.
- The *objective* of statistics is to make inferences about a population based on information contained in a sample.
- *Statistical inference* not only enables us to make a good inference, but it also gives us a measure of its goodness.

SUPPLEMENTARY EXERCISES

PSY **2.10** A study suggests that left-handed people are more subject to accidents than right-handed people (Coren, S., "Left-Handedness and Accident-Related Injury Risk," *American Journal of Public Health*, Vol. 79, No. 8, August 1989). According to the article, Dr. Coren, a psychologist, sampled 1,896 college students at the University of British Columbia. Each person was tested to determine whether they were left- or right-handed and each was asked (among a number of questions) whether they had sustained at least one injury in the last 2 years. The data showed that 51.7% of 180 left-handed persons and 36.1% of 1,716 right-handed persons sustained at least one injury in the last 2 years.

a. Identify the elements observed in this sample.
b. Speculate on the universe that was of interest to Coren.
c. Describe the variable(s) observed and their type.
d. What do you think was the purpose of collecting the data on the 1,896 college students?

BIO **2.11** Table 2.5 gives the diameter and height (in millimeters) of 10 fossil specimens of the shellfish *Rotularia* (*Annelida*) *fallax* that were discovered near the Antarctic Peninsula (Macellari, C. E., "Revisions of Serpulids of the Genus *Rotularia* (*Annelida*) at Seymour Island (Antarctic Peninsula) and Their Value in Stratigraphy," *Journal of Paleontology*, Vol. 58, No. 4, July 1984).

a. Identify the elements observed in this sample.
b. Speculate on the universe that was of interest to Macellari.
c. Describe the variables observed and their type.
d. What do you think was the purpose of collecting the sample data of Table 2.5?

TABLE 2.5 Diameters and Heights of 10 Fossilized Shellfish

Specimen	Diameter	Height
1	185	78
2	194	65
3	173	77
4	200	76
5	179	72
6	213	76
7	134	75
8	191	77
9	177	69
10	199	65

GEN **2.12** *The New York Times* (August 22, 1989) published an extensive article on the nature of cats, their sensitivity to changes in their surroundings, their unusual ability to adjust to a wide variety of circumstances, and their predatory and social characteristics. One bit of research that was reported concerns the cat's ability to survive a fall. Dr. Wayne Whitney and Dr. Cheryl Metcalf of the Animal Medical Center in Manhattan recorded data on 132 cats that fell from the windows of multistory buildings in New York. Among the variables recorded for each case (cat fall) were the number of floors the cat fell and whether the cat survived. They found that 122 of the cats survived. The cat that fell the farthest, 32 floors, was one of the survivors. In addition, they report that data seem to show that the higher the fall, the greater the chance of survival. Only 1 of 22 cats that fell from above the 7th floor died. Interesting?

a. Identify the elements observed in this sample.
b. How many elements were included in the sample?
c. Identify a universe that might have been of interest to the researchers.
d. Describe the variable(s) observed and their type.
e. What do you think was the purpose of collecting the sample data?

References

1. Freedman, D., Pisani, R., Purves, R., and Adhikari, A., *Statistics*, 2nd edition. New York: W. W. Norton & Co., 1991.

2. Huff, D., *How to Lie with Statistics.* New York: W. W. Norton & Co., 1954.

3. McClave, J. and Dietrich, F., *Statistics*, 5th edition. San Francisco: Dellen Publishing Co., 1991.

4. Mendenhall, W. and Beaver, R., *Introduction to Probability and Statistics*, 8th edition. Boston: PWS-Kent, 1991.

5. Moore, D. and McCabe, G., *Introduction to the Practice of Statistics*, 2nd edition. New York: W. H. Freeman & Co., 1993.

6. Scheaffer, R., Mendenhall, W., and Ott, L., *Elementary Survey Sampling*, 4th edition. Boston: PWS-Kent, 1990.

7. Sincich, T., *Statistics by Example*, 4th edition. San Francisco: Dellen Publishing Co., 1990.

8. Tanur, J. M., Mosteller, F., Kruskal, W. H., Lehmann, E. L., Link, R. F., Pieters, R. S., and Rising, G. R. (eds.), *Statistics: A Guide to the Unknown*, 3rd edition. Pacific Grove, California: Wadsworth & Brooks/Cole, 1989.

three

Graphical Descriptions of Data

In a Nutshell

To draw an inference from a sample, we must be able to describe it. The easiest way is to "see" it, to grasp it visually—to describe it graphically. This chapter will show you how to use graphics to describe the sample data and to gain insight into the nature of the sampled population.

3.1 Describing Qualitative Data

3.2 Relative Frequency Distributions for Quantitative Data

3.3 Graphical Descriptions of the Cholesterol Data Set, Case 2.3

3.4 Stem-and-Leaf Displays

3.5 Key Words and Concepts

3.1 Describing Qualitative Data

Recall that qualitative data cannot be measured on a numerical scale. They can only be categorized.

Categorical data can be easily summarized and displayed graphically. To see how this is done, let us return to the fish species data described in Case 2.2 and presented in Table 2, Appendix 1.

You will recall that Table 2, Appendix 1, gives data on four variables—species, length, weight, and DDT concentration—observed for each of 144 fish sampled from the Tennessee River in Alabama during the summer of 1980. Of these, the species of a fish is a qualitative observation that can fall into one and only one of three categories—channel catfish, large-mouth bass, and small-mouth buffalo.

To summarize the species data, we count the number of observations that fall in each of the three species categories. **The count for a given category is called the *category frequency*.** Table 3.1 shows the category frequency for each of the three species of fish collected in the study. Of the total of 144 fish, 96 were channel catfish, 12 were large-mouth bass, and 36 were small-mouth buffalo. The category frequencies are summarized in Table 3.1.

TABLE 3.1 Summary of Fish Species Data: Category Frequencies

Category	Species of Fish	Frequency
1	Channel catfish	96
2	Large-mouth bass	12
3	Small-mouth buffalo	36
Total		144

In addition to the actual category frequencies, we also like to know the proportion or the percentage of the total number of fish species falling in each category. **The proportion of the total number of observations falling in a particular category is called the *category relative frequency*.** For example, the category relative frequency for category 1 (channel catfish) is 96/144 or .67. The relative frequency for category 2 (large-mouth bass) is $12/144 = .08$. For category 3 (small-mouth buffalo), it is $36/144 = .25$. The category frequencies for the species data are shown in Table 3.2.

The ***category percentages***, obtained by multiplying each category relative frequency by 100, are shown in Table 3.3. Note that each of the 144 fish falls into one

TABLE 3.2 Summary of Fish Species Data: Category Relative Frequencies

Category	Species of Fish	Freqency	Relative Frequency
1	Channel catfish	96	.67
2	Large-mouth bass	12	.08
3	Small-mouth buffalo	36	.25
Total		144	1.00

TABLE 3.3 Summary of Fish Species Data: Category Percentages

Category	Species of Fish	Frequency	Relative Frequency	Percentage
1	Channel catfish	96	.67	67
2	Large-mouth bass	12	.08	8
3	Small-mouth buffalo	36	.25	25
Total		144	1.00	100

and only one of the three species categories. Therefore, the sum of the category frequencies must equal 144, the sum of the category relative frequencies must equal 1, and the sum of the category percentages must equal 100.

Notation Used in Graphical Description of Data

In general, we use the following symbols and formulas to calculate category relative frequencies:

Symbols

n = Total number of observations in a data set

f_i = Number of observations in category i

$\frac{f_i}{n}$ = Relative frequency for category i

$\frac{(100)f_i}{n}$ = Percentage of observations in category i

The sum of the category frequencies will always equal *n*. From this fact it follows that the sum of the category relative frequencies will always equal 1 and the sum of the category percentages will always equal 100.

Bar Graphs and Pie Charts

Qualitative data are graphically described using either a *bar graph* or a *pie chart*. A bar graph for the data of Table 3.1, typical of the ones shown in magazines and newspapers, is shown in Figure 3.1. The computer-generated bar graph for the same data, using the SAS software package, is shown in Figure 3.2 (page 30). The graphs display one bar for each category. The height of a category bar is proportional to the category frequency (or relative frequency or percentage), which is measured along the vertical axis of the graph. The width of the bars has no meaning.

Figures 3.3 and 3.4 (page 31) show pie charts, one artist-drawn and one computer-generated, for the data of Table 3.1. The whole pie represents the 144 observations of the complete data set. The pie is divided into slices, one for each category. The size of a category pie slice is proportional to the relative frequency (or, most often, the percentage) of observations falling in that category. The pie chart makes it easy to compare the proportions of fish in the three species categories.

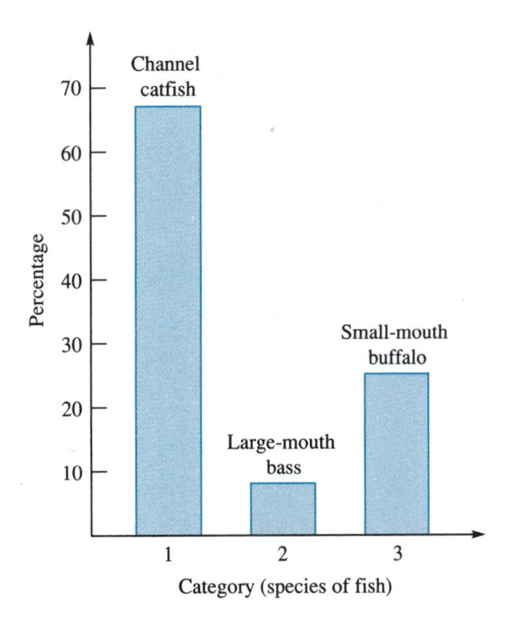

FIGURE 3.1 A bar graph for the species data of Table 3.1

FIGURE 3.2 A computer-generated bar graph for the data of Table 3.1 using the SAS software package

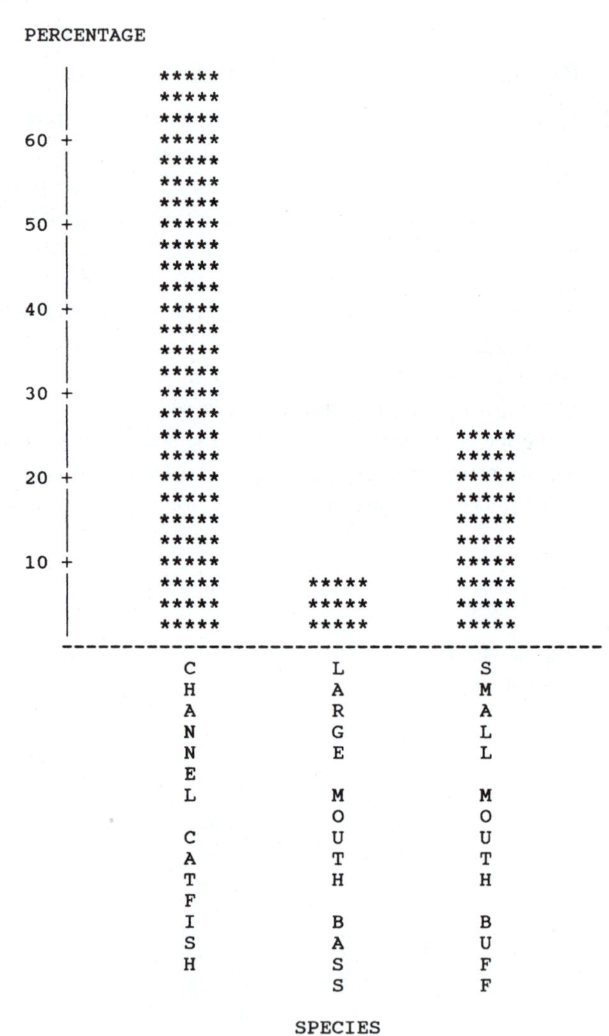

FIGURE 3.3 A pie chart for the data of Table 3.1

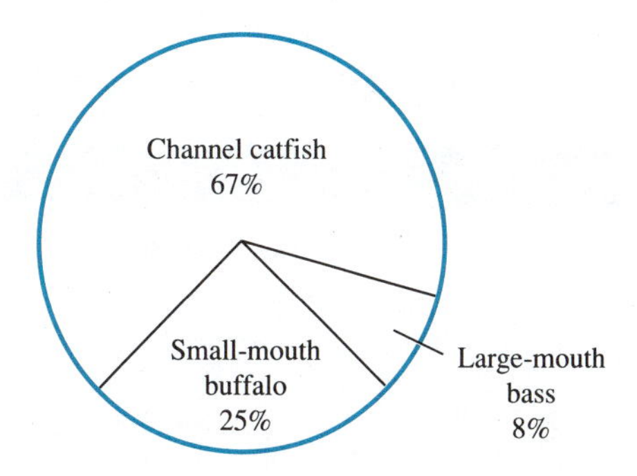

FIGURE 3.4 A computer-generated pie chart for the data of Table 3.1 using the SAS software package

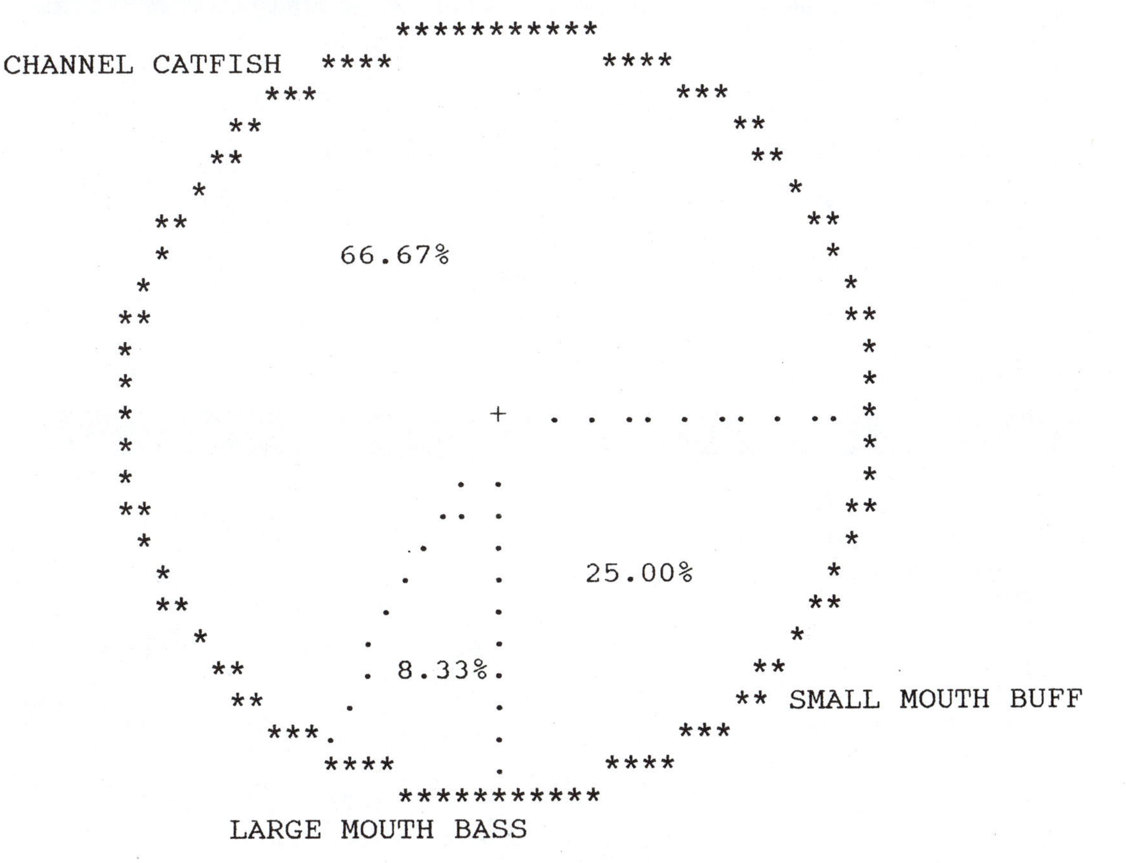

FIGURE 3.5 Graphical descriptions of the key indicators of a united Germany: Exercise 3.1

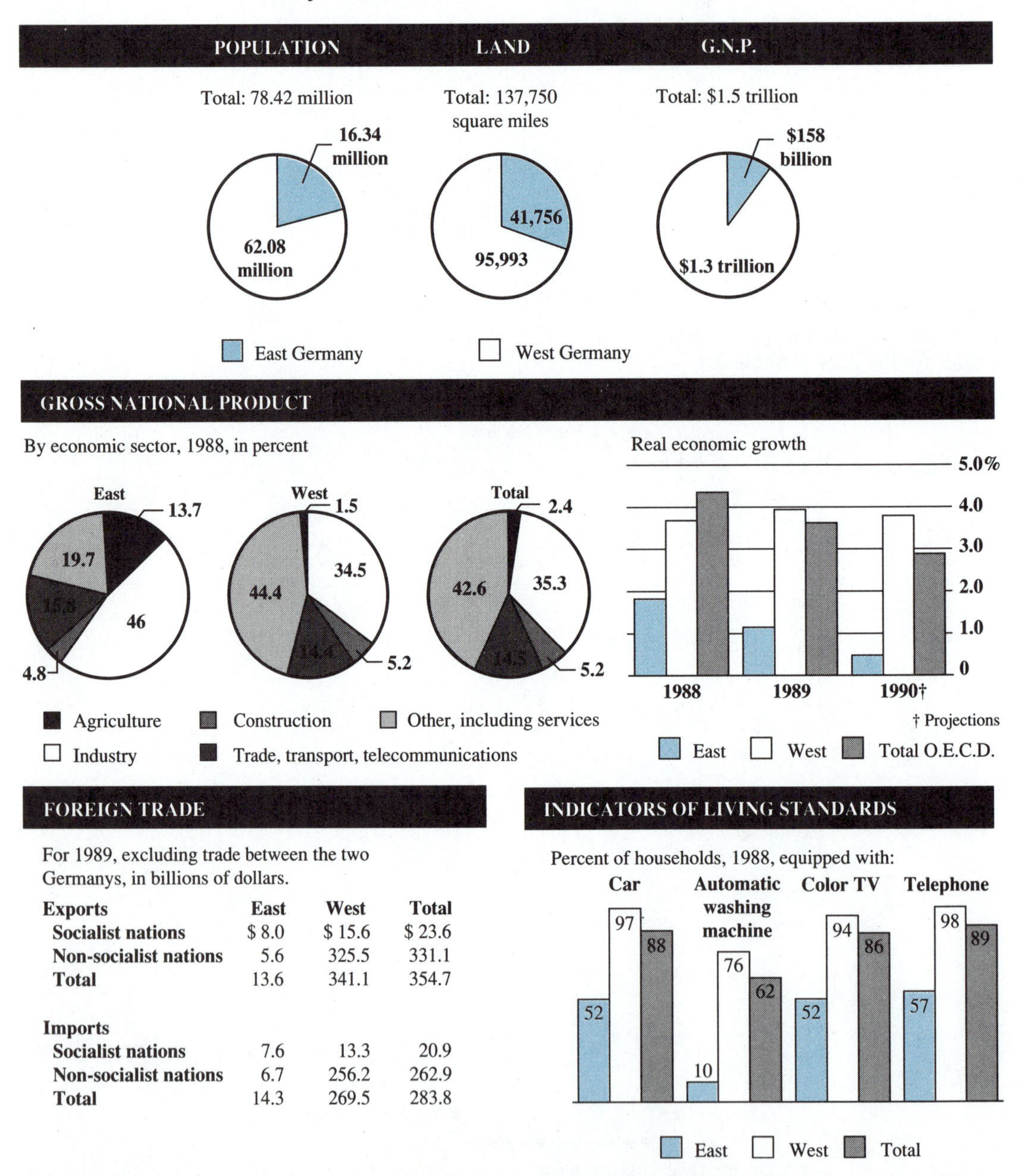

	East	West	Total
Exports			
Socialist nations	$ 8.0	$ 15.6	$ 23.6
Non-socialist nations	5.6	325.5	331.1
Total	13.6	341.1	354.7
Imports			
Socialist nations	7.6	13.3	20.9
Non-socialist nations	6.7	256.2	262.9
Total	14.3	269.5	283.8

Sources: Central Intelligence Agency, Organization for Economic Cooperation and Development, PlanEcon.

EXERCISES

ECON **3.1** The pie charts and bar graphs in Figure 3.5 show Central Intelligence Agency estimates of some key economic indicators of a united Germany (*The New York Times*, June 24, 1990). Examine Figure 3.5 and explain the consequences of unification on the population and land for a united Germany.

3.2 Refer to Exercise 3.1. Discuss the current makeup of the Gross National Product for both East and West Germany and the consequences of unification.

3.3 Refer to Exercise 3.1. Use the pie charts and bar graphs to show the effects of unification on foreign trade.

3.4 Explain how the bar charts in Figure 3.5 show the projected changes in standard of living in a united Germany.

FIGURE 3.6 Pie chart showing how farm debt was funded by type of lender

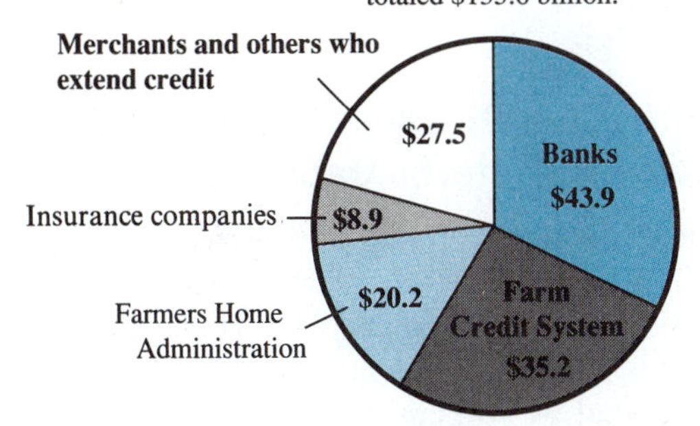

Source: Agriculture Department. Copyright © 1990 by The New York Times Company. Reprinted by permission.

BUS **3.5** The pie chart in Figure 3.6 (*The New York Times*, April 24, 1990) shows how farm debt in the United States is distributed, in billions of dollars, among available lenders.

a. Approximately what proportion of the total amount of loans is provided by banks?

b. Approximately what proportion is provided by the Farmers Home Administration (FHA)?

c. The total debt in 1989 was $135.6 billion. Use the loan amounts shown on the pie chart to calculate the exact proportions of the total loaned by (1) banks and by (2) FHA. Compare your answers to the answers that you gave for parts **a** and **b**.

HLTH **3.6** The outbreak of gastroenteritis associated with eating shellfish reached epidemic proportions in New York State in 1982 (Morse, D. L., et al., "Widespread Outbreaks of Clam- and Oyster-Associated Gastroenteritis," *New England Journal of Medicine*, Vol. 314, No. 11, March 1986). The paper by Morse et al. contains the bar graph of Figure 3.7 (page 34), which shows the number of well-documented cases of shellfish-associated gastroenteritis by month in 1982. What does this graphical description of cases per month tell you about the distribution of cases throughout 1982?

3.7 A consumer preference survey found that of 1,200 consumers questioned, 671 persons preferred product A to product B, 339 preferred B to A, and 190 had no preference. Identify the qualitative variable measured in the survey, identify the categories, and construct a relative frequency table for the data. Construct a bar graph showing the proportions of responses falling in the categories. Describe the data with a pie chart.

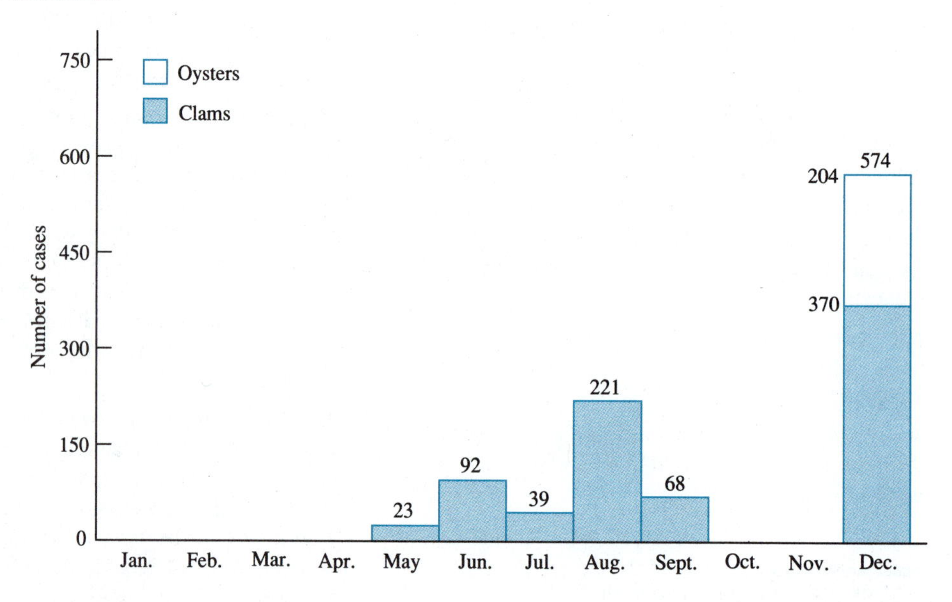

FIGURE 3.7 Distribution of shellfish-related cases of gastroenteritis: Exercise 3.7

3.2 Relative Frequency Distributions for Quantitative Data

A Quantitative Data Set: Some TCHOL Measurements from the University of Florida Cholesterol Data

To demonstrate how to graphically describe a set of quantitative measurements, we will use a sample of $n = 30$ total cholesterol (TCHOL) measurements selected from the set of 107 TCHOL measurements of Table 3, Appendix 1. These measurements, given in milligrams per deciliter (mg/dl), are shown in Table 3.4. The graphical description of the 30 measurements that we will construct is called a *relative frequency distribution.*

Relative Frequency Distributions

A ***relative frequency distribution*** **(also called a** ***relative frequency histogram*****) is to quantitative data what a bar graph is to qualitative data.** Total cholesterol is measured along the horizontal axis of the graph. This axis is divided into con-

TABLE 3.4 A Sample of 30 Total Cholesterol Measurements

236	267	230	161	202	159
237	289	200	131	176	153
170	272	337	224	190	269
278	269	203	312	169	123
235	215	295	235	209	168

FIGURE 3.8 Locating the classes for a relative frequency distribution

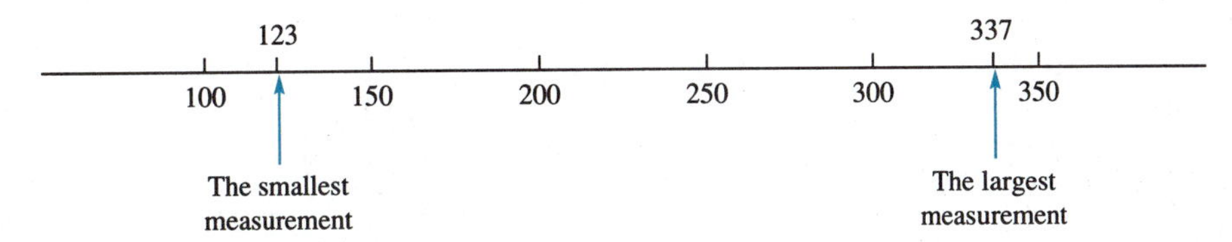

necting, nonoverlapping intervals of equal width, starting below the smallest measurement, 123, in the data set and ending above the largest, 337. See Figure 3.8. **The intervals, called *classes*, correspond to the categories of a qualitative variable.**

For example, the first class includes 100 and goes to but does not include 150. The second, third, fourth, and fifth class intervals are 150 to 200, 200 to 250, 250 to 300, and 300 to 350. The class interval width is 50. **Note that the classes were chosen so that they contain all of the measurements and it is impossible for an observation to fall on a point that divides two classes.**

The next step is to determine the class frequency for each class. **The *class frequency* for a particular class is the number of measurements falling in that class.** Examining Table 3.4, you can see that two measurements fall in the first class. Therefore, the class frequency for the first class is 2. Counting the number of measurements in the second, third, fourth, and fifth classes, we find that the class frequencies are 8, 11, 7, and 2, respectively.

The *class relative frequency* is the class frequency divided by the number n of measurements in the sample. For example, the class relative frequency for class 1 is $2/30 = .067$. It is $8/30 = .267$ for class 2, $11/30 = .367$ for class 3, $7/30 = .233$ for class 4, and $2/30 = .067$ for class 5. The sum of all class relative frequencies is always equal to 1. For this example, the sum of class relative frequencies is $.067 + .267 + .367 + .233 + .067 = 1.001$ (differs from 1.000 only because of rounding in the computation of relative frequencies).

The final step in constructing a relative frequency distribution is to locate a bar over each interval. **The height of the bar is proportional to the *class relative frequency*** (or, depending on your preference, the ***class frequency*** or ***class percentage***, the percentage of the measurements falling in an individual class). The completed relative frequency distribution for the data of Table 3.4 is shown in Figure 3.9.

FIGURE 3.9 The relative frequency distribution for the sample of cholesterol measurements, Table 3.4

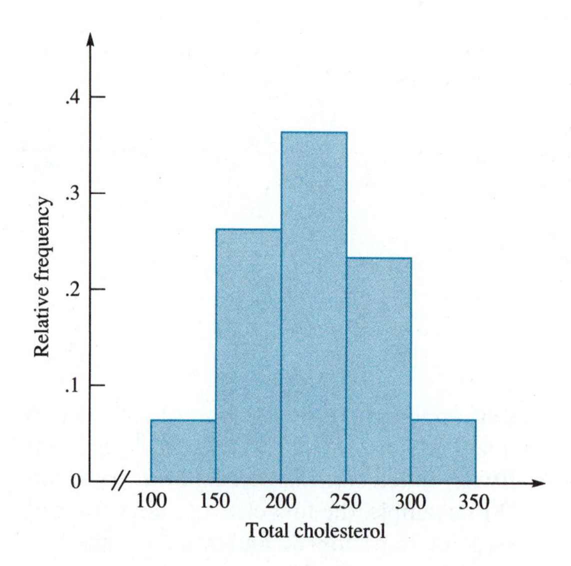

Interpreting a Relative Frequency Distribution

What does Figure 3.9 tell us about the 30 TCHOL measurements of Table 3.4? You can see at a glance that all of the measurements fall between 100 and 350. Most fall between 150 and 300. We know this because the area (as well as the height) of the bar over a particular interval is proportional to the number of measurements falling in that interval.

For example, the relative frequency for class 1 is .067. Therefore, the area of the bar over class 1 is .067 of the total area of the bars in the distribution. Similarly, the proportion of people with TCHOL scores less than 200 is the sum of the relative frequencies of classes 1 and 2, i.e., .067 + .267 = .334. Therefore, the sum of the areas of the bars over classes 1 and 2 is equal to .334 (i.e., 33.4%) of the total area of the bars in the distribution (see Figure 3.10). Summarizing, **if we let the total area of**

FIGURE 3.10 The area over classes 1 and 2 represents .334 of the total area

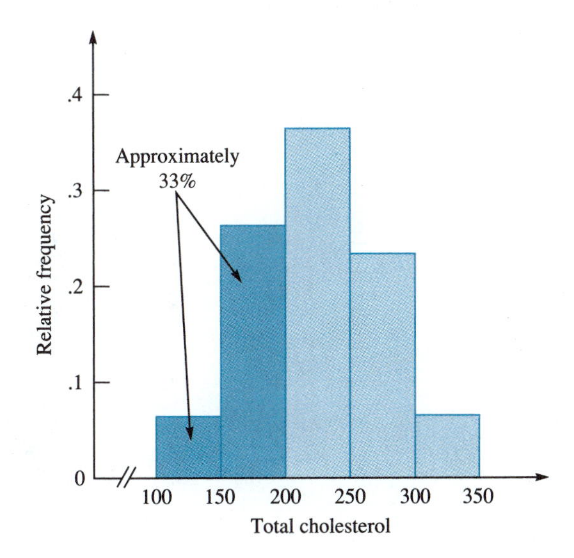

the bars in the relative frequency distribution equal 1, then the area of the bar lying over a particular interval is equal to the proportion of the total number of observations that fall in that interval.

Now let us apply what we have learned. An article in *U.S. News & World Report* (June 18, 1990) states that a body cholesterol level of less than 200 mg/dl is considered to be satisfactory; levels of 200–239 indicate "borderline" or moderate risk of heart disease. People with TCHOL levels above 239 are considered to be at risk of atherosclerosis, heart disease, and stroke.

Examining Figure 3.9, we can quickly obtain a picture of the cholesterol levels of the people in our sample. Visually approximating the area over the interval corresponding to "satisfactory" levels (i.e., 199 or less) of cholesterol, we can see that approximately 1/3 of the area of the distribution lies to the left of 200. That tells us that only 1/3 or 33% of the people in the sample of 30 had TCHOL measurements in the satisfactory zone.

Approximately 30% of the total area appears to lie to the right of 240. Therefore, approximately 30% of the people in the sample had TCHOL measurements equal to 240 or more and are at risk of heart disease at some time in the future (see Figure 3.11 on page 38). Approximately 2/3 of the total area of the bars lies to the right of 200. Therefore, approximately 2/3 of the people in the sample had TCHOL measurements that classify them as borderline or at risk.

FIGURE 3.11 The proportion of people who are "at risk"

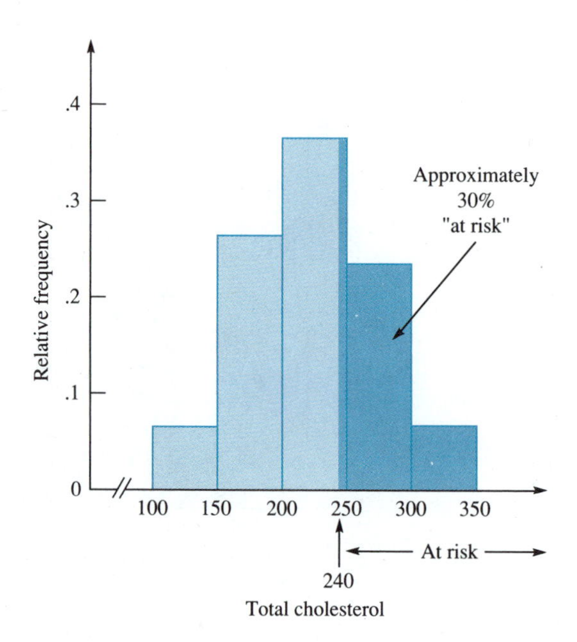

Probabilistic Interpretations of Areas Under a Relative Frequency Distribution

Recall that the total number of observations in the data set in Table 3.4 was $n = 30$. Suppose that you were to write these numbers on poker chips, one to each chip, thoroughly mix the chips, and then draw one. What is the chance that you will draw a number less than 200?

We know that 1/3 of the area of the relative frequency distribution lies to the left of 200 or, equivalently, 1/3 of the measurements are less than 200. Therefore, the chance of selecting a TCHOL measurement less than 200 from the sample of 30 is 1 in 3. The probability that the chip will show a number larger than 350 is 0 because there is no area under the distribution (and, therefore, no measurement) above 350.

We will have more to say about the meaning of probability in Chapter 5. For the moment, it is sufficient to note that the area over an interval under a relative frequency distribution has a probabilistic interpretation. Suppose that you were to draw a single measurement from the data set. The area over a particular interval is equal to the probability that the measurement will fall in that interval.

The Sample and Population Relative Frequency Distributions

Every population consists of a set of data. Therefore, it too can be described by a relative frequency distribution. Of course, we never have the entire population in hand, so its distribution exists in our minds.

Most populations contain a large number of measurements. Therefore, we envision a very narrow class interval and a distribution that appears approximately as a smooth curve. The relative frequency distribution for the sample of 30 TCHOL measurements, and our conception of the relative frequency distribution for the population from which the sample was selected, are shown in Figure 3.12.

FIGURE 3.12 The sample and our conception of the population relative frequency distribution

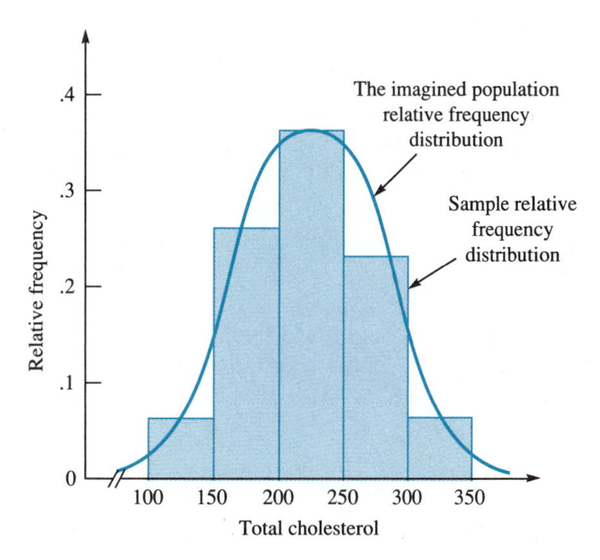

The sample relative frequency distribution is important because it gives us our first best guess as to the nature of the population relative frequency distribution. It suggests the interval within which most of the data lie. It gives us information concerning the likely shape of the distribution, something that may affect the results of the statistical methods we will learn in Chapters 6–11.

Comparing Our Sample with Data from the Minnesota Heart Survey

We cannot say how similar the sample relative frequency distribution for the 30 TCHOL measurements, Figure 3.9, is to the population relative frequency distribution, but we can compare it with a relative frequency distribution based on a very large study.

The Minnesota Heart Survey collected cholesterol data on residents of the Twin Cities (Minneapolis–St. Paul) area in the time periods 1980–1982 and 1985–1987. One objective of the study was to see whether trends in public awareness and treatment of high cholesterol produced a decrease in the TCHOL levels of residents in that area from 1980–1982 to 1985–1987. An article by L. Burke Gregory et al. summarizes the results of the survey ("Trends in Serum Cholesterol Levels from 1980 to 1987, The Minnesota Heart Survey," *New England Journal of Medicine*, Vol. 324, No. 14, April 1991).

Gregory's article gives four relative frequency distributions, one for males and one for females for each of two time periods, 1980–1982 and 1985–1987. Figure 3.13a combines the information in the two 1985–1987 distributions to show a TCHOL relative frequency distribution for the total of 4,545 men and women tested during that period. Let's compare this distribution with the distribution for our sample of 30 TCHOL measurements, Figure 3.9.

The relative frequency distribution for the 30 TCHOL measurements of Figure 3.9 is reproduced in Figure 3.13b. It appears to have approximately the same shape and spread as the Minnesota distribution, but its center appears to be shifted approximately 30 mg/dl to the right. This difference could be due to the fact that the Florida distribution is based on a small number (30) of measurements or it may be that the TCHOL level for Florida residents is slightly higher than the level in the Twin Cities area. Despite this difference, the distribution shown in Figure 3.13b does provide insight into the nature of the distribution given in Figure 3.13a. In a similar manner, it provides insight into the nature of the relative frequency distribution for its population.

Types of Relative Frequency Distributions

Relative frequency distributions can assume many different shapes. We will often talk about distributions that are mound-shaped, symmetric, or skewed.

A *mound-shaped distribution* of data appears as a hump or mound, as shown in Figure 3.14a (page 42).

A *symmetric distribution* is a distribution that is symmetric about a vertical axis, as shown in Figure 3.14b. The portion of the distribution to the left of the vertical axis is the mirror image of the portion to the right of the axis. You can see that the symmetric distribution in Figure 3.14b is mound-shaped, but the mound-shaped distribution in Figure 3.14a is not symmetric.

A *skewed distribution* is one that tails off in one direction, to the right (or left), as shown in Figure 3.14c.

FIGURE 3.13 Comparing the sample TCHOL relative frequency distribution with a distribution based on the Minnesota Heart Survey

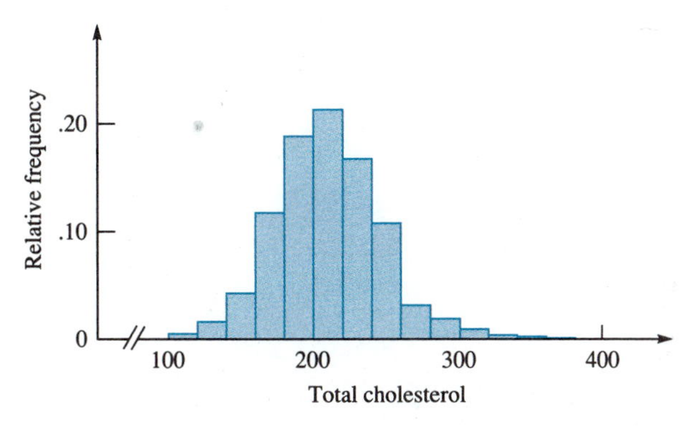

a. Minnesota Heart Survey

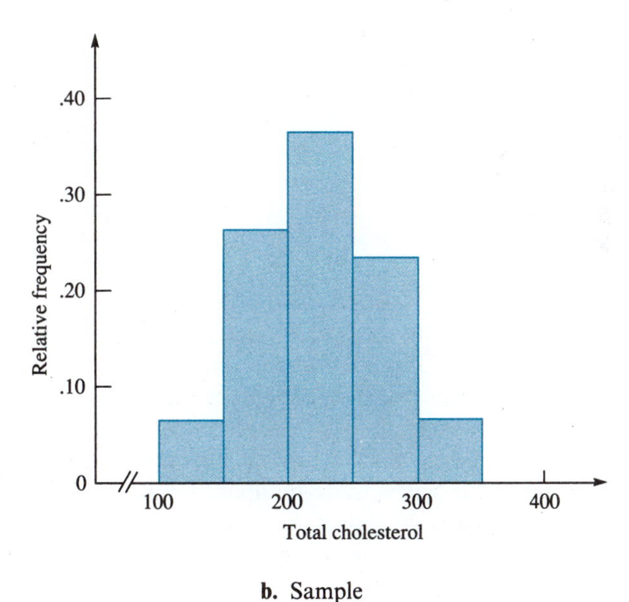

b. Sample

Constructing a Relative Frequency Distribution

The reason for constructing the relative frequency distribution for a sample is to gain insight into the nature of the population relative frequency distribution. Is it skewed, relatively symmetric, or what? To answer this question, we

FIGURE 3.14 Characteristics of distributions

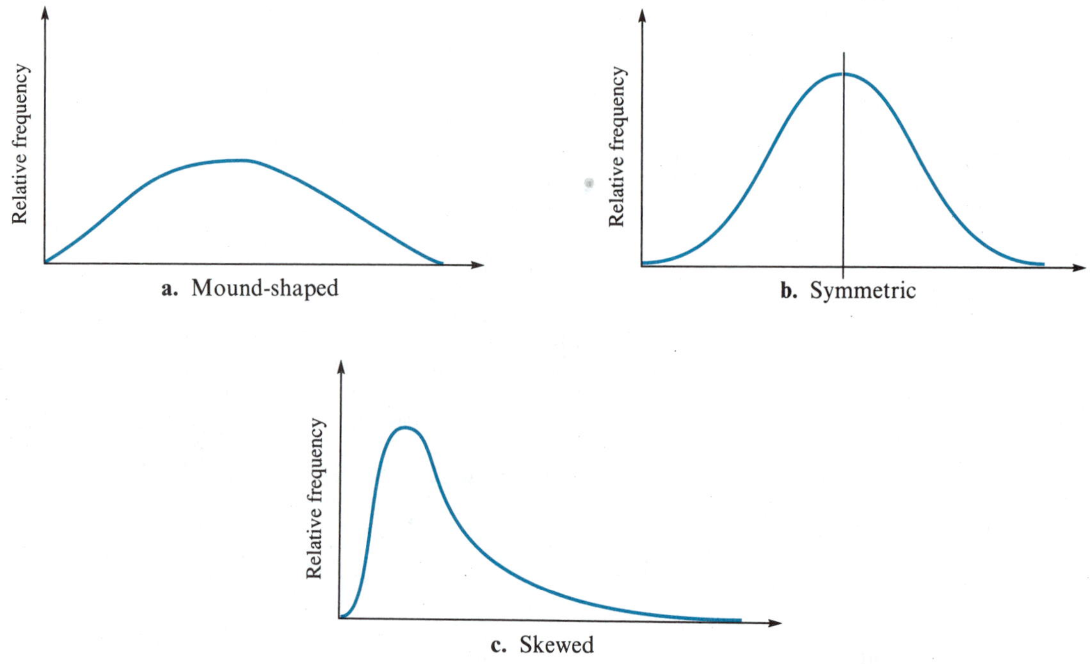

Relative frequency

a. Mound-shaped

Relative frequency

b. Symmetric

Relative frequency

c. Skewed

need to select the class width (and, consequently, the number of classes) so that the outline of the graphed distribution will reveal these properties.

The problem of selecting the number of classes will usually be solved for us. Rough drafts (and sometimes the final versions) of relative frequency distributions are almost always done on a computer and the computer will be programmed to select the appropriate number of classes for the data set. Usually, we will use between 5 and 20 classes, a small number of classes for a small data set (say, 10 to 15 measurements) and a larger number for a larger data set.

Computer-Generated Relative Frequency Distributions

The format of a computer-generated graphical display of a distribution of data will vary from one computer software package to another, but all of the outputs will be recognizable as relative frequency (frequency or percentage) distributions. Some packages print figures that are identical to hand-drawn versions. Other printouts produce primitive replicas of these figures. Printouts of relative frequency (frequency, or percentage) distributions for various computer packages appear in examples and exercises that follow. For information about the printouts for other software packages, see the instruction booklets that accompany the packages.

For example, Figure 3.15 (page 44) shows the computer-generated relative frequency (frequency, or percentage) distributions for the 30 total cholesterol (TCHOL) measurements of Table 3.4 using the Minitab and SAS software packages for personal computers.

The Minitab printout of the frequency distribution for the data of Table 3.4 varies substantially from the hand-drawn version, Figure 3.9, as well as the SAS version of Figure 3.15b. In Minitab, the distribution is rotated through 90° so that the horizontal axis of Figure 3.9 is now in a vertical position. If you rotate the Minitab figure 90° in a counterclockwise direction, you obtain a distribution with a shape similar to that of Figure 3.9. In Minitab, midpoints of the class intervals are shown in the column at the left. The second column gives the "count," which is what we have called *class frequency*. The bars that are typically used to represent the class frequencies are replaced by a horizontal run of stars, one for each measurement. For example, the frequency in the class with midpoint 160 contains 5 measurements (count = 5) and is represented by five stars.

The SAS percentage distribution, Figure 3.15b, is similar in construction to the typical hand-drawn distribution, except that the bars are formed by a rectangular arrangement of stars that do not completely cover the class intervals. The midpoints of the class intervals, 125, 175, ..., 325, are shown along the horizontal axis. Since these midpoints are 50 units apart, the class intervals used for this SAS distribution are 100 to 150, 150 to 200, 200 to 250, ..., 300 to 350. The height of the bar over a particular class interval is proportional to the percentage of the total number of observations falling in that class. This percentage is equal to 100 times the class frequency.

3.3 Graphical Descriptions of the Cholesterol Data Set, Case 2.3

Background Information on Body Cholesterol

In Case 2.3 we briefly explained the importance of body cholesterol and the relevance of the variables by which it is measured. Before we examine the complete University of Florida cholesterol data set (Table 3, Appendix 1), we will review and add to this background information.

FIGURE 3.15 Relative frequency distributions produced by the Minitab and SAS statistical software packages

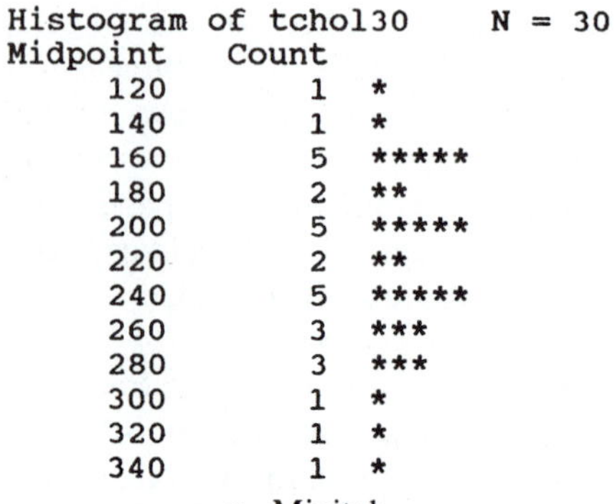

a. Minitab

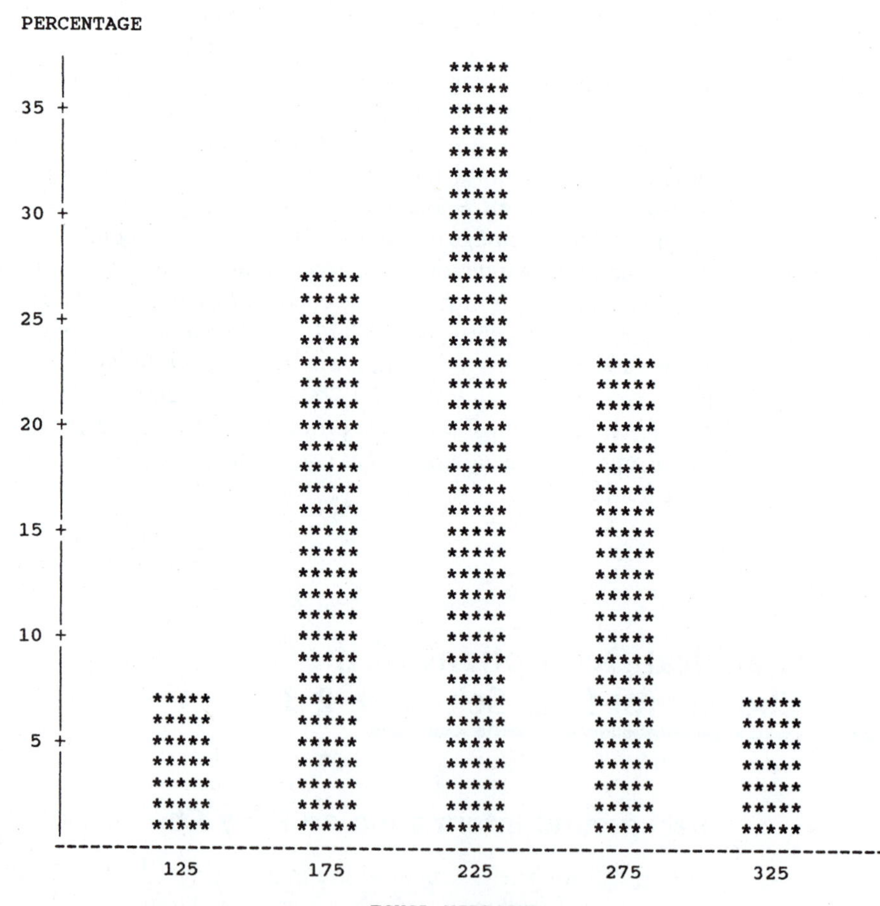

b. SAS

We have heard much from the news media over the last 10 years about the harmful effects of high blood cholesterol levels and the fact that cholesterol starts building up in the walls of the blood vessels at an early age.

Cholesterol has three components: low-density lipids (LDL), high-density lipids (HDL), and very low-density lipids (VLDL). LDL is often called "bad cholesterol" because high levels of LDL, and possibly VLDL, are associated with a higher risk of coronary heart disease. In contrast, HDL is a "good cholesterol." The higher the level of HDL, the lower the risk of heart disease. High levels of triglycerides, fatty molecules in the blood plasma, may also be associated with a higher risk of heart disease.

Physicians use several different variables to measure a person's propensity for developing coronary heart disease. The most common measure is a person's total cholesterol (TCHOL), where TCHOL = HDL + LDL + VLDL. Some physicians prefer the ratios, TCHOL/HDL or LDL/HDL, while others look at the individual levels of LDL and HDL.

So, how do we know whether we are prone to heart disease? Levels of total cholesterol or of the ratio, LDL/HDL, that present a risk of heart disease are age-dependent. Values of TCHOL and LDL/HDL that locate levels of elevated risk are lower at younger ages and increase with age. As a general rule, desirable and elevated levels are as follows (Grundy, Scott M., et al., "Cardiovascular and Risk Factor Evaluation of Healthy American Adults, A Statement for Physicians by an Ad Hoc Committee Appointed by the Steering Committee, American Heart Association," *Circulation*, Vol. 75, No. 6, 1987):

	Desirable	**Elevated**
Total cholesterol	Less than 200	240 or above
Ratio, LDL/HDL	Less than 3.0	

The Cholesterol Data Set

Case 2.3 describes the cholesterol data of Table 3, Appendix 1. Table 3 gives the age, sex, and measurements on the total cholesterol (TCHOL), triglycerides (TRI), high-density lipids (HDL), low-density lipids (LDL), very low-density lipids (VLDL), and the ratio, LDL/HDL, for 107 people. The universe from which the 107 people (elements of the universe) were selected is the set of all apparently healthy humans (aged 17 to 62), none of whom were suspected of having high blood cholesterol levels at the time the data were collected. Consequently, the 107 measurements on each of the eight variables, AGE, SEX, TCHOL, TRI, HDL, LDL, VLDL, and LDL/HDL, can be viewed as samples from the populations that would be collected if we were able (which we are not) to take measurements on every member of the universe. (For purposes of illustration, we chose a smaller and more manageable sample of $n = 30$ TCHOL measurements from this data set to obtain the TCHOL measurements of Table 3.4.)

We will confine our attention to the SEX, AGE, TCHOL, and RATIO data of Table 3, Appendix 1. Figure 3.16 gives the SAS bar chart for (a) SEX and the frequency distributions for (b) AGE, (c) TCHOL, and (d) the ratio, LDL/HDL. Notice how the graphics summarize the information in the data sets.

The bar chart in Figure 3.16a shows that the persons included in the sample were almost equally balanced between males and females, with slightly more males. The mound-shaped frequency distribution for age (Figure 3.16b) is centered over the class with midpoint 44, and the ages range from 16 (the lower class bound) to 72 (the upper class bound).

The frequency distribution for TCHOL (Figure 3.16c) is roughly mound-shaped but there appear to be a few very large measurements in the class that goes from 380 to 420. Approximately one-third of the area under the distribution lies to the left of 200, the midpoint of the third class, which tells us that only one-third

FIGURE 3.16 SAS graphical descriptions of SEX, AGE, TCHOL, and RATIO data on 107 people, from Table 3 of Appendix 1

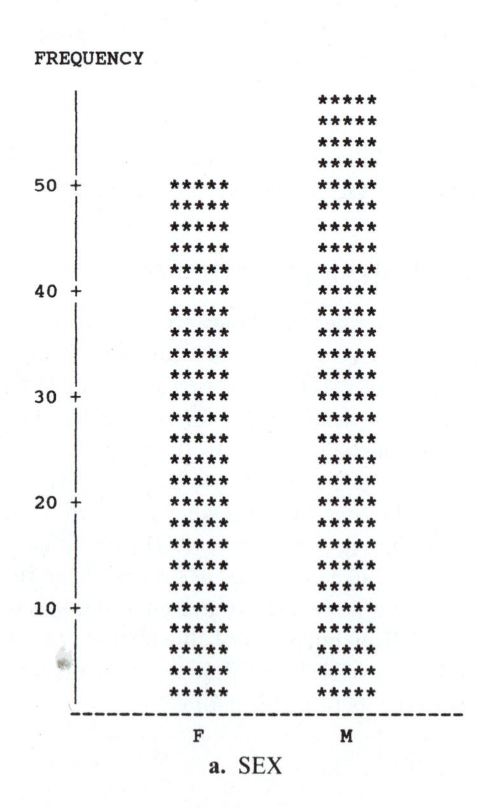

a. SEX

FIGURE 3.16 ***Continued***

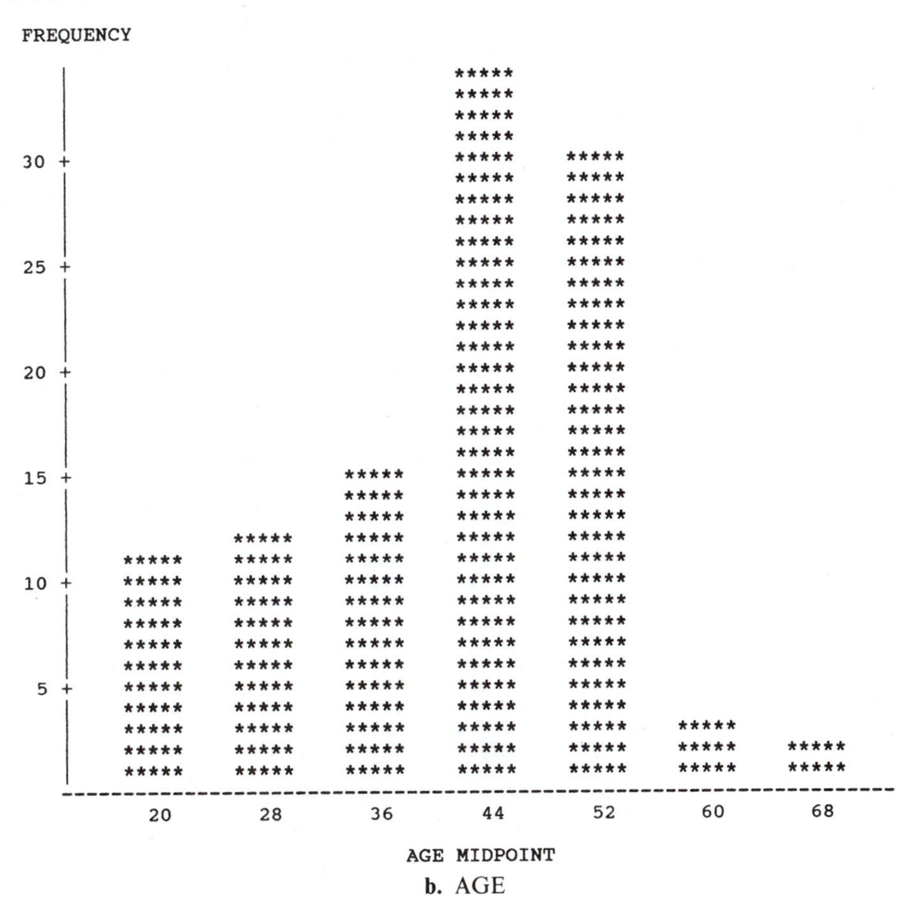

b. AGE

of the people included in the study had total cholesterol measurements less than 200—i.e., levels of TCHOL that fall in the desirable range.

The distribution of the ratio LDL/HDL, a second measure of the risk of heart disease, is slightly skewed to the right (see Figure 3.16d). A glance at the area under the distribution confirms what we learned from our examination of the TCHOL distribution. Approximately one-third of the area under the distribution is less than 3.0, the upper limit on the desirable level of RATIO. This tells us that only one-third of the people in the study had desirable levels of LDL/HDL.

If you have had a complete physical examination recently, you have probably had your cholesterol level tested. If you have, you may want to compare your level with the sampling of other healthy adults contained in Table 3, Appendix 1. Where do your TCHOL and RATIO scores fall in the distributions shown in Figures 3.16c and 3.16d?

FIGURE 3.16 ***Continued***

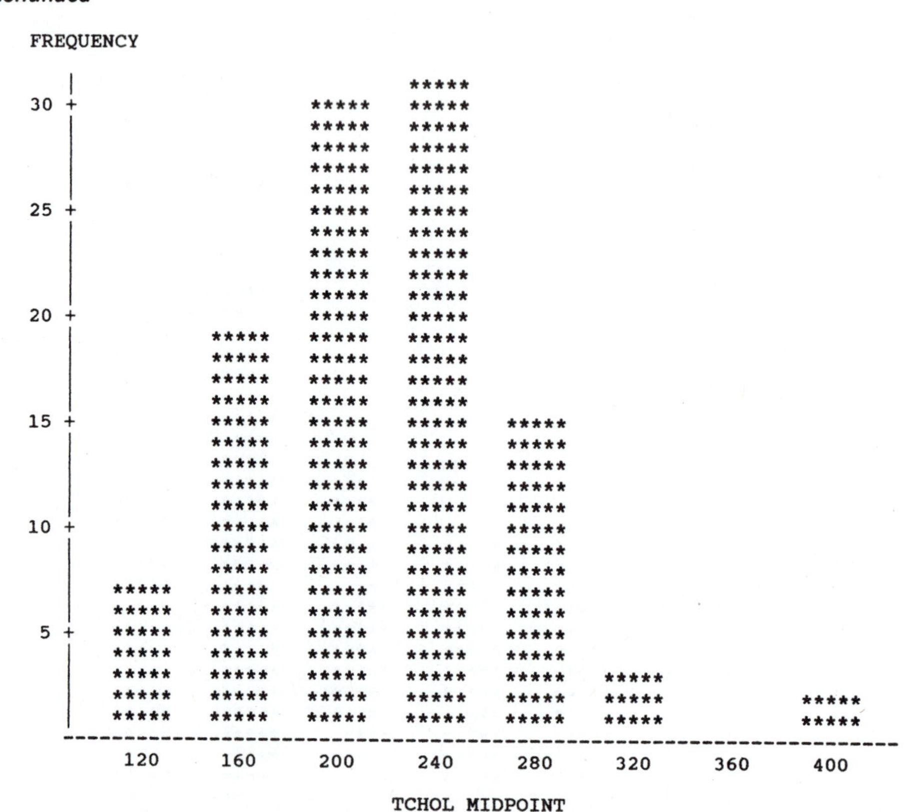

c. TCHOL

FIGURE 3.16 ***Continued***

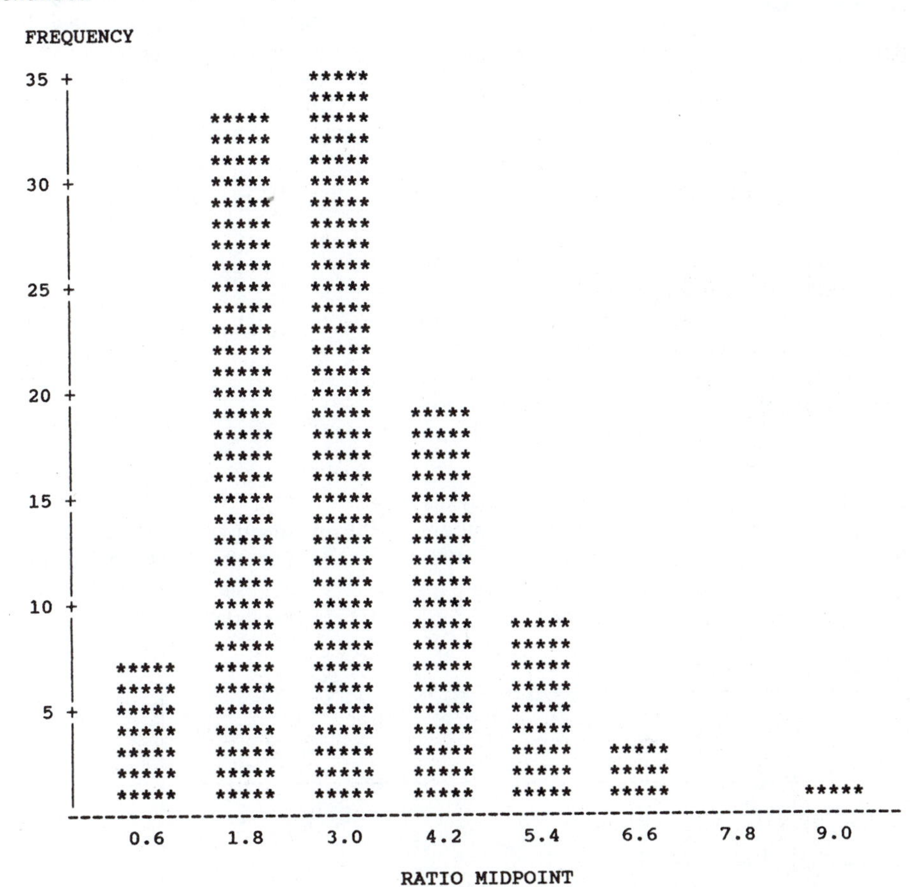

d. RATIO

EXERCISES

3.8 What is a relative frequency distribution?

3.9 What characteristics of a data set can you see with a relative frequency distribution?

3.10 Of what relevance is the area under a relative frequency distribution?

GEN **3.11** An article in *Risk Management* (February 1986) states that most deaths due to fires in compartmentalized fire-resistant buildings (e.g., hotels, motels, apartments) occur when the occupants attempt to evacuate. Table 3.5 gives the number of victims for each of 14 major fires and Figure 3.17 shows a frequency distribution for the data produced by the Data Desk computer software package.

a. What is the difference between a frequency distribution and a relative frequency distribution?
b. Which interval contains the measurement 10?
c. What class intervals did Data Desk choose for the relative frequency distribution?

FIGURE 3.17 A frequency distribution for hotel fire victims produced by Data Desk software package

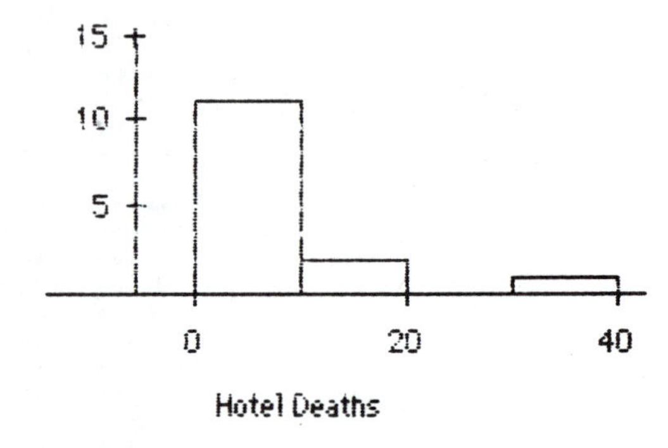

d. Identify the first class and give its frequency.
e. Describe the shape, location, and spread of the data set. What conclusions can you draw from the data?

TABLE 3.5 Fire Victims

Fire	Number of Victims
Las Vegas Hilton (Las Vegas)	5
Inn on the Park (Toronto)	5
Westchase Hilton (Houston)	8
Holiday Inn (Cambridge, Ohio)	10
Conrad Hilton (Chicago)	4
Providence College (Providence)	8
Baptist Towers (Atlanta)	7
Howard Johnson (New Orleans)	5
Cornell University (Ithaca, New York)	9
Wesport Central Apartments (Kansas City, Missouri)	4
Orrington Hotel (Evanston, Illinois)	0
Hartford Hospital (Hartford, Connecticut)	16
Milford Plaza (New York)	0
MGM Grand (Las Vegas)	36

Source: Macdonald, J. N. "Is Evacuation a Fatal Flaw in Fire Fighting Philosophy?" *Risk Management*, Vol. 33, No. 2, Feb. 1986, p. 37.

BIO **3.12** Refer to Case 2.2 in Chapter 2. Table 2 of Appendix 1 gives the species, length (in centimeters), weight (in grams), and DDT concentration (in parts per million) for each of 144 fish sampled from the Tennesee River in Alabama during the summer of 1980 (*Source:* Sincich, T., *Statistics by Example*, 4th edition, Dellen Publishing Co., San Francisco, 1990). Figures 3.18a, 3.18b, and 3.18c give, respectively, the SAS computer printouts of the frequency distributions for the length, weight, and DDT content for the 144 fish.

a. Explain what the distribution of fish lengths tells you about that data set.

b. Use areas under the frequency distribution to estimate the proportion of fish between 30 and 45 centimeters in length.

c. Explain what the distribution of fish weights tells you about that data set.

d. Use areas under the frequency distribution to estimate the proportion of fish weighing more than 800 grams.

FIGURE 3.18 Graphical descriptions of the fish data from Table 2, Appendix 1

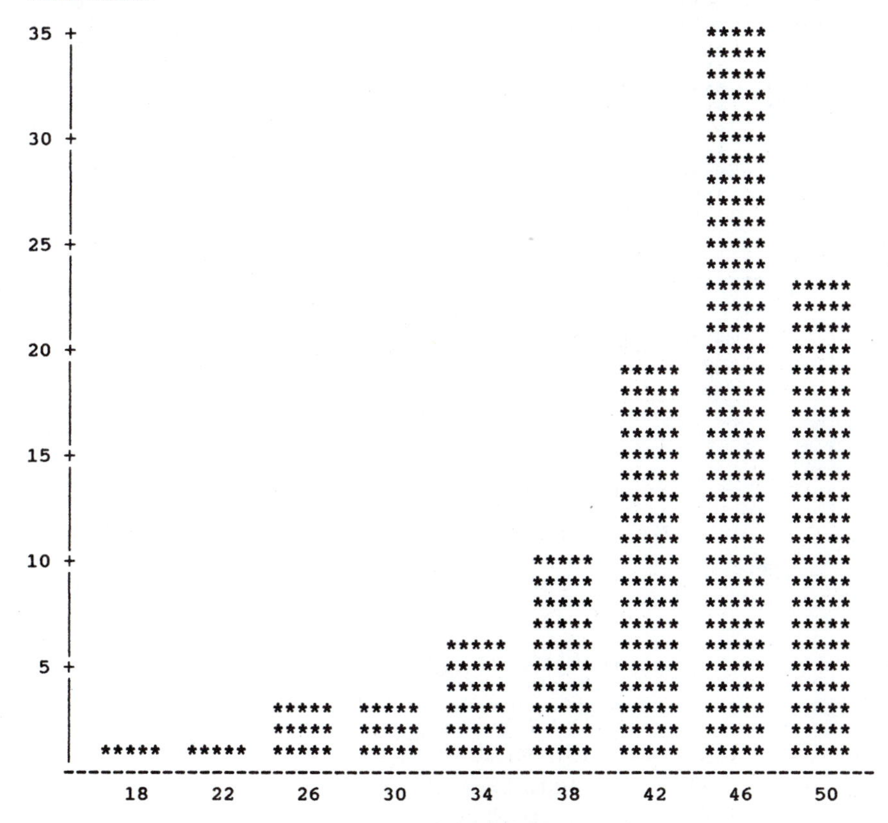

a. Length

FIGURE 3.18 ***Continued***

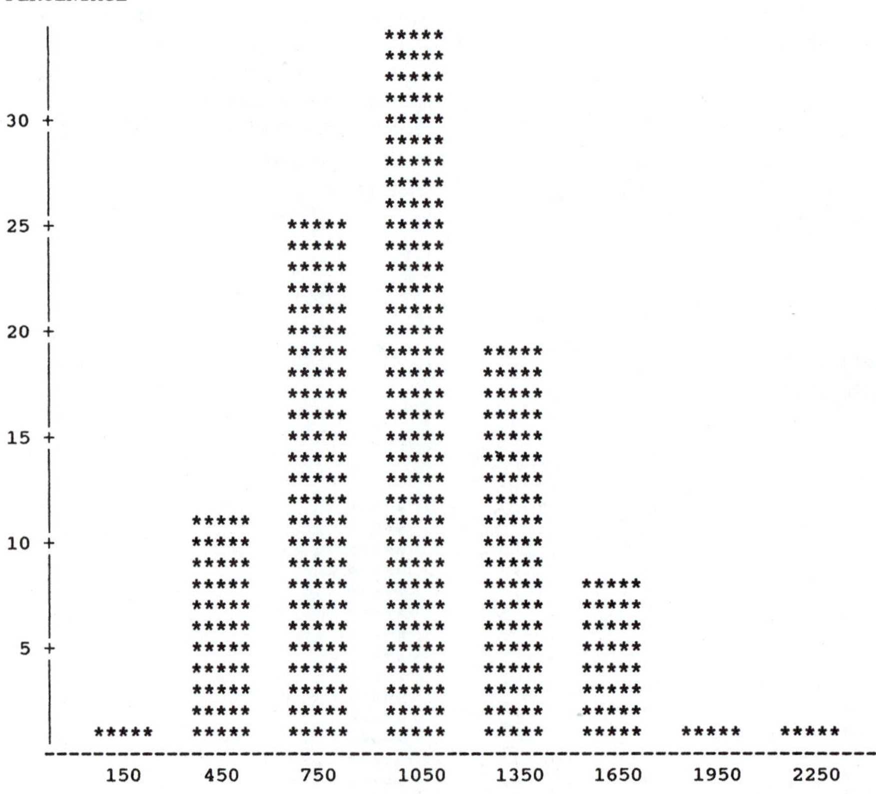

b. Weight

FIGURE 3.18 *Continued*

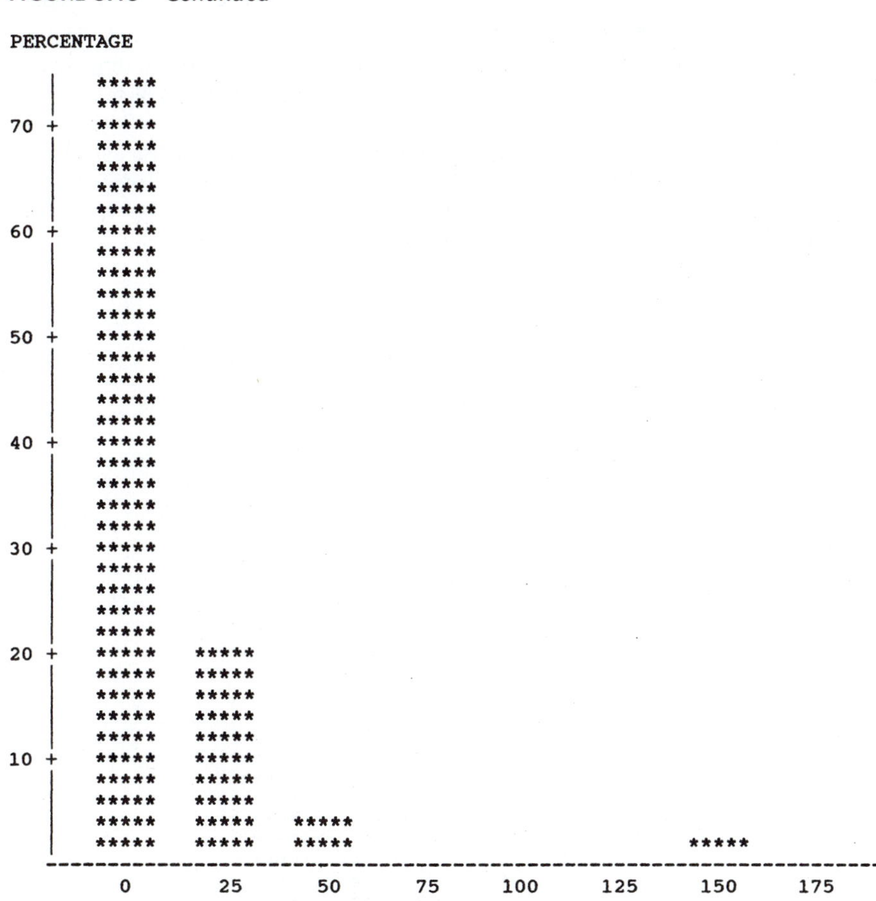

c. DDT concentration

e. Explain what the distribution of DDT concentrations tells you about that data set.
f. Use areas under the frequency distribution to estimate the proportion of fish containing more than 5 parts per million of DDT. (*Note*: 5.0 ppm is the maximum safe amount as specified by the U.S. Food and Drug Administration.)

BIO **3.13** The 144 fish of Table 2, Appendix 1, consist of 96 channel catfish, 12 large-mouth bass, and 36 small-mouth buffalo. Figure 3.19 gives the SAS computer printout of the frequency distribution of fish lengths (in centimeters) for each of these species.

a. Examine the distributions for the three species and describe how they differ. Which species of fish tends to be the longest? Which species shows the greatest variation in fish lengths?
b. Use areas under the frequency distributions to estimate the proportion of fish for each species that exceed 30 centimeters in length.

FIGURE 3.19 Frequency distributions of fish lengths for three species of fish

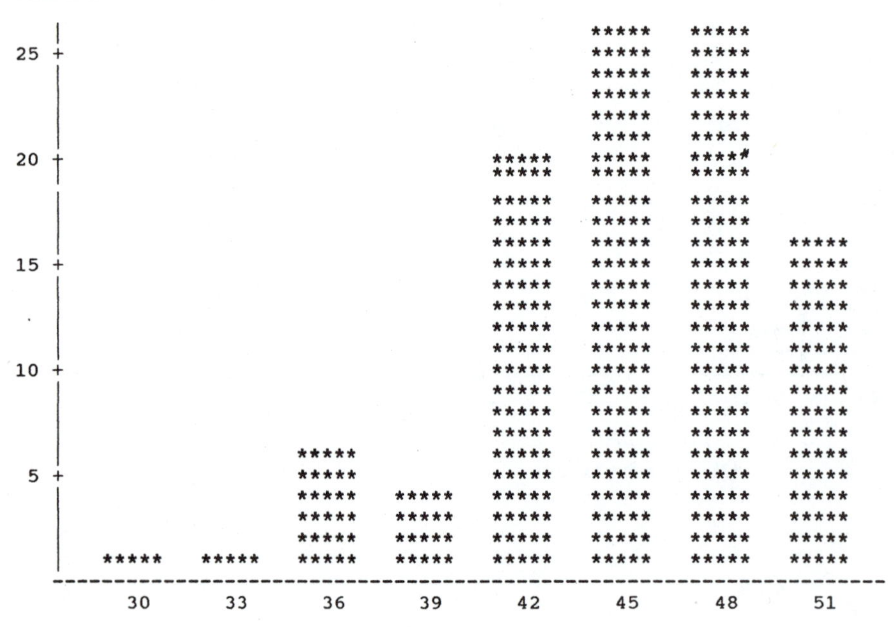

a. Channel catfish

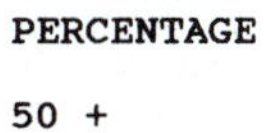

FIGURE 3.19 ***Continued***

PERCENTAGE

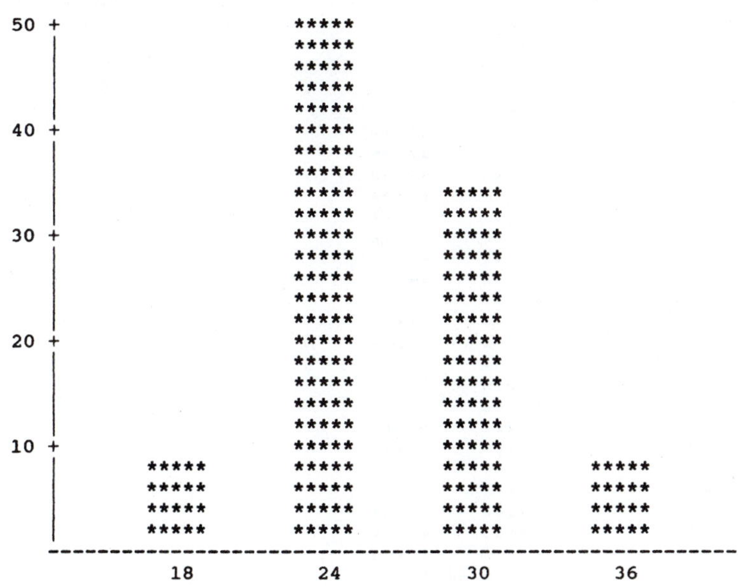

LENGTH MIDPOINT

b. Large-mouth bass

FIGURE 3.19 ***Continued***

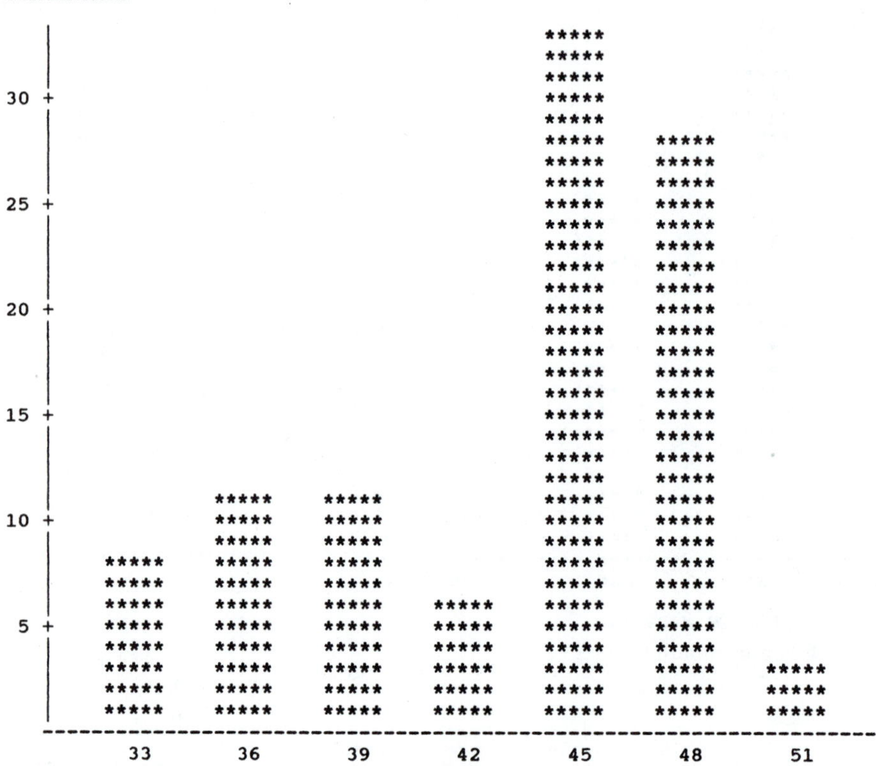

c. Small-mouth buffalo

BIO **3.14** Figure 3.20 shows the SAS printout of the frequency distribution of fish weights (in grams) for each of the three fish species of Table 2, Appendix 1.

a. Examine the distributions for the three species and describe how they differ. Which species of fish tends to weigh the least? Which species shows the least variation in fish weights?

b. Use areas under the frequency distributions to estimate the proportion of fish for each species that exceed 800 grams in weight.

FIGURE 3.20 Frequency distributions of fish weights for three species of fish

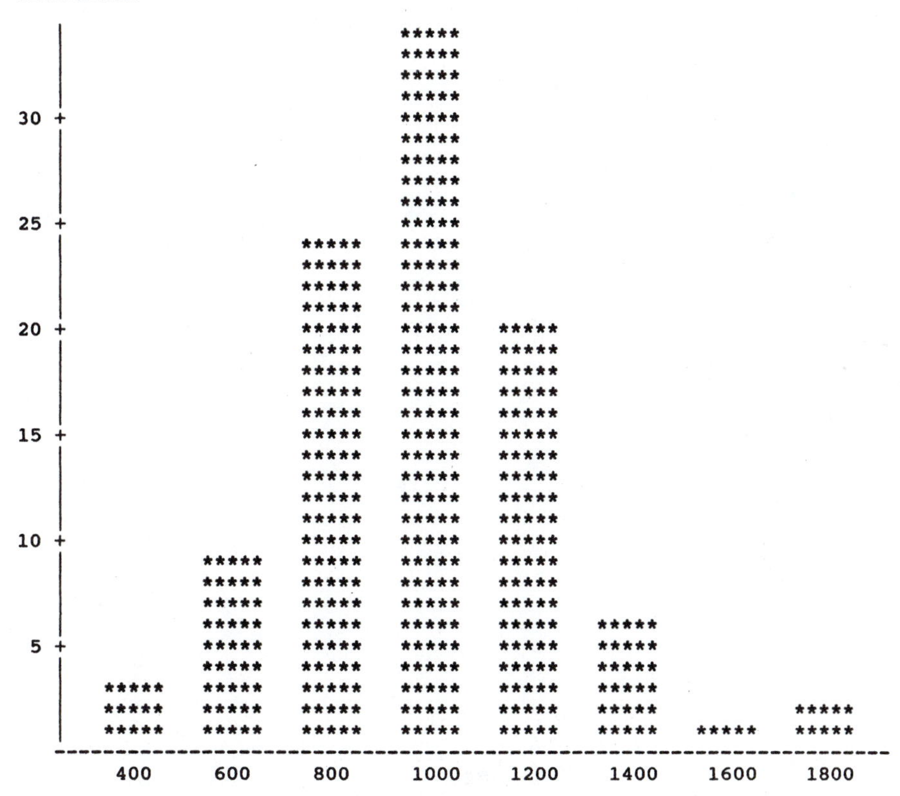

a. Channel catfish

FIGURE 3.20 ***Continued***

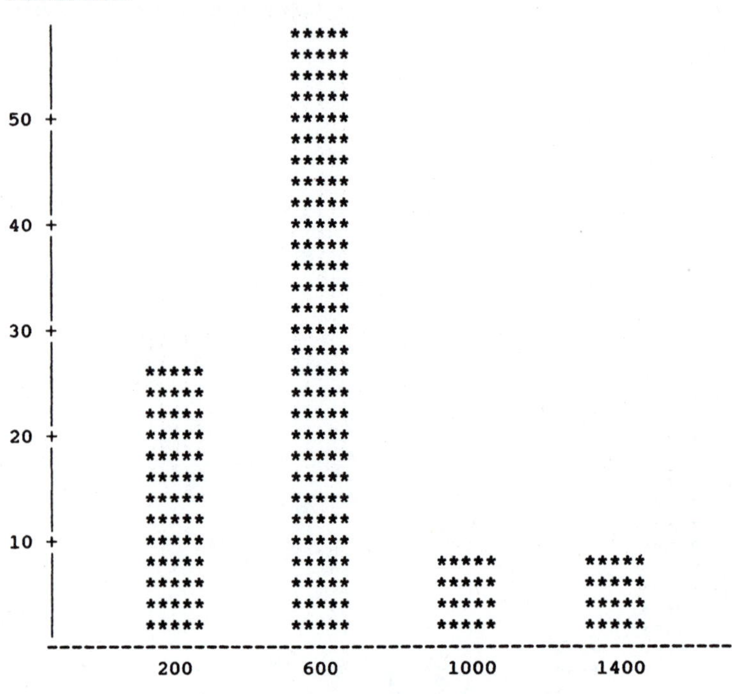

b. Large-mouth bass

FIGURE 3.20 ***Continued***

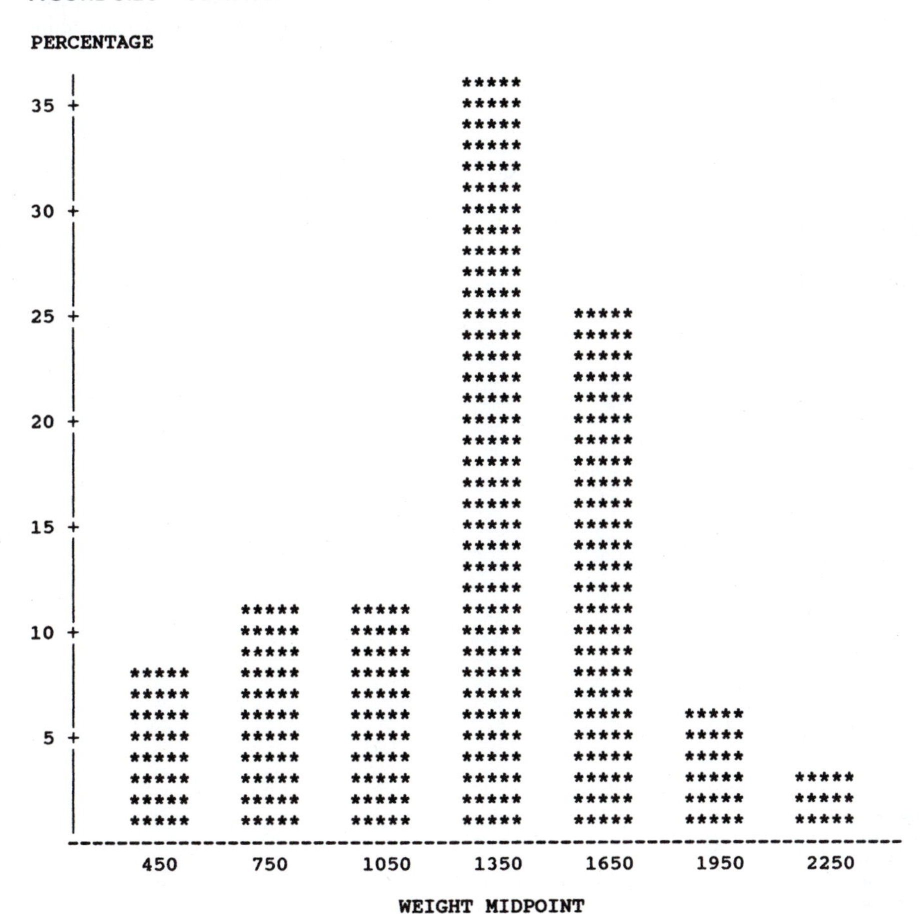

c. Small-mouth buffalo

3.15 Figure 3.21 shows the frequency distribution of concentrations (ppm) of DDT in fish for each of the three fish species of Table 2, Appendix 1.

BIO

a. Examine the distributions for the three species and describe how they differ.

b. Use areas under the frequency distributions to estimate the proportion of fish for each species that have concentrations of DDT exceeding 5.0 ppm.

FIGURE 3.21 Frequency distributions of the concentrations of DDT in three species of fish

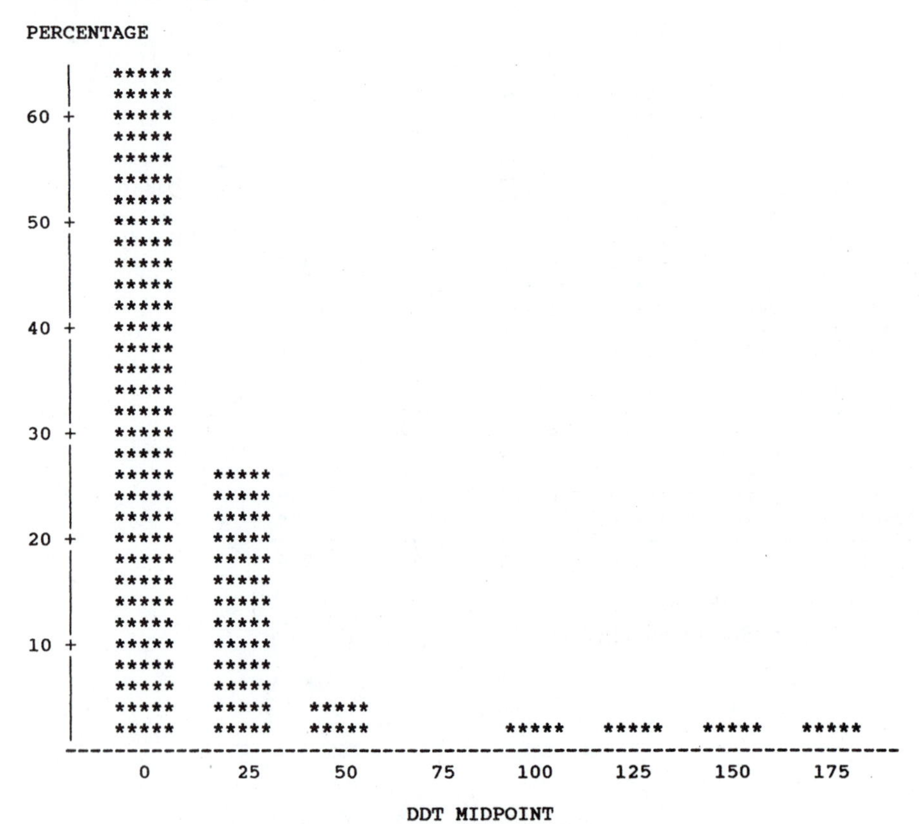

a. Channel catfish

FIGURE 3.21 ***Continued***

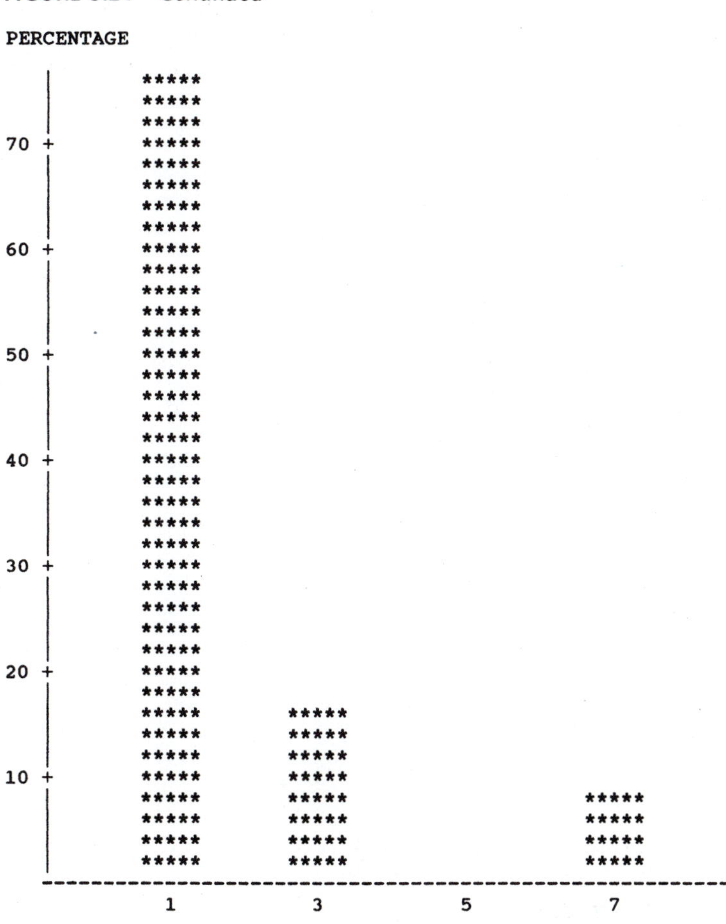

b. Large-mouth bass

FIGURE 3.21 ***Continued***

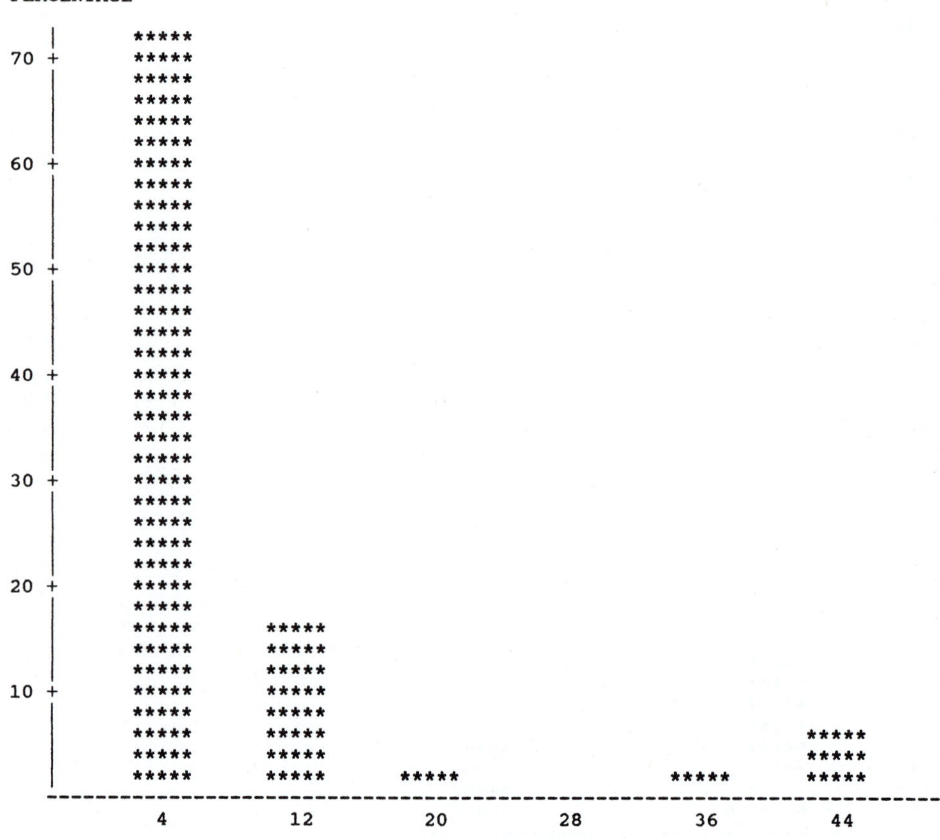

c. Small-mouth buffalo

GEN **3.16** Table 3.6 gives the basic subway fare for 10 major cities of the world (*U.S. News & World Report*, June 25, 1990).

TABLE 3.6 Basic Subway Fares (in dollars)

City	Fare	City	Fare
Moscow	.01	Mexico	.11
Tokyo	.78	Osaka	.78
New York	1.15	Leningrad	.01
Paris	.87	Seoul	.28
London	1.17	Hong Kong	.32

a. Construct (manually or with the aid of a computer) a relative frequency distribution to portray the distribution of basic subway fares. Use five classes to span the range, 0 to .25, .26 to .50, .51 to .75, etc.

b. Explain why we have chosen to use 5 classes rather than 20 to span the data set.

c. Is the data set mound-shaped? Is it symmetric? Is it skewed?

d. Give the class interval that contains the smallest measurement.

e. Give the class interval that contains the largest measurement.

f. Where, approximately, does the distribution appear to be centered?

DEM **3.17** Table 3.7 (*The New York Times*, April 7, 1990) gives the average life expectancy for people living in 33 developed nations. Construct, manually or with the use of a computer, a relative frequency distribution of these 33 life expectancies and explain what this graphical description tells you about life expectancy in the 33 countries.

TABLE 3.7 Life Expectancies in 33 Developed Nations

Country	Life Expectancy	Country	Life Expectancy
Australia	76.3	Italy	75.5
Austria	75.1	Japan	79.1
Belgium	74.3	Luxembourg	74.1
Britain	75.3	Malta	74.8
Bulgaria	71.5	The Netherlands	76.5
Canada	76.5	New Zealand	74.2
Czechoslovakia	71.0	Norway	76.3
Denmark	74.9	Poland	71.0
East Germany	73.2	Portugal	74.1
West Germany	75.8	Rumania	69.9
Finland	74.8	Soviet Union	69.8
France	75.9	Spain	76.6
Greece	76.5	Sweden	77.1
Hungary	69.7	Switzerland	77.6
Iceland	77.4	United States	75.0
Ireland	73.5	Yugoslavia	71.0
Israel	75.2		

3.4 Stem-and-Leaf Displays

A *stem-and-leaf display*, another method for describing a set of quantitative data, is a variation on the concept of a relative frequency distribution. It provides a quick and easy method to describe a small set of sample data.

Figure 3.22 shows a stem-and-leaf display for the 30 TCHOL measurements of Table 3.4.

FIGURE 3.22 Stem-and-leaf display for the 30 total cholesterol measurements of Table 3.4

Stem	Leaf
1	23, 31
1	53, 59, 61, 68, 69, 70, 76, 90
2	00, 02, 03, 09, 15, 24, 30, 35, 35, 36, 37
2	67, 69, 69, 72, 78, 89, 95
3	12, 37

Stem-and-Leaf Displays: How to Construct Them

To construct a stem-and-leaf display, we divide each measurement into two parts as shown in Figure 3.23. **The portion of the number lying to the left of the vertical divider is called the *stem* of the number; the portion to the right is called the *leaf*.** The location of the divider between stems and leaves is at the same location for all measurements—between a specific pair of digits.

FIGURE 3.23 Dividing measurements into stems and leaves

2 | 36   2 | 67   2 | 30

Figure 3.23 shows the stem and leaf for each of the first three measurements in the TCHOL data set of Table 3.4. For example, for the measurement 236, the stem is 2 and the leaf is 36. This particular example of a stem-and-leaf display places the divider between the hundreds digit and the tens digit.

The stems of a stem-and-leaf display are the same as the classes of a relative frequency distribution. The location of the divider between stems and leaves is up to us. When choosing the point of division between a stem and a leaf, the objective is to arrive at a number of stems that will show a piling up of the data in the stems.

One choice for the stems for the TCHOL data would be to locate the point of division between the hundreds digit and the tens digit, as was done in Figure 3.23. But, since the measurements range from the smallest, 123, to the largest, 337, we would have only three stems corresponding to the three hundreds digits, 1, 2, and 3. A better choice would be to split the tens digit into two sets: 0 to 4 and 5 to 9. This would create six stems. Each of the hundreds digits (1, 2, 3) would appear twice—the first time for measurements for which the tens digit is 0 to 4, and the second for measurements with tens digit of 5 to 9.

The complete set of data is then arranged in rows, one row for each of the six stems described above (see Figure 3.22). The stems are arranged in order, usually from the top to the bottom, along the vertical axis of the display. (Some stem-and-leaf displays arrange the stems from bottom to top.)

Each leaf is then placed in the row corresponding to its stem. For example, the leaf in the measurement 236 is 36. Since 236 contains a 3 in the tens digit, it is placed in the first "2" stem row. Similarly, the leaf for the number 267 is 67. It contains a 6 in the tens digit, so it is placed in the second "2" stem row. **This process is continued until all of the leaves are recorded. Then the measurements within each stem row are arranged in order, from the smallest to the largest.** (Note that the second stem row for the hundreds digit 3 has been eliminated, because there were no measurements in that row.)

The resulting data display, Figure 3.22, looks like a frequency distribution tipped on its side. In addition, it presents the measurements in order and in a manner so that an observer of the display can recover the original data set. For example, starting at the top of the display, the measurements are 123, 131, 153, 159, and so on.

Modified Stem-and-Leaf Displays

Some statisticians prefer to simplify the stem-and-leaf display by retaining only the first digit in a leaf. Thus, for example, the leaf 23 would become a 2 and the leaf 31 would become a 3. The disadvantage of this type of modification, shown in Figure 3.24, is that you can no longer reconstruct the data set from the display.

A stem-and-leaf display is especially useful for describing a small set of data. The display not only shows how the data are distributed, but it also lets you recombine the stems with the leaves and thereby reconstruct the data set. Another

FIGURE 3.24 A modification of the stem-and-leaf display of Figure 3.22

Stem	Leaves
1	2, 3
1	5, 5, 6, 6, 6, 7, 7, 9
2	0, 0, 0, 0, 1, 2, 3, 3, 3, 3, 3
2	6, 6, 6, 7, 7, 8, 9
3	1, 3

advantage is that it arranges the data in order, which is of value when using certain statistical methods. A disadvantage of stem-and-leaf displays is that they do not show relative frequencies. They are also awkward to use if the data set is large.

Computer-Generated Stem-and-Leaf Displays

Most of the stem-and-leaf displays that you will see are computer-generated. The computer will choose the points of division between stems and leaves so that they create intervals of equal width. The procedure for choosing the stems and the resulting printout of stems and leaves vary from one software package to another. Some, including Minitab, print the stem, from the smallest in the top row of the diagram to the largest at the bottom (as shown in Figure 3.25a). Others, including SAS, reverse the order, as shown in Figure 3.25b. Some print the entire leaf for a measurement, whereas others (such as Minitab) drop digits from the leaves so that it is impossible to reconstruct the data.

FIGURE 3.25 Computer-generated stem-and-leaf displays for the 30 total cholesterol measurements of Table 3.4

```
Stem-and-leaf of tchol30   N  = 30
Leaf Unit = 10


    2      1 23
    4      1 55
    9      1 66677
   10      1 9
   15      2 00001
   15      2 233333
    9      2
    9      2 66677
    4      2 89
    2      3 1
    1      3 3
```

a. Minitab

```
Stem Leaf                      #
  32 7                         1
  30 2                         1
  28 95                        2
  26 79928                     5
  24
  22 405567                    6
  20 02395                     5
  18 0                         1
  16 18906                     5
  14 39                        2
  12 31                        2
     ----+----+----+----+
Multiply Stem.Leaf by 10**+1
```

b. SAS

Figure 3.25 gives the Minitab and SAS versions of the stem-and-leaf display that we show in Figure 3.22. In order to understand the printouts of the stem-and-leaf displays for these or any of the other software packages, see the manual that accompanies your particular package.

EXERCISES

3.18 Construct a stem-and-leaf display for the subway fare data set of Exercise 3.16. Describe the shape, location, and spread of the data set.

3.19 Construct a stem-and-leaf display for the life expectancy data set of Exercise 3.17. Describe the shape, location, and spread of the data set.

3.20 Figure 3.26 shows a stem-and-leaf display for a data set.

a. How many measurements are in the data set?
b. Find the smallest measurement in the data set.
c. Find the largest measurement in the data set.
d. Locate the third largest measurement in the data set and identify its stem and its leaf.
e. What percentage of the measurements in the data set exceed 899?

FIGURE 3.26 Stem-and-leaf display for Exercise 3.20

Stem	Leaves
1	09
2	12, 54
3	23, 59, 71
4	18, 25, 45, 62, 95
5	16, 39, 57, 80
6	21, 67, 92
7	11, 31, 42
8	19, 58
9	04
10	
11	26

3.21 Refer to Exercise 3.20 and reconstruct the data set.

3.5 Key Words and Concepts

- ***Qualitative data sets*** are described using ***pie charts*** or ***bar graphs***.
- Pie charts and bar graphs show graphically the proportions (***relative frequencies***) of the total number of observations that fall in the various categories.
- ***Quantitative data sets*** are described using ***relative frequency distributions*** and ***stem-and-leaf displays***.
- Both relative frequency distributions and stem-and-leaf displays show how data are distributed over contiguous ***class intervals*** of equal width.
- Relative frequency distributions also show the ***class relative frequency*** (or class frequency or class percentage) for each class. In addition, the area of a bar over an interval enables us to visually approximate the proportion of the total number of measurements that fall in that interval.
- A stem-and-leaf display provides a quick and easy way to describe a small data set. It provides a satisfactory outline of the shape of the frequency distribution and makes it relatively easy to calculate a class frequency. It also presents the data in order, and allows us to reconstruct the data set. These advantages tend to diminish as the sample size increases and it is awkward to use a stem-and-leaf display for large data sets.

SUPPLEMENTARY EXERCISES

HLTH **3.22** The water that you drink may be harmful to your health, particularly if you live in an older city that is still using lead-lined pipes that were installed in some of the early water systems. Table 3.8 gives measurements of the average amounts (in milligrams per liter) of lead, copper, and iron in samples of water taken each day for 23 days from the Boston water system. A SAS relative frequency distribution for the lead concentration measurements is shown in Figure 3.27.

a. What does the relative frequency distribution tell you about the data set?

b. The Environmental Protection Agency (EPA) specifies that the maximum amount of lead allowed in drinking water is .05 milligram/liter. Use areas under the relative frequency distribution to estimate the proportion of water samples that have concentrations of lead that exceed .05 milligram per liter.

TABLE 3.8 Lead, Copper, and Iron Concentrations in Boston Water, 1983

Lead		Copper		Iron	
.035	.015	.12	.04	.20	.14
.060	.015	.18	.04	.33	.12
.055	.022	.10	.05	.22	.12
.035	.043	.07	.07	.17	.16
.031	.030	.08	.10	.15	.17
.039	.019	.09	.04	.19	.13
.038	.021	.16	.08	.17	.15
.049	.036	.14	.05	.17	.13
.073	.016	.07	.05	.23	.14
.047	.010	.07	.04	.18	.11
.031	.020	.08	.04	.25	.11
.016		.07		.14	

Source: Karalekas, P. C., Jr., Ryan, C. R., and Taylor, F. B. "Control of Lead, Copper, and Iron Pipe Corrosion in Boston," *Journal of the American Water Works Association*, Vol. 75, No. 2, February 1983.

FIGURE 3.27 A SAS relative frequency distribution of lead concentrations in Boston water, 1983

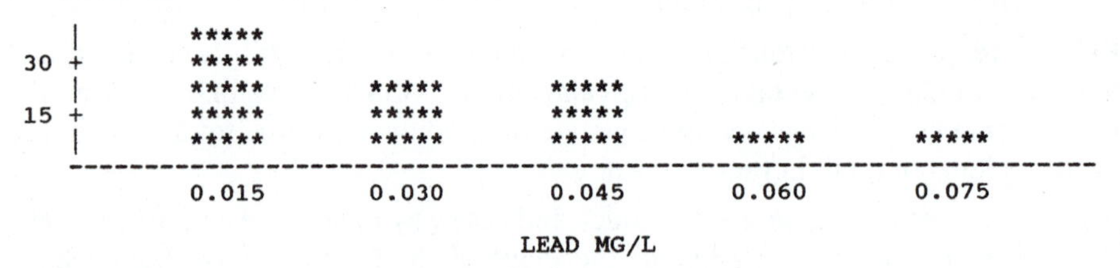

HLTH **3.23** Figure 3.28 gives the SAS printout of the relative frequency distribution for the copper concentration measurements of Table 3.8.

a. What does the relative frequency distribution tell you about the data set?

b. The EPA specifies that the maximum amount

PERCENTAGE

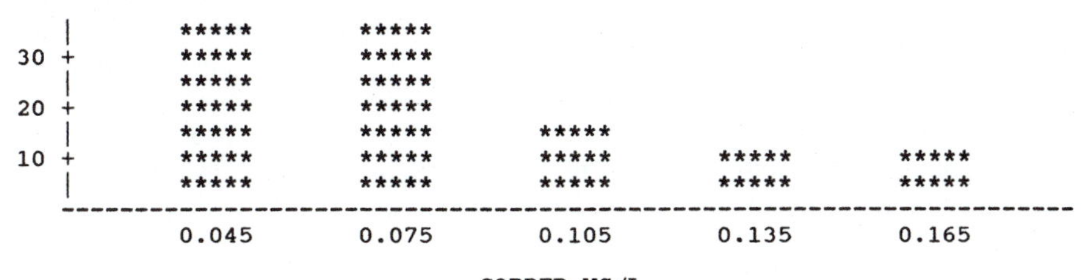

FIGURE 3.28 A SAS relative frequency distribution of copper concentrations in Boston water, 1983

of copper allowed in drinking water is 1.0 milligram/liter. Use areas under the relative frequency distribution to estimate the proportion of water samples that have concentrations of copper that exceed 1.0 milligram per liter.

HLTH **3.24** Figure 3.29 gives the SAS printout of the relative frequency distribution for the iron concentration measurements of Table 3.8.

a. What does the relative frequency distribution tell you about the data set?

b. The EPA specifies that the maximum amount of iron allowed in drinking water is .3 milligram per liter. Use areas under the relative frequency distribution to estimate the proportion of water samples that have concentrations of iron that exceed .3 milligram per liter.

PERCENTAGE

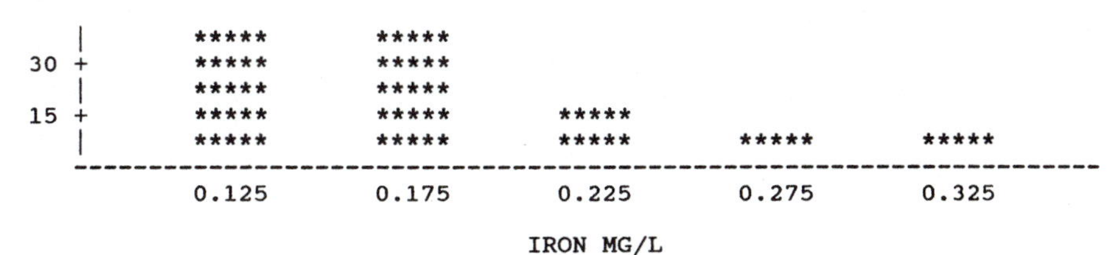

FIGURE 3.29 A SAS relative frequency distribution of iron concentrations in Boston water, 1983

EXERCISES FOR YOUR COMPUTER

BUS **3.25** Refer to Case 2.1 and the percentage iron data given in Table 1, Appendix 1. Use the data disk supplied with this text and a statistical software package to construct a relative frequency distribution for the 390 percentage iron measurements.

a. Explain what the relative frequency distribution tells you about the sample data.
b. Describe the population of data from which the sample was selected.
c. Why would anyone be interested in knowing about the population?
d. What does the sample relative frequency distribution tell you about the population relative frequency distribution?

3.26 Refer to Case 2.3 and the cholesterol data given in Table 3, Appendix 1. Use a statistical software package to construct a relative frequency distribution for the 107 LDL measurements. In the context of cholesterol levels, explain what the relative frequency distribution tells you about the sample data.

3.27 Refer to Case 2.3 and the cholesterol data given in Table 3, Appendix 1. Use your computer to construct a relative frequency distribution of the TCHOL measurements for all female members of the data set. Construct a distribution for all male members. Compare the two sample relative frequency distributions. Do the two distributions suggest differences in the cholesterol levels of men and women?

References

1. Freedman, D., Pisani, R., Purves, R., and Adhikari, A., *Statistics*, 2nd edition. New York: W. W. Norton & Co., 1991.
2. McClave, J. and Dietrich, F., *Statistics*, 5th edition. San Francisco: Dellen Publishing Co., 1991.
3. Mendenhall, W. and Beaver, R., *Introduction to Probability and Statistics*, 8th edition. Boston: PWS-Kent, 1991.
4. Moore, D. and McCabe, G., *Introduction to the Practice of Statistics*, 2nd edition. New York: W. H. Freeman & Co., 1993.
5. Sincich, T., *Statistics by Example*, 4th edition. San Francisco: Dellen Publishing Co., 1990.

four

1 2 3 4 5 6 7 8 9 10 11

Numerical Descriptions of Quantitative Data

In a Nutshell

A graph allows us to visualize the distribution of a set of sample data but it does not tell us by how much it may differ from the distribution of the sampled population. To measure that difference, we need to quantify the characteristics of a data set.

4.1 Why We Need a Numerical Method for Describing a Data Set

4.2 Measures of Central Tendency

4.3 Measures of Variation or Spread

4.4 Applying the Empirical Rule: Approximating a Standard Deviation

4.5 Applying the Empirical Rule: Processes and Statistical Control

4.6 Measures of Relative Standing

4.7 Looking for Outliers

4.8 Numerical Descriptions of TCHOL and RATIO, Case 2.3

4.9 Statistics and Parameters

4.10 Key Words and Concepts

4.1 Why We Need a Numerical Method for Describing a Data Set

We learned in Chapters 2 and 3 that we sample because we want to make an inference about the population from which the sample was selected. The set of 30 TCHOL measurements given in Table 3.4 is a sample selected from among the millions of TCHOL measurements associated with all "healthy" people, aged 17 to 62, in Florida. The researcher did not sample to learn about the TCHOL values of the 30 people. His objective was more general: He wanted to use what he learned about the sample to deduce (i.e., infer) the nature of the TCHOL measurements for all people in the sampled universe.

One way to infer the nature of a population of quantitative data is to construct a relative frequency distribution for a sample of data and use it to help us envision the unknown relative frequency distribution for the population.

For example, Figure 4.1 reproduces the relative frequency distribution for the sample of 30 TCHOL measurements given in Table 3.4. We would imagine that the relative frequency distribution for the population would be similar to Figure 4.1. It is also possible that it looks quite different. That, of course, is the problem. If we use the irregular shape of the sample relative frequency distribution to infer the

FIGURE 4.1 The relative frequency distribution for the sample of $n = 30$ total cholesterol measurements

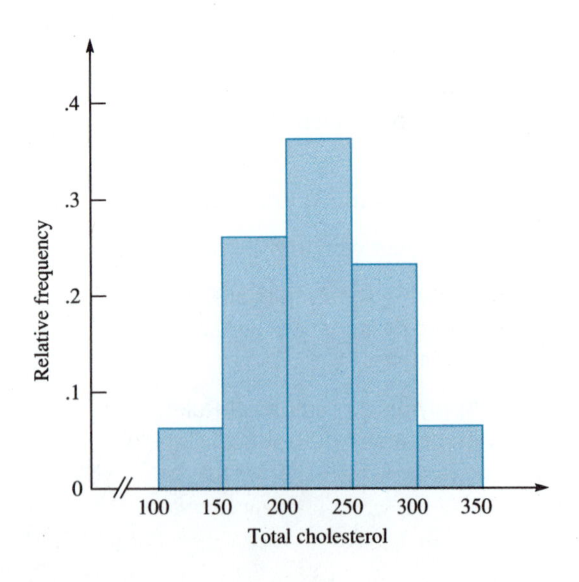

shape of the population distribution, how much in error are we likely to be? And, how do we measure that error? How do we measure the difference between two irregular figures, the sample and the population relative frequency distributions?

One way to solve this problem is to look for some numbers that measure the essential features of a relative frequency distribution. Then we could use the sample descriptive numbers to infer the values of the population numbers.

For example, we would guess that the location of the center of the population distribution would be close to the center of the sample distribution shown in Figure 4.1. We would also guess that the spread of the two distributions, from the smallest to the largest measurements, would be similar.

Noting that the sample relative frequency distribution in Figure 4.1 centers near 225 and that the measurements vary from 100 (the lower class boundary), to 350 (the upper class boundary), we would imagine that the population relative frequency distribution is centered at or near 225 and is spread over an interval equal to or slightly larger than the spread of the sample. The population relative frequency distribution might appear as shown in Figure 4.2.

FIGURE 4.2 How the TCHOL population relative frequency distribution might look

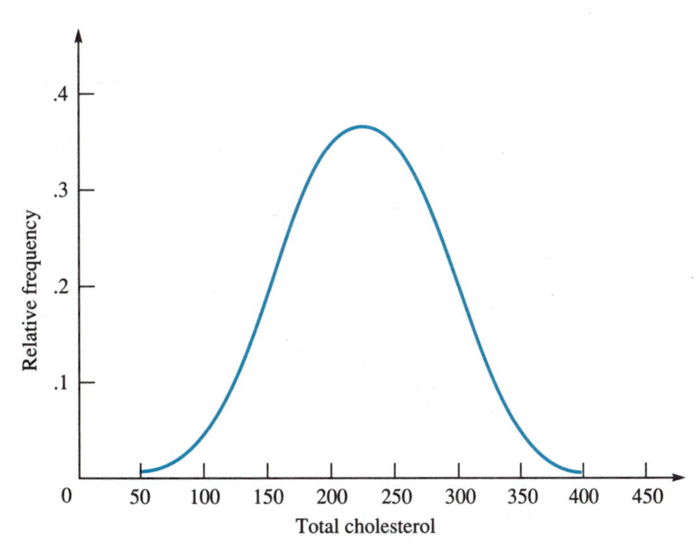

A ***numerical descriptive method*** **for describing a set of data involves the selection of two numbers that enable us to create a mental image of the relative frequency distribution for that data set.** The first number locates the center of the distribution; the second measures its spread. See Figure 4.3 on page 74.

FIGURE 4.3 Measuring the location and spread of a distribution

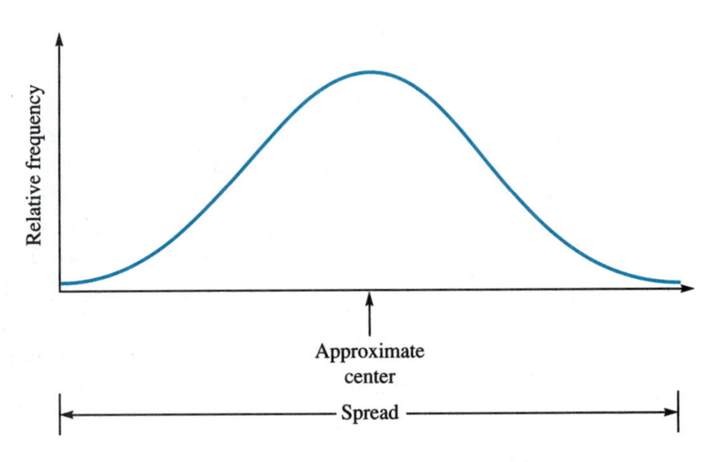

4.2 Measures of Central Tendency

Suppose that you were asked to give a single number that would be most typical of the 30 total cholesterol measurements in Table 3.4. What number would you choose? We think that you would pick a number near the center of the relative frequency distribution shown in Figure 4.1, say 225.

A number that locates the center of a relative frequency distribution is called a *measure of central tendency*.

We will describe two measures of central tendency: the mean and the median. Each is useful, in a different way, as will be seen below.

A Measure of Central Tendency: The Mean

The mean of a set of data is very familiar to us. It is what, in ordinary usage of the word, we refer to as an "average." When we speak of the average length of time it takes to complete a homework assignment, we have in mind the mean of a sample of times to complete an assignment or the mean of some conceptual population of times.

The *mean of a set of measurements* is their arithmetic average, i.e.,

$$\text{Mean of a set of } n \text{ measurements} = \frac{\text{Sum of the measurements}}{n}$$

Some Essential Notation

If we let the symbol x represent the variable that we are measuring, then the *sample mean*, represented by the symbol $\bar{x}$, is given by the formula

$$\bar{x} = \frac{\text{Sum of the sample measurements}}{n}$$

or, in summation notation,*

$$\bar{x} = \frac{\Sigma x}{n}$$

Example 4.1

Find the mean of the following sample of $n = 5$ measurements: 3.1, 4.6, 2.9, 6.0, and 4.7.

Solution

The mean of the sample of $n = 5$ measurements is

$$\bar{x} = \frac{3.1 + 4.6 + 2.9 + 6.0 + 4.7}{5} = \frac{21.3}{5} = 4.26$$

Does this sample mean, $\bar{x} = 4.26$, locate the center of this set of five measurements? The smallest measurement is 2.9, the largest is 6.0. Therefore, we see that the sample mean, $\bar{x} = 4.26$, falls near the center of this set of data.

The mean is the point at which measurements to the left of the mean balance those to the right (see Figure 4.4). If the largest measurement in the sample in Example 4.1 shifts to the right, then the mean (the balance point) must shift with it.

FIGURE 4.4 The mean as a balance point

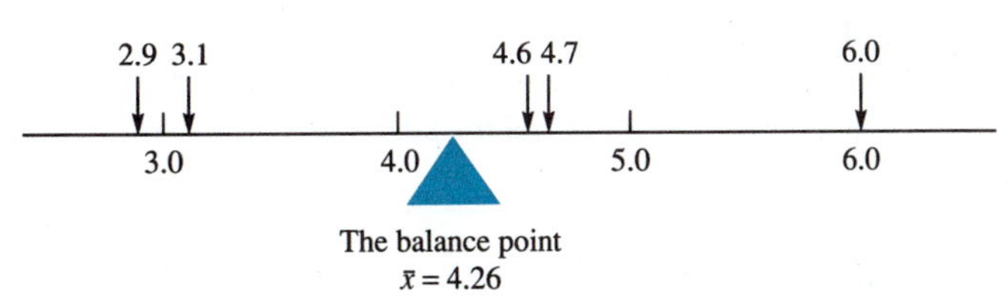

* The Greek symbol Σ (sigma) is mathematical notation for the word *sum*. Σx tells us to sum all x values in the sample.

A Measure of Central Tendency: The Median

A second measure of the central tendency of a set of data is the median. **The *median*, *m*, of a set of *n* measurements is the value of *x* that falls in the middle when the measurements are ranked in order from the smallest to the largest. If *n* is odd, the median is the midddle measurement; if *n* is even, the median is halfway between the two middle ranked measurements.**

Half of the measurements lie to the left of the median; half lie to the right. Since the total area under a relative frequency distribution is equal to 1, the median is the value of x that divides the area under the relative frequency distribution in half, with .5 to the left of the median and .5 to the right. See Figure 4.5.

FIGURE 4.5 Interpretation of the median

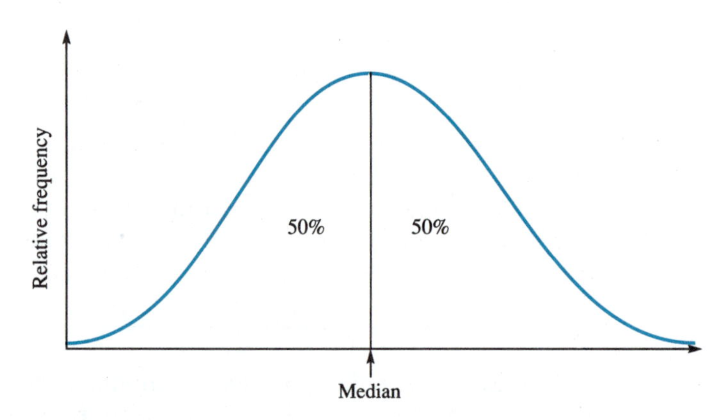

Example 4.2

Find the median for the following sample of $n = 5$ measurements: 3.1, 4.6, 2.9, 6.0, and 4.7.

Solution

The $n = 5$ measurements, ranked in order, are 2.9, 3.1, 4.6, 4.7, and 6.0. Since $n = 5$ is odd, the median is the measurement in the middle or, $m = 4.6$.

Example 4.3

Find the median for the following sample of $n = 6$ measurements: 15, 9, 12, 6, 4, and 29.

Solution

The ranked measurements are 4, 6, 9, 12, 15, and 29. Since $n = 6$ is even, the median m is halfway between the two middle measurements, or $m = (9 + 12)/2 = 10.5$.

The median is less affected than the mean when the data include a few measurements that deviate greatly from the others. To illustrate, in Example 4.1, we found that the mean of the set of five measurements, 2.9, 3.1, 4.6, 4.7, and 6.0, was equal to 4.26. The median for the same five measurements was found in Example 4.2 to equal 4.6. Suppose that we replace the largest measurement, 6.0, by the much larger measurement 20.0. Because $\bar{x}$ is a balance point, the mean of the sample shifts to the right to $\bar{x} = 7.06$ (see Figure 4.6). In contrast, the median for this new set of measurements is still 4.6.

FIGURE 4.6 The effect of a large measurement on the mean

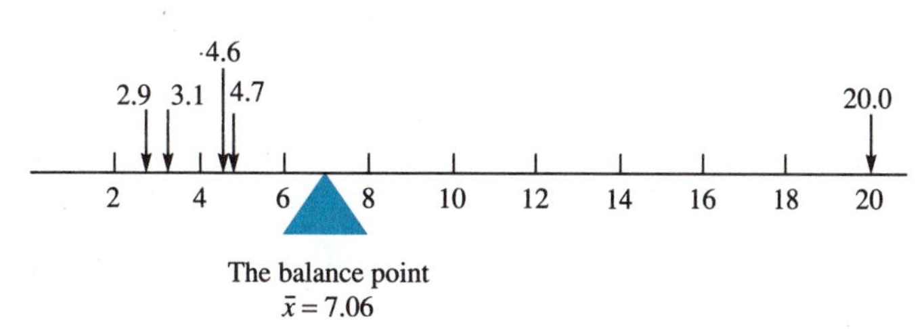

The median is often preferred to the mean when describing distributions of data in the social sciences, because we do not want a "typical" income, net worth, family expenditure, etc., to be influenced by a few extremely large (or small) measurements. For example, if we wanted to give a measure of a typical family's net worth in Arkansas, we would not want the enormous net worth of a very wealthy person (e.g., the late Sam Walton, the founder of Wal-mart and one of the wealthiest people in the United States) to overly inflate the value. The median would be less affected by a few large values than would the mean.

The mean is more commonly used than the median to typify the center of most data sets because we are accustomed to thinking in terms of "averages"—i.e., means. In addition, the mean is easier to use for making inferences. We will show you in Chapter 6 how to use a sample mean to estimate the value of a population mean. When we do, we will also place a bound, i.e., an upper limit, on our error of estimation.

4.3 Measures of Variation or Spread

Now that we know how to locate the center of a distribution, we want to know something about its variation or spread: Is it concentrated and thus easy to visualize, or widely dispersed and thus more difficult to analyze? For example, do most of the measurements in the total cholesterol relative frequency distribution shown in Figure 4.1 fall near the center of the distribution, or do they vary greatly, say from 100 to 350? The most widely used measures of variation are the *range*, the *variance*, and the *standard deviation* of a set of measurements.

A Measure of Variation: The Range

The simplest measure of spread or variation in a set of data is the range. **The *range* is the difference between the largest and the smallest members of the data set.**

Example 4.4

Find the range of the $n = 30$ measurements in the TCHOL data set given in Table 3.4.

Solution

The smallest member of the set of $n = 30$ measurements in Table 3.4 is 123 and the largest is 337. Therefore, the range of the set of data is (see Figure 4.7)

$$\text{Range} = \text{Largest} - \text{Smallest} = 337 - 123 = 214$$

FIGURE 4.7 The range for the 30 TCHOL measurements of Table 3.4

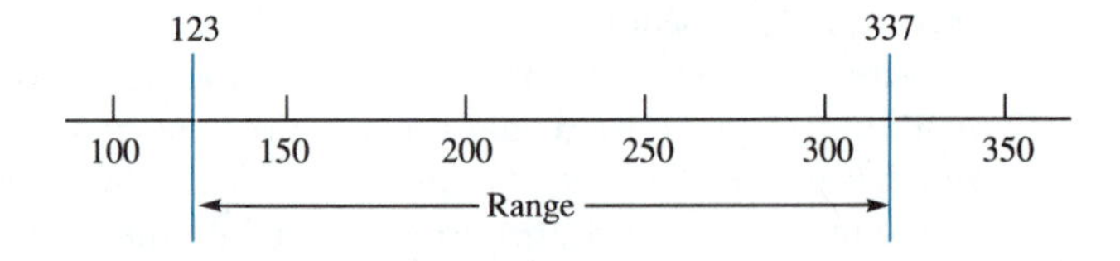

A Measure of Variation: The Variance

A second and more sensitive measure of variation is based on the deviations of the measurements from their mean.

The *deviation* of a measurement x from a sample mean $\bar{x}$ is equal to $(x - \bar{x})$. For example, if a measurement is $x = 2$ and the mean is $\bar{x} = 4.1$, then the deviation of $x = 2$ from its mean is $(x - \bar{x}) = (2 - 4.1) = -2.1$ (see Figure 4.8). **Note that if a deviation is negative, x lies to the left of the mean. If the deviation is positive, x lies to the right of the mean.**

FIGURE 4.8 The deviation of $x = 2$ from the sample mean, $\bar{x} = 4.1$

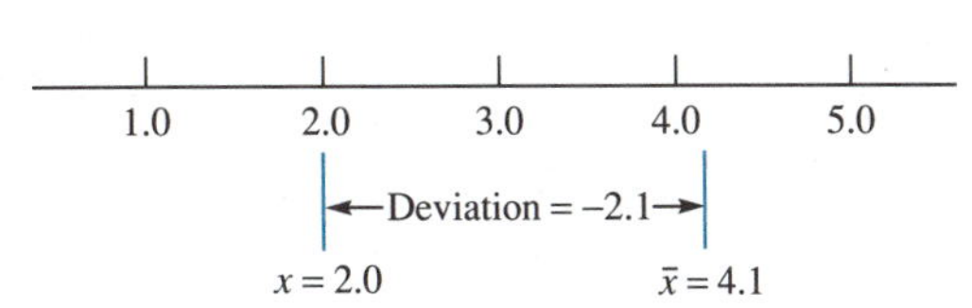

The larger the deviations of a set of measurements from their mean, the greater will be the variation of the data. There are a number of quantities that we could concoct (e.g., the average of the unsigned deviations) to measure the variability of a data set. The problem is that only one of these measures can be easily interpreted. It is called the *standard deviation* of a data set.

To find the standard deviation of a set of measurements, we must first calculate a quantity known as the variance. **The *variance* of a sample of n measurements, denoted by the symbol s^2, is equal to the sum of the squares of the deviations of the measurements from their mean divided by $(n - 1)$, i.e.,**

$$s^2 = \frac{\Sigma(x - \bar{x})^2}{n - 1}$$

Knowing the variance of a set of measurements will not help us picture their spread. However, the square root of the variance, called the standard deviation, will.

A Measure of Variation: The Standard Deviation

The *sample standard deviation*, denoted by the symbol s, is equal to the positive square root of the variance, i.e., $s = \sqrt{s^2}$.

The Empirical Rule

A rule of thumb explains how a standard deviation measures the variation in a data set or, equivalently, the spread in a relative frequency distribution. It is called the ***Empirical Rule*. It tells us that approximately 68% of the measurements will lie within one standard deviation, 95% within two standard deviations, and almost all of the measurements will lie within three standard deviations of their mean** (see Figure 4.9 on page 80).

FIGURE 4.9 The Empirical Rule: Percentages of measurements within one, two, and three standard deviations of the mean

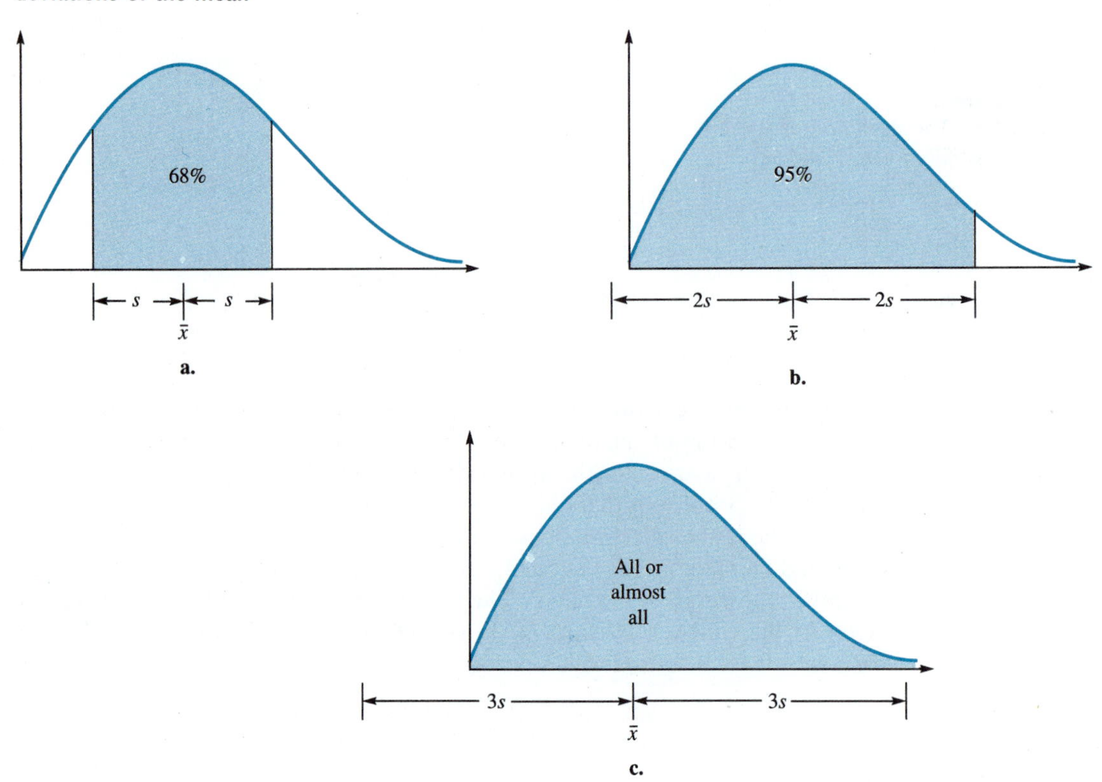

The Empirical Rule provides a good description for most sets of data observed in the real world. The percentages of the data in the intervals will not agree exactly with the percentages given by the Empirical Rule but they will be close enough, particularly for the intervals $\bar{x} \pm 2s$ and $\bar{x} \pm 3s$, to be of practical value.

Figure 4.10 shows frequency distributions for three data sets, each containing 16 measurements. Each data set contains a combination of the integers 1, 2, 3, 4, ... For example, the data set shown in Figure 4.10a contains eight 1's, three 2's, two 3's, one 4, one 5, and one 6. The mean and standard deviation are shown above each distribution, and the interval $\bar{x} \pm 2s$ is located below the horizontal axis of each graph. You can see that the range of each data set is approximately equal to 4s and that the percentage of observations falling in the interval $\bar{x} \pm 2s$ agrees reasonably well with the 95% given in the Empirical Rule.

FIGURE 4.10 Three distributions and their standard deviations

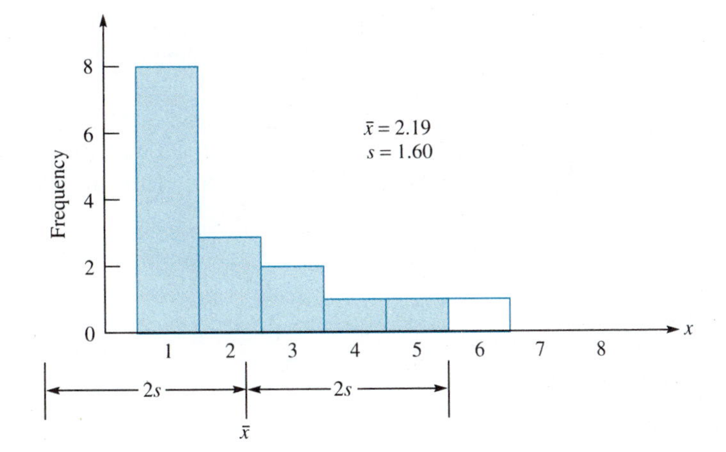

a. $\bar{x} \pm 2s$ contains 93.75% of the observations.

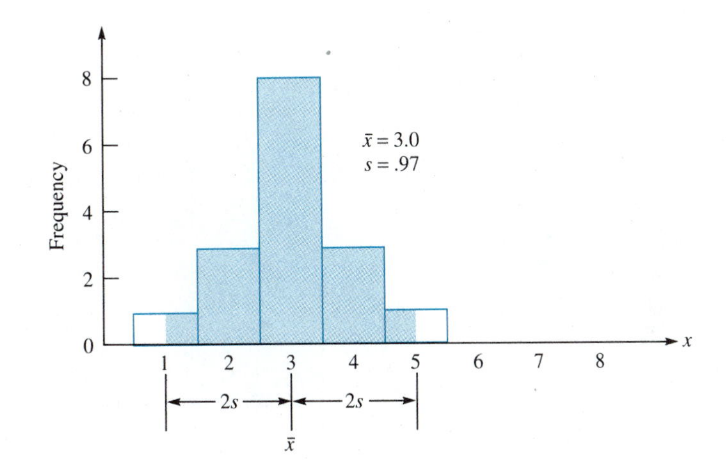

b. $\bar{x} \pm 2s$ contains 87.5% of the observations.

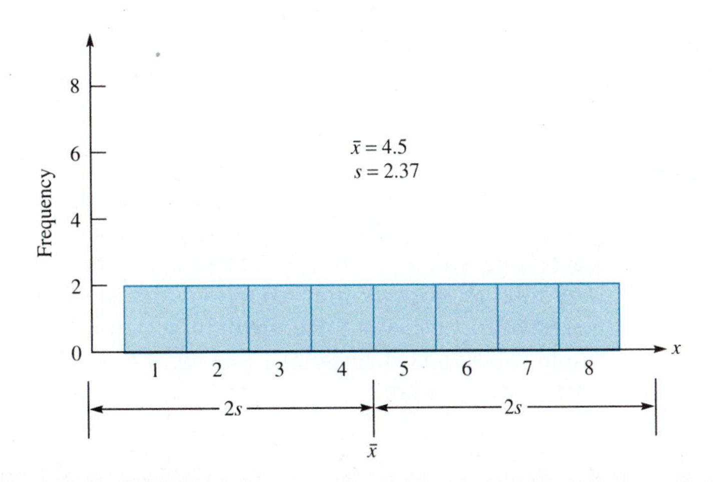

c. $\bar{x} \pm 2s$ contains 100% of the observations.

How well do the standard deviation and the Empirical Rule describe the variation of the 30 total cholesterol measurements of Table 3.4? A computer printout (to be shown at the end of this section) gives the mean and standard deviation for the 30 total cholesterol measurements as $\bar{x} = 220.5$ and $s = 54.9$. Figure 4.11 shows the relative frequency distribution for the data with the intervals $\bar{x} \pm s$, $\bar{x} \pm 2s$, and $\bar{x} \pm 3s$ marked along the horizontal axis of the distribution.

FIGURE 4.11 Relative frequency distribution for the 30 total cholesterol measurements of Table 3.4

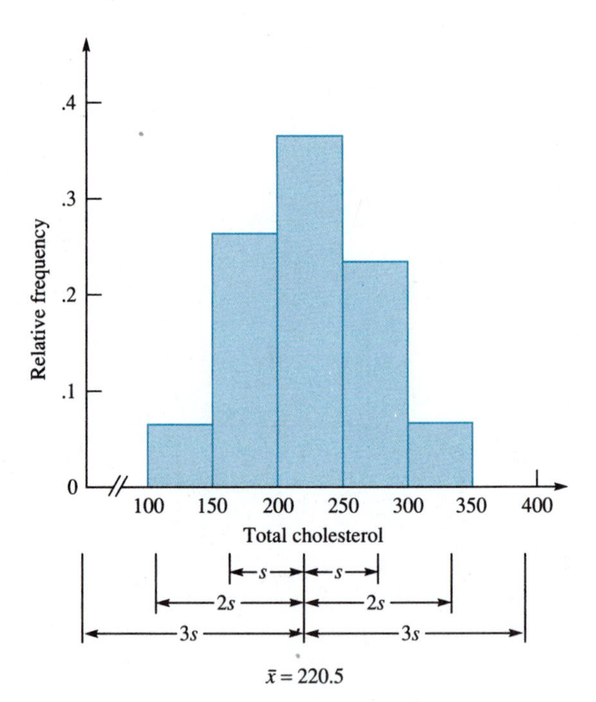

Table 4.1 shows the number and percentages of measurements falling in the intervals $\bar{x} \pm s$, $\bar{x} \pm 2s$, and $\bar{x} \pm 3s$. You can see that the proportion of measurements lying within one, two, and three standard deviations of the mean agree quite well with the proportions given in the Empirical Rule. Approximately 68% of the measurements lie within one standard deviation of the mean, approximately 95% within two standard deviations, and all or almost all lie within three standard deviations of the mean.

TABLE 4.1 Demonstrating the Empirical Rule

	Interval	Frequency TCHOL Data	Relative Frequency TCHOL Data	Empirical Rule
$(\bar{x} \pm s)$	165.6 to 275.4	20	.67	.68
$(\bar{x} \pm 2s)$	110.7 to 330.3	29	.97	.95
$(\bar{x} \pm 3s)$	55.8 to 385.2	30	1.00	All or almost all

Example 4.5

The scores on a national standard achievement test have a mean equal to 810 and a standard deviation equal to 107. Describe the distribution of scores.

Solution

The distribution of scores might appear as shown in Figure 4.12. The distribution is centered over the mean, 810. According to the Empirical Rule, approximately 68% of the scores should lie in the interval $\bar{x} \pm s = 810 \pm 107$, or 703 to 917. Most of the scores, close to 95%, should lie in the interval $\bar{x} \pm 2s = 810 \pm 2(107)$, or 596 to 1,024. All or almost all of the scores should lie in the interval $\bar{x} \pm 3s = 810 \pm 3(107)$, or 489 to 1,131.

FIGURE 4.12 Visual reconstruction of the relative frequency distribution, Example 4.5

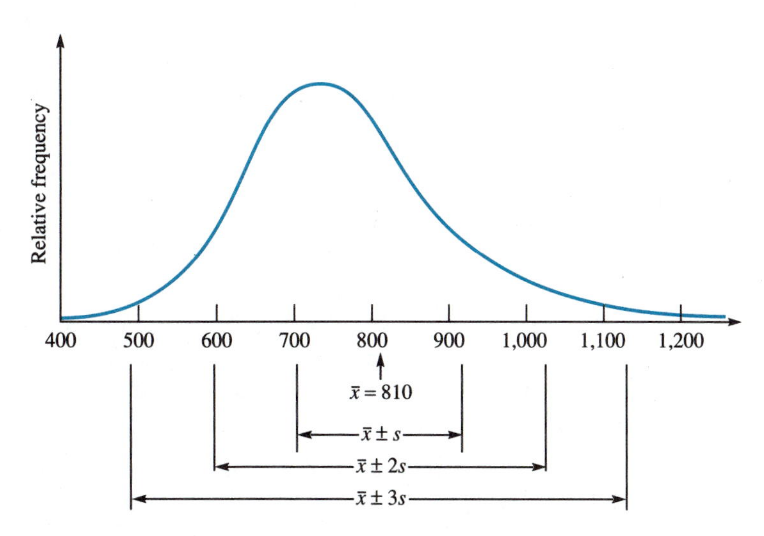

Calculating the Variance and Standard Deviation for a Data Set

In a practical situation, you will not calculate the variance and standard deviation of a set of data. It will be done on a computer. You only need to have a clear understanding of what the variance and standard deviation of a set of measurements really are, how to find the standard deviation on a computer printout, and, most important, how it measures the variation in a set of data. But, just to be sure that you really understand how a standard deviation measures variation, we will calculate one for a small set of data.

Example 4.6

Calculate the variance and the standard deviation for the following sample containing $n = 5$ measurements: 4, 0, 5, 7, and 2.

Solution

Recall that the sample variance s^2 is a function of the deviations of the measurements from their mean. The formulas for s^2 and for the standard deviation s are

$$s^2 = \frac{\Sigma(x - \bar{x})^2}{n - 1}$$

and

$$s = \sqrt{s^2}$$

1. The first step in calculating s^2 is to calculate the mean $\bar{x}$ of the five measurements:

$$\begin{aligned} \bar{x} &= \frac{\Sigma x}{n} \\ &= \frac{4 + 0 + 5 + 7 + 2}{5} \\ &= \frac{18}{5} \\ &= 3.6 \end{aligned}$$

Figure 4.13 is a **dot diagram** that shows the location of the mean, $\bar{x} = 3.6$, relative to the five measurements in the data set. Each of the five measurements is represented on the diagram by a dot. Notice the deviations of the five measurements from their mean.

2. The second step in calculating s^2 is to construct Table 4.2.

The five measurements are listed in the first column of Table 4.2. We next calculate the deviation of each measurement from the mean, $\bar{x} = 3.6$. The deviation,

FIGURE 4.13 A dot diagram for the $n = 5$ measurements, $x = 4, 0, 5, 7,$ and 2

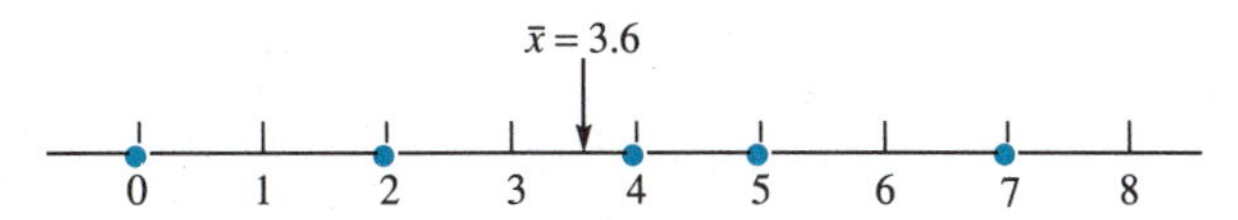

TABLE 4.2 Table for Calculating s

	x	$(x - \bar{x})$	$(x - \bar{x})^2$
	4	.4	.16
	0	−3.6	12.96
	5	1.4	1.96
	7	3.4	11.56
	2	−1.6	2.56
Totals	18	0	29.20

$(x - \bar{x})$, for each of the five measurements is shown in column 2. For example, the first value of x is 4 and its deviation from the mean, $(x - \bar{x})$, is equal to $(4 - 3.6) = .4$.

The third column of Table 4.2 contains the squares, $(x - \bar{x})^2$, of the deviations. The square of the deviation corresponding to the first measurement, $x = 4$, is $(x - \bar{x})^2 = (.4)^2 = .16$.

3. The last step in constructing Table 4.2 is to obtain the column totals.

The sum of the x values, needed to calculate $\bar{x}$, appears at the bottom of column 1. The sum of the deviations, shown at the bottom of column 2, is always equal to 0. **The sum of squares of the deviations, shown at the bottom of column 3, is what we need to calculate the sample variance s^2.** The variance is

$$s^2 = \frac{\Sigma(x - \bar{x})^2}{n - 1} = \frac{29.20}{4} = 7.30$$

and the sample standard deviation is $s = \sqrt{s^2} = \sqrt{7.30} = 2.7$.*

---

* The procedure followed in Table 4.2 to calculate the sum of squares of deviations will help you to understand what a standard deviation is, but it is also subject to rounding errors. A less intuitive but faster and more accurate shortcut procedure is used by computer software packages. For that formula, see Mendenhall and Beaver (1991).

Computer Printouts of the Mean and Measures of Variation

Figure 4.14 shows the Minitab and SAS computer printouts of the numerical descriptive measures for the 30 total cholesterol measurements of Table 3.4. The mean and standard deviation for the data of Table 3.4 are shaded on both printouts. The variance is also shaded on the SAS printout. Some of the other numerical descriptive measures are relevant to our discussion; others are not. We will discuss the relevant measures in the sections that follow.

FIGURE 4.14 Minitab and SAS printouts of numerical descriptive measures for the 30 total cholesterol measurements of Table 3.4

```
                N     MEAN   MEDIAN   TRMEAN   STDEV   SEMEAN
tchol30        30    220.5    219.5    219.7    54.9     10.0

              MIN      MAX       Q1       Q3
tchol30     123.0    337.0    169.7    269.0
```

a. Minitab

```
                        Moments

N                  30   Sum Wgts          30
Mean         220.4667   Sum             6614
Std Dev      54.94306   Variance     3018.74
Skewness     0.182297   Kurtosis    -0.68574
USS           1545710   CSS         87543.47
CV           24.92126   Std Mean    10.03119
T:Mean=0     21.97813   Prob>|T|      0.0001
Sgn Rank        232.5   Prob>|S|      0.0001
Num ^= 0           30
```

b. SAS

EXERCISES

4.1 What is the mean of a data set? How does it help to describe the data set?

4.2 What is the median of a data set? How does it help to describe the data set?

4.3 When is the mean preferred to the median to describe the central tendency of a data set?

4.4 When is the median preferred to the mean to describe the central tendency of a data set?

4.5 What is the range of a data set? How does it help to describe the data set?

4.6 Make up a sample of five numbers such that four of the five sample measurements are larger than the sample mean.

4.7 What is the standard deviation of a data set? How does it help to describe the data set?

4.8 State the Empirical Rule.

4.9 What is the purpose of the Empirical Rule?

4.10 What is the variance of a data set? How are the variance and the standard deviation of a set of measurements related?

4.11 A sample of $n = 50$ measurements has a mean of $\bar{x} = 89.7$ and a standard deviation of $s = 10.4$.

a. Find the variance of the sample.
b. Describe the distribution.
c. Within what limits would you expect approximately 95% of the measurements to fall?

MFG **4.12** A sample of the lengths of life of $n = 1{,}000$ auto batteries has a mean of $\bar{x} = 2.07$ years and a standard deviation of $s = .4$ year.

a. Describe the distribution of times to failure for the 1,000 batteries.
b. Within what limits would you expect approximately 95% of the battery lifetimes to fall?

GEN **4.13** A sample of the percentage of seats occupied on a scheduled airline flight has a mean equal to 75 and a standard deviation equal to 10. What percentage of observations would you expect to fall in the interval from 55 to 95? Would you expect many flights to have an occupancy of less than 40%? Why?

4.14 A relative frequency distribution for a data set is approximately symmetric. If the minimum and maximum observations in the data set are 28 and 110, respectively, approximately where is the median located? Explain your reasoning.

4.15 Refer to Exercise 4.14. Suppose the distribution is skewed to the left. Where, relative to your answer to Exercise 4.14, would you expect the median to be located? Explain.

4.16 If you were told that the mean of a distribution is much larger than the median, what would you conclude about the distribution?

4.17 Suppose that a distribution of 100 measurements has mean equal to 10.5 and standard deviation equal to 1.8. Describe your mental image of the distribution.

BUS **4.18** *The New York Times* (November 6, 1991) states, "The median price for an existing home was $99,200 (in November), down five-tenths of 1 percent from September's median price of $99,700." What do these numbers tell us about the price for an existing home?

4.19 If a distribution of 25 measurements has mean equal to 152.1 and standard deviation equal to 20.7, describe your mental image of the distribution.

4.20 If a distribution of 30 measurements has mean equal to 1.05 and standard deviation equal to .62, describe your mental image of the distribution.

GEN **4.21** Table 4.3 reproduces the data of Table 3.5, which gives the number of deaths incurred in evacuating fires in compartmentalized fire-resistant buildings. Figures 4.15 and 4.16 (page 88) give the SAS printouts of the frequency distribution and the

TABLE 4.3 Fire Deaths

Fire	Number of Victims
Las Vegas Hilton (Las Vegas)	5
Inn on the Park (Toronto)	5
Westchase Hilton (Houston)	8
Holiday Inn (Cambridge, Ohio)	10
Conrad Hilton (Chicago)	4
Providence College (Providence)	8
Baptist Towers (Atlanta)	7
Howard Johnson (New Orleans)	5
Cornell University (Ithaca, New York)	9
Wesport Central Apartments (Kansas City, Missouri)	4
Orrington Hotel (Evanston, Illinois)	0
Hartford Hospital (Hartford, Connecticut)	16
Milford Plaza (New York)	0
MGM Grand (Las Vegas)	36

Source: Macdonald, J. N. "Is Evacuation a Fatal Flaw in Fire Fighting Philosophy?" *Risk Management*, Vol. 33, No. 2, Feb. 1986, p. 37.

FIGURE 4.15 SAS frequency distribution for fire deaths

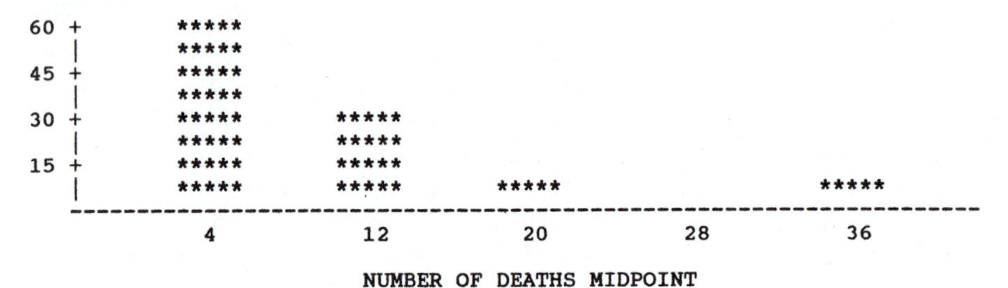

FIGURE 4.16 SAS printout of numerical descriptive measures for fire deaths

```
                       Moments

N                     14   Sum Wgts           14
Mean            8.357143   Sum               117
Std Dev         8.940893   Variance     79.93956
Skewness         2.52694   Kurtosis     7.614333
USS                 2017   CSS          1039.214
CV               106.985   Std Mean     2.389554
T:Mean=0        3.497365   Prob>|T|       0.0039
Sgn Rank              39   Prob>|S|       0.0005
Num ^= 0              12

                  Quantiles(Def=5)

 100% Max             36          99%         36
  75% Q3               9          95%         36
  50% Med              6          90%         16
  25% Q1               4          10%          0
   0% Min              0           5%          0
                                   1%          0

 Range                36
 Q3-Q1                 5
 Mode                  5
```

numerical descriptive measures, respectively, for the data.

a. Find the mean and standard deviation on the printout shown in Figure 4.16.

b. Calculate the variance of the measurements and compare with the value shown on the printout.

c. Find the smallest, the largest, and the range of the number of deaths on the printout.

d. Calculate the intervals $\bar{x} \pm s$, $\bar{x} \pm 2s$, and $\bar{x} \pm 3s$, and mark the intervals along the horizontal axis of the distribution shown in Figure 4.15.

e. Based on the areas under the frequency distribution, estimate the proportion of the total number of measurements that fall in each of the intervals of part **d**. Explain how these proportions agree or disagree with those given by the Empirical Rule.

f. Calculate the exact number and the proportion of the total number of measurements that fall in the intervals of part **d**. Explain how these proportions agree or disagree with those given by the Empirical Rule.

BUS **4.22** Figures 4.17 and 4.18 give the Minitab printouts of the frequency distribution and the numerical descriptive measures, respectively, for the 390 percentage iron ore measurements of Table 1, Appendix 1.

a. Find the mean and standard deviation on the printout in Figure 4.18.

b. Calculate the variance of the measurements.

c. Find the smallest and the largest measurements on the printout. Calculate the range.

d. Calculate the intervals $\bar{x} \pm s$, $\bar{x} \pm 2s$, and $\bar{x} \pm 3s$, and mark the intervals along the vertical axis of the distribution shown in Figure 4.17.

e. Based on the areas under the frequency distribution, estimate the proportion of the total number of measurements that fall in each of the intervals of part **d**. Explain how these proportions agree or disagree with the Empirical Rule.

f. To obtain a more accurate estimate of the proportions found in part **d**, sum the relative frequencies of the classes whose midpoints fall in each of the intervals of part **d**. Explain how these proportions agree or disagree with the Empirical Rule.

HLTH **4.23** Table 4.4 (page 90) gives the white blood cell count (WBC) and the lymphocyte count (LYMPHO)

FIGURE 4.17 Minitab frequency distribution for percentage iron ore

```
Histogram of %iron    N = 390
Each * represents 5 obs.

Midpoint   Count
    63.0       1  *
    63.5       7  **
    64.0      10  **
    64.5      12  ***
    65.0      42  *********
    65.5      88  ******************
    66.0     139  ****************************
    66.5      76  ****************
    67.0      15  ***
```

FIGURE 4.18 Minitab printout of numerical descriptive measures for percentage iron ore

```
                N      MEAN    MEDIAN    TRMEAN     STDEV    SEMEAN
%iron         390    65.743    65.830    65.794     0.694     0.035

              MIN       MAX        Q1        Q3
%iron      62.770    66.860    65.410    66.230
```

TABLE 4.4 Lymphocyte Count and White Blood Cell Count Data for Exercise 4.23

Case Number	WBC	LYMPHO	Case Number	WBC	LYMPHO
1	4,100	14	26	4,300	9
2	5,000	15	27	5,200	16
3	4,500	19	28	3,900	18
4	4,600	23	29	6,000	17
5	5,100	17	30	4,700	23
6	4,900	20	31	7,900	43
7	4,300	21	32	3,400	17
8	4,400	16	33	6,000	23
9	4,100	27	34	7,700	31
10	8,400	34	35	3,700	11
11	5,600	26	36	5,200	25
12	5,100	28	37	6,000	30
13	4,700	24	38	8,100	32
14	5,600	26	39	4,900	17
15	4,000	23	40	6,000	22
16	3,400	9	41	4,600	20
17	5,400	18	42	5,500	20
18	6,900	28	43	6,200	20
19	4,600	17	44	4,900	26
20	4,200	14	45	7,200	40
21	5,200	8	46	5,800	22
22	4,700	25	47	8,400	61
23	8,600	37	48	3,100	12
24	5,500	20	49	4,000	20
25	4,200	15	50	6,900	35

Source: Royston, J. P. "Some Techniques for Assessing Multivariate Normality Based on the Shapiro–Wilk *W*." *Applied Statistics*, Vol. 32, No. 2, 1983, pp. 121–133.

for each of $n = 50$ West Indian or African workers. Figures 4.19 and 4.20 give the SAS printouts of the percentage distribution and the numerical descriptive measures, respectively, for the white blood cell count data.

a. Find the mean and standard deviation on the printout in Figure 4.20.

b. Calculate the variance of the measurements and compare with the value shown on the printout.

c. Find the smallest measurement, the largest measurement, and the range on the printout.

d. Calculate the intervals $\bar{x} \pm s$, $\bar{x} \pm 2s$, and $\bar{x} \pm 3s$, and mark the intervals along the horizontal axis of the distribution shown in Figure 4.19.

e. Based on the areas under the distribution, estimate the proportion of the total number of measurements that fall in each of the intervals of part **d**. Explain how these proportions agree or disagree with the Empirical Rule.

FIGURE 4.19 SAS percentage distribution for white blood cell count data

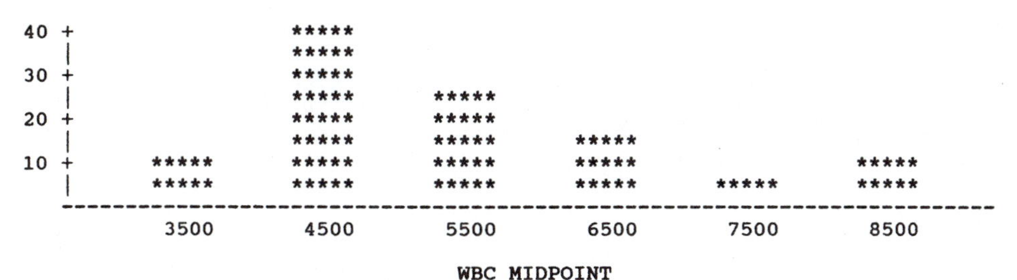

FIGURE 4.20 SAS printout of numerical descriptive measures for white blood cell count data

Moments

N	50	Sum Wgts	50
Mean	5334	Sum	266700
Std Dev	1387.791	Variance	1925963
Skewness	0.87264	Kurtosis	0.120791
USS	1.517E9	CSS	94372200
CV	26.01783	Std Mean	196.2633
T:Mean=0	27.17778	Prob>|T|	0.0001
Sgn Rank	637.5	Prob>|S|	0.0001
Num ^= 0	50		

Quantiles(Def=5)

100% Max	8600	99%	8600
75% Q3	6000	95%	8400
50% Med	5050	90%	7800
25% Q1	4300	10%	3950
0% Min	3100	5%	3400
		1%	3100
Range	5500		
Q3-Q1	1700		
Mode	6000		

f. Calculate the exact number and the proportion of the total number of measurements that fall in each of the intervals of part **d**. Explain how these proportions agree or disagree with the Empirical Rule.

HLTH **4.24** Figures 4.21 and 4.22 (page 92) give the SAS printouts of the percentage distribution and the numerical descriptive measures, respectively, for the lymphocyte count data of Table 4.4. Follow the instructions of Exercise 4.23.

FIGURE 4.21 SAS percentage distribution for lymphocyte count data

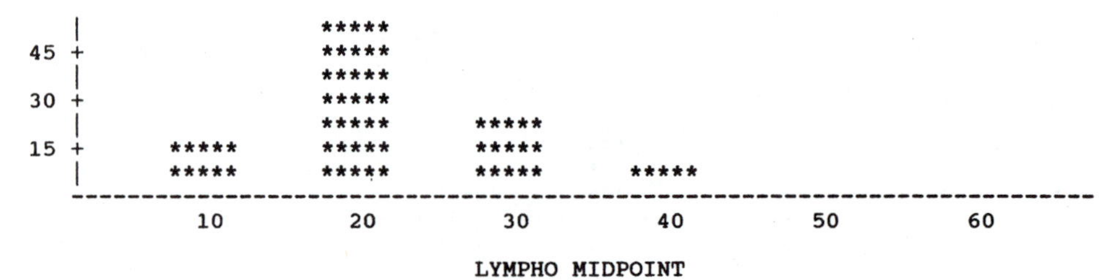

FIGURE 4.22 SAS printout of numerical descriptive measures for lymphocyte count data

Moments

N	50	Sum Wgts	50
Mean	22.68	Sum	1134
Std Dev	9.541403	Variance	91.03837
Skewness	1.547021	Kurtosis	4.302262
USS	30180	CSS	4460.88
CV	42.06968	Std Mean	1.349358
T:Mean=0	16.80799	Prob>|T|	0.0001
Sgn Rank	637.5	Prob>|S|	0.0001
Num ^= 0	50		

Quantiles(Def=5)

100% Max	61	99%	61
75% Q3	26	95%	40
50% Med	20.5	90%	34.5
25% Q1	17	10%	13
0% Min	8	5%	9
		1%	8
Range	53		
Q3-Q1	9		
Mode	20		

DEM **4.25** Table 4.5 reproduces the data of Table 3.7, which gives the average life expectancy for each of 33 developed countries. Figures 4.23 and 4.24 give the Data Desk printouts of the frequency distribution and the numerical descriptive measures, respectively.

a. Find the mean and standard deviation on the printout shown in Figure 4.24.

b. Calculate the variance of the measurements and compare with the value shown on the printout.

c. Find the smallest measurement, the largest measurement, and the range on the printout.

d. Calculate the intervals $\bar{x} \pm s$, $\bar{x} \pm 2s$, and $\bar{x} \pm 3s$, and mark the intervals along the horizontal axis of the distribution shown in Figure 4.23.

e. Based on the areas under the frequency distribution, estimate the proportion of the total number of measurements that fall in each of the intervals of part **d**. Explain how these proportions agree or disagree with the Empirical Rule.

FIGURE 4.23 Data Desk frequency distribution for life expectancies

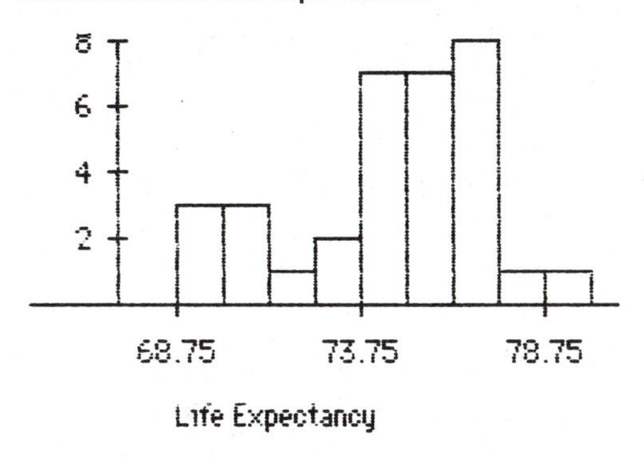

TABLE 4.5 Average Life Expectancies for 33 Developed Countries

Country	Life Expectancy	Country	Life Expectancy
Australia	76.3	Italy	75.5
Austria	75.1	Japan	79.1
Belgium	74.3	Luxembourg	74.1
Britain	75.3	Malta	74.8
Bulgaria	71.5	The Netherlands	76.5
Canada	76.5	New Zealand	74.2
Czechoslovakia	71.0	Norway	76.3
Denmark	74.9	Poland	71.0
East Germany	73.2	Portugal	74.1
West Germany	75.8	Rumania	69.9
Finland	74.8	Soviet Union	69.8
France	75.9	Spain	76.6
Greece	76.5	Sweden	77.1
Hungary	69.7	Switzerland	77.6
Iceland	77.4	United States	75.0
Ireland	73.5	Yugoslavia	71.0
Israel	75.2		

FIGURE 4.24 Data Desk printout of numerical descriptive measures for life expectancies

```
Summary statistics for    Life Expectancy
NumNumeric = 33
Mean = 74.530
Median = 75
Standard Deviation = 2.4412
Range = 9.4000
Variance = 5.9597
Minimum = 69.700
Maximum = 79.100
```

f. Calculate the exact number and the proportion of the total number of measurements that fall in the intervals of part **d**. Explain how these proportions agree or disagree with the Empirical Rule.

BIO **4.26** The SAS percentage distribution of the DDT concentration (ppm) of 144 fish sampled from the Tennessee River, Figure 3.18c, is reproduced in Figure 4.25 and the SAS printout of the numerical descriptive measures for the data is shown in Figure 4.26 (page 96).

a. Find the mean and standard deviation on the printout shown in Figure 4.26.

b. Calculate the variance of the measurements and compare with the value shown on the printout.

c. Find the smallest measurement, the largest measurement, and the range on the printout.

d. Calculate the intervals $\bar{x} \pm s$, $\bar{x} \pm 2s$, and $\bar{x} \pm 3s$, and mark the intervals along the horizontal axis of the distribution shown in Figure 4.25.

e. Based on the areas under the distribution, estimate the proportion of the total number of measurements that fall in each of the intervals of part **d**. Explain how these proportions agree or disagree with the Empirical Rule.

f. To obtain a more accurate estimate of the proportions found in part **d**, sum the relative frequencies of the classes whose midpoints fall in each of the intervals of part **d**. Explain how these proportions agree or disagree with the Empirical Rule.

FIGURE 4.25 SAS percentage distribution for the DDT content of 144 fish

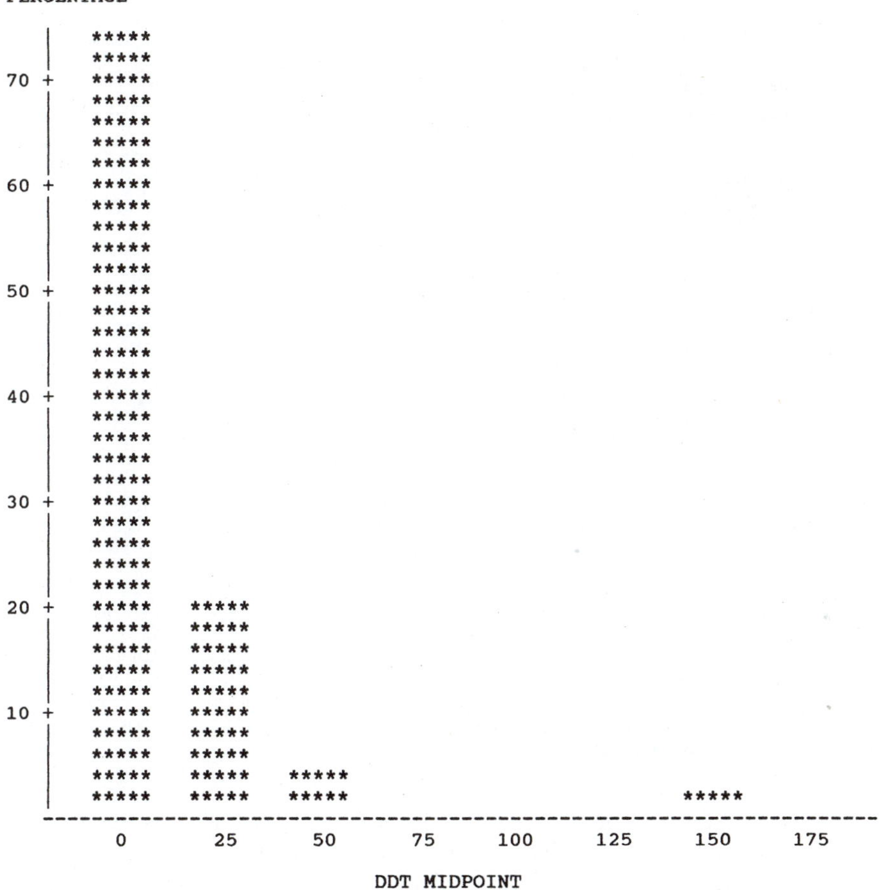

FIGURE 4.26 SAS printout of numerical descriptive measures for the DDT content of 144 fish

```
                          Moments

N                144   Sum Wgts          144
Mean          24.355   Sum           3507.12
Std Dev     98.37859   Variance     9678.346
Skewness     9.61166   Kurtosis     102.2388
USS          1469419   CSS           1384003
CV          403.9359   Std Mean     8.198215
T:Mean=0    2.970768   Prob>|T|       0.0035
Sgn Rank        5220   Prob>|S|       0.0001
Num ^= 0         144


                    Quantiles(Def=5)

 100% Max        1100          99%          360
  75% Q3           13          95%           61
  50% Med        7.15          90%           28
  25% Q1         3.35          10%          1.2
   0% Min        0.11           5%         0.43
                                1%         0.18

 Range        1099.89
 Q3-Q1           9.65
 Mode              12
```

4.4 Applying the Empirical Rule: Approximating a Standard Deviation

Often we need to approximate the standard deviation of a set of measurements. In one situation, we have the data but we want an approximate value of the standard deviation s to use as a rough check on a value calculated from the formula given in Section 4.3. In a situation we will discuss in Chapters 6 and 7, we do not have the data set, but we have an idea about the approximate values of the smallest and the largest members. Then we will find an approximation to the standard deviation s based on the range of a data set.

The easiest way to approximate the standard deviation of a data set is to apply the Empirical Rule in reverse. Since almost all of the measurements in a data set will lie within three standard deviations of the mean, it follows that the range of a moderately large to large data set will be approximately equal to six standard deviations (three standard deviations above and three below the mean). Consequently, the standard deviation of a data set should be approximately equal to the range divided by 6. See Figure 4.27.

FIGURE 4.27 The relationship between the range and s for large data sets

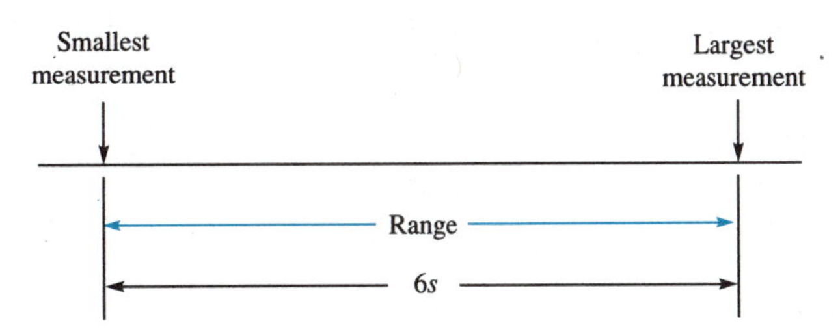

Large data sets are usually more variable than small ones because they are likely to contain one or more extreme values. A more accurate range approximation for s, adjusted for sample size, would use the divisors of the range given in Table 4.6.

TABLE 4.6 Divisors of the Range to Approximate s

Sample Size n	5	10	25	100
Divisor	2.5	3	4	6

Example 4.7

Your data set contains $n = 20$ measurements with a range $r = 87$. Find an approximate value for the standard deviation s of the data set.

Solution

From Table 4.6, you can see that the sample size $n = 20$ is between the tabulated values 10 and 25. Therefore, the range divisor should be between 3 and 4. We could interpolate to find the value of the divisor, but we will obtain an adequate approximation to s using a divisor equal to 4. Then

$$\text{Range approximation to } s = \frac{\text{Range}}{4} = \frac{87}{4} = 21.75$$

If the computer printout value for s differs greatly from this value, your printout value may be in error. You may have made one or more incorrect data entries.

EXERCISES

4.27 If a data set contains $n = 16$ measurements and its range equals 36, what would you expect the standard deviation of the data set to be, approximately? Explain your reasoning.

4.28 If a data set contains $n = 8$ measurements and its range equals 95, what would you expect the standard deviation of the data set to be, approximately? Explain your reasoning.

4.29 You believe that the smallest measurement in a population is 2 and the largest is 18. What is your best guess as to the value of the population standard deviation? Explain your reasoning.

GEN **4.30** The smallest and largest numbers of deaths in the set of $n = 14$ fires, Table 4.3, are 0 and 36, respectively. Give an approximation to the standard deviation for the data set and justify your answer. Compare your answer with the computed value shown in the printout in Figure 4.16.

BUS **4.31** The smallest and largest values of the $n = 390$ percentage iron ore measurements in Table 1, Appendix 1, are 62.77 and 66.86, respectively. Give an approximation to the standard deviation for the data set and justify your answer. Compare your answer with the computed value shown in the printout, Figure 4.18.

HLTH **4.32** The smallest and largest values of the $n = 50$ white blood cell counts on West Indian and African workers, Table 4.4, are 3,100 and 8,600, respectively. Give an approximation to the standard deviation for the data set and justify your answer. Compare your answer with the computed value shown in the printout in Figure 4.20.

HLTH **4.33** The smallest and largest values of the $n = 50$ lymphocyte counts on West Indian and African workers, Table 4.4, are 8 and 61, respectively. Give an approximation to the standard deviation for the data set and justify your answer. Compare your answer with the computed value shown in the printout in Figure 4.22.

DEM **4.34** The smallest and largest members of the data set on average life expectancies for 33 developed countries, Table 4.5, are 69.7 and 79.1, respectively. Give an approximation to the standard deviation for the data set and justify your answer. Compare your answer with the computed value shown in the printout in Figure 4.24.

4.5 Applying the Empirical Rule: Processes and Statistical Control

A Case

A management problem associated with a motor oil can-filling operation—the 1-quart cans that you purchase at your local filling station—was described by V. Filimon et al. (Filimon, V., Maggass, D. Frazier, and Klingel, L., "Some Applications of Quality Control Techniques in Motor Oil Can-Filling," *Industrial Quality Control*, Vol. 12, No. 2, 1955). Company accountants discovered that the company received 10,000,000 quarts of oil stock per month but shipped considerably less. If the company received 10,000,000 cans in a given month and shipped only 9,700,000, what happened to the lost 300,000 cans of oil?

The search for the missing oil focussed on the filling machine. Does each loading spindle discharge exactly 1 quart of oil into the waiting can? Filimon et al., by experimentation, showed that the answer is no. The amount of oil discharged from a single spindle varies from one discharge to another and it varies from one spindle to another. This variation in fill amounts led Filimon and colleagues to conclude that the missing oil stocks left the plant in overfilled cans.

The oil can-filling operation that we have just described is typical of many other business or manufacturing operations—operations that emit a continuous stream of elements produced over time. **An operation that emits a continuous stream of product over time is sometimes called a *process*.** The quality of these product elements is measured by one or more variables.

A product can be tangible or intangible. For example, a can of oil is a tangible product. The exact amount of oil loaded into a 1-quart can is a measure of the "quality" of that single can of oil. In contrast, the entry of deposits and withdrawals into a bank's computer is a continuing operation that produces an intangible product, a service. The observed element—the product—is the collection of entries made for a particular day, and the "quality" of a day's entries is measured by the number of incorrect entries.

Management, government agencies, etc., specify that the variables that measure the quality of a product must satisfy certain specifications. (Specifications are what they want, not necessarily what they get.) For example, either management or government might specify that a can of oil must contain at least 1 quart of oil. Ideally, the oil company would like the distribution of oil discharges per can to be centered at 32 ounces and have a standard deviation equal to 0 (see Figure 4.28). That way, they would satisfy specifications but they would never overfill cans. Practically, the company would like to satisfy these specifications and, at the same time, keep the overfill per can to a minimum.

To accomplish this, the company would want the distribution of oil discharges to the cans to lie mostly to the right of 1 quart (see Figure 4.28b) and to possess the smallest standard deviation possible. You can see from Figure 4.28b that only a

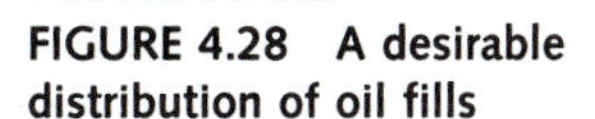

FIGURE 4.28 A desirable distribution of oil fills

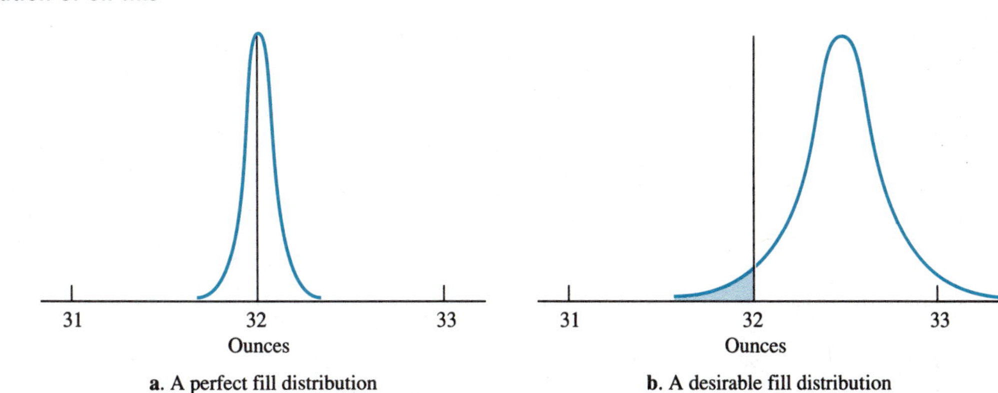

a. A perfect fill distribution

b. A desirable fill distribution

small percentage of cans (those in the tail of the distribution to the left of 32 ounces) will contain less than 1 quart of oil. The smaller the spread of the distribution, the smaller will be the overfill of the cans.

From this discussion, you can see that the manager of an operation wants to control the mean and minimize the standard deviation of the distribution for a quality variable. At the very least, the manager wants the distribution of quality measurements to be stable—i.e., remain unchanged from one day (hour, or minute) to another.

FIGURE 4.29 Processes in (a) and out (b and c) of control

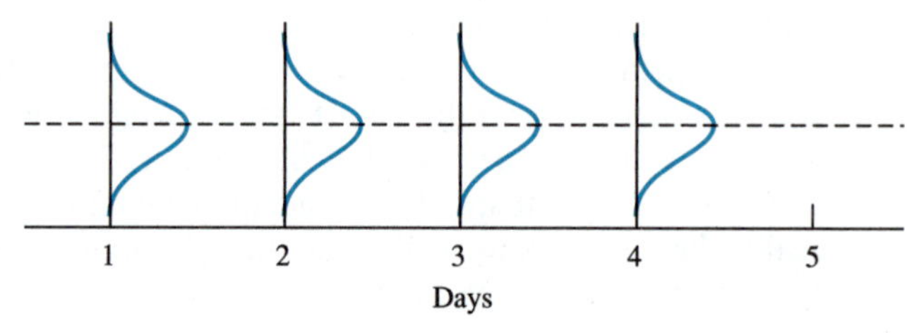

a. In control

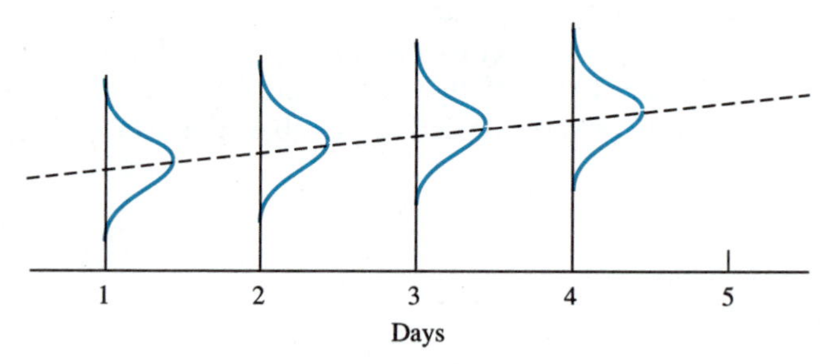

b. Out of control

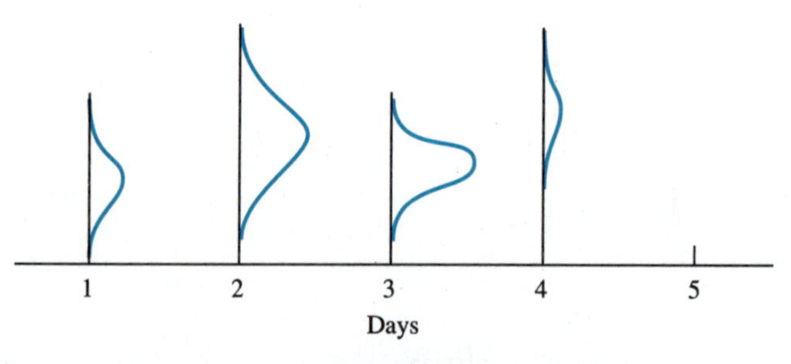

c. Out of control

Statistical Control

A quality variable whose distribution remains unchanged over time is said to be *in statistical control*. Figure 4.29a shows the distribution of a quality variable for each of four days. If the process is in control, the distribution of the measurements of the quality variable should be exactly the same on days 2, 3, and 4, as it is on day 1.

In contrast, the distributions shown in Figures 4.29b and 4.29c would correspond to processes out of control. The variability of the distributions shown in Figure 4.29b remains unchanged over time, but the mean is shifting upward. In Figure 4.29c, both the mean and the variability of the distributions appear to change over time.

Control Charts

Quality variables for an ongoing operation are monitored using *control charts*. A control chart for the oil can-filling operation might appear as shown in Figure 4.30.

FIGURE 4.30 A control chart for the oil can-filling operation

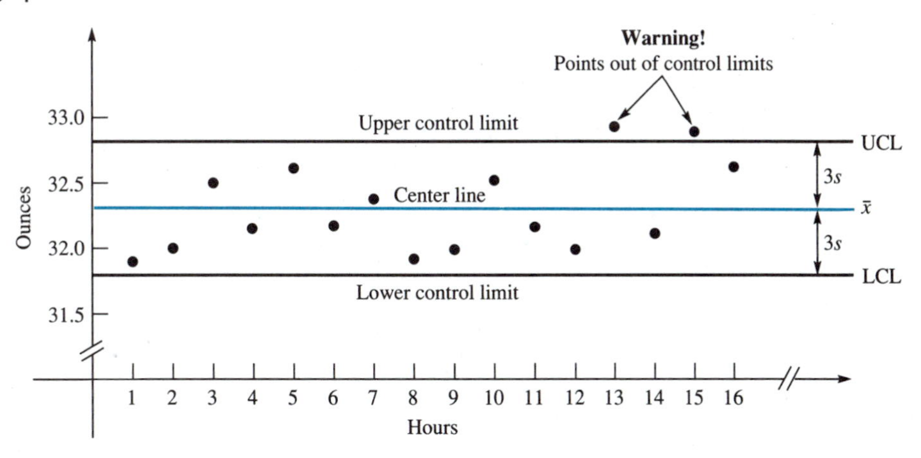

To construct the control chart shown in Figure 4.30, we need to collect data to determine the mean and standard deviation of the distribution of discharges when the process is "in control." Periodically, say once each hour, we take an accurate measurement of the amount of oil discharged into a can. After a large number, say 100, of hours, we calculate the mean $\bar{x}$ and standard deviation s for the 100 measurements.

The vertical axis at the left of the chart in Figure 4.30 is marked off in the units of measurement (for our example, in ounces) of the quality variable. **The horizontal line through the center of the chart, called the *center line*, is located at $\bar{x}$.** It locates the center of the "in-control" distribution.

Finally, we apply the Empirical Rule. Since all or almost all of a set of measurements should fall within three standard deviations of their mean, we construct horizontal lines, one at a distance equal to $3s$ above and one $3s$ below the center line of the chart. **These lines are called the upper (UCL) and lower (LCL) control limits.**

To use the chart, the operator of the can-filling machine removes a single filled can from the machine at equally spaced intervals of time and measures the amount of oil that it contains. This amount is plotted on the chart. If the process remains in control, all or almost all of the plotted points should stay within the control limits. If a point falls outside of the control limits, it is a warning that the process may be going out of control, and the machine operator should begin to look for the cause of the out-of-limits measurement. Figure 4.31 shows segments of two control charts

FIGURE 4.31 Two control charts for can-filling process

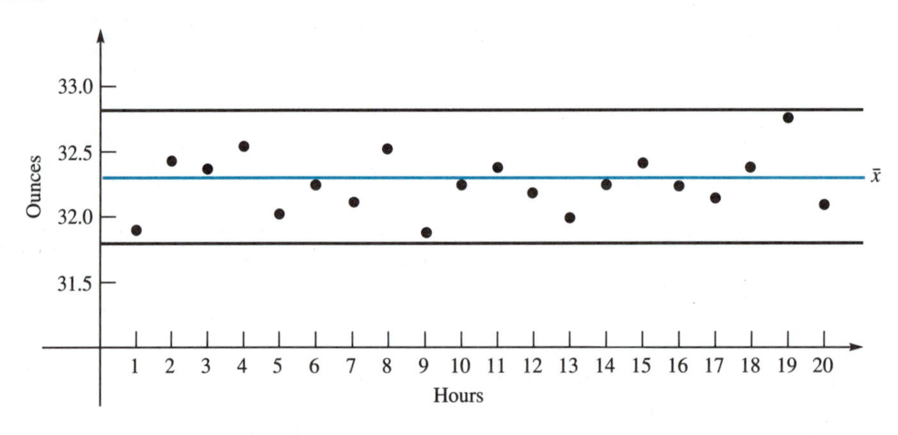

a. Process in control

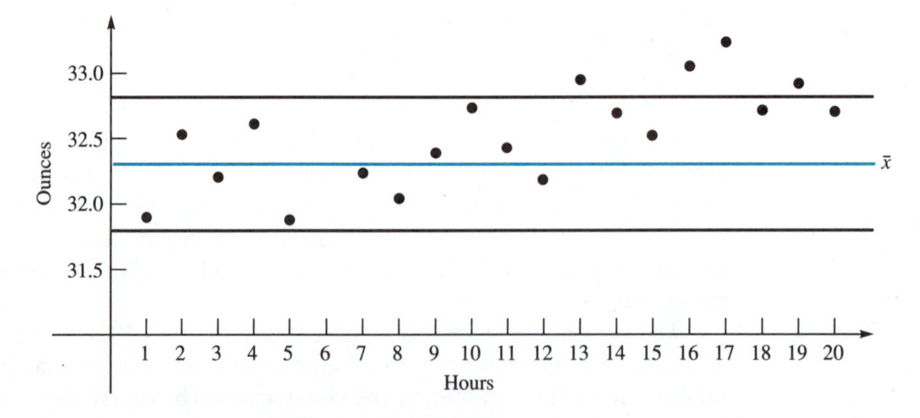

b. Process out of control

for the oil can-filling machine. The segment in Figure 4.31a indicates a process in control, whereas Figure 4.31b indicates a process out of control.

Most control charts are based on small samples (rather than individual measurements) collected at equally spaced intervals of time. For example, if the manager wants to detect changes in the mean fill of the cans, it would be better to select $n = 3$ cans each hour, calculate the sample mean, and construct an $\bar{x}$-chart. Such a chart would show a plot of the hourly sample means. The location of the center line and control limits would be based on the mean and standard deviation of a large number of $\bar{x}$'s.

Control charts for a single process variable and for a sample mean are only two of a number of different control charts that can be constructed to monitor quality variables. Each chart is based on the same concept: We identify the interval within which all or almost all of the plotted points should fall; points that fall outside of these limits are highly improbable if the process is in control and thus suggest the possibility that the process is out of control.

EXERCISES

4.35 What do we mean by an ongoing operation or process?

4.36 What is a quality variable?

4.37 What do we mean when we say that a process is in control?

4.38 Describe a control chart for a quality variable.

4.39 What is the objective of a control chart?

4.6 Measures of Relative Standing

The Terminology

A *measure of relative standing* locates the position of a measurement, in rank, relative to other members of the data set. For example, you might be interested in the position of your test score relative to the test scores of other members of your class.

Relative standing is described in terms of percentiles. **The *pth percentile* of a data set is the value of x such that $p\%$ of the measurements are less than x and $(100 - p)\%$ are larger than x.**

Example 4.8

Suppose that you scored at the 68th percentile on a class examination. How did you score relative to the rest of your class?

Solution

Your score is such that 68% of the test scores are less than it and 32% are larger than it. Your score is also the value of x such that 68% of the area under the relative frequency distribution lies to the left of x and 32% will lie to its right. See Figure 4.32.

FIGURE 4.32 Location of the 68th percentile of a data set

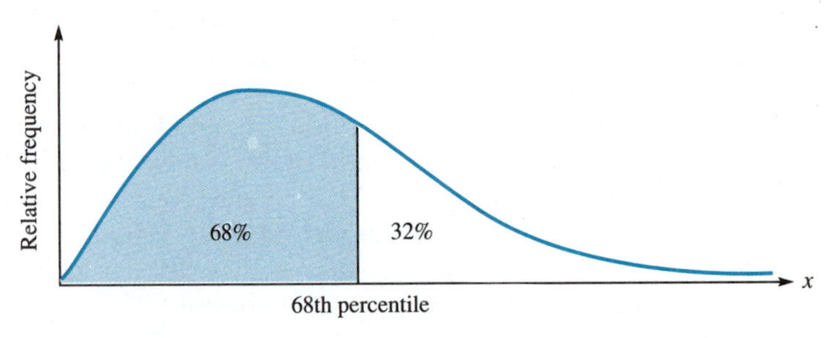

The 25th percentile is called the *lower quartile* and is designated Q_1. Twenty-five percent of the measurements in the data set are less than Q_1 and 75% are larger than Q_1.

The 75th percentile is called the *upper quartile* and is designated Q_3. Seventy-five percent of the measurements in the data set are less than Q_3 and 25% are larger than Q_3.

The median is the middle quartile, Q_2. The quartiles divide the ranked measurements into four groups of equal numbers of measurements. They also divide the area under the relative frequency distribution into four equal areas. See Figure 4.33.

FIGURE 4.33 Location of the quartiles for a distribution

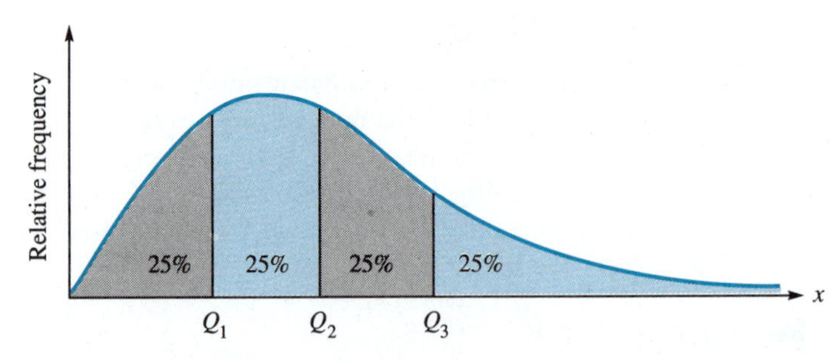

Example 4.9

The grade on your last examination was less than the lower quartile. How did you fare relative to other members of your class?

Solution

Since you scored below the lower quartile, 75% or more of the other students exceeded your score.

***Deciles* divide the ranked scores into groups, each containing 10% of the scores.** The rth decile is the value of x such that $10r\%$ of the measurements are less than x and $(100 - 10r)\%$ of the measurements are larger than x. For example, the third decile is the value of x such that 30% of the measurements are less than x and 70% are larger than x.

The Location of the Median and Skewness

The location of the median, relative to the lower (Q_1) and upper (Q_3) quartiles of a data set, provides a rough measure of the skewness of a relative frequency distribution. If the distribution is symmetric, the median will fall midway between Q_1 and Q_3. If the distribution is skewed to the right, the median will be shifted to the left of the midpoint. If the distribution is skewed to the left, the median will be shifted to the right of the midpoint (see Figure 4.34).

FIGURE 4.34 Using the median to detect skewness

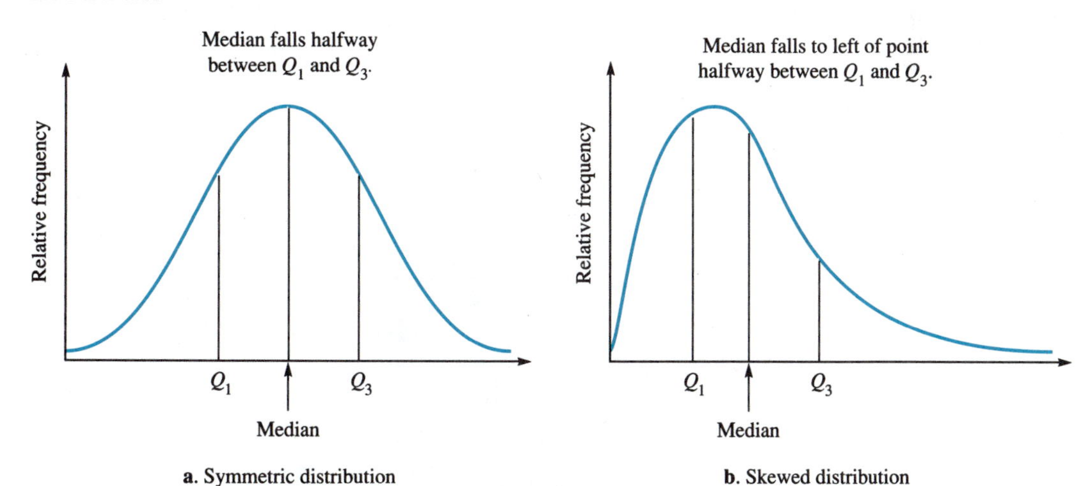

Figure 4.35a (page 106) shows the Minitab printout of the percentiles, quartiles, and other related statistics for the 30 total cholesterol measurements of Table 3.4. Figure 4.35b shows the corresponding SAS printout (SAS refers to the percentiles as

FIGURE 4.35 Minitab and SAS computer printouts of useful percentiles

	N	MEAN	MEDIAN	TRMEAN	STDEV	SEMEAN
tchol30	30	220.5	219.5	219.7	54.9	10.0

	MIN	MAX	Q1	Q3
tchol30	123.0	337.0	169.7	269.0

a. Minitab

Quantiles(Def=5)

100% Max	337	99%	337
75% Q3	269	95%	312
50% Med	219.5	90%	292
25% Q1	170	10%	156
0% Min	123	5%	131
		1%	123
Range	214		
Q3-Q1	99		
Mode	235		

b. SAS

FIGURE 4.36 Relative frequency distribution for the $n = 30$ total cholesterol measurements of Table 3.4

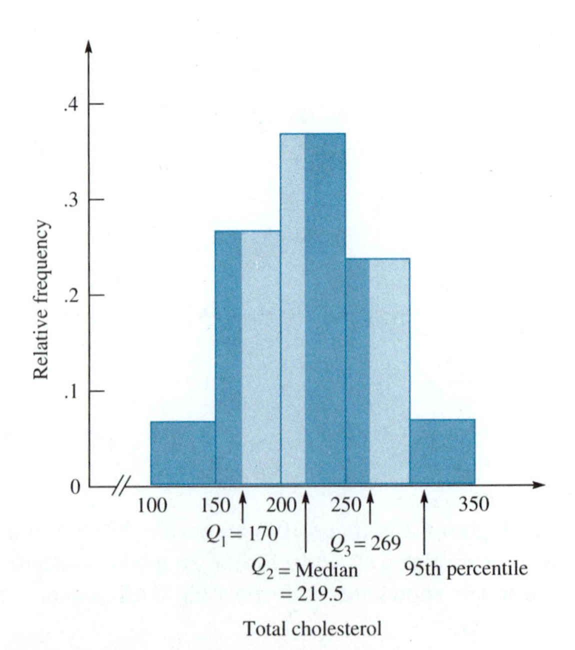

quantiles). Both Minitab and SAS show the smallest (MIN) and the largest (MAX) measurements, as well as the lower and upper quartiles and the median. In addition, SAS gives the 1st, 5th, 10th, 90th, 95th, and 99th percentiles. For the TCHOL data set, MIN = 123, Q_1 = 170, median = 219.5, Q_3 = 269, and MAX = 337. (*Note*: SAS values for Q_1 and Q_3 are rounded to the nearest integer.)

Figure 4.36 shows the relative frequency distribution for the 30 measurements with the quartiles marked along the horizontal axis. Vertical lines located over the quartiles divide the total area under the distribution into four equal areas. The 95th percentile, 312, is also located on the axis. Ninety-five percent of the total cholesterol measurements lie below 312 and 5% lie above.

4.7 Looking for Outliers

What Is an Outlier?

A measurement that appears to be improbably large (or small) relative to other measurements in a data set is called an *outlier*. For example, Figure 4.37 (page 108) reproduces the relative frequency distribution for the 107 TCHOL measurements described in Case 2.3 (see also Table 3, Appendix 1). All of the measurements fall in the interval from 100 to 340, except for two: 399 and 414. Is it improbable that two measurements from the same population could differ so much from the others? In other words, are the two measurements outliers?

An outlier may be a valid member of a data set (improbable values do occur occasionally!) or it may have occurred due to an incorrect measurement, a faulty recording, unsuitable experimental conditions, etc. If the latter, you will want to remove the outlier from the data set. The first step in dealing with this problem is to decide whether an observation is an outlier—i.e., a measurement that is highly improbable given the other measurements in the data set.

Detecting Outliers: A Box Plot for the 30 TCHOL Measurements

One method of detecting outliers is a *box plot*. Figure 4.38 shows a box plot for the $n = 30$ total cholesterol (TCHOL) measurements of Table 3.4. The axis of measurement is a horizontal line. The lower and upper quartiles (from Figure 4.35), $Q_1 = 170$ and $Q_3 = 269$, are marked on the axis and a box is constructed over the interval from Q_1 to Q_3.

The width of the box, the distance between the upper and lower quartiles, is called the *interquartile range*. Since this distance covers two quartiles, 50% of the measurements in the data set fall in this interval.

$$\text{Interquartile range (IQR)} = Q_3 - Q_1$$

FIGURE 4.37 A SAS relative frequency distribution for the 107 TCHOL measurements of Table 3, Appendix 2

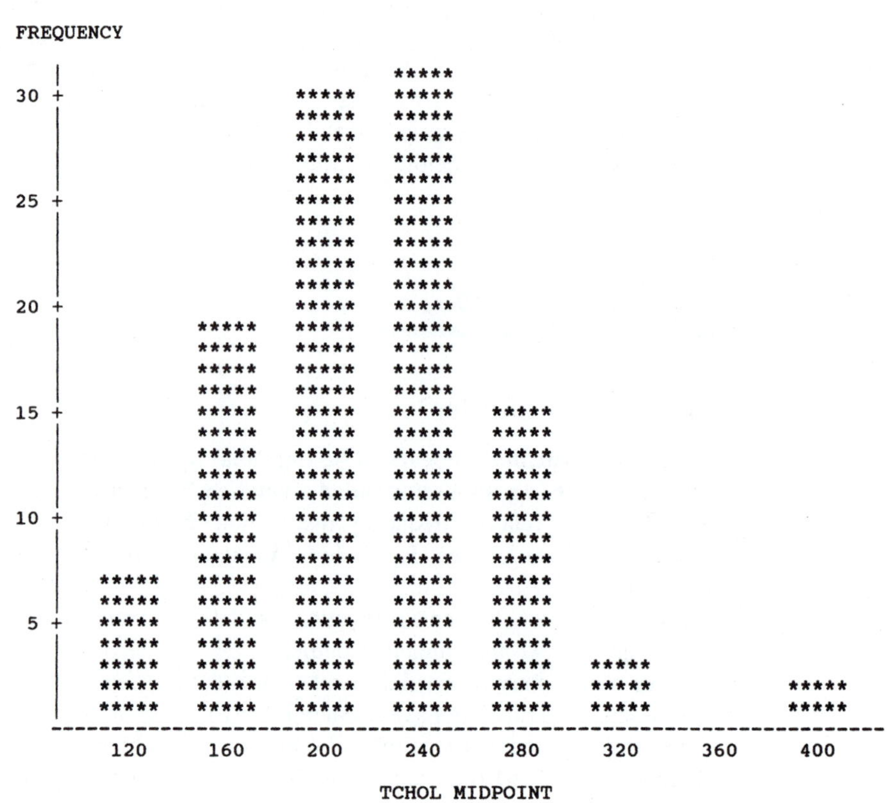

FIGURE 4.38 A box plot for the data of Table 3.4

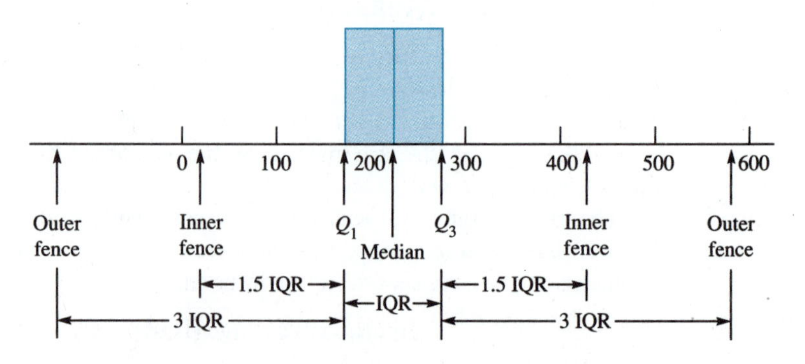

The interquartile range for the $n = 30$ total cholesterol measurements of Table 3.4 is

$$\text{IQR} = Q_3 - Q_1 = 269 - 170 = 99 \text{ mg/dl}$$

(Note that the interquartile range is shown on the SAS printout in Figure 4.35.)

***Inner fences* are located a distance of 1.5(IQR) below Q_1 and above Q_3. *Outer fences* are located 3(IQR) below Q_1 and above Q_3 (see Figure 4.38).** Measurements that fall outside of the inner fences are plotted along the axis of measurement. **Those measurements that fall outside of the outer fences are identified as outliers** that should be checked to determine whether they are faulty observations. Measurements falling outside of the inner fences but inside of the outer fences are suspect outliers, indicating possible disagreement with other members of the data set.

The smallest and largest measurements in the total cholesterol data set of Table 3.4 are 123 and 337, respectively. Neither lies outside of the inner fences of the box plot shown in Figure 4.38. Therefore, there are no outliers or suspect outliers in the 30 measurements in the TCHOL data set.

Outliers: What to Do with Them

If you are able to identify the cause of an outlier (an error in measurement, a recording error, or whatever), eliminate the measurement from the data set. If you are unable to find a cause, do not eliminate the measurement. **It may be the most important measurement in the data set.**

For example, suppose that you were to draw a sample of $n = 4$ measurements from a population that has the relative frequency distribution shown in Figure 4.39 and that the four measurements assumed values indicated by the four dots. Looking only at the four dots, it appears that the dot at the far right is an outlier. However, if you look at the location of the dots relative to their population relative frequency distribution, you can see that the three dots on the left—the ones that at first glance appeared to be most representative of the population—fall below the population mean. If the fourth measurement were tossed out without sufficient reason, the three measurements would underestimate the mean. Therefore, the fourth observation is

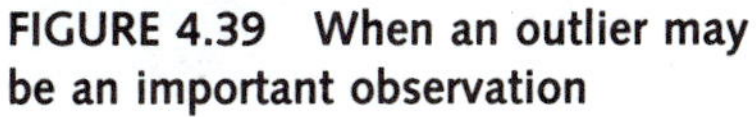

FIGURE 4.39 When an outlier may be an important observation

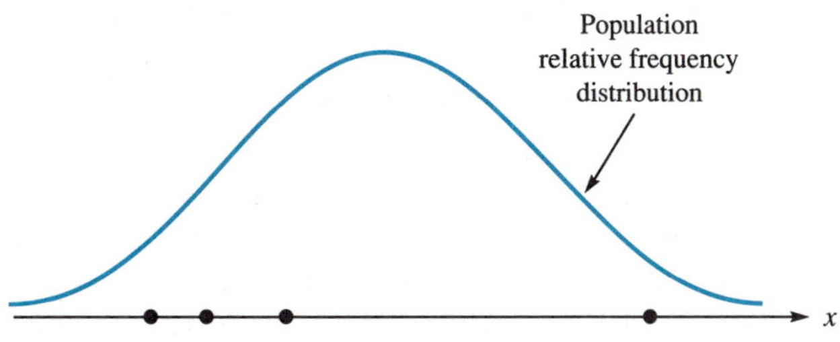

the most important member of this sample. Its inclusion will provide a better estimate of the population mean because it balances the three observations that fall below the mean.

Computer-Generated Box Plots

The appearance of a computer-generated box plot for a set of data varies from one software package to another, but all will perform their intended job—the identification of outliers. Figure 4.40 gives the Minitab and SAS box plots for the 30 TCHOL measurements of Table 3.4. The horizontal line segment below the box is the interval that lies within the inner fences. Suspect outliers and outliers are indicated by asterisks or dots. Compare these two computer-drawn box plots with the hand-drawn box plot shown in Figure 4.38, for the same set of data. None of the plots indicate suspect outliers or outliers in the data set.

FIGURE 4.40 Minitab and SAS box plots for the 30 TCHOL measurements of Table 3.4

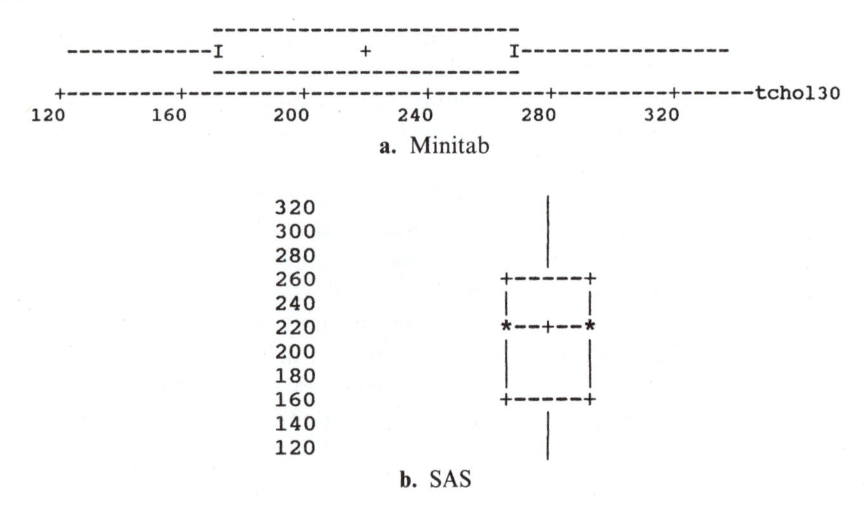

4.8 Numerical Descriptions of TCHOL and RATIO, Case 2.3

Recall Case 2.3 and Table 3 of Appendix 1, which contains cholesterol data on 107 "healthy" adults, aged 17 to 62. In addition to the sex (SEX) and age (AGE) of each individual, the table gives readings on total cholesterol (TCHOL), triglycerides (TRI), high-density lipids (HDL), low-density lipids (LDL), very low-density lipids

(VLDL), and the ratio (RATIO), LDL/HDL. While all of these variables are related to the risk of heart disease, the two variables commonly used to measure risk are TCHOL and RATIO.

Figure 4.41 shows the SAS computer printouts of the relative frequency distributions, numerical descriptive statistics, and box plots for these two important measures of risk. The numerical descriptive measures assume values in agreement with the SAS distributions that we discussed in Section 3.3. In particular, note that the standard deviation for each distribution is approximately equal to one-sixth of the range. The most outstanding feature of Figure 4.41, however, is that both box plots indicate the presence of outliers; the two largest TCHOL measurements (414 and 399) and the three largest RATIO measurements (8.57, 6.49, and 6.22). Each of these measurements should be checked to determine whether any was the result of incorrect measurement or calculation. Any measurement found to be faulty should be eliminated from the data sets.

FIGURE 4.41 **SAS numerical descriptive measures for the TCHOL and RATIO data, Table 3, Appendix 1**

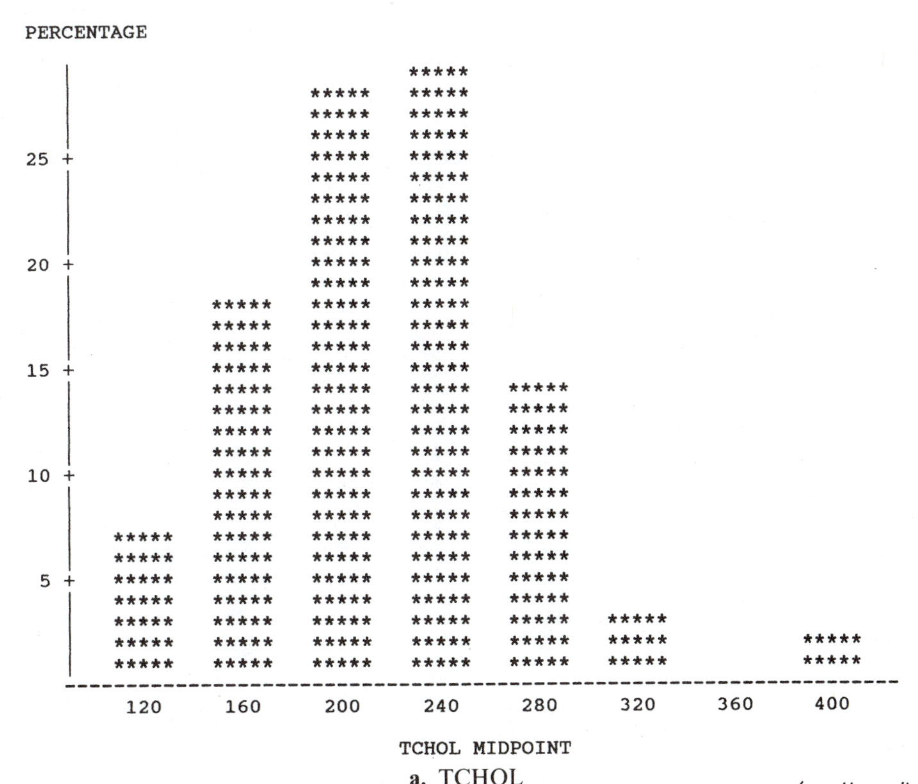

a. TCHOL

(continued)

FIGURE 4.41 *Continued*

Moments

N	107	Sum Wgts	107
Mean	217.7935	Sum	23303.9
Std Dev	53.20961	Variance	2831.263
Skewness	0.771339	Kurtosis	1.678801
USS	5375551	CSS	300113.8
CV	24.43123	Std Mean	5.143967
T:Mean=0	42.33959	Prob>|T|	0.0001
Sgn Rank	2889	Prob>|S|	0.0001
Num ^= 0	107		

Quantiles(Def=5)

100% Max	414	99%	399
75% Q3	249	95%	298
50% Med	217	90%	279
25% Q1	181	10%	155
0% Min	120	5%	131
		1%	123
Range	294		
Q3-Q1	68		
Mode	205		

Extremes

Lowest	Obs	Highest	Obs
120(	49)	311(	20)
123(	81)	312(	75)
126(	76)	337(	51)
126(	14)	399(	40)
130(	83)	414(	103)

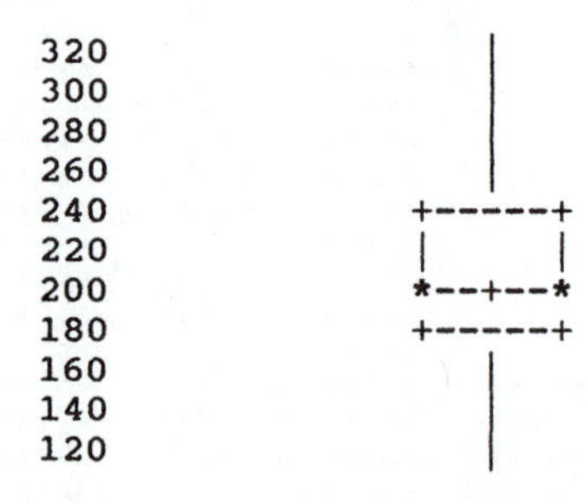

FIGURE 4.41 *Continued*

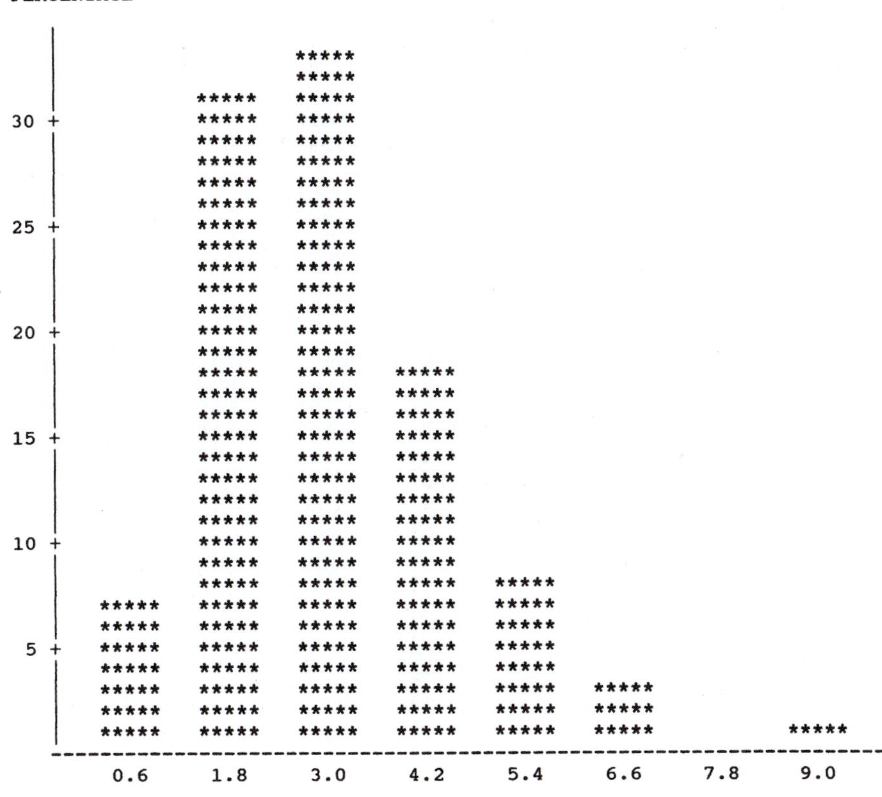

Moments

N	107	Sum Wgts	107
Mean	3.008777	Sum	321.9391
Std Dev	1.455412	Variance	2.118224
Skewness	0.997354	Kurtosis	1.291836
USS	1193.175	CSS	224.5317
CV	48.37222	Std Mean	0.1407
T:Mean=0	21.38434	Prob>|T|	0.0001
Sgn Rank	2889	Prob>|S|	0.0001
Num ^= 0	107		

(continued)

FIGURE 4.41 *Continued*

```
                 Quantiles(Def=5)

100% Max   8.573529          99%   6.487121
 75% Q3    3.683871          95%   5.874286
 50% Med   2.815385          90%   5.405882
 25% Q1    2.014286          10%   1.330233
  0% Min   0.463918           5%   1.149474
                              1%   0.725424
Range      8.109612
Q3-Q1      1.669585
Mode           3.65


                     Extremes

  Lowest       Obs       Highest      Obs
0.463918(       41) 5.897143(         48)
0.725424(       81) 6.045872(         51)
0.935556(      105) 6.215789(         27)
1.025974(       11) 6.487121(        103)
 1.09375(       98) 8.573529(         20)
```

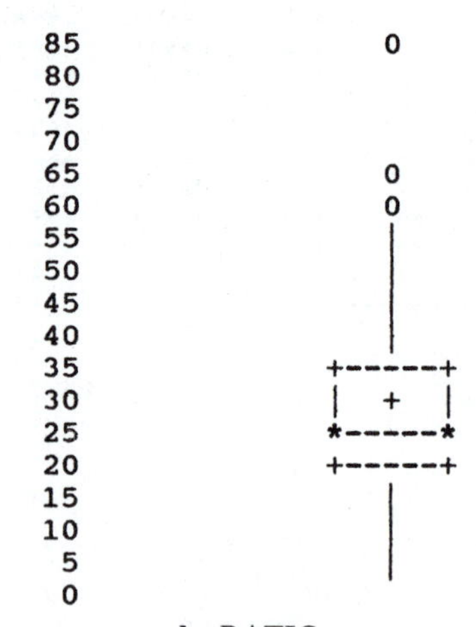

b. RATIO

EXERCISES

4.40 What is a measure of relative standing?

4.41 What is the pth percentile of a relative frequency distribution?

4.42 What are the quartiles of a relative frequency distribution?

4.43 How is the median of a relative frequency distribution related to the quartiles of the distribution?

4.44 How is the third decile related to percentiles?

4.45 Suppose that you scored at the 84th percentile in your class. How did you score relative to the other members of your class?

HLTH **4.46** *U.S.A. Today* (July 7, 1989) states that a study, published in the *Journal of the American Medical Association*, suggests that 60 million U.S. adults have cholesterol levels that are too high. It goes on to say that 43% of the 11,864 adults in the study had cholesterol levels below 200, 30% between 200 and 240, and 27% above 240. They conclude that, overall, 36% are at high risk for developing heart disease.

a. Give the 43rd percentile for the cholesterol levels.
b. Give the 73rd percentile for the cholesterol levels.
c. What can you say about the location of the 30th percentile?
d. If 36% of the people in the study are at risk of developing heart disease, at approximately what cholesterol level does this occur?

EDUC **4.47** The lower, middle, and upper quartiles for a distribution of test scores are 460, 580, and 780, respectively.

a. Suppose that you scored 470 on the test. Approximately what percentage of the total number of scores exceed your score?
b. What can you say about the skewness of the distribution? Explain.

4.48 What is the interquartile range for the data set in Exercise 4.47?

4.49 What is the purpose of a box plot?

4.50 Draw a box plot for the data set described in Exercise 4.47.

GEN **4.51** Refer to the $n = 14$ measurements on deaths due to fires in compartmentalized fire-resistant buildings, Table 4.3. The Data Desk computer printout of the numerical descriptive statistics for the data is shown in Figure 4.42.

FIGURE 4.42 Data Desk printout of numerical descriptive measures for fire death data, Table 4.3

```
Summary statistics for      Hotel Deaths
NumNumeric = 14
NumCases = 14
Mean = 8.3571
Median = 6
Standard Deviation = 8.9409
Interquartile range = 5.2500
Range = 36
Variance = 79.940
Minimum = 0
Maximum = 36
25-th %ile = 4
75-th %ile = 9.2500
```

a. Find the lower and upper quartiles on Figure 4.42.
b. Find the median.
c. Calculate the value of the interquartile range and compare it with the value shown on the printout.
d. Construct a box plot for the data.
e. Check the individual measurements against the box plot and plot any that fall outside of the inner fences.
f. Are there any outliers? If so, what should you do about them?
g. Based on your box plot, what can you say about the skewness of the relative frequency distribution for the data?

DEM **4.52** Refer to the measurements on average life expectancies for $n = 33$ developed countries, Table 4.5. The Data Desk computer printout of the numerical descriptive statistics for the data is shown in Figure 4.43.

a. Find the lower and upper quartiles on Figure 4.43.

b. Find the median.

c. Calculate the value of the interquartile range and compare it with the value shown on the printout.

d. Construct a box plot for the data.

e. Check the individual measurements against the box plot and plot any that fall outside of the inner fences.

f. Are there any outliers? If so, what should you do about them?

g. Based on your box plot, what can you say about the skewness of the relative frequency distribution for the data?

FIGURE 4.43 Data Desk printout of numerical descriptive measures for average life expectancies

```
Summary statistics for     Life Expectancy
NumNumeric = 33
Mean = 74.530
Median = 75
Standard Deviation = 2.4412
Interquartile range = 3.0500
Range = 9.4000
Variance = 5.9597
Minimum = 69.700
Maximum = 79.100
25-th %ile = 73.350
75-th %ile = 76.400
```

4.9 Statistics and Parameters

Throughout this chapter, we have presented symbols for sample numerical descriptive measures such as the sample mean, variance, and standard deviation. These sample numerical descriptive measures are called **statistics**.

The population has corresponding numerical descriptive measures but we do not calculate them. (That is what we are attempting to avoid by sampling.) These measures are called **parameters**. We represent population parameters by symbols so that we can talk about them, and we use the sample statistics to estimate their values. This procedure is part of what we call *statistical inference*.

A numerical descriptive measure *for a sample* is called a *statistic*.

A numerical descriptive measure *for a population* is called a *parameter*.

Symbols for population parameters are usually letters of the Greek alphabet. The population mean is represented by the symbol μ (mu), which is the

Greek letter "m". Similarly, the population standard deviation is represented by the symbol σ (sigma), the Greek letter "s". The sample and population symbols for the mean, standard deviation, and variance are shown in the box.

Symbols for Some Sample and Population Numerical Descriptive Measures

	Sample	**Population**	
Mean	$\bar{x}$	μ	**(the Greek letter "mu")**
Standard deviation	s	σ	**(the Greek letter "sigma")**
Variance	s^2	σ^2	**("sigma-squared")**

As you might suspect, we will use the sample mean $\bar{x}$ to estimate the value of the population mean μ, and the sample standard deviation s to estimate the population standard deviation σ. We will be able to use these estimates of μ and σ to form a mental picture of the population relative frequency distribution.

Measuring Error and Goodness

Why is a number more useful than a figure (the shape of a sample relative frequency distribution) for making inferences? The answer is that we can easily give a numerical measure of error associated with a numerical inference. For example, a newspaper report on a public opinion poll might state that "70% of the public favor no smoking in public places, with a sampling error of plus or minus 3%." The 70% is an estimate (a sample statistic) of the true but unknown percentage (the parameter) of the public that favor no smoking in public places. We cannot say how much the estimate, 70%, will deviate from the true percentage, but we are told that the ***sampling error*** (or ***error of estimation***), **the deviation between the estimate and the true percentage**, will be less than 3%. This published bound, plus or minus 3%, on the error of estimation gives us a meaningful numerical measure of how good the estimate is likely to be.

EXERCISES

4.53 How are the range and the standard deviation of a data set related?

4.54 What is a parameter? Give an example of a parameter.

4.55 What is a statistic? Give an example of a statistic.

4.56 If 10% of the members of this class are left-handed, is the 10% a statistic or a parameter? Explain.

4.10 Key Words and Concepts

- ***Numerical descriptive measures*** not only provide meaningful descriptions of a set of data, **but they also are better suited to making inferences** than the graphical methods.
- Measures of location, the ***mean*** and the ***median***, **locate the approximate center** of the measurements in a data set.
- The ***range*** and the ***standard deviation*** **measure the variation or dispersion** of a data set.
- ***Percentiles***, ***deciles***, and ***quartiles*** are used to **locate the relative position** of a measurement within a data set.
- **Numerical descriptive measures calculated from sample data** are called ***statistics***.
- **Population numerical descriptive measures** are called ***parameters***.
- Sample statistics are used to make inferences about population parameters.
- **The numerical difference between an estimate and the estimated parameter** is called the ***sampling error*** or the ***error of estimation***.
- Numerical descriptive measures allow us to measure the "goodness" of an inference, whereas graphical methods of description do not.

SUPPLEMENTARY EXERCISES

MFG **4.57** Although a coffee company ships coffee in cans indicating a content of 11.5 ounces, the actual weight loaded by a filling machine varies from can to can. A sample of 100 filled cans was selected from the filling operation. The mean and standard deviation of their weights were $\bar{x} = 11.60$ ounces and $s = .15$ ounce. Construct a mental image of the distribution of the 100 fill weights.

BIO **4.58** The SAS relative frequency distribution of the DDT content (ppm) of 144 fish sampled from the Tennessee River is shown in Figure 4.44 and the SAS printout of the numerical descriptive measures for the data is shown in Figure 4.45 (page 120).

a. Find the lower and upper quartiles on Figure 4.45. What is the practical interpretation of the lower quartile? The upper quartile?

b. Find the median. What is the practical interpretation of the median?

c. Calculate the value of the interquartile range and compare it with the value shown on the printout.

d. Construct a box plot for the data.

e. Check the individual measurements against the box plot and plot any that fall outside of the inner fences.

f. Are there any outliers? If so, what should you do about them?

g. Based on your box plot, what can you say about the skewness of the relative frequency distribution for the data?

BUS **4.59** Refer to the 390 percentage iron ore measurements of Table 1, Appendix 1. The Minitab printout of the numerical descriptive statistics for the data is shown in Figure 4.46 (page 120).

a. Find the lower and upper quartiles on Figure 4.46. What is the practical interpretation of the lower quartile? The upper quartile?

FIGURE 4.44 SAS relative frequency distribution for the DDT content of 144 fish

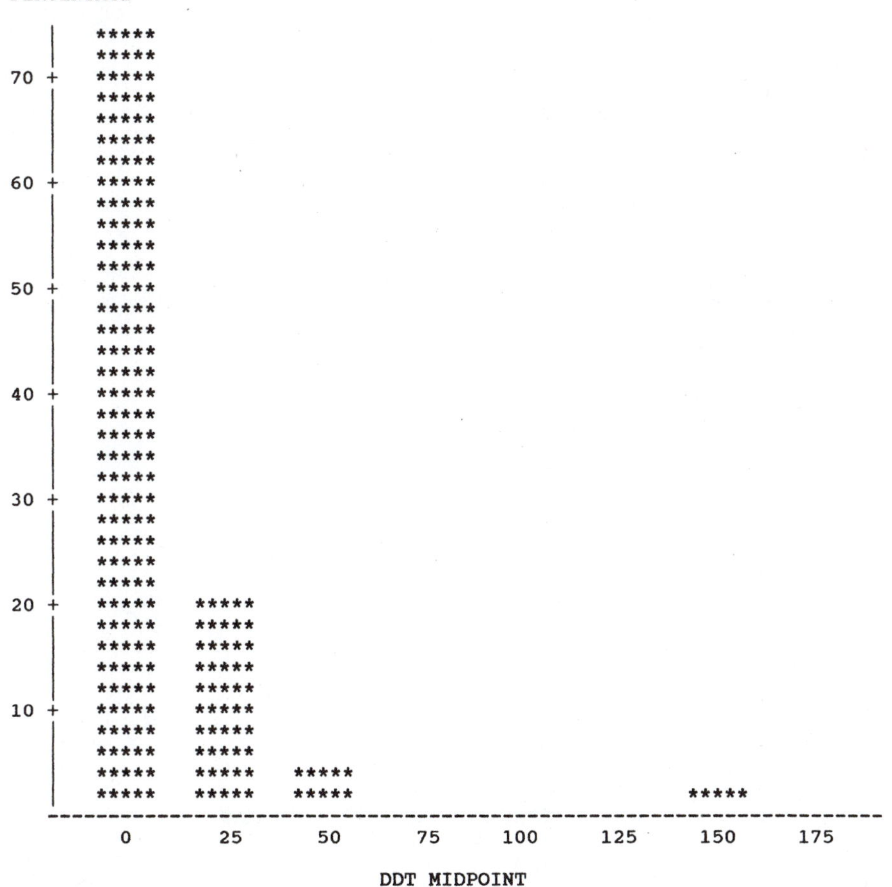

FIGURE 4.45 SAS printout of numerical descriptive measures for the DDT content of 144 fish

```
                    Moments

N               144   Sum Wgts        144
Mean         24.355   Sum         3507.12
Std Dev    98.37859   Variance   9678.346
Skewness    9.61166   Kurtosis   102.2388
USS         1469419   CSS         1384003
CV         403.9359   Std Mean   8.198215
T:Mean=0   2.970768   Prob>|T|     0.0035
Sgn Rank       5220   Prob>|S|     0.0001
Num ^= 0        144

               Quantiles(Def=5)

 100% Max      1100          99%       360
  75% Q3         13          95%        61
  50% Med      7.15          90%        28
  25% Q1       3.35          10%       1.2
   0% Min      0.11           5%      0.43
                              1%      0.18

 Range      1099.89
 Q3-Q1         9.65
 Mode            12
```

FIGURE 4.46 Minitab printout of the numerical descriptive measures for the iron ore data

```
              N     MEAN   MEDIAN   TRMEAN    STDEV   SEMEAN
%iron       390   65.743   65.830   65.794    0.694    0.035

            MIN      MAX       Q1       Q3
%iron    62.770   66.860   65.410   66.230
```

b. Find the median. What is the practical interpretation of the median?

c. Calculate the value of the interquartile range and compare it with the value shown on the printout.

d. Construct a box plot for the data.

e. Check the individual measurements against the box plot and plot any that fall outside of the inner fences.

f. Are there any outliers? If so, what should you do about them?

g. Based on your box plot, what can you say about the skewness of the relative frequency distribution for the data?

4.60 If a data set contains $n = 100$ measurements and its range equals 10.2, what would you expect the standard deviation of the data set to be, approximately? Explain your reasoning.

HLTH **4.61** The smallest and the largest of the 107 TCHOL measurements of Table 3, Appendix 1, are 120 and 414, respectively. Use these values to approximate the standard deviation of the data set. Is this approximation of the same magnitude as the standard deviation ($s = 53.20961$) shown on the printout, Figure 4.41a?

HLTH **4.62** The smallest and the largest of the 107 RATIO measurements of Table 3, Appendix 1, are .463918 and 8.573529, respectively. Use these values to approximate the standard deviation of the data set. Is this approximation of the same magnitude as the standard deviation ($s = 1.455412$) shown on the printout, Figure 4.41b?

EXERCISE FOR YOUR COMPUTER

GEN **4.63** A common misunderstanding of the Empirical Rule is that it applies only to a symmetric, bell-shaped distribution of data (which we will discuss in Chapter 5). The percentages given in the Empirical Rule apply, with a reasonable degree of approximation, to a wide variety of distributions, skewed as well as symmetric. Pick any real data set (biological, sociological, economic, engineering, or whatever) containing at least 20 measurements (the more, the better). Use a computer to calculate the numerical descriptive measures for the data set. I would bet that all or almost all of the measurements fall within $3s$ of the mean $\bar{x}$. Close to 95%, or perhaps all, of the measurements will fall within $2s$ of the sample mean.

References

1. Freedman, D., Pisani, R., Purves, R., and Adhikari, A., *Statistics*, 2nd edition. New York: W. W. Norton & Co., 1991.
2. McClave, J. and Dietrich, F., *Statistics*, 5th edition. San Francisco: Dellen Publishing Co., 1991.
3. Mendenhall, W. and Beaver, R., *Introduction to Probability and Statistics*, 8th edition. Boston: PWS-Kent, 1991.
4. Moore, D. and McCabe, G., *Introduction to the Practice of Statistics,* 2nd edition. New York: W. H. Freeman & Co., 1993.
5. Sincich, T., *Statistics by Example,* 4th edition. San Francisco: Dellen Publishing Co., 1990.

How Probability Helps Us Make Inferences

In a Nutshell

We learned in Chapters 2 and 4 that the objective of statistics is to make an inference about a population of data based on the limited information contained in a sample. Probability is the tool that makes statistical inference possible.

5.1 Shifting Gears

5.2 Probability and Inference

5.3 Some Concepts of Probability

5.4 Random Samples

5.5 Probability Distributions

5.6 The Binomial Probability Distribution

5.7 The Normal Probability Distribution

5.8 Sampling Distributions

5.9 Making Statistical Inferences: Estimation

5.10 Making Statistical Inferences: Tests of Hypotheses

5.11 Key Words and Concepts

5.1 Shifting Gears

An introductory course in statistics is served in three portions. The first, consisting of Chapters 1–4, is earthy and real. It gives you an overall view of statistics, what it is intended to accomplish, and why it plays an important role in the modern world. We see real sets of data and learn how to describe them. And we begin to see how sample statistics might be used to deduce (infer) the nature of a sampled population.

The second portion of the course represents a shift in gears. We leave the practical familiarity of data and turn to the less familiar abstract concepts of probability. Now we must learn how we make the deductive leap from a sample to a population. Thus, Chapter 5 is a foundations chapter. It sets the stage for the statistical applications to come in the third portion of the course, Chapters 6–11.

The third part of the course requires a second shift in gears. Now we go back to the practical and consider some real-life sampling and inference-making problems. Woven into a discussion of the data sets, almost all derived from the news media or research literature, are the basic inferential concepts of Chapter 5. Thus, our understanding of the concepts of Chapter 5 will be reinforced and developed as we see how they are applied in Chapters 6–11.

5.2 Probability and Inference

What Is Probability?

Before we examine the relationship between probability and statistical inference, we need to clarify the meaning of the word *probability*.

For a meaningful definition of *probability*, let's see how we use the word. Most observations can result in one of a number of outcomes. For example, a tossed coin can result in either a head or a tail. We do not know which of the two outcomes will occur when we make a particular observation, but we say that the odds of a head or a tail are 50–50 or that the probability of observing a head is $\frac{1}{2}$. By these statements, we mean that we believe that we are as likely to observe a head as a tail. If we say that the probability of observing a head is .2, we mean that an observation would be much less likely to result in a head than a tail. If the probability of a head is 1.0, we believe that we will be certain to observe a head. Thus, we think of the probability of the occurrence of a particular outcome as a number between 0 and 1. What does it measure? **Probability measures our belief that an observation will result in some particular outcome.**

How Probability Is Used to Make Statistical Inferences

Statistical inference works the same way as your personal intuitive inference-making system works. If, as you approach a bank, a man with a stocking mask runs out of the door, satchel in hand, followed by the bank guard wielding a gun, what do you infer? You will probably infer that the bank is being robbed. How did you arrive at that conclusion? You considered the possible set of circumstances that might have produced the sample of events that you observed. Was the masked man trying to avoid getting sunburned? Was the masked man robbing the bank and trying to hide his identity? Was the masked man simply the bank manager traveling incognito? Of these and many other justifications for the events that you observed, you decided that the outcome that you observed was most probable if the bank was being robbed. Thus, you based your inference, a decision, on your assessment of a set of probabilities.

Statistical inference uses the same type of reasoning. For example, consider the oil can-filling operation of Section 4.5. Suppose that you collected 25 filled cans, measured the amount of oil in each, and calculated the mean of the sample of 25 fills to be 32.4 ounces. Suppose also that you wanted to use this sample mean to estimate (infer) the value of the mean μ of the population of fills for all cans filled during the past month. Intuitively, what would be your estimate of μ?

Figure 5.1 shows three hypothetical population relative frequency distributions: one with mean $\mu = 16.5$, one with mean $\mu = 32.4$, and one with mean $\mu = 37.3$. If the mean of the sample of 25 can fills is 32.4, for which population was a sample mean of 32.4 most likely? Intuitively, it seems reasonable (see Figure 5.1) to assume that the probability that you will draw a sample with mean close to $\bar{x} = 32.4$ is greater if $\mu = 32.4$ than if $\mu = 16.5$ or 37.3! And, in fact, our intuition is correct. Theoretical statistics shows that a distribution of fills with a mean of $\mu = 32.4$ has a higher probability of producing a sample mean of $\bar{x} = 32.4$ than a distribution of fills with a different mean. Thus, if we choose $\bar{x} = 32.4$ as our best estimate of the population mean μ, we are making an inference about something (a population parameter μ) based on our intuitive evaluations of probabilities. We decided that it was most likely that $\bar{x} = 32.4$ if $\mu = 32.4$.

FIGURE 5.1 Hypothetical population relative frequency distributions. For which is $\bar{x}$ = 32.4 most probable?

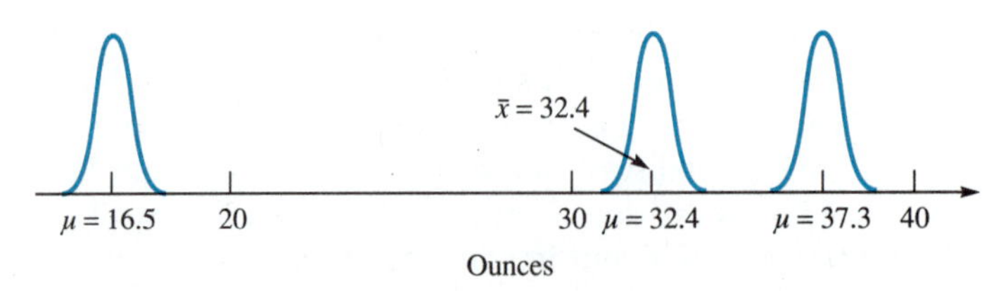

These two examples, the bank episode and the sample of can fills, show that inference, both personal and statistical, relies on an assessment of the probabilities of observed sample outcomes. **The difference between personal and statis-**

tical inferential methods is that statistical inference does not rely on intuitive assessments of probabilities. It bases its inferences on the probabilities of sample outcomes that have been calculated using the mathematical theory of probability. We will not have to perform these calculations. These probabilities have been calculated years ago by theoretical statisticians and are presented in tables.

5.3 Some Concepts of Probability

The *theory of probability* (as opposed to the word *probability*) is a mathematical theory (a model of reality) that enables us to calculate the probability of one outcome based on its relationship to other outcomes and their probabilities. To commence, we need to quantify what we mean by the probability of an outcome. We have said that the probability of an outcome is a number that measures our belief that a single observation will result in that outcome. To most of us, the exact number that measures the probability of a specific outcome is based on the relative frequency concept of probability.

The Relative Frequency Concept of Probability

If an observation is repeated a very large number, N, times and r observations result in outcome A, then the *probability that a single observation results in outcome A*, denoted by the symbol $P(A)$, is equal to

$$P(A) = \frac{r}{N}$$

For example, suppose that records show that of 10,000 one-quart oil cans filled by a filling machine, 3,470 contain 32 or more ounces of oil. Suppose that we were to draw a single can from production. We would expect the probability that the can contains 32 or more ounces to equal (or be very close to)

$$P(\text{32 or more ounces}) = \frac{r}{n} = \frac{3{,}470}{10{,}000} \approx .35$$

We would prefer to have a much larger number of observations on which to base our calculation. But even with n as small as 10,000 cans, I think we would agree that the probability that the can will contain 32 ounces or more is close to .35.

Mutually Exclusive Outcomes and the Additive Property of Probabilities

Two different outcomes of an observation, call them A and B, are said to be *mutually exclusive* if a single observation cannot result in both outcomes A and B. For example, suppose that we observe the content of a single can of oil selected from production.

Let A be the outcome that the can contains less than 32 ounces and let B be the outcome that the can contains between 32 and 33 ounces. Then A and B are mutually exclusive outcomes. If A occurs (that is, the can contains less than 32 ounces), then B (the can contains 32 to 33 ounces) cannot occur, and vice versa.

The concept of mutually exclusive outcomes is useful when we want to find the probability that *either* outcome A *or* outcome B occurs. We use the following rule:

If outcomes A and B are mutually exclusive, then the probability that a single observation results in either A or B is equal to $P(A) + P(B)$, i.e.,

Additive Rule of Probability for Mutually Exclusive Outcomes

$$P(\text{observe either } A \text{ or } B) = P(A) + P(B)$$

Example 5.1

Suppose that we draw an oil can from a filling machine and observe its fill weight. We will define two outcomes, call them A and B, as follows:

A: The can contains less than 32 ounces

B: The can contains at least 32 but less than 33 ounces

If $P(A) = .65$ and $P(B) = .20$, find the probability that either A or B will occur—i.e., the probability that the can contains less than 33 ounces.

Solution

A and B are mutually exclusive outcomes because if A occurs, B cannot occur. Thus,

$$\begin{aligned} P(\text{either } A \text{ or } B \text{ occurs}) &= P(\text{the weight is less than 33 ounces}) \\ &= P(A) + P(B) = .65 + .20 = .85 \end{aligned}$$

Therefore, the probability that a can drawn from production will contain less than 33 ounces is .85.

The Additive Rule applies to any number of mutually exclusive outcomes. For example, if A, B, and C are mutually exclusive outcomes, then the probability that either A, B, or C occurs is equal to $P(A) + P(B) + P(C)$.

Example 5.2

If you toss a balanced die, the chances are equal that you will observe a 1, 2, 3, 4, 5, or 6. That is, the probability of observing either a 1, 2, 3, 4, 5, or 6 is $\frac{1}{6}$. Find the probability of tossing a die and observing an odd number.

Solution

The probability of observing an odd number is the probability of observing either a 1, 3, or 5. Since all three outcomes are mutually exclusive (e.g., if you observe a 1, you could not possibly observe a 3 or a 5 at the same time),

$$P(\text{observing either a 1, 3, or 5}) = P(1) + P(3) + P(5)$$
$$= \frac{1}{6} + \frac{1}{6} + \frac{1}{6} = \frac{3}{6} = \frac{1}{2}$$

Independent Outcomes and the Multiplicative Property of Probabilities

Two outcomes, *A* and *B*, are said to be *independent* if the probability that outcome *A* will occur is unaffected by the fact that outcome *B* has or has not occurred, i.e., what happens to *A* is independent of what happens to *B*.

For example, suppose that we make an observation each morning. Define two of the many possible outcomes, A and B, as follows:

A: The sun rises that day

B: I have eggs for breakfast that day

Since the probability that the sun rises on any given morning is always 1, regardless of whether I do or do not have eggs for breakfast, the outcomes A and B are by definition independent outcomes.

As another example, suppose that we toss a coin two times and observe how it lands. Let A be the observance of a head on the first toss and B a head on the second toss. Then A and B are independent outcomes because the probability of a head on the second toss (outcome B) is $\frac{1}{2}$ regardless of the outcome of the first toss (whether A does or does not occur).

In contrast, suppose that we observe a basketball player toss two "free throws" (instead of two coins) and that two outcomes, A and B, are A: a basket on the first shot and B: a basket on the second shot. Then A and B are not independent outcomes because the player will use what he or she learns from the first shot to adjust the distance for the second shot. Therefore the probability that B occurs depends on whether A does or does not occur.

If we know that two outcomes are independent, we can use the definition of independence to find the probability that both of the outcomes occur in a single repetition of an experiment, as follows:

The probability that both of two independent outcomes, *A* and *B*, occur in a single observation is equal to *P*(*A*) times *P*(*B*), i.e.,

The Multiplicative Rule of Probability for Independent Outcomes

$$P(\text{both outcomes } A \text{ and } B \text{ occur}) = P(A)P(B)$$

Example 5.3

Suppose that we toss two coins and observe the upper faces of the coins. Define the following outcomes:

A: Coin #1 results in a head

B: Coin #2 results in a head

Find the probability that the tosses result in two heads.

Solution

The probability of tossing a head in a single toss of a coin is $\frac{1}{2}$. Therefore, $P(A) = \frac{1}{2}$ and $P(B) = \frac{1}{2}$. Since outcomes A and B are independent, i.e., the probability that B will occur is independent of whether A has or has not occurred,

$$P(\text{both } A \text{ and } B \text{ occur}) = P(A)P(B) = \left(\frac{1}{2}\right)\left(\frac{1}{2}\right) = \frac{1}{4}$$

The Multiplicative Rule applies to any number of independent outcomes. Thus, if A, B, and C are independent outcomes, the probability that all of the outcomes, A, B, and C, occur is

$$P(A, B, \text{and } C \text{ occur}) = P(A)P(B)P(C)$$

For example, the probability of observing three heads in the toss of three coins is

$$P(\text{tossing three heads}) = \left(\frac{1}{2}\right)\left(\frac{1}{2}\right)\left(\frac{1}{2}\right) = \frac{1}{8}$$

The Complement of an Outcome

The *complement* of an outcome A is the outcome that A does not occur. The complement of an outcome A is represented by the symbol $\bar{A}$. For example, suppose that we select a person and measure their total cholesterol (as in Case 2.3). Let A be the outcome that the person's total cholesterol is less than 200. Then the complement $\bar{A}$ of A is the outcome that A does not occur, i.e., $\bar{A}$ is the outcome that the person's total cholesterol is 200 or more. Note that if A occurs, $\bar{A}$ cannot occur (and vice versa).

The probability of the complement of an outcome A is equal to

$$P(\bar{A}) = 1 - P(A)$$

For example, if an outcome is A: a person's total cholesterol is less than 200, its complement is the outcome $\bar{A}$: a person's total cholesterol is 200 or more. If $P(A) = .35$, then the probability that a person's cholesterol is 200 or more is

$$P(\bar{A}) = 1 - P(A) = 1 - .35 = .65$$

5.4 Random Samples

The bank robbery and the oil can-filling examples of Section 5.2 taught us that statistical inference is based on probability. **The outcomes that we will be interested in are the numbers that appear in a sample. In order to make a statistical inference, we need to be able to assess the probabilities of all of these possible sample outcomes, that is, we need to select a *probability sample*.**

A Probability Sample

A *probability sample* is selected in such a way that the probabilities of the various sample outcomes can be determined. To give a simple example, suppose that the universe consists of all of the students in this class and that I want to select a sample of $n = 2$ students from this universe. I decide to select the sample by placing each student's name on a card, shuffling the deck thoroughly, and then selecting two cards (two names). It is possible (proof omitted) to calculate the probability that any pair of students, say Susan and John, will be selected and I can do this for every different pair of students in the universe. Therefore, the pair selected would represent a probability sample. This sample can be used for statistical inference.

In contrast, suppose that I decide to select the two persons for the sample by choosing the first two students who walk into the next class meeting. There is no way that I can calculate the probability of selection for every possible pair of students. John, for the first time this semester, may come early. Therefore, the pair selected would not represent a probability sample and the sample would be unsuitable for statistical inference.

There are many different sampling procedures that will produce probability samples. The most frequently employed method is called *simple random sampling* or just *random sampling*.

A Random Sample

A *random sample* is one that is selected in such a way that, for a given sample size, every different sample in the population has an equal probability of being selected. The number n of observations in a sample is called the *sample size*.

Suppose that a population contains only the four measurements 1, 3, 7, and 2. If we wish to draw a sample of $n = 2$ measurements from this population, there are only six different samples that can be drawn: the combinations (1, 3), (1, 7), (1, 2), (3, 7), (3, 2), and (7, 2). The sampling will result in a random sample if the selection is performed in such a way that the probability of selecting any one of these samples is equal to $\frac{1}{6}$. This could be done by numbering six poker chips, one for each of the six samples, putting them in a hat, mixing them, and drawing one chip at random. Or it could be done by placing four chips in the hat, one for each of the numbers, 1, 3, 7, and 2, dipping into the hat, and selecting two of the four chips at random.

How to Draw a Random Sample

The best way to select a random sample is to use a *random number table* such as the one shown in Table 1 of Appendix 2. A portion of Table 1 of Appendix 2 is reproduced in Table 5.1. Table 5.1 is constructed in such a way that every number in the table occurs with equal frequency, i.e., the observation of any one number is as likely as any other. We will illustrate the use of Table 5.1 with an example.

TABLE 5.1 Partial Reproduction of the Table of Random Numbers, Table 1 of Appendix 2

Row \ Column	1	2	3	4	5	6	7	8	9	10
1	10480	15011	01536	02011	81647	91646	69179	14194	62590	36207
2	22368	46573	25595	85393	30995	89198	27982	53402	93965	34095
3	24130	48360	22527	97265	76393	64809	15179	24830	49340	32081
4	42167	93093	06243	61680	07856	16376	39440	53537	71341	57004
5	37570	39975	81837	16656	06121	91782	60468	81305	49684	60672
6	77921	06907	11008	42751	27756	53498	18602	70659	90655	15053
7	99562	72905	56420	69994	98872	31016	71194	18738	44013	48840
8	96301	91977	05463	07972	18876	20922	94595	56869	69014	60045
9	89579	14342	63661	10281	17453	18103	57740	84378	25331	12566
10	85475	36857	53342	53988	53060	59533	38867	62300	08158	17983
11	28918	69578	88231	33276	70997	79936	56865	05859	90106	31595
12	63553	40961	48235	03427	49626	69445	18663	72695	52180	20847
13	09429	93969	52636	92737	88974	33488	36320	17617	30015	08272
14	10365	61129	87529	85689	48237	52267	67689	93394	01511	26358
15	07119	97336	71048	08178	77233	13916	47564	81056	97735	85977
16	51085	12765	51821	51259	77452	16308	60756	92144	49442	53900

Example 5.4

Suppose that you want to sample the opinions of the 4,547 adults living in a small town. Use the portion of the random number table reproduced in Table 5.1 to select a random sample of 400 people from this universe.

Solution

The first step in selecting a random sample is to list and number the names of the 4,547 adults. **The listing of all elements in the universe is called a *frame*.** Then go to Table 5.1 and randomly select a starting point, anywhere. Suppose that we decide to start with the number in row 5 and column 4, 16656. Since the adults in our universe are numbered from 1 to 4,547, we will use only the first four digits of our five-digit numbers and will discard 0 and all numbers larger than 4,547. Since the first number chosen from the table is 16656, the first person to enter our sample is the one numbered 1,665. The next number you select is the one in the next row (or column). We will go down the column to the next row and read 42751. Using the first four digits of this number, we select the person numbered 4,275 as the second member of our sample. The next random number is 69994. Discard this number because 6,999 exceeds 4,547; go on to 07972. Then the third member of our sample is the person numbered 797. Continue this procedure until you have selected 400 people for your sample. If the same random number is drawn more than once, discard it for all except the first draw.

Difficulties in Drawing Random Samples

Drawing random samples (and other probability samples) can be difficult or impossible because we may not be able to construct a frame. Thus there is no possible way to select a random sample from a conceptual population of data,—a population that exists in our minds. The best that we can do in sampling the results of a new laboratory experiment is to repeat the experiment a number n of times and hope that this sample is representative of what we might observe if the experiment were repeated, over and over again, a very large number of times. At best, we hope that it represents a good approximation to a random sample!

Similarly, there is no way to list all elements in the universe of all sea bass in the western Atlantic Ocean. You can select the fish from different areas of the ocean and thereby achieve an element of randomness, but you cannot be certain that you have a random sample.

It is also impossible to select a random sample in a national opinion poll because it is impossible to obtain a list of all adults in the United States; you cannot obtain a complete frame. (Not even the U.S. Bureau of the Census can obtain the frame! It was accused in 1990 of major omissions in listing all residents of the United States.) The best that the pollsters can do is to try to obtain a frame that appears to represent the sampled population and to select a probability sample from it. A description of how a November 1991 *New York Times*/CBS News Poll was conducted is shown in Figure 5.2 (*The New York Times*, November 26, 1991).

FIGURE 5.2 A description of how *a New York Times*/CBS News Poll was conducted

How the Survey Was Taken

The latest New York Times/CBS News Poll is based on telephone interviews conducted Nov. 18 to 22 with 1,106 adults around the United States, excluding Alaska and Hawaii.

The sample of telephone exchanges called was selected by a computer from a complete list of exchanges in the country. The exchanges were chosen so as to assure that each region of the country was represented in proportion to its population. For each exchange, the telephone numbers were formed by random digits, thus permitting access to both listed and unlisted numbers. The numbers were then screened so that calls would be placed to residences only. Within each household one adult was designated by a random procedure to be the respondent for the poll.

The results have been weighted to take account of household size and number of telephone lines into the residence and to adjust for variations in the sample relating to region, race, sex, age and education.

In theory, in 19 cases out of 20, the results based on such samples will differ by no more than three percentage points in either direction from what would have been obtained by seeking out all American adults.

The potential sampling error for smaller subgroups is larger. For example, for those who say they vote in Democratic primaries or caucuses, it is plus or minus five percentage points.

In addition to sampling error, the practical difficulties of conducting any survey of public opinion may introduce other sources of error into the poll.

5.5 Probability Distributions

A *probability distribution* is a theoretical model for a population's relative frequency distribution. The variable that it describes is called a *random variable*. The purpose in constructing these models is to enable us to calculate the probabilities of various sample outcomes.

There are two types of random variables, *discrete* and *continuous*.

Discrete Random Variables and Their Probability Distributions

***Discrete random variables* are those that assume a countable number of values.** The number of values that x can assume is usually finite but it can be infinitely large as long as the values are countable. Typical discrete random variables are counts of things, as the following examples illustrate:

1. The number x of occupied beds in a hospital on a given day: If the hospital contains 260 beds, x can assume any one of the values, $x = 0, 1, 2, 3, 4, \ldots, 259, 260$.
2. The number x of voters in a sample of 400 who favor a particular candidate: x can assume the values $x = 0, 1, 2, 3, 4, \ldots, 399, 400$.
3. The number x of auto accidents per day at a specific intersection: Although we would expect x to be small, there is no limit to the number of accidents that could occur on a given day. Therefore (in theory!), x could assume any one of the countable values, $x = 0, 1, 2, 3, 4, 5, \ldots, \infty$. The symbol ∞ means infinitely large—that is, large beyond all bound.

The *probability distribution for a discrete random variable* x is a table, graph, or formula that gives the probability $p(x)$ for each value of x. The probability $p(x)$ will always assume a value in the interval from 0 to 1, and the sum of the values of $p(x)$ for all values of x will always equal 1.

Figure 5.3 shows the probability distribution for a discrete random variable x that can assume any one of the values $0, 1, 2, 3, 4, \ldots, 19, 20$. The probability for a value of x, say $x = 5$, is represented by a bar centered over $x = 5$. The height of the bar is equal to $p(x)$ and the bars are of equal width. Therefore, the area of a bar over a value of x is proportional to $p(x)$. For example, the height of the bar over $x = 5$ is approximately .17 and the area of the bar is .17 or 17% of the total area of all of the bars in the distribution. Therefore, $p(5) = .17$. The sum of the values of $p(x)$ over all values of x is equal to 1.

The values that a discrete random variable can assume are mutually exclusive outcomes. For example, suppose that x is the total number of rooms vacant per day at a small hotel. If $x = 1$ on a given day, x could not, for the same

FIGURE 5.3 An example of a discrete probability distribution

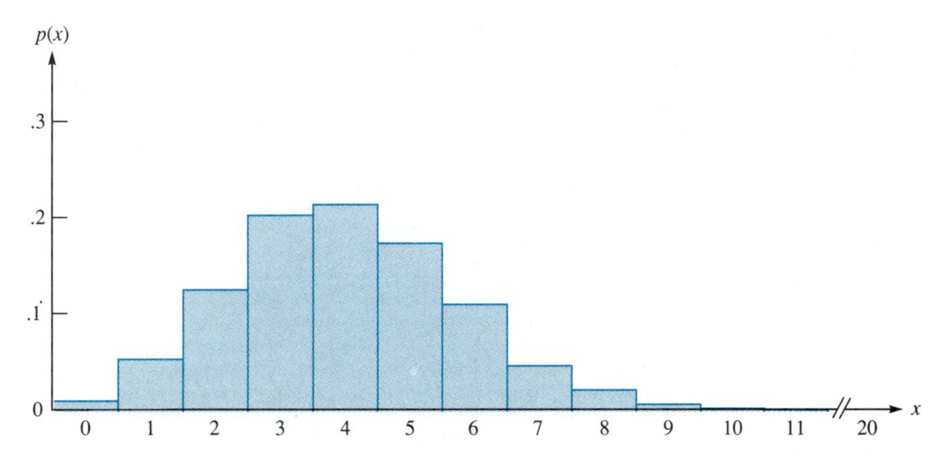

day, equal 2. Therefore the values $x = 1$ and $x = 2$ represent mutually exclusive outcomes. In fact, any pair of x values will always be mutually exclusive outcomes.

Since two values of x, say a and b, are mutually exclusive outcomes, then (according to the Additive Rule of Probability) the probability that x equals either a or b is equal to $p(a) + p(b)$. If $p(1) = .1$ and $p(2) = .3$, then the probability that the hotel will have either $x = 1$ or $x = 2$ vacant rooms on a given day is equal to $p(1) + p(2) = .1 + .3 = .4$.

Continuous Random Variables and Their Probability Distributions

A *continuous random variable* is one that can assume the infinitely large number of values associated with the points on a line interval. Examples of continuous random variables are the following:

1. The length x of a beetle
2. The amount x of blood in a human
3. The length x of time required to complete an examination

The *probability distribution for a continuous random variable* is a smooth curve as shown in Figure 5.4. The total area under the curve is equal to 1. The probability that x falls in the interval between two points, say a and b, is equal to the area under the relative frequency distribution over that interval (see Figure 5.4) and is represented by the symbol $P(a < x < b)$. The probability that x equals a specific value, say $x = a$, is always equal to 0. This last statement is the reason that we have to have two different types of probability distributions—one for discrete and one for continuous random variables. It is impossible to assign

FIGURE 5.4 An example of a continuous probability distribution

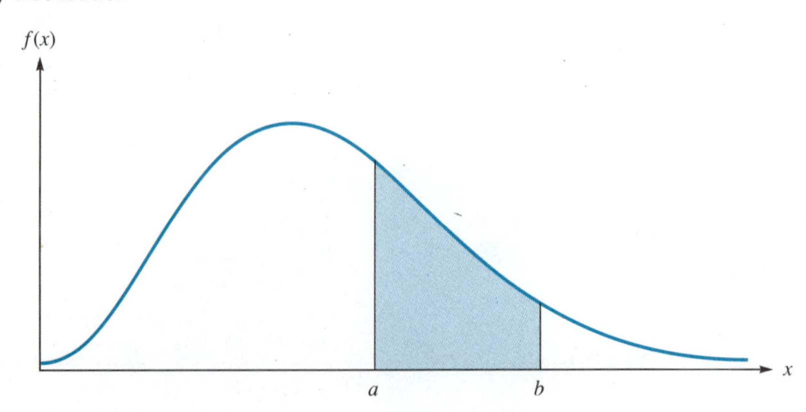

a finite amount of probability to each of the infinitely large number of points in a line interval and have the sum of the probabilities equal 1.

The probability that a continuous random variable will fall in a specific interval is tabulated for certain population relative frequency distributions that occur frequently in nature. One of the most important is discussed in Section 5.7.

EXERCISES

5.1 What is the objective of statistics?

5.2 What does probability measure?

5.3 How does probability play a role in statistical inference?

5.4 Explain the relative frequency concept of probability.

5.5 What do we mean when we say that two outcomes of an observation are mutually exclusive?

5.6 What do we mean when we say that two outcomes of an observation are independent?

5.7 Give the Additive Rule of Probability for mutually exclusive outcomes.

5.8 Give the Multiplicative Rule of Probability for independent outcomes.

5.9 What is the complement of an outcome?

GEN **5.10** Suppose that you observe the status of a legal trial at the end of a day. Consider the following outcomes: (1) the jury returns a conviction and (2) the trial is continued to the next day. State whether these two outcomes are mutually exclusive and explain why or why not.

5.11 Refer to Exercise 5.10. State whether the two outcomes are independent and explain why or why not.

MFG **5.12** An electrical fuse box contains four fuses—one bad and three good. Two fuses are removed from the box. Let A represent the outcome that the first fuse is defective and let B represent the outcome that the second fuse is defective. Are A and B independent outcomes? Explain.

GEN **5.13** Jones and Smith are two of five candidates running for mayor of a city. The odds-makers say that Jones will win with probability equal to .3 and that Smith will win with probability equal to .4. What is the probability that either Jones or Smith will win the election?

5.14 Refer to Exercise 5.13. What is the probability that neither Jones nor Smith wins the election?

5.15 The probability of observing a 6 in the toss of a single die is $\frac{1}{6}$. What is the probability of observing a pair of 6's when tossing a pair of dice?

5.16 What is a probability sample?

5.17 Why is probability sampling important when making statistical inferences?

5.18 Is a random sample a probability sample? Explain.

5.19 What is a random sample?

GEN **5.20** A public opinion poll of a city's adult population was conducted by including in the sample every tenth adult that passed a corner in the downtown section of the city until the sample of 1,000 adults was selected. Would this procedure produce a sample that is, for all practical purposes, a random sample? Explain.

BUS **5.21** Refer to Case 2.1. Explain why you think that the 390 1.5-kg samples of ore do or do not represent a random sample from the 20,000-ton consignment of ore.

5.22 Use the random number table (Table 1 of Appendix 2) to draw a random sample of $n = 5$ elements from a universe that contains 973 elements. Explain your procedure.

5.23 Use the random number table (Table 1 of Appendix 2) to draw a random sample of $n = 6$ elements from a universe that contains 10,589 elements. Explain your procedure.

GEN **5.24** A city council wishes to select a subcommittee of three members from among its total of 20 members. Use the random number table (Table 1 of Appendix 2) to identify the three members to be included on the subcommittee and explain how you arrived at your selection.

5.25 Suppose that you want to select a random sample of $n = 2$ elements from a universe that contains $N = 5$ elements. What is the probability that one particular sample will be selected?

5.26 What is a discrete random variable?

5.27 What is a continuous random variable?

5.28 State whether each of the following random variables is discrete or continuous.

a. Daily rainfall at a particular location
b. Number of transactions performed per hour at a bank
c. Number of hits in a baseball game
d. Weight of a fish
e. Length of time required for you to run a 100-yard dash

5.29 Follow the instructions of Exercise 5.28 for the following random variables:

a. Your cholesterol level
b. Your blood pressure
c. Number of workers injured per day at a factory
d. Number of eligible voters in a sample of 1,000 who favor a particular candidate
e. Number of chicken eggs produced per year in the United States

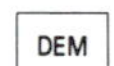

5.30 Let x equal the number of auto accidents per day at a particular intersection, and consider the outcomes $x = 0$ and $x = 1$. Are these mutually exclusive outcomes? Explain.

5.6 The Binomial Probability Distribution

Binomial Experiments

A *New York Times*/CBS News Poll (*The New York Times*, May 30, 1990) was conducted to investigate the opinions of American adults on their attitudes toward the former Soviet Union. The sampled universe consisted of all adults living in the United States. Each adult represented an element of the universe and each possessed or did not possess a particular characteristic—they either agreed or did not agree with a statement posed in the survey. A sample of $n = 1{,}140$ adults was selected from the universe and the number x of people in the sample favoring a statement was recorded. For example, $x = 684$, or 60% of all people in the sample, agreed that "it is very important to U.S. interests that Gorbachev remain in power in the Soviet Union." The objective of this sampling was to use the sample data, the number x in the sample who respond "yes" to a particular question, to estimate the proportion (or percentage) of all Americans who possess the same opinion.

The *New York Times*/CBS News Poll is typical of a whole class of similar statistical problems. All involve **sampling n elements from a universe where each element can elicit one, and only one, of two responses—yes or no, red or white, agree or disagree, and so forth. Sampling that satisfies these conditions is called a *binomial experiment* and the random variable x is called a *binomial random* variable.** The characteristics of a binomial experiment are listed in the box.

Characteristics of a Binomial Experiment

1. **The experiment consists of n identical trials.**
2. **Each trial can result in one and only one of two possible outcomes. We will denote one outcome by the symbol S (for success) and the other by the symbol F (for failure).**
3. **The probability of success S on a single trial is equal to p and remains the same from trial to trial. The probability of a failure F on a single trial is equal to $q = (1 - p)$. Note that $p + q = 1$.**
4. **The trial outcomes are independent.**
5. **The binomial random variable x is the number of successes in n trials.**

To determine whether a particular discrete random variable is a binomial random variable, check the experiment that gave rise to the variable to see whether it satisfies the five characteristics of a binomial experiment.

Example 5.5

Flip a coin $n = 10$ times and observe the number x of heads. Explain why x is or is not a binomial random variable.

Solution

1. The experiment consists of $n = 10$ identical tosses of a coin. Each toss is a trial.
2. Each toss results in one of two possible outcomes: a head, which we will denote as a success S, or a tail, which we will denote as a failure F.
3. The probability of a head on a single trial is $\frac{1}{2}$ (assuming that the coin is balanced).
4. The trials are independent (i.e., the outcome of any one trial is independent of the outcome of any other).
5. The variable x is the number of heads (successes) in $n = 10$ tosses of the coin.

Since the experiment satisfies the five characteristics of a binomial experiment, x is a binomial random variable.

Example 5.6

Why is the *New York Times*/CBS News Poll described at the beginning of this section a binomial experiment?

Solution

Checking the experiment against the five characteristics of a binomial experiment, we find that:

1. The experiment consists of $n = 1{,}140$ identical trials (each trial is the selection of a single adult from the universe of all adults in the United States).
2. Each trial can result in one of two outcomes: agree with a statement (a success) or do not agree (a failure).
3. If the sample is a random sample, the probability that a given trial will result in a success (a person responding "yes" to a question) is equal to p, the unknown proportion of adults in the universe who would respond "yes" to the question. Actually, this probability will change from trial to trial depending on the answers of those adults drawn in previous trials. However, because the number of adults in the universe is so large relative to the sample size, the change from trial to trial in the probability of drawing a person who responds "yes" will be negligible. Therefore, we can assume that, for all practical purposes, the probability of a success on any given trial is equal to p.
4. The trials are independent (except for the point noted in step 3).
5. We are interested in the number x of successes (adults responding "yes" to a question) in the $n = 1{,}140$ trials.

Since the experiment satisfies the five characteristics of a binomial experiment, x is a binomial random variable.

Example 5.7

A manufacturer of electrical toasters ships them in boxes, with 24 in each box. Periodically, three toasters are selected from a box and each is examined for defects. Let x equal the number of defective toasters in the sample of $n = 3$. Explain why x is or is not a binomial random variable.

Solution

1. The experiment consists of $n = 3$ identical trials (a trial involves selecting a single toaster from the box).
2. Each trial results in a success (a defective) or a failure (a nondefective).
3. What happens to the probability of success (drawing a defective) from trial to trial? For example, suppose that 2 of the 24 toasters in the box are defective. Then, the probability of selecting a defective toaster on the first draw is $\frac{2}{24}$. The probability of selecting a defective toaster on the second draw depends on the outcome of the first draw. It is equal to $\frac{1}{23}$ if the first draw was a defective and $\frac{2}{23}$ if the first draw was nondefective. Since the probability of a success does not remain the same from trial to trial, x is not a binomial random variable.

Notice that both Examples 5.6 and 5.7 involve the selection of n elements from a universe containing N elements and counting the number x of elements in the sample that possess some characteristic. Public opinion and preference polls are typical examples of this type of experiment. Why, then, is Example 5.6 a binomial experiment and Example 5.7 is not? The answer is that the number N of elements in the universe was so large in Example 5.6 that the probability of success remained approximately the same from trial to trial. In contrast, N was so small in Example 5.7 that the probability of success for a given trial depended on the outcomes of previous trials. Therefore, the trial outcomes were dependent.

The Binomial Probability Distribution

The binomial probability distribution $p(x)$ depends on the number n of trials and the probability p of success. We do not need to know the formula for $p(x)$ because the probabilities have been tabulated for small samples and they can be approximated for large samples using an approximation given in Section 5.7. Figure 5.5 (page 140) shows graphs of binomial probability distributions for $n = 10$ and $p = .1$, $.5$, and $.9$.

Of What Relevance Is the Binomial Probability Distribution in Statistics?

The binomial probability distribution makes it possible for us to calculate the probabilities of various sample outcomes, and thereby to draw inferences, when we conduct a binomial experiment—say, a public opinion poll. The results of opinion polls appear almost daily in our newspapers and in telecasts. The binomial distribution (and its properties which follow) will help us (in Chapter 9) evaluate the results of a poll.

The Mean and Standard Deviation of a Binomial Probability Distribution

The formulas for calculating the mean μ and the standard deviation σ for a particular binomial distribution (that is, for given values of n and p) are shown in the box.

Mean and Standard Deviation of a Binomial Probability Distribution

Mean: $\mu = np$

Standard deviation: $\sigma = \sqrt{np(1 - p)}$

For example, the binomial probability distribution for $n = 10$ and $p = .5$ (see Figure 5.5b) has a mean

$$\mu = np = 10(.5) = 5$$

FIGURE 5.5 Binomial probability distributions for n = 10

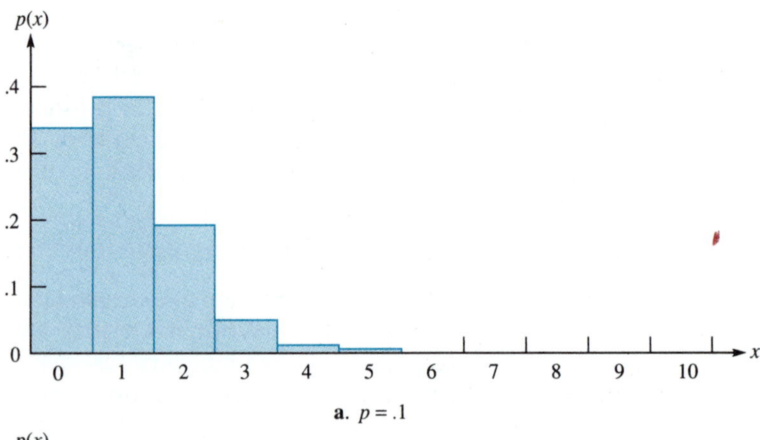

a. $p = .1$

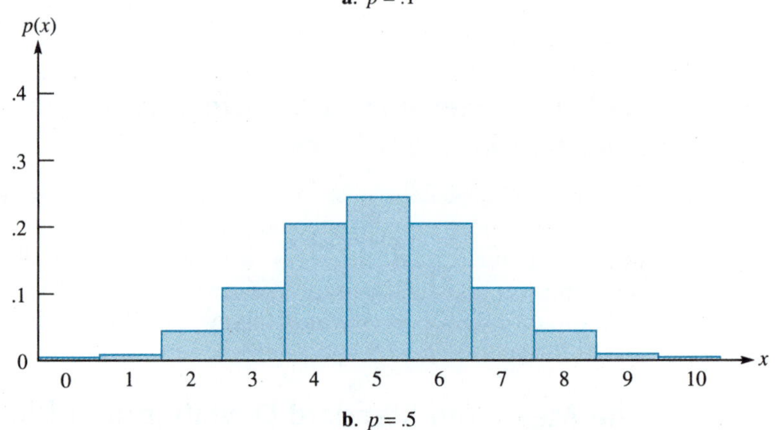

b. $p = .5$

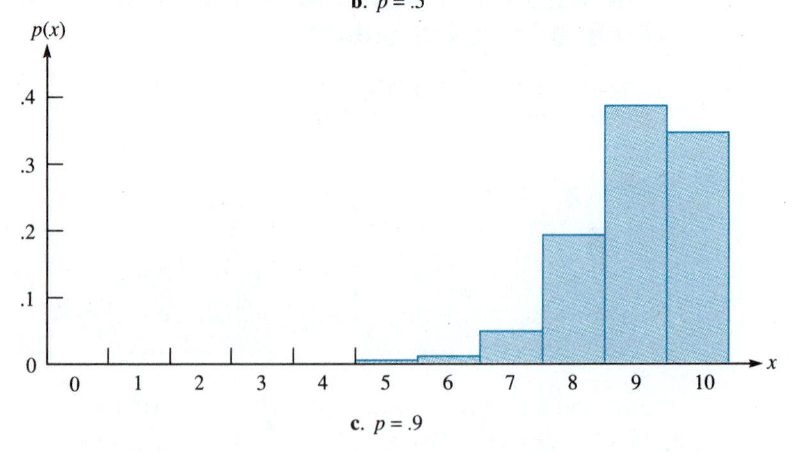

c. $p = .9$

and standard deviation

$$\sigma = \sqrt{np(1-p)} = \sqrt{10(.5)(.5)} = \sqrt{2.5} = 1.58$$

Note that when $p = .5$, the binomial probability distribution is symmetric about its mean.

When $p = .1$ (Figure 5.5a),

$$\mu = np = 10(.1) = 1 \quad \text{and} \quad \sigma = \sqrt{np(1-p)} = \sqrt{10(.1)(.9)} = \sqrt{.9} = .95$$

and the distribution is skewed to the right. When $p = .9$ (Figure 5.5c),

$$\mu = np = 10(.9) = 9 \quad \text{and} \quad \sigma = \sqrt{np(1-p)} = \sqrt{10(.9)(.1)} = .95$$

and the distribution is skewed to the left.

When the sample size n is large enough so that the interval $\mu - 3\sigma$ to $\mu + 3\sigma$ falls in the interval $x = 0$ to $x = n$, the binomial probability distribution will be nearly symmetric about its mean. If $\mu - 3\sigma$ is less than 0, the distribution will tend to be skewed to the right. If $\mu + 3\sigma$ is larger than n, the distribution will tend to be skewed to the left. As we might expect, the Empirical Rule will provide a good description of the binomial probability distribution, particularly when it is symmetric about its mean.

Example 5.8

A political analyst claims that 60% of the eligible voters in Florida favor an income tax to solve the state's budgetary problems. Suppose that a random sample of 1,000 voters is selected from the universe of all voters in the state.

a. Describe the probability distribution of the number x of voters in the sample who favor a state income tax if, in fact, the analyst's claim is true.

b. Find the interval $\mu - 3\sigma$ to $\mu + 3\sigma$. What does this interval tell us about the distribution?

Solution

a. The sampling described in this experiment is a poll that satisfies the requirements of a binomial experiment (refer to Example 5.6). Drawing a voter favoring a state income tax will be defined as a success. Therefore, $p(x)$ will be a binomial probability distribution with $n = 1{,}000$ and $p = .6$. The distribution will possess a mean

$$\mu = np = 1{,}000(.6) = 600$$

and standard deviation

$$\sigma = \sqrt{np(1-p)} = \sqrt{1{,}000(.6)(.4)} = 15.5$$

b. The interval $\mu - 3\sigma$ to $\mu + 3\sigma$, or $600 - 3(15.5)$ to $600 + 3(15.5)$, or 553.5 to 646.5, falls within the interval from 0 to $n = 1{,}000$, so we would expect the distribution to be nearly symmetric about its mean. We would also expect μ, σ, and the Empirical Rule to provide a good description of $p(x)$. In particular, we would expect the probability distribution $p(x)$ to be centered over $\mu = 600$ and all or almost all of $p(x)$ to fall in the interval from 553.5 to 646.5. Since x can assume so many values, the probability rectangles will be so narrow that $p(x)$ will look like a smooth curve, as shown in Figure 5.6.

FIGURE 5.6 A sketch of $p(x)$ for Example 5.8

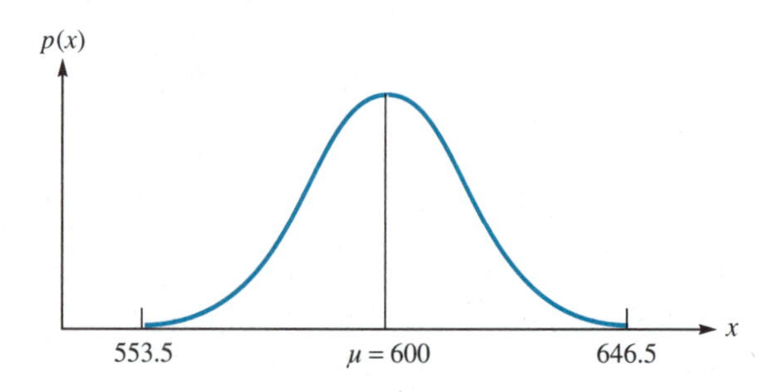

An Example of Statistical Inference Using the Binomial Probability Distribution

Example 5.9

An automobile manufacturer claims that fewer than 5% of its car buyers complain of lack of service from its dealers. A survey of 500 buyers of the manufacturer's new cars found that 39 new car buyers were dissatisfied with their dealer's service. Is this sample outcome inconsistent with the manufacturer's claim? (Attention, readers! This problem is asking us to make an inference about a binomial parameter p. It is our first attempt to employ statistical inference. Note how we do it.)

a. Suppose that the manufacturer's claim is true; that is, only 5% (or less) of its car buyers complain of a lack of service. Describe the probability distribution of the number x of buyers in a sample of 500 who complain about service.

b. Why will answering the question "Is this sample outcome inconsistent with the manufacturer's claim?" result in an inference about a population parameter?

c. If the manufacturer's claim is true, is it likely that the number x of dissatisfied customers is as large as 39?

d. What can we conclude about the manufacturer's claim?

Solution

a. We will assume that the total number of new cars sold by the manufacturer is large relative to the sample size, $n = 500$, and, therefore, that the customer survey represents a binomial experiment. A single trial involves interviewing a single new car buyer and a success is drawing a buyer who is dissatisfied with service. Then, if the manufacturer's claim is true, the proportion of successes in the population is no more than $p = .05$. The mean of the probability distribution of x successes (dissatisfied customers) in a sample of n trials is

$$\mu = np = 500(.05) = 25$$

and the standard deviation is

$$\sigma = \sqrt{np(1-p)} = \sqrt{500(.05)(.95)} = 4.87$$

Three standard deviations is equal to $3(4.87) = 14.61$, and the interval $\mu - 3\sigma$ to $\mu + 3\sigma$ is $25 - 14.61$ to $25 + 14.61$, or 10.39 to 39.61. Since this interval falls within the interval from 0 to 500, we would expect μ, σ, and the Empirical Rule to provide a very good description of $p(x)$. We would imagine that the probability distribution $p(x)$ might look like the distribution shown in Figure 5.7. (To simplify the artist's task, we show the distribution as a smooth curve.)

FIGURE 5.7 Sketch of $p(x)$ for Example 5.9

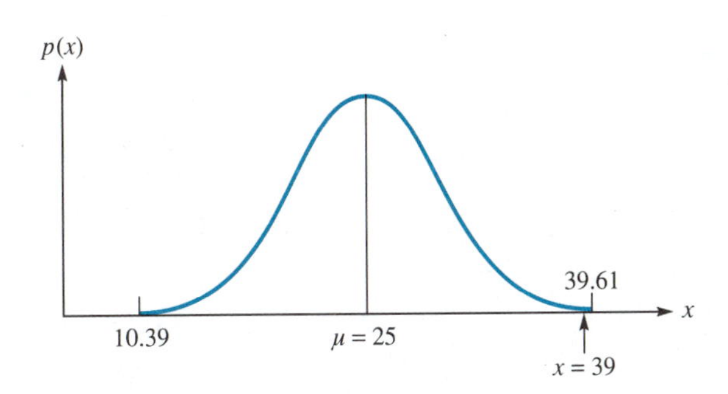

b. The proportion p of owners who are dissatisfied with customer service is the parameter for a binomial population of "yes" and "no" responses. Asking us to decide whether the manufacturer's claim is correct (i.e., whether $p = .05$ or less) is asking us to make an inference about a binomial population parameter p.

c. To decide whether a value of x as large as 39 is likely, assuming $p = .05$, see Figure 5.7. You can see that $x = 39$ falls approximately 3σ above the mean $\mu = 25$. Therefore, the probability of observing a value of x as large as or larger than $x = 39$ is very small.

d. So what do we conclude? If we agree with the manufacturer and conclude that $p = .05$, then we must believe that miracles do happen! Highly improbable outcomes can occur but they are, by definition, not likely. Therefore, if x is as large as or larger than 39, we would be inclined to believe that the manufacturer's claim is incorrect and that the proportion p of owners who are dissatisfied with service exceeds .05.

Example 5.9 shows how probability is used to make statistical inferences. **We rejected the manufacturer's claim that no more than 5% of new car buyers were dissatisfied with their dealer's services because the extraordinarily large number of dissatisfied buyers in the sample was improbable if, in fact, the manufacturer's claim was true.** Therefore, we made an inference about the binomial parameter p. We inferred that p was larger than .05.

Binomial Probability Tables

Table 2 of Appendix 2 gives values of the binomial probability distribution $p(x)$ for values of $n = 5$, 10, 15, 20, and 25, and $p = .01, .05, .10, .20, \ldots, .90, .95, .99$. More extensive tables are listed in the references. A reproduction of Table 2a, for $n = 5$, is shown in Table 5.2.

TABLE 5.2 Binomial Probabilities for $n = 5$: A Reproduction of Table 2a of Appendix 2

	p												
x	.01	.05	.1	.2	.3	.4	.5	.6	.7	.8	.9	.95	.99
0	.9510	.7738	.5905	.3277	.1681	.0778	.0313	.0102	.0024	.0003	.0000	.0000	.0000
1	.0480	.2036	.3280	.4096	.3601	.2592	.1563	.0768	.0283	.0064	.0005	.0000	.0000
2	.0010	.0214	.0729	.2048	.3087	.3456	.3125	.2304	.1323	.0512	.0081	.0011	.0000
3	.0000	.0011	.0081	.0512	.1323	.2304	.3125	.3456	.3087	.2048	.0729	.0214	.0010
4	.0000	.0000	.0004	.0064	.0283	.0768	.1563	.2592	.3601	.4096	.3280	.2036	.0480
5	.0000	.0000	.0000	.0003	.0024	.0102	.0313	.0778	.1681	.3277	.5905	.7738	.9510

The values of p appear across the top of Table 5.2. The probabilities for a binomial probability distribution with $p = .4$ appear in the column under .4. The values of x are shown at the left side of the table. To find the value of $p(x)$ for $x = 3$ and $p = .4$, we read the tabulated number in the row corresponding to $x = 3$ and the column for $p = .4$. This value, shaded in Table 5.2, is $p(3) = .2304$.

Example 5.10

Find the probability that x is less than or equal to 3 for a binomial random variable with $n = 10$ and $p = .5$.

Solution

Table 2b of Appendix 2 gives the values of $p(x)$ for $n = 10$. Therefore, to find the probability that x is less than or equal to 3, we need to find the sum of $p(x)$ from $x = 0$ to $x = 3$. See Figure 5.8. From Table 2b,

$$\begin{aligned} P(x \leqslant 3) &= p(0) + p(1) + p(2) + p(3) \\ &= .0010 + .0098 + .0439 + .1172 \\ &= .1719 \end{aligned}$$

FIGURE 5.8 Binomial distribution for Example 5.10

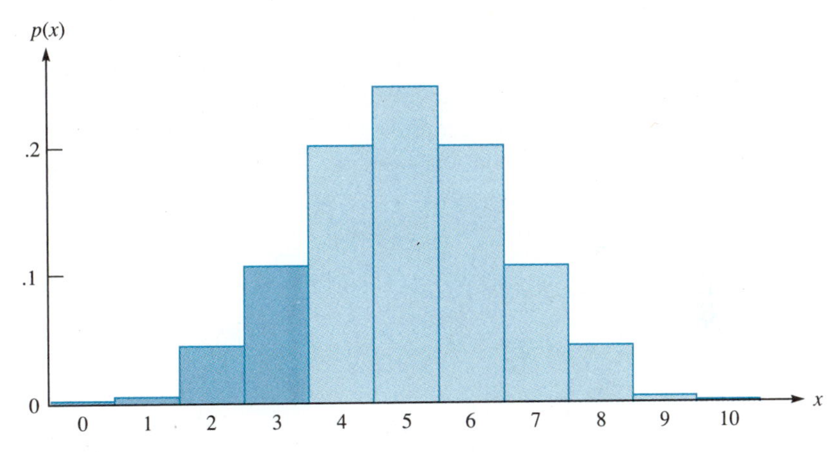

Example 5.11

A manufacturer of cameras claims that no more than 1% of its cameras are defective.

a. Suppose that the manufacturer's claim is true and that the proportion defective in the universe is $p = .01$. If you receive a shipment of $n = 25$ cameras, what is the probability that more than one will be defective?

b. If the number x of defective cameras in a sample is 2, what would you conclude about the manufacturer's claim?

Solution

a. This experiment satisfies the five characteristics of a binomial experiment. A trial consists of drawing a camera from production, inspecting it, and deciding whether it is defective or nondefective. A success is a defective, the proportion p of defectives in the population is .01, and the experiment contains $n = 25$ trials. The outcome, observing more than one defective in the sample of 25, is the observance of $x = 2, 3, 4, 5, \ldots, 23, 24$, or 25 defectives. Since the sum of all values of $p(x)$ is equal to 1, the sum of $p(x)$ from $x = 2$ to $x = 25$ is equal to $1 - [p(0) + p(1)]$. See Figure 5.9. The probabilities $p(0)$ and $p(1)$ are given in Table 2e as $p(0) = .7778$ and $p(1) = .1964$. Therefore, if the proportion of defective cameras emerging from production is only .01, the probability of observing more than one defective camera in a shipment of 25 is

$$P(x \geqslant 2) = 1 - [p(0) + p(1)] = 1 - [.7778 + .1964]$$
$$= .0258$$

b. If a shipment of 25 cameras contained 2 defective cameras, we would conclude that either we have observed a highly improbable (.0258) outcome, assuming that the manufacturer's claim ($p = .01$) is true, or else the manufacturer's claim is false (and $p > .01$). Because the observance of 2 or more defective cameras in a shipment is so improbable (.0258), we would conclude that the proportion of defective cameras produced by the manufacturer exceeds .01. As in Example 5.9, this example shows how probability is used to make an inference about a population based on the information contained in a sample.

FIGURE 5.9 Binomial distribution for Example 5.11

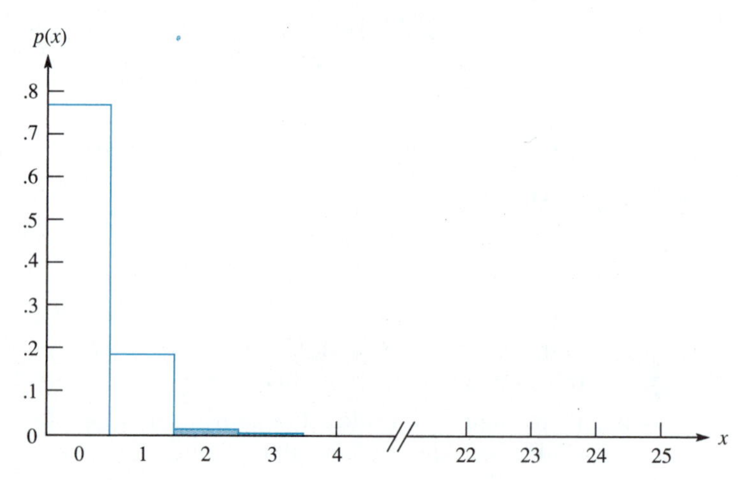

Other Discrete Probability Distributions

The binomial probability distribution is not the only discrete probability distribution used to model the relative frequency distributions of populations of data. Others, such as the Poisson and the hypergeometric probability distributions (see the references), are useful for specific applications but we will not refer to them in the remainder of the text. In contrast, the frequent use of public opinion polls in social research and the occurrence of similar populations of data in medicine, engineering, and business will require many references to the binomial probability distribution.

EXERCISES

BUS **5.31** Suppose that you were to select a random sample of chemical plants in the United States in order to estimate the percentage that are not in compliance with Environmental Protection Agency regulations. Explain why this is or is not (to a reasonable degree of approximation) a binomial experiment.

MFG **5.32** A random sample of 10 electrical motors was selected from a shipment of 40. Each motor was tested and the number x of defective motors was recorded. Explain why this experiment is or is not (to a reasonable degree of approximation) a binomial experiment.

BUS **5.33** An electric power company plans to observe the number x of electrical outages per month for each of the next 36 months. Explain why x is or is not (to a reasonable degree of approximation) a binomial random variable.

MFG **5.34** A manufacturer of radios buys transistors in large lots. A random sample of 25 transistors is selected from each lot and the number x of defective transistors is recorded. Explain why x is or is not (to a reasonable degree of approximation) a binomial random variable.

5.35 Give the mean and standard deviation for a binomial probability distribution with $n = 2{,}000$ and $p = .4$.

5.36 Refer to the binomial experiment of Exercise 5.35. Would you expect to observe a value of x as small as or smaller than 700? Explain.

5.37 Give the mean and standard deviation for a binomial probability distribution with $n = 400$ and $p = .1$. Sketch the distribution showing the interval within which most of the distribution will lie. Would you expect the distribution to be highly skewed? Explain.

GEN **5.38** A baseball player has a .300 batting average, i.e., he gets a hit 30% of the times at bat. If the player comes to bat 100 times, what is the mean number of hits that you would expect? With a reasonable probability, what is the maximum number of hits that you would expect? The minimum number? Explain.

GEN **5.39** Approximately 40% of television viewers are tuned in during the Super Bowl (*The New York Times Magazine*, July 23, 1989). Suppose that you were to randomly sample 1,000 TV viewers during the Super Bowl. Describe the distribution of the number x of viewers who will be watching the Super Bowl.

5.40 Refer to Exercise 5.39. Will this distribution be relatively symmetric? Explain.

5.41 Suppose that x has a binomial distribution with $n = 10$ and $p = .6$. Find:

a. The probability that x is equal to 5
b. The probability that x is less than 5
c. The probability that x is larger than 5

5.42 See Table 5.2. Graph $p(x)$ for a binomial probability distribution with $n = 5$ and $p = .5$.

5.43 Repeat Exercise 5.42 for $n = 5$, $p = .1$.

5.44 Repeat Exercise 5.42 for $n = 5$ and $p = .9$.

5.45 Suppose that x has a binomial probability distribution with $n = 20$ and $p = .2$. Find:

a. The mean value of x
b. The standard deviation of x
c. The probability that x will exceed 4
d. The interval $\mu \pm 2\sigma$
e. The probability that x will fall within the interval $\mu \pm 2\sigma$. How does your answer compare with the Empirical Rule?

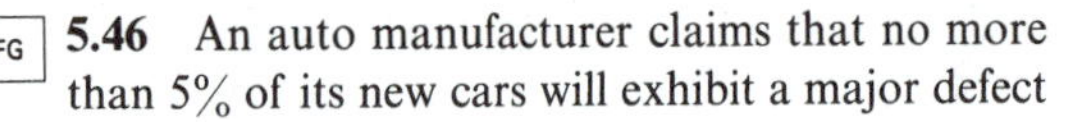

MFG **5.46** An auto manufacturer claims that no more than 5% of its new cars will exhibit a major defect during the first 3 years of operation. A random sample of 25 owners of the cars found that 6 had experienced a major defect. What can you say about the manufacturer's claim? Explain.

MFG **5.47** A manufacturer ships fuses in lots of 1,200 and guarantees that a lot will contain no more than 5% defective. A purchaser has decided to sample 25 fuses from each lot and accept a lot only if the number of defective fuses in the sample is 4 or less. What is the probability of:

a. Accepting a lot if it contains 5% defective
b. Accepting a lot if it contains 10% defective
c. Rejecting a lot if it contains 5% defective

5.7 The Normal Probability Distribution

The normal probability distribution is regarded as the most important probability distribution in statistics. It is important because it provides a good approximation to so many population relative frequency distributions of data—discrete as well as continuous. It also provides, when certain conditions are satisfied, a good approximation to some probability distributions. One of these, as you will subsequently learn, is the binomial probability distribution of Section 5.6.

The *normal probability distribution* is a symmetric, bell-shaped, continuous distribution with mean μ and standard deviation σ. See Figure 5.10. **In theory, x can assume any value from $-\infty$ to ∞. In practice, almost all values of x will fall within 3σ of the population mean μ.**

FIGURE 5.10 The normal probability distribution

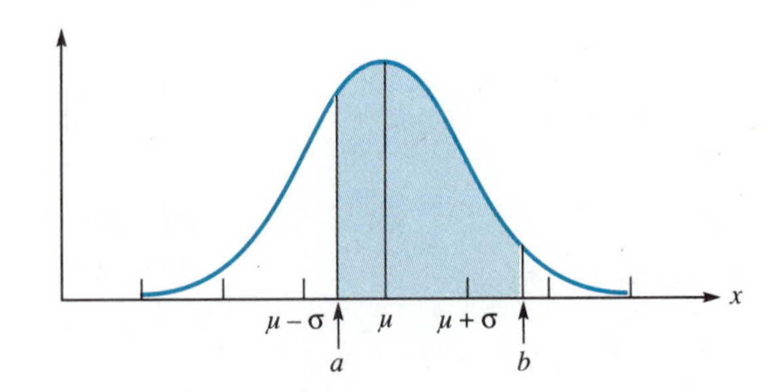

Because the normal distribution is continuous, we can only give the probability that the random variable x falls in an interval. **The probability that x falls in the interval from $x = a$ to $x = b$ is the area under the normal curve over the interval from $x = a$ to $x = b$.** This is the shaded area in Figure 5.10. The probability that x is equal to a specific value, say $x = a$, is equal to 0 because there is no area above a single point. **The total area under the normal curve is equal to 1. Half of the area under the normal distribution curve lies to the left of the mean; half lies to the right.**

Finding Areas Under the Normal Curve

If a sample statistic has a normal probability distribution, we must use the normal distribution to decide whether the observed value of the statistic is probable or improbable. Thus, we must be able to calculate the probability that a normally distributed variable will assume values within a particular interval.

Areas under the normal curve, which represent probabilities, are tabulated and appear in Table 3 of Appendix 2. Since we will need to know only whether an observed value of x is improbable, we will use only a few entries from Table 3—the areas under the normal curve that are shown in Table 5.3.

TABLE 5.3 Areas* Under the Normal Curve

z	Area
.52	.20
.67	.25
.84	.30
1.00	.34
1.28	.40
1.65	.45
1.96	.475
2.33	.49
2.58	.495

* The areas shown here are rounded from the values given in Table 3 of Appendix 2.

Table 5.3 gives areas under the normal curve between the mean and a point located z standard deviations to the right of the mean (see Figure 5.11 on page 150). (Areas to the left of the mean are not tabulated because the normal curve is

FIGURE 5.11 Areas recorded in Table 5.3

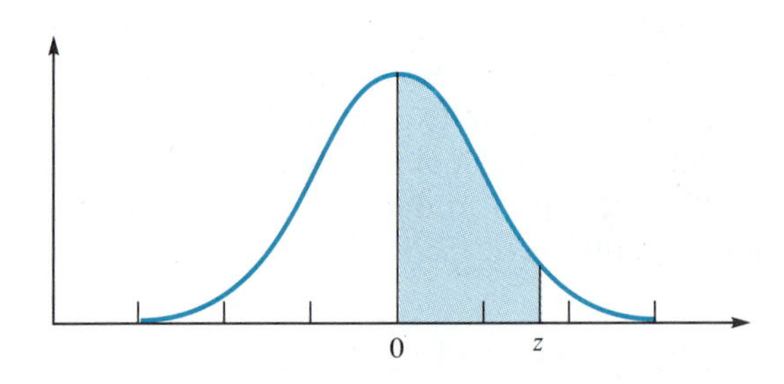

symmetric and, therefore, areas to the left of the mean are equal to the corresponding areas to the right of the mean.)

The Standard Normal z Distribution

All of the standard tables give areas under the normal curve for values of z, the distance that x deviates from its mean μ expressed in units of its standard deviation σ; that is,

$$z = \frac{x - \mu}{\sigma}$$

The probability distribution of the normal random variable z, called a *standard normal distribution*, has a mean $\mu = 0$ and standard deviation $\sigma = 1$.

Example 5.12

Find the total area under the normal curve to the right of the mean.

FIGURE 5.12 Normal curve: Example 5.12

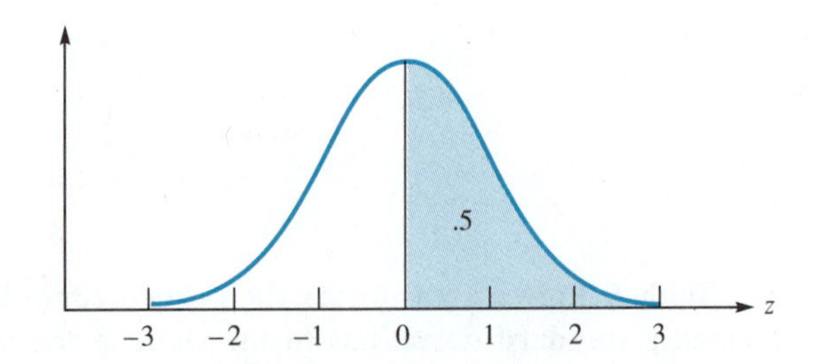

Solution

The total area under the normal curve is equal to 1. Because the normal curve is symmetric about its mean, one-half (.5) of the area will lie to the left of the mean and one-half (.5) to the right. See Figure 5.12.

Example 5.13

Give the z-value corresponding to the 80th percentile for the normal curve.

Solution

The z-value for the 80th percentile is the value of z such that 80% of the area under the normal curve lies to its left (see Figure 5.13). Since the area to the left of the mean is .5, the 80th percentile is the value of z corresponding to an area to the right of the mean equal to .3. This value of z is given in Table 5.3 as .84. Therefore the 80th percentile is located a distance of $z = .84$ standard deviation to the right of the mean.

FIGURE 5.13 Normal curve: Example 5.13

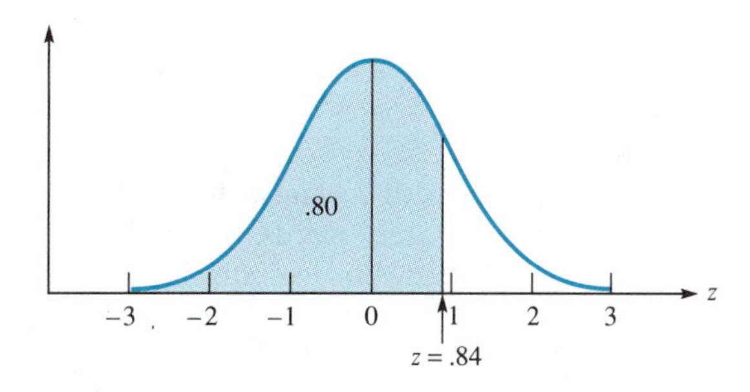

Example 5.14

Find the probability that a normally distributed random variable x will lie more than 1.96σ away from its mean μ.

Solution

The probability that x lies more than 1.96σ away from its mean is the sum of the shaded tail areas under the curve shown in Figure 5.14 (page 152). The area between the mean and a point $z = 1.96$ standard deviations to the right of the mean (see Table 5.3) is .475. Since the total area to the right of the mean is equal to .5, the shaded area in the upper tail is equal to $.5 - .475 = .025$. Because the normal curve is symmetric, the shaded area in the lower tail is also equal to .025. Therefore, the total

FIGURE 5.14 Normal curve: Example 5.14

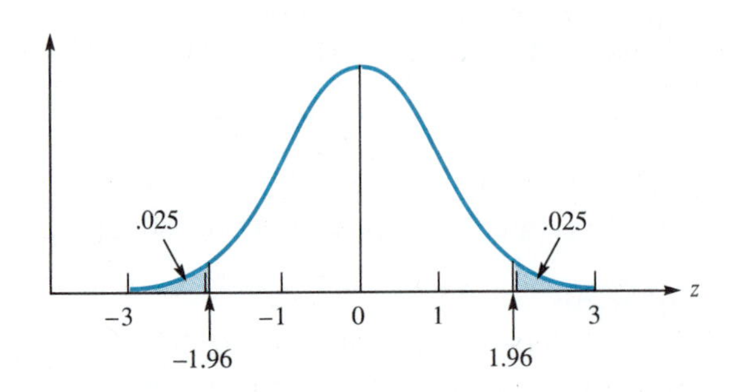

shaded area under the normal curve shown in Figure 5.14—that is, the probability that x lies more than 1.96σ away from its mean, is $.025 + .025 = .05$.

Example 5.15

A large set of test grades was approximately normally distributed with mean equal to 712 and standard deviation equal to 120. If your grade on the test was 920, how did your grade rank relative to the other grades? To answer this question, give the approximate percentile score for your grade.

Solution

As a first step, we need to know how many standard deviations your grade lies above the mean; in other words, we need to find the z-value that corresponds to your grade of 920. Substituting into the formula for z, we obtain

$$z = \frac{x - \mu}{\sigma} = \frac{920 - 712}{120} = 1.73$$

FIGURE 5.15 Normal curve: Example 5.15

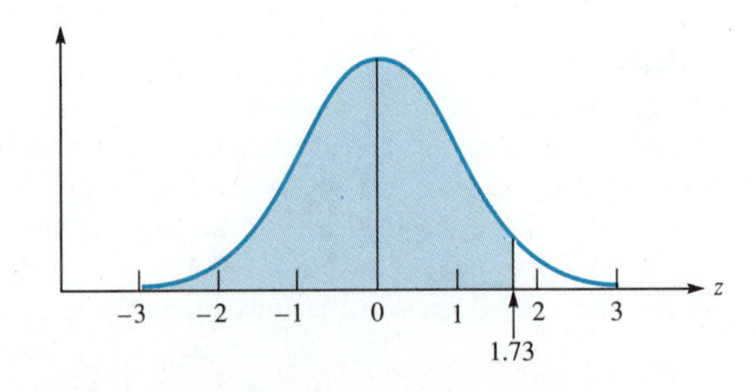

The percentile for your grade is the area under the normal curve, to the left of $z = 1.73$, times 100 (see Figure 5.15). The area to the left of the mean is .5. The approximate area to the right of the mean can be found in Table 5.3. You can see in Table 5.3 that $z = 1.65$ corresponds to an area equal to .45 and $z = 1.96$ corresponds to .475. Since your z-value falls between $z = 1.65$ and $z = 1.96$, the area under the normal curve to the left of your z-value, $z = 1.73$, is between $(.5 + .45)$ and $(.5 + .475)$, or between .95 and .975. Therefore, you scored between the 95th and the 97.5th percentiles on the test.

Example 5.16

Refer to Example 5.15 and give the percentile score for a test grade equal to 552.

Solution

The z-value corresponding to $x = 552$ is

$$z = \frac{x - \mu}{\sigma} = \frac{552 - 712}{120} = -1.33$$

The fact that z is *negative* tells us that x lies 1.33 standard deviations *below* the mean, $\mu = 712$. The area to the left of $z = -1.33$ is the shaded area in Figure 5.16. Since the total area below the mean is equal to .5, we must find the area corresponding to $z = -1.33$ and subtract it from .5. You can see in Table 5.3 that $z = 1.28$ corresponds to the area .40, and $z = 1.65$ corresponds to .45. Therefore, the area to the left of $z = -1.33$ is between $(.5 - .40)$ and $(.5 - .45)$, or between .05 and .10. Therefore, a grade of 552 is between the 5th and the 10th percentiles.

FIGURE 5.16 Normal curve: Example 5.16

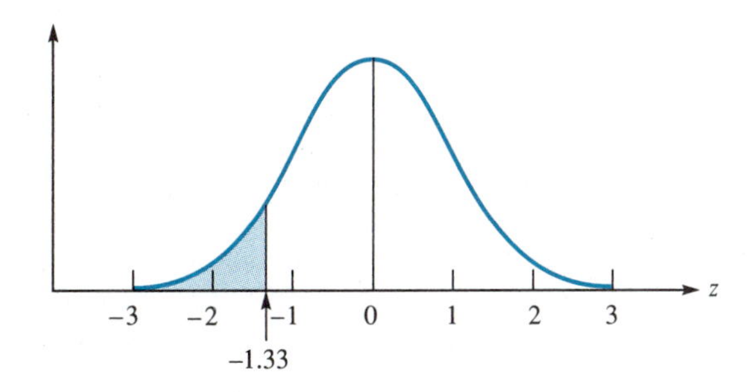

Example 5.17

The mean and standard deviation of the scores on a national achievement test were 655 and 146, respectively. If you were informed that you scored at the 90th percentile, what was your test score?

Solution

Since the area to the left of the mean is .5, the z-value for the 90th percentile is the value that corresponds to an area to the right of the mean equal to .4 (see Figure 5.17) in Table 5.3. This z-value is 1.28. (Note that the z-value takes a plus sign because the test score x lies to the right of the mean. If the test score fell below the mean, the z-value would be negative.) To find the test score corresponding to this value of z, we substitute $\mu = 655$, $\sigma = 146$, and $z = 1.28$ into the formula and solve for x:

$$z = \frac{x - \mu}{\sigma}$$

$$1.28 = \frac{x - 655}{146}$$

Solving for x gives your test score:

$$x = (146)(1.28) + 655 = 842$$

FIGURE 5.17 Normal curve: Example 5.17

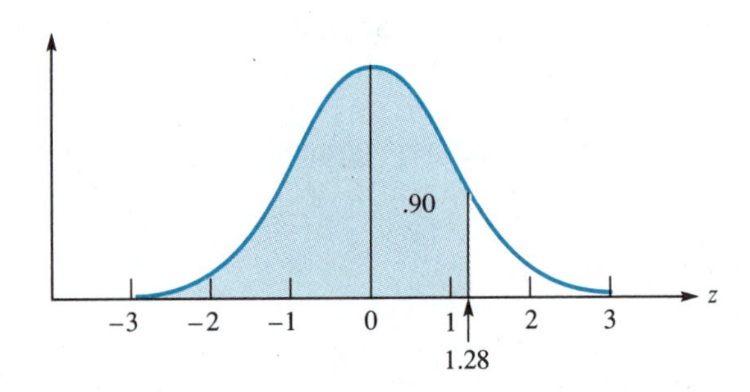

Keep in mind that the normal probability distribution, like all other probability distributions, is a model for a population relative frequency distribution. Theoretically, the systolic blood pressure for a human is a continuous random variable that can assume any one of the infinitely large number of values in the interval from 0 to ∞. In practice, blood pressure is not a large number and it is read to the nearest integer. So, for example, the infinitely large number of values of x between a pair of integers, say 72 and 73, never appear in a real population of data. Despite this discrepancy between theory and reality, the normal probability distribution provides a very good model for many population relative frequency distributions, particularly measurements on natural phenomena. In addition, it enables us to calculate the probability of sample outcomes and to use them to make inferences about the sampled populations.

The Normal Approximation to the Binomial and Other Probability Distributions

The normal probability distribution is not only a good model for many population relative frequency distributions, it is also a good *approximation*, under certain conditions, for many other probability distributions (discrete as well as continuous).

The binomial probability distribution can be approximated by a normal distribution when the interval $\mu \pm 3\sigma$ [i.e., $np \pm 3\sqrt{np(1-p)}$] falls within the interval from $x = 0$ to $x = n$. The larger the value of n, the better will be the approximation. For example, for a binomial probability distribution with $n = 100$ and $p = .3$,

$$\mu = np = (100)(.3) = 30 \quad \text{and} \quad \sigma = \sqrt{np(1-p)} = \sqrt{100(.3)(.7)} = \sqrt{21} = 4.6$$

Therefore, the interval $\mu \pm 3\sigma$ is 16.2 to 43.8. Since this interval is included in the interval from 0 to 100, this binomial probability distribution can be approximated very well by a normal distribution with mean $\mu = 30$ and standard deviation $\sigma = 4.6$.

EXERCISES

5.48 Describe a normal probability distribution.

5.49 What is a standard normal distribution?

5.50 Suppose that x is a normally distributed random variable. Find the probability that x will fall within:

a. 2.58 standard deviations of its mean
b. 2.33 standard deviations of its mean
c. 1.96 standard deviations of its mean
d. 1.65 standard deviations of its mean
e. 1.28 standard deviations of its mean

Prepare a table showing the results of this exercise. We will use it in the chapters that follow.

5.51 Suppose that x is a normally distributed random variable. Find the probability that x will fall:

a. More than 2.58 standard deviations away from its mean
b. More than 2.33 standard deviations away from its mean
c. More than 1.96 standard deviations away from its mean
d. More than 1.65 standard deviations away from its mean
e. More than 1.28 standard deviations away from its mean

Prepare a table showing the results of this exercise. We will use it in the chapters that follow.

5.52 Suppose that x is a normally distributed random variable with mean equal to 60 and standard deviation equal to 5. How many standard deviations above or below the mean are the following values of x?

a. 53.6
b. 63
c. 47
d. 36
e. 69.5

5.53 Refer to Exercise 5.52. Give the percentile location of $x = 53.6$.

BIO **5.54** The quartiles divide the area under the normal curve into four equal parts. If the heights of male adults of an ancient Indian tribe are normally distributed with mean equal to 5.7 feet and a standard deviation equal to .4 foot, find the heights that locate the three quartiles. If a height falls between the first and the second quartiles, between what two heights will it fall?

5.55 When can you assume that a binomial probability distribution will be approximately normal?

5.56 Will a normal distribution provide a good approximation to a binomial distribution if $n = 25$ and $p = .3$?

5.57 If x is normally distributed with mean equal to 4 and standard deviation equal to .5, which of the following observed values of x might be regarded as "rare events"? Explain.

a. 3.4
b. 4.4
c. 2.7
d. 5.0

5.58 Suppose that x has a binomial probability distribution with $n = 1{,}000$ and $p = .01$. Give the mean and standard deviation of x. Will this distribution be approximately normal? Why or why not?

5.59 Suppose that x has a binomial probability distribution with $n = 100$ and $p = .01$. Give the mean and standard deviation of x. Will this distribution be approximately normal? Why or why not?

5.8 Sampling Distributions

We said in Chapter 4 that sample statistics are used to make inferences—estimates or decisions—about population parameters. Since probability is the mechanism that enables us both to make an inference and to evaluate its goodness, we need to be able to make probability statements about sample statistics—that is, we need to know their probability distributions.

The probability distribution for a statistic is called its *sampling distribution.* Remember that a probability distribution is a theoretical model for a population relative frequency distribution. Mathematical statisticians use analytical methods to derive the sampling (i.e., probability) distribution for a sample statistic but it can also be found by simulating repeated sampling.

To illustrate, suppose that we want to find the sampling distribution for the sample mean $\bar{x}$. Select a random sample of $n = 5$ measurements from a population and calculate the sample mean, $\bar{x}$. Throw those measurements back into the population, draw another sample of 5 measurements, and calculate $\bar{x}$ for this sample. Repeat this sampling process an infinitely large number of times until we have an infinitely large "population" of $\bar{x}$'s. The relative frequency distribution for these $\bar{x}$'s is the sampling distribution of $\bar{x}$. It enables us to determine whether an observed value of $\bar{x}$ is probable or improbable.

The standard deviation of the sampling distribution of a statistic is called the *standard error* of the statistic and is represented by the symbol SE( ). For example, the standard error of the sampling distribution of $\bar{x}$ is represented by the symbol SE($\bar{x}$). The standard error of a statistic measures its variability. Since we will use sample statistics to estimate and make decisions about population parameters, we would like the standard error of a statistic to be as small as possible.

The Sampling Distribution of the Sample Mean

It can be shown (proof omitted) that the sampling distributions for many statistics are approximately normal. Perhaps the most important of these statistics is the sample mean, $\bar{x}$. The following theorems, which can be proved mathematically, give the sampling distribution of a sample mean $\bar{x}$ when certain conditions are satisfied.

Theorem 1

If a random sample of n measurements is drawn from a population that is normally distributed with mean μ and standard deviation σ, the sampling distribution of the sample mean $\bar{x}$ will be normally distributed with mean μ and standard error $SE(\bar{x}) = \sigma/\sqrt{n}$.

Theorem 2
The Central Limit Theorem

If a random sample of n measurements is drawn from a nonnormal population with a finite mean μ and standard deviation σ, then, when n is large, the sampling distribution of the sample mean $\bar{x}$ will be approximately normally distributed with mean equal to μ and standard error $SE(\bar{x}) = \sigma/\sqrt{n}$. The approximation will become more and more accurate as n becomes larger and larger. The sampling distribution for the sample mean is shown in Figure 5.18.

FIGURE 5.18 Sampling distribution of $\bar{x}$

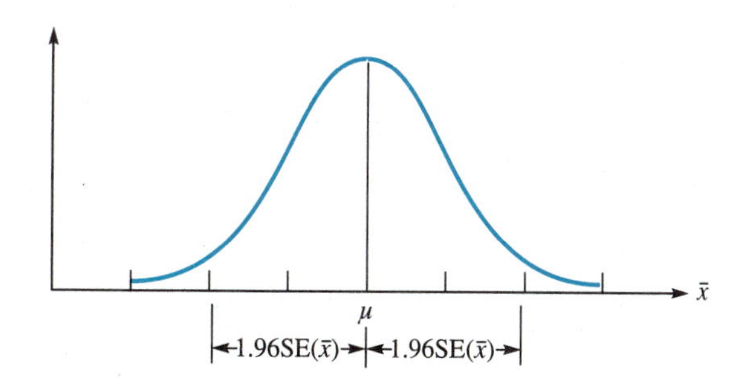

What the Theorems Tell Us About the Sampling Distribution of $\bar{x}$

Theorem 1 and Theorem 2 (the Central Limit Theorem) tell us that for most sampling situations, the sampling distribution of $\bar{x}$ will be approximately normally distributed—i.e., possess a distribution that is approximately normal. The mean of the sampling distribution will be the same as the mean of the sampled population, and the standard error of the sampling distribution will be $\sigma/\sqrt{n}$, where σ is the standard deviation of the sampled population and n is the sample size. See Figure 5.18.

If the population that you are sampling from has a normal distribution, then the sampling distribution of $\bar{x}$ will be exactly normal. Even if the population is not normally distributed, the sampling distribution of $\bar{x}$ will be approximately normally distributed (because of the Central Limit Theorem) when the sample size is large. The greater the skewness of the population distribution, the larger the sample size will have to be to achieve normality in the sampling distribution of $\bar{x}$. Regardless, we will subsequently show that the sample size need not be very large even when the population distribution is highly skewed.

Applying Theorems 1 and 2 to a Sampling Problem

What are the practical consequences of Theorems 1 and 2? To answer that question, consider the following example. Let us pretend that the set of 107 total cholesterol measurements in Table 3 of Appendix 1 is a population with mean μ and standard deviation σ. In a practical situation we would not know μ and σ, but we know their values for this example because we calculated them in Chapter 4 to be $\mu = 217.8$ and $\sigma = 53.2$. Now suppose that we were to draw a sample of $n = 5$ measurements from this "population." Theorems 1 and 2 tell us that the sampling distribution of $\bar{x}$ will be approximately normally distributed with mean $\mu = 217.8$ (the same as the mean of our sampled population) and standard error equal to $SE(\bar{x}) = \sigma/\sqrt{n} = 53.2/\sqrt{5} = 23.8$ (see Figure 5.19).

FIGURE 5.19 A sketch of the sampling distribution of $\bar{x}$ for $n = 5$, $\mu = 217.8$, and $\sigma = 53.2$

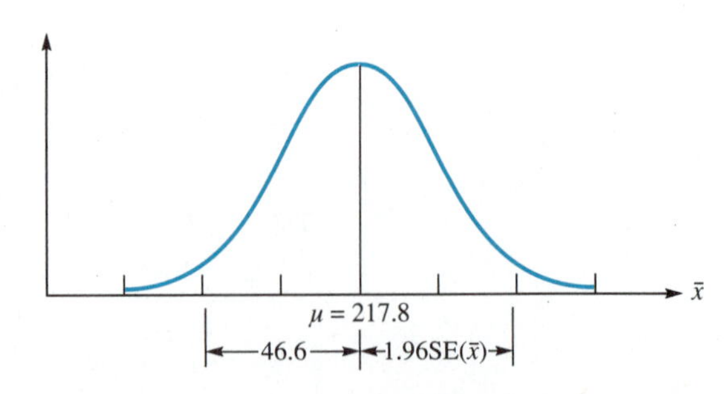

If you were to draw a random sample of $n = 5$ measurements from the population of cholesterol measurements in Table 3 of Appendix 1, the probability (from the table of areas under the normal curve, Table 5.3) is .95 that your sample mean $\bar{x}$ will fall within 1.96 standard errors, i.e.,

$$(1.96)\ \mathrm{SE}(\bar{x}) = 1.96(23.8) = 46.6$$

of the population mean $\mu = 217.8$. Try it! Draw a random sample of $n = 5$ measurements from our "population" of cholesterol measurements in Table 3 of Appendix 1, calculate the sample mean $\bar{x}$, and see where it falls. We think that it will fall within $1.96\ \mathrm{SE}(\bar{x}) = 46.6$ of the population mean, $\mu = 217.8$. The odds are 19 to 1!

When Will the Sample Size Be Large Enough for the Sampling Distribution of $\bar{x}$ to Be Approximately Normal?

How large must the sample size n be in order that the sampling distribution of $\bar{x}$ be approximately normally distributed? The answer depends on the shape of the population relative frequency distribution. We conducted an experiment to give you a partial answer. We selected 1,000 random samples of $n = 5$ measurements from each of two different types of population relative frequency distributions (shown in Figure 5.20a). The first is a uniform relative frequency distribution, which is perfectly flat and symmetric about its mean. The second is a negative exponential distribution, which is highly skewed to the right. We then

FIGURE 5.20 Approximations to the sampling distributions of $\bar{x}$ for two different population relative frequency distributions

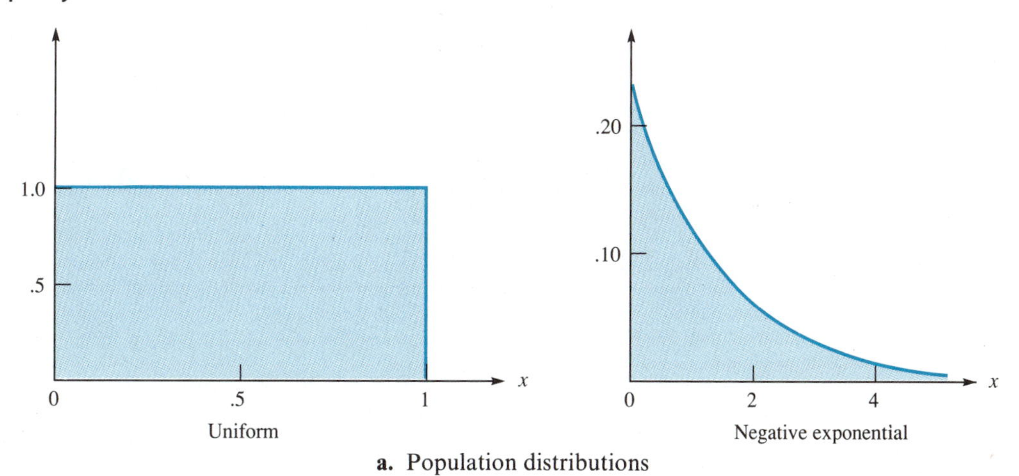

a. Population distributions

FIGURE 5.20 ***Continued***

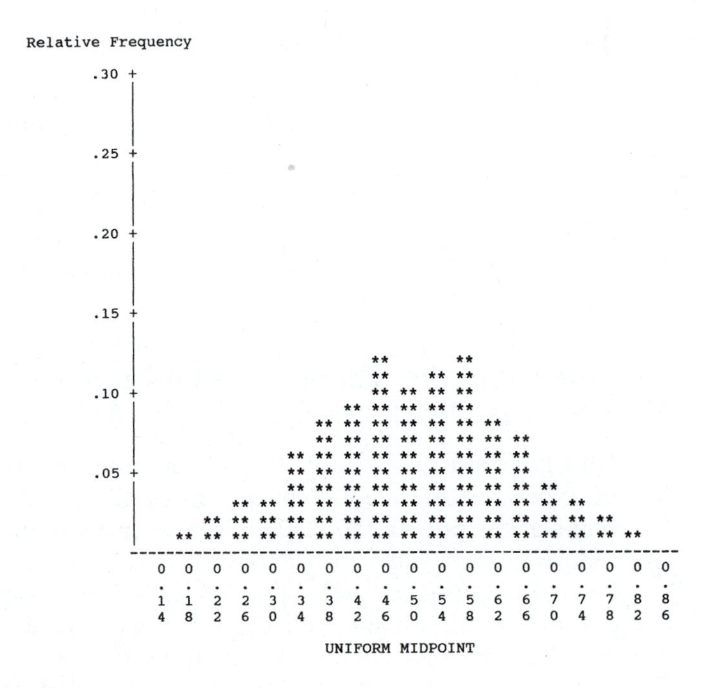

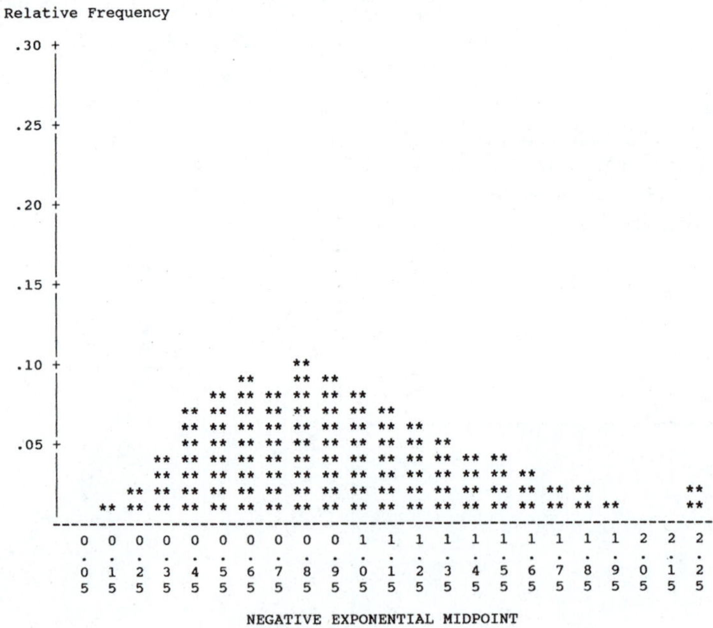

b. Distributions of $\bar{x}$ for $n = 5$

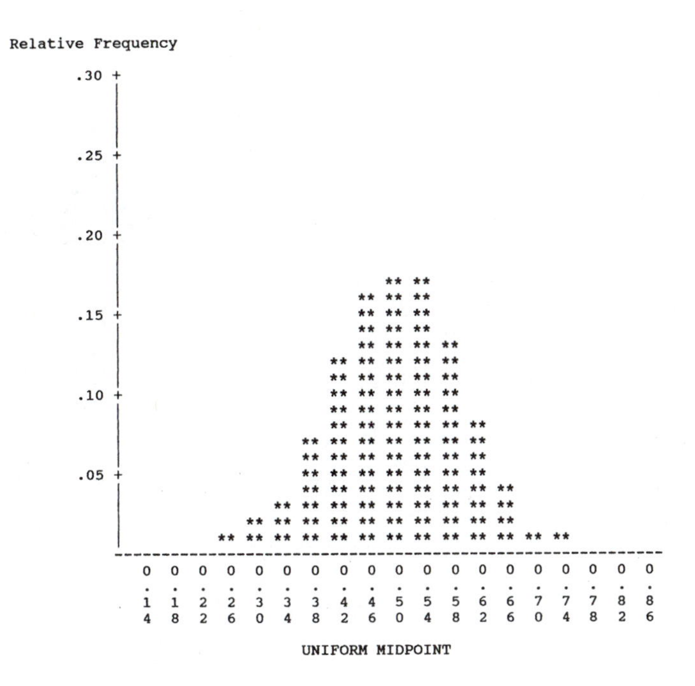

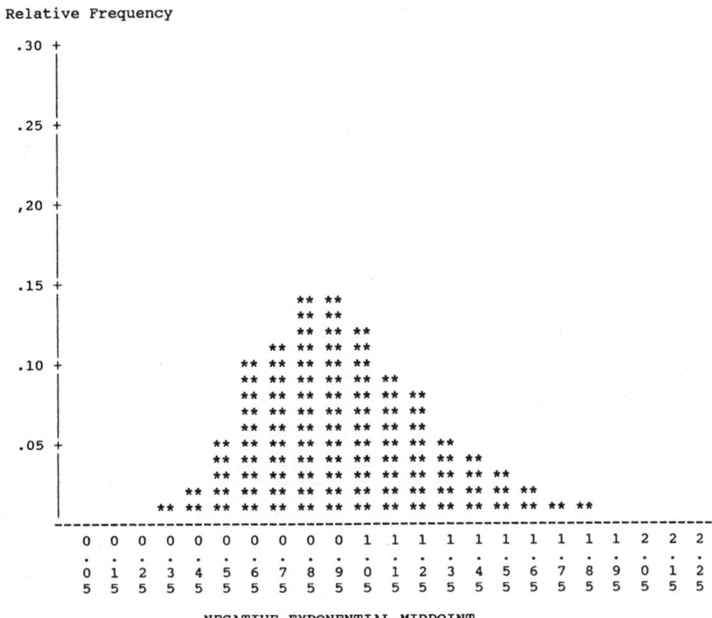

c. Distributions of $\bar{x}$ for $n = 10$

FIGURE 5.20 ***Continued***

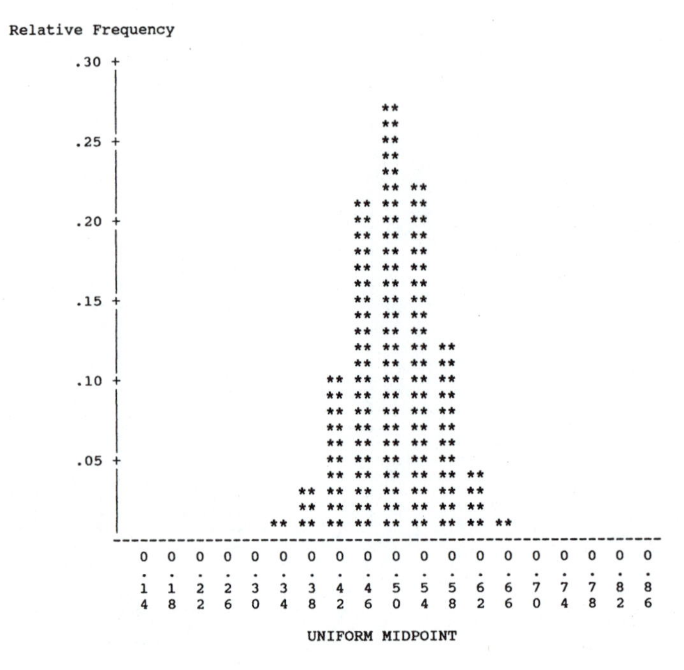

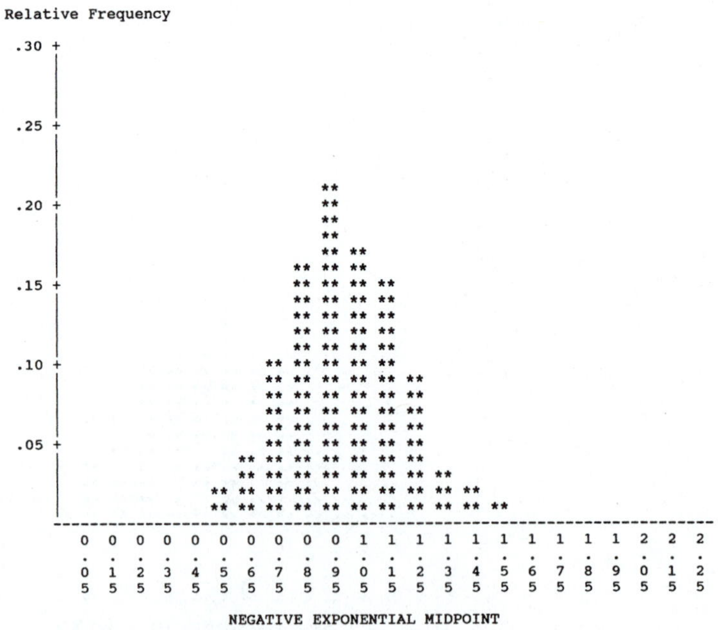

d. Distributions of $\bar{x}$ for $n = 25$

calculated $\bar{x}$ for each of the 1,000 samples and constructed relative frequency distributions of the sample means for each of the two populations. These distributions, shown in Figure 5.20b, are approximations to the sampling distributions of $\bar{x}$ for random samples of $n = 5$ from the two populations.

Similarly, Figures 5.20c and 5.20d give approximations to the sampling distribution of $\bar{x}$ for the two populations for $n = 10$ and $n = 25$. Notice that all of the sampling distributions tend to be mound shaped, even for n as small as $n = 5$. The shapes of the distributions tend to become more and more bell shaped as n becomes larger. What happens to the standard error of the sampling distribution as the sample size n increases? It decreases! Note how the values of $\bar{x}$ tend to group closer and closer to the population mean as n becomes larger.

The implications of the Central Limit Theorem are amazing when you realize that it applies to any population, regardless of the shape of its relative frequency distribution. All that we need to know is that μ and σ exist and that n is large. How large? Not very large at all (say, as small as $n = 5$ or $n = 10$) if the population relative frequency distribution is not too badly skewed!

EXERCISES

5.60 What is the sampling distribution of a statistic?

5.61 What is the standard error of a statistic?

5.62 State the Central Limit Theorem. What is the practical significance of the Central Limit Theorem?

HLTH **5.63** Use the random number table in Table 1 of Appendix 2 to draw a random sample of $n = 10$ measurements from among the 107 TCHOL measurements of Table 3 of Appendix 1. (Explain precisely how you arrived at your specific set of sample measurements.) The mean and standard deviation of this "population" of $N = 107$ measurements are $\mu = 217.8$ and $\sigma = 53.2$. Calculate $SE(\bar{x})$. Calculate $\bar{x}$ and see whether it falls within the interval $\mu \pm 1.96SE(\bar{x})$. What is the probability that $\bar{x}$ should fall within the interval $\mu \pm 1.96SE(\bar{x})$?

5.64 Have each member of the class perform Exercise 5.63. Calculate the proportion of sample means that fall within the interval $\mu \pm 1.96SE(\bar{x})$. This proportion should be close to what value (assuming that the class is reasonably large)?

5.65 Of what practical significance is Theorem 1 of Section 5.8?

5.9 Making Statistical Inferences: Estimation

Suppose that we conducted a survey to determine the proportion p of students who favor Jones as the next student body president. Of the 500 students randomly sampled in the poll, 230 favor Jones. Based on the sample data, we would likely choose the sample proportion, $230/500 = .46$, as our "best guess" as to the value of the population proportion p. This best guess, based on the sample data, is called an *estimate* of the population parameter p.

The material in this section is intended to introduce you to the *concepts* of statistical estimation. More to follow! In the succeeding chapters, we will present some very practical problems that will require estimates for their solution. We will give you the formula used to calculate the estimate for each population parameter and then show you, by example, how it is applied. Now let us consider some new words and concepts used in statistical estimation.

An *estimator* is a rule that tells us how to calculate an estimate of a parameter based on sample observations. For example, the sample mean $\bar{x}$ is an estimator of the population mean μ. It is expressed as a formula that tells us exactly how to calculate the value of $\bar{x}$ from the sample measurements. If a sample contains $n = 3$ measurements, 5, 9, and 2, then our estimate of μ is

$$\bar{x} = \frac{5 + 9 + 2}{3} = 5.33$$

There are two types of estimates of population parameters: *point estimates* and *interval estimates*.

Point Estimators

A *point estimator* uses the sample data to produce a single number (a point) that is an estimate of a population parameter. For example, $\bar{x}$ is a point estimator of the population mean μ and the sample standard deviation s is a point estimator of the population standard deviation σ.

Point estimation is analogous to firing a rifle at a target. You, the marksperson, are the estimator. A single shot is an estimate and the bull's-eye of the target is the parameter that you want to hit. To evaluate your marksmanship, you fire hundreds of shots at the target. The distribution of shots on the target is analogous to a sampling distribution. A quick glance at the distribution can tell us how close to the bull's-eye you are likely to come when you fire your next shot.

A good point estimator of a population parameter is one that has a sampling distribution with a mean equal to the estimated parameter and as little variation as possible. In the context of target practice, we would like the shots (estimates) to be centered about the bull's-eye (the population parameter) and to fall as close to it as possible. For example, $\bar{x}$ is a good estimator of the population mean μ. Its sampling distribution, shown in Figure 5.21, has a mean equal to μ and a standard error $SE(\bar{x}) = \sigma/\sqrt{n}$ that is small relative to the standard error of other estimators of μ.

If the mean of an estimator (i.e., the mean of its sampling distribution*) is equal to the estimated parameter, the estimator is said to be *unbiased*. Other-

* The expression "mean (or standard error) of a statistic" is an abbreviation of the expression "mean (or standard error) *of the sampling distribution* of the statistic."

FIGURE 5.21 The sampling distribution of the sample mean $\bar{x}$

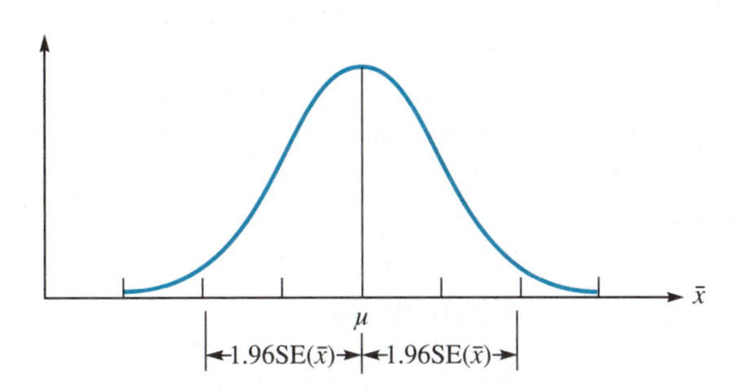

FIGURE 5.22 Sampling distributions for unbiased and biased estimators

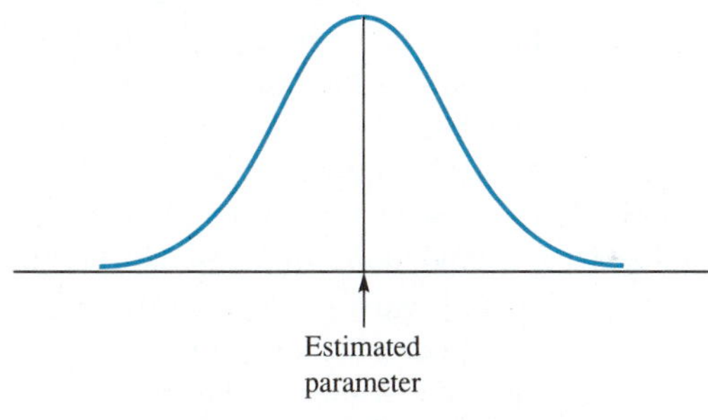

a. Unbiased sampling distribution

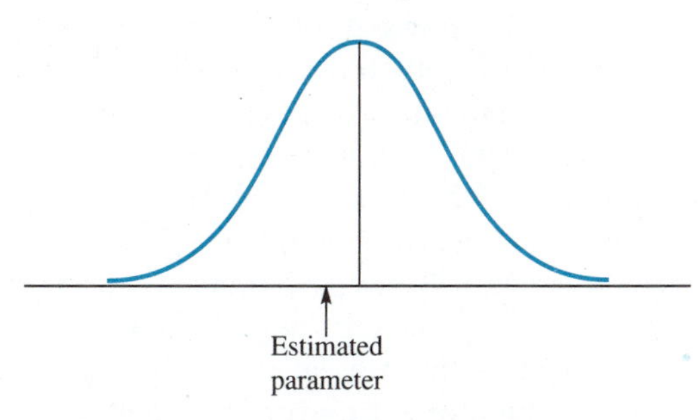

b. Biased sampling distribution

wise, the estimator is said to be *biased* (see Figure 5.22). For example, it can be shown (proof omitted) that $\bar{x}$ is an unbiased estimator of the population mean μ and s^2 is an unbiased estimator of the population variance σ^2. In contrast, the sample standard deviation s is a biased estimator of σ.

For the rifle firing analogy, the target for an unbiased marksperson (i.e., estimator) shows a distribution of shots that is centered over the bull's-eye. Some shots hit to the left of the bull's-eye and some to the right, but the mean of the distribution is equal to the estimated parameter. In contrast, the distribution of shots for a biased marksperson (i.e., estimator) tends to be shifted to either the right or the left of the bull's-eye. The mean of the distribution is not equal to the estimated parameter.

Confidence Intervals

An *interval estimator* is a rule that tells us how to calculate two points—an interval—that will enclose the estimated parameter with a specified probability. The interval is called a *confidence interval.* The lower point is called the *lower confidence limit* (LCL) and the upper point is called the *upper confidence limit* (UCL). The probability that the interval encloses the estimated parameter is called the *confidence coefficient.* For example, suppose that someone calculated a 95% confidence interval for a population mean μ and reported that the interval was from 1.1 to 1.3. This tells us that we estimate that the interval from 1.1 to 1.3 straddles the population mean. The lower confidence limit is 1.1, and the upper confidence limit is 1.3. The confidence coefficient for a 95% confidence interval is .95.

How good is this interval estimate? Does the interval really enclose the parameter μ? We do not know; it may or it may not. What we do know is that if we were to draw many, many different samples from the population and calculate a 95% confidence interval for each, approximately 95% of the intervals would enclose μ. Thus, the confidence coefficient measures the likelihood (and our confidence) that a particular interval will enclose the estimated parameter. This concept is illustrated in Figure 5.23.

Figure 5.23 shows 20 of a large number of confidence intervals, each represented by a horizontal line segment. The vertical line locates the population mean μ. If a horizontal line segment crosses the vertical line, then the confidence interval, which it represents, encloses μ. Note that the confidence intervals vary in width and location. If we were to sample over and over again a large number of times and calculate a confidence interval for each sample, 95% of the intervals (for a 95% confidence interval) will enclose μ. In other words, if we were to draw a single random sample from the population and calculate a 95% confidence interval for μ, the probability is .95 that the interval will enclose μ.

Which type of estimator should you use—point or interval? It depends on which you like. Most people understand the concept of a point estimate, so that is the type discussed in newspapers and magazines. Newswriters often give, along with an estimate, a **margin of error.** The stated margin of error is usually equal to two or three standard errors of the estimate or equal to the half-width of a 95% or a 99% confidence interval. Confidence intervals are often used in scientific articles.

FIGURE 5.23 Representation of 20 confidence intervals

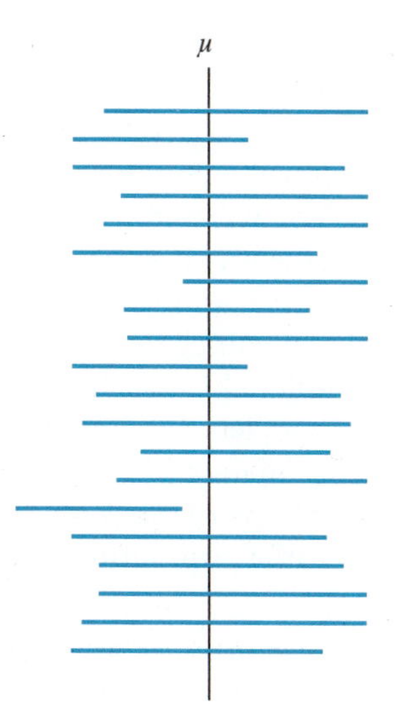

How Does Probability Play a Role in Statistical Estimation?

Theoretical statisticians use the sampling distribution of an estimator—i.e., its probability distribution—to determine how far a point estimate is likely to deviate from the estimated parameter. The sampling distribution is also used to construct a confidence interval and to determine the confidence coefficient. Thus, probability enables us to evaluate the goodness of an estimate.

5.10 Making Statistical Inferences: Tests of Hypotheses

A second method for making statistical inferences is called a *statistical test of an hypothesis*. For example, suppose that a federal agency specifies that the amount of oil in 32-ounce cans of motor oil "must average" (i.e., the mean content of the cans must equal) 32 ounces. To determine whether the fill operation is operating in accordance with the regulation, the manager of the fill operation (or an inspector for the federal agency) would want to sample n cans and measure their fill weights. Then, based on the sample fill weights, the manager would like to decide whether the fill machine is operating according to regulations; that is, the manager must decide whether the mean fill weight μ is equal to 32 ounces or more. Further, the manager would like to know the chances of having made a wrong decision.

A statistical test has a different objective than estimation. Estimation asks the question,"What is μ and how large is the error of estimation?" A statistical test asks the question, "Is μ greater than (or less than) 32 ounces and what is the probability that our sample leads us to the wrong decision?"

This section will present the basic *concepts* associated with a statistical test of an hypothesis. You will find that the line of reasoning employed in a statistical test runs counter to the way that most of us arrive at a decision. To help you adjust to this new perspective, we introduce you to the new concepts and language in this section and then reinforce them as we repeatedly apply our techniques to the practical problems that follow in Chapters 6–11. The applications in those chapters not only help you to understand how a test works; they also help you to see why a test is preferred to estimation in certain practical situations.

A *statistical test of an hypothesis* is a procedure for making a decision about the value of a population parameter. The test is composed of four parts:

1. **The *alternative* (or research) *hypothesis* about the population parameter, denoted by the symbol H_a, is the one that you wish to support.** For example, suppose that a political candidate wants to infer the proportion p of eligible voters who will vote for her in the next election. The candidate is hoping that a sampling of n voters from this universe will provide evidence to show that the candidate will be a winner, i.e., that $p > .5$. Thus, the candidate hopes that the sample data support the alternative hypothesis, H_a: $p > .5$. Note that p is the parameter of a binomial population of data.

2. **The *null hypothesis* about the population parameter, denoted by the symbol H_0, is the opposite of the alternative hypothesis. It is the hypothesis that you wish to reject.** The null hypothesis for the political candidate's test is that the candidate will receive no more than 50% of the votes, i.e., H_0: $p = .5$ (or less). If the data provide evidence to indicate that $p = .5$ is false—i.e., that p is larger than .5—then we reject the null hypothesis, H_0: $p = .5$, and conclude that the alternative hypothesis, H_a: $p > .5$, is true—i.e., that the political candidate will win the election.

3. **The *test statistic* is a statistic, calculated from the sample data, that depends on the value of the parameter that you are testing.** For example, suppose that we were to sample 400 voters from the population of all eligible voters. The number x of voters in the sample who favor our candidate is related to p, the proportion of all voters in the population who favor her. The larger the value of p, the larger x is likely to be. **You do not have to worry about choosing a good test statistic for testing a particular hypothesis. We will tell you the test statistic to use for each type of test.**

4. **The rejection region for the test is the set of values of the test statistic that are contradictory to the null hypothesis and that favor the alternative hypothesis.** For example, large values of x, the number of voters in the sample who favor the political candidate, are contradictory to the null hypothesis, H_0: $p = .5$.

The larger the value of x, the greater the contradiction. If 300 of the 400 voters in the sample favor the candidate, it appears that there is some evidence to indicate that p is larger than .5. If $x = 350$, the evidence is even greater.

The logic behind a statistical test is based on the "rare event" concept employed in reaching decisions in Examples 5.9 and 5.11, Section 5.6. The "rarity" of the test result, assuming H_0 to be true, is measured by what is called the p-value for the test. **The *p-value* for a statistical test is the probability of observing a value of the test statistic as contradictory (or more) to the null hypothesis as the computed value of the test statistic.***

The decision to reject the null hypothesis and to accept the alternative hypothesis depends on the extent to which the value of the test statistic contradicts the null hypothesis. **The more improbable the value of the test statistic, assuming the null hypothesis to be true, the smaller will be the p-value. Therefore, we reject H_0 when the p-value for a test is small.**

Example 5.18

Suppose that 400 eligible voters were sampled and that $x = 225$ favor the candidate. Is $x = 225$ large enough to lead us to reject the null hypothesis and conclude that the candidate will be a winner? Do the data provide sufficient evidence to indicate that the political candidate will win the election?

Solution

To answer this question, we need to find the p-value for the test. Since large values of x are contradictory to the null hypothesis, the p-value is the probability of observing a value of x as large as or larger than the observed value, $x = 225$. The sampling distribution for x, given that the null hypothesis is true (i.e., that $p = .5$), is a binomial probability distribution with $n = 400$ and $p = .5$. When n is large (see Section 5.7), the sampling distribution for x will be approximately normally distributed with mean and standard deviation

$$\mu = np = (400)(.5) = 200$$

and

$$\sigma = \sqrt{np(1-p)} = \sqrt{(400)(.5)(.5)} = \sqrt{100} = 10$$

The sampling distribution for x is shown in Figure 5.24 (page 170).

The p-value for the test, the probability that x is greater than or equal to 225, is equal to the shaded area under the sampling distribution to the right of 225. To find this p-value, we need to find A, the area between the mean $\mu = 200$ and $x = 225$ under the normally distributed sampling distribution (Figure 5.24). To find this area, we use the method described in Section 5.7.

*Do not confuse the p in "p-value" with the binomial proportion p. We employ this confusing double use of the symbol p to be consistent with other literature on this subject.

FIGURE 5.24 The sampling distribution for the number x of voters who favor the candidate

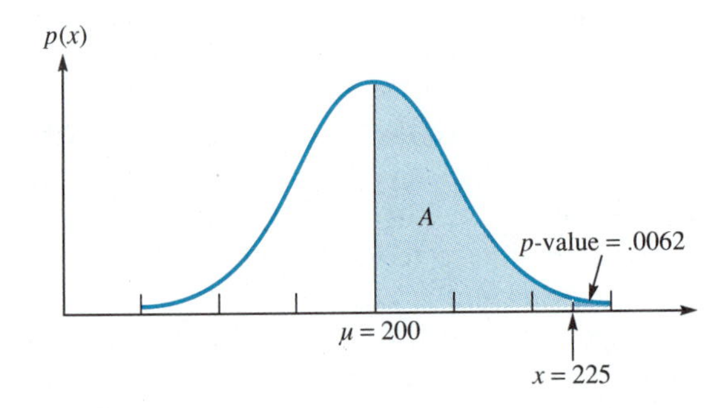

The z-value corresponding to $x = 225$ is

$$z = \frac{x - \mu}{\sigma} = \frac{225 - 200}{10} = 2.5$$

Then the area A under the standard normal curve (Table 3 of Appendix 2) is .4938. Since the total area under a normal curve lying to the right of the mean is equal to .5, the area to the right of $z = 2.5$, the p-value for the test, is

$$p\text{-value} = .5 - A = .5 - .4938 = .0062$$

What does this mean? It means that the probability of observing as many as 225 voters who favor the political candidate in a sample of 400, assuming she will not win, is only .0062. Thus either we have witnessed a very rare event or else our null hypothesis, $p = .5$, is false. Rare events occur but not very often! Consequently, we conclude that the null hypothesis is false and that the proportion of voters favoring our candidate is larger than .5.

Measuring Risk: The Probability of Making the Wrong Decision

There is always a risk of making a mistake when you make a decision. **You may reject the null hypothesis when it is true (called a *Type I error*) or you may fail to reject it when it is false (called a *Type II error*).**

The probability of rejecting the null hypothesis when it is true, i.e., making a Type I error, is denoted by the Greek letter α. It is the probability that the test statistic will fall in the rejection region when H_0 is true.

You control the risk of making a Type I error because you choose the value of α for your test. For example, suppose that you were willing to take the risk of mistakenly rejecting the null hypothesis 5% of the time. Then you would reject H_0

whenever the p-value for a test was less than or equal to .05. If you wanted to subject yourself to less risk, say being wrong only 1% of the time, you could choose $\alpha = .01$ and reject H_0 only when the p-value for the test was less than .01.

Choosing the Value of α

Why not choose a very small value for α and make the risk of making a Type I error (rejecting the null hypothesis when it is true) very small? The answer is that the less risk you take of making a Type I error (rejecting the null hypothesis when it is true), the greater the risk you take of making a Type II error (failing to reject H_0 when it is false). **As you decrease the probability of making one type of error, you increase the probability of making the other.**

To solve this problem, use a value of α that gives you adequate but not extreme protection against making a Type I error. If you are trying to detect a change in some population parameter, choose α equal to .10 or .05. If you wish to decrease the risk of making a Type I error, choose a smaller value for α, say .01. If you want to decrease the probabilities of making both types of error, increase the sample size.

An Analogy

If you are struggling with the new concepts employed in a test of an hypothesis, the following analogy may help. The prosecutor in a murder trial plays the role of the scientist. Although the defendant is assumed innocent (the null hypothesis), the prosecutor believes in her heart (so did the grand jury) that the defendant is guilty (the alternative hypothesis) and *wants to prove it*. Evidence is presented to the jury. If the weight of evidence contradicts the assumption of innocence—i.e., the p-value for the evidence is very small—the jury finds the defendant guilty.

The outcome of a trial is subject to the same errors encountered in a statistical test. The defendant can be convicted (rejecting H_0) when he is innocent (a Type I error) or can be found not guilty (not rejecting H_0) when in fact he is guilty (a Type II error). The difference between most statistical tests and the judicial system is that the judicial system attempts to make α, the risk of convicting an innocent person, as small as possible. This increases the probability of failing to convict a defendant who is guilty.

Reading Computer Printouts

Most computer printouts allow us to conduct a statistical test of an hypothesis very easily. They give the computed value of the test statistic and usually give the p-value for a test. If we are given only the value of the test statistic, we will obtain the exact or approximate p-value for the test by consulting a table that gives tail areas under the sampling distribution of the test statistic. If the printout gives the p-value for the test, we can very quickly evaluate the test results.

In the chapters that follow, we will review the concepts covered in this section and will learn how to conduct some very useful statistical tests. We will assume that the computing has been (or could be) done on a computer and will stress the logic involved in the test and the practical interpretation of the results.

EXERCISES

5.66 What is an estimator?

5.67 What are the two types of estimators?

5.68 What is a point estimator?

5.69 How does the sampling distribution of a point estimator help to provide a measure of its goodness?

5.70 What is an interval estimator?

5.71 What is a confidence interval?

5.72 How do you measure the goodness of a confidence interval?

5.73 Suppose that a 95% confidence interval for a population mean is given as 8.1 to 9.8. What does the 95% imply?

5.74 What are the four parts of a statistical test of an hypothesis?

5.75 Which hypothesis do you wish to support?

5.76 What is a Type I error? A Type II error?

5.77 What is the symbol for the probability of making a Type I error?

5.78 Why do we not protect ourselves against rejecting H_0 when it is true by choosing a very small value for α?

5.79 What is the *p*-value for a statistical test?

5.80 How do you use a *p*-value to decide whether to reject the null hypothesis and accept the alternative hypothesis?

5.11 Key Words and Concepts

- ***Probability*** is a measure of our belief that an observation will result in some particular outcome.
- The ***relative frequency concept of probability*** is as follows: If an observation is repeated a very large number, N, of times and r observations result in outcome A, then the probability that a single observation results in outcome A, denoted by the symbol $P(A)$, is equal to

$$P(A) = \frac{r}{N}$$

- In order to make a statistical inference, we need to be able to assess the probability of observing a particular sample outcome.
- A ***probability sample*** is selected in such a way that the probabilities of the various sample outcomes can be determined.
- A ***random sample*** is one that is selected in such a way that every different sample of fixed size in the population has an equal probability of being selected.
- There are two kinds of random variables: discrete and continuous.

- A ***discrete random variable*** can assume a countable number of values. Its probability distribution gives the probability $p(x)$ for each value of x. Typical is the binomial random variable.
- A ***continuous random variable*** can assume any of the infinitely large number of points contained in a line interval. The probability distribution for a continuous random variable is usually a smooth curve. The area under the curve over an interval gives the probability that x will fall within that interval. Typical is the normal probability distribution.
- Sample statistics are used to estimate and to make decisions about population parameters. The properties of a statistic are completely described by its sampling distribution.
- The ***sampling distribution*** of a statistic is its probability distribution. The standard deviation of a sampling distribution is called the standard error of the statistic.
- The sampling distribution shows how close to the estimated parameter an estimate is likely to fall. In choosing an estimator, we select the one with the best sampling distribution, the one that is centered over the estimated parameter and has the smallest standard error.
- Decisions about the values of parameters are made using tests of hypotheses. The goodness of a decision is measured by the probability we have made an error.
- You will learn more about point estimates, confidence intervals, and tests of hypotheses in the chapters that follow as we apply these inferential tools to answer some practical questions.

SUPPLEMENTARY EXERCISES

5.81 Suppose that a sample statistic is used to estimate a population parameter. How can we tell how close to the parameter the estimate is likely to fall? Explain.

5.82 Where in this chapter (not including Section 5.3) did we need to know the Additive Rule for mutually exclusive outcomes?

5.83 Where in this chapter (not including Section 5.3) did we need to know the concept of independent outcomes?

5.84 A telephone survey is conducted by contacting a random sample of listed telephone subscribers in the United States. Explain why this process will or will not produce a random sample of all adults in the United States.

EXERCISES FOR YOUR COMPUTER

5.85 A random number generator is a statistical software package that will produce a list of random numbers. If your software package includes a random number generator, use it to generate some random numbers. Then use those numbers to identify the first 10 persons to be included in a random sample of 500 voters from a universe of 215,245 registered voters in a community. Explain how you selected your sample.

5.86 Some software packages will generate a

random sample of n measurements from a normal distribution (usually, the standard normal z distribution). If you have access to such a package, draw a random sample of $n = 25$ measurements from the distribution. Calculate $\bar{x}$. Calculate the mean and the standard error of $\bar{x}$ (based on the mean and standard deviation of the sampled normal population) and describe the sampling distribution of $\bar{x}$. Within what limits would you expect your sample mean to fall? Does it? Explain.

References

1. Beyer, W. C., *Handbook of Tables for Probability and Statistics,* 2nd edition. Cleveland, Ohio: The Chemical Rubber Co., 1968.
2. Feller, W. *An Introduction to Probability Theory and Its Applications,* Vol. 1, 3rd edition. New York: Wiley, 1968.
3. Freedman, D., Pisani, R., Purves, R., and Adhikari, A., *Statistics,* 2nd edition. New York: W. W. Norton & Co., 1991.
4. McClave, J. and Dietrich, F., *Statistics,* 5th edition. San Francisco: Dellen Publishing Co., 1991.
5. Mendenhall, W. and Beaver, R., *Introduction to Probability and Statistics,* 8th edition. Boston: PWS-Kent, 1991.
6. Mendenhall, W., Scheaffer, R. L., and Wackerly, D., *Mathematical Statistics with Applications,* 3rd edition. Boston: Duxbury Press, 1986.
7. Moore, D. and McCabe, G., *Introduction to the Practice of Statistics,* 2nd edition. New York: W. H. Freeman & Co., 1993.
8. Mosteller, F., Rourke, R. E. K., and Thomas, G. B., *Probability with Statistical Applications,* 2nd edition. Reading, Massachusetts: Addison-Wesley, 1970.
9. National Bureau of Standards, *Tables of the Binomial Probability Distribution.* Washington, D.C.: Government Printing Office, 1949.
10. Sincich, T., *Statistics by Example,* 4th edition. San Francisco: Dellen Publishing Co., 1990.

six

Inferences About a Population Mean

In a Nutshell

Chapter 5 lays the foundation. It tells how probability enables us to make the intellectual jump from a sample to a population. We will reinforce and develop these ideas in Chapter 6 and use them to make inferences about the mean of a population.

6.1 The Problem

6.2 The *z* and the *t* Statistics

6.3 Confidence Intervals for a Population Mean

6.4 Another Way to Report an Estimate

6.5 Why Test Hypotheses About a Population Mean?

6.6 Tests of Hypotheses About a Population Mean

6.7 Choosing the Sample Size

6.8 Why Assumptions Are Important

6.9 Key Words and Concepts

6.1 The Problem

How much frozen yogurt does a TCBY store sell per day? There is no unique answer. As we learned in Chapter 2, daily sales of yogurt, like the characteristics of most things observed in nature, vary from one observation to another and are best described by a relative frequency distribution.

So, let us try again. What are the "typical" frozen yogurt sales per day at your local TCBY store? When we ask that question, we are asking for a measurement in the middle of a population distribution of daily sales—a measure of central tendency such as the mean or median. In fact, most often (without saying so) we are asking for the "average" daily sales, or what we call in statistics the *population mean.*

This chapter explains how to make inferences about a population mean, how to find confidence intervals for it, and how to test hypotheses about its value.

6.2 The *z* and the *t* Statistics

Recall Case 2.1 and the sampling of the 20,000-ton consignment of iron ore. Three hundred ninety 1.5-kg specimens of ore were selected from the consignment, tested, and the percentage of iron was recorded for each specimen. The objective of the sampling was to estimate the amount of iron in the 20,000-ton consignment. The mean $\bar{x}$ of the sample of 390 percentage iron measurements was 65.74%. How far from the true mean percentage of iron in the consignment is the sample mean $\bar{x}$ likely to be? For example, is it probable that $\bar{x} = 65.74\%$ deviates from the unknown population percentage by as much as one-tenth of 1% (i.e., .1%)? How large is the sampling error likely to be? To answer this and other questions about $\bar{x}$, we must know its sampling distribution.

The Sampling Distribution of $\bar{x}$

We know from Section 5.8 that if we have drawn a random sample of n measurements from a population with mean μ and standard deviation σ, the sampling (i.e., probability) distribution of $\bar{x}$ will be, to a reasonable degree of approximation, normally distributed. The mean of the sampling distribution of $\bar{x}$ will equal the mean μ of the sampled population. The standard error (i.e., the standard deviation) of the sampling distribution will equal $\text{SE}(\bar{x}) = \sigma/\sqrt{n}$.

The sampling distribution of the sample mean for samples of 390 percentage iron measurements is shown in Figure 6.1. Notice that the distribution is normally distributed and that the unknown mean μ of the 20,000-ton consignment is shown in the center of the distribution. The standard error of the mean (shown on Fig-

FIGURE 6.1 The sampling distribution of $\bar{x}$ for the sample of 390 percentage iron measurements

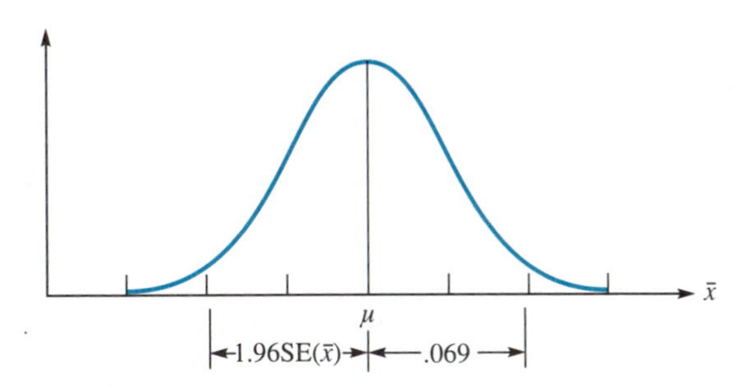

ure 4.18, the Minitab printout of numerical descriptive measures for the percentage iron data) is $SE(\bar{x}) = .035$. We learned in Section 5.7 that 95% of the area under the normal distribution lies within 1.96 standard deviations of its mean. The standard deviation of the sampling distribution of $\bar{x}$ is $SE(\bar{x}) = .035$ and $1.96SE(\bar{x}) = 1.96(.035) = .069$. This distance is measured off on each side of the unknown mean μ. The result, Figure 6.1, is a good description of the sampling distribution of $\bar{x}$ for a sample of 390 percentage iron measurements.

The Standard Normal *z* Statistic

Example 6.1

Now let us go back to our original question: What is the probability that a sample mean of 390 percentage iron measurements deviates from the mean percentage iron in the whole consignment by as much as .1% iron?

Solution

To answer this question, we have to know how many standard errors the deviation $(\bar{x} - \mu) = .1\%$ represents. This deviation, expressed in units of the standard error, $SE(\bar{x}) = .035$, is the standard normal z statistic of Section 5.7:

$$z = \frac{\bar{x} - \mu}{SE(\bar{x})}$$

Substituting the deviation $(\bar{x} - \mu) = .1\%$ and $SE(\bar{x}) = .035$ into the formula for z, we find that

$$z = \frac{\bar{x} - \mu}{SE(\bar{x})} = \frac{.1}{.035} = 2.86$$

This tells us that the deviation $(\bar{x} - \mu) = .1\%$ represents 2.86 standard errors of $\bar{x}$.

FIGURE 6.2 Locating the deviation $(\bar{x} - \mu) = .1\%$

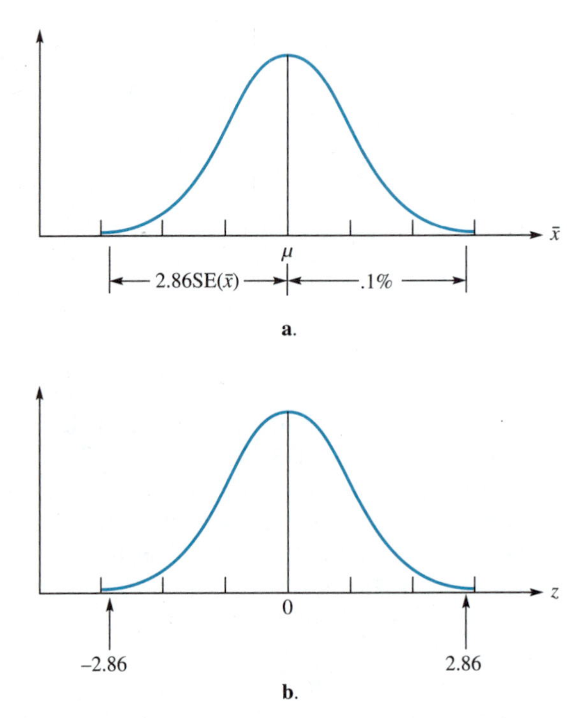

Figure 6.2a shows the deviation, $(\bar{x} - \mu)$, marked off below the sampling distribution of $\bar{x}$ and Figure 6.2b shows the corresponding standard normal z distribution with $z = \pm 2.86$ measured on each side of its mean 0. You can see that $z = 2.86$ and $z = -2.86$ fall far out in the tails of the z distribution. Therefore, the probability that an estimate $\bar{x}$, based on 390 percentage iron measurements, deviates from μ by as much as .1%, i.e., .1% or more, is almost 0.

Upper-Tail Areas for the z Distribution

Finding the probability in Example 6.1 that an estimate $\bar{x}$ deviates from μ by as much as .1% was relatively easy because the deviation, .1%, was so large in terms of SE($\bar{x}$). The Empirical Rule tells us that a deviation as large as 2.86 standard errors is improbable. Suppose, instead, that we had asked for the probability that the error of estimation, the deviation of $\bar{x}$ from the mean percentage iron in the 20,000-ton consignment, was as large as .05%. To make this or other probability statements about a particular value of $\bar{x}$, we first need to find the z-value corresponding to $\bar{x}$—that is, we need to find how many SE($\bar{x}$)'s the sample mean $\bar{x}$ deviates from μ—and then use (as we did in Section 5.7) a table of areas under the z distribution to find the desired probability.

Finding the Probability That $\bar{x}$ Deviates from μ by a Specified Amount

The probability that z is larger than a value z_a is the area a in the upper tail of the z distribution to the right of z_a. The location of z_a and the area a are shown in Figure 6.3a. These areas can be calculated from Table 5.3 in Section 5.7, or from Table 3 in Appendix 2, both of which give areas under the normal curve between the mean 0 and a value of z. For example, the tabulated area between the mean 0 and $z = 1.96$ (from Table 5.3) is .475 and the total area to the right of the mean is .5. Therefore, the area to the right of $z_{.025} = 1.96$ is $.5 - .475 = .025$. See Figure 6.3b. This and other important z-values and corresponding upper-tail areas are given in Table 6.1 (page 180).

The following examples will show how we can use Table 6.1 and Table 3 of Appendix 2 to calculate the probability of observing values of $\bar{x}$ that deviate greatly from the population mean μ.

FIGURE 6.3

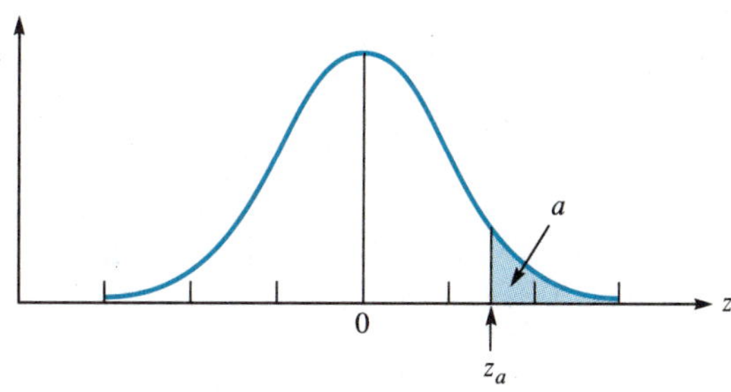

a. A z distribution showing z_a and a

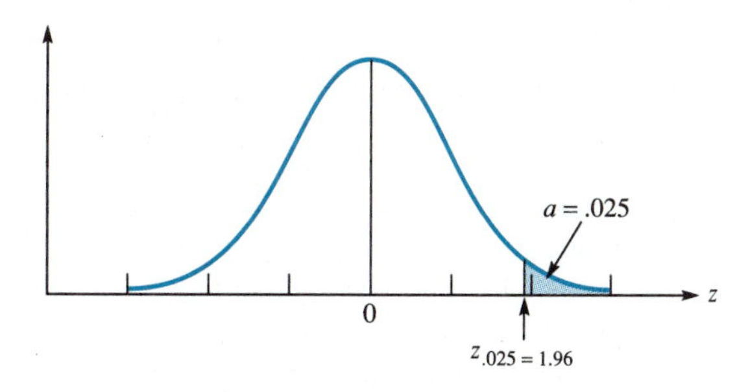

b. Area under the normal curve to the right of $z_a = 1.96$

TABLE 6.1 Some Upper-Tail Areas for the Standard Normal Distribution

z_a	Upper-Tail Area a
.84	.20
1.00	.16
1.28	.10
1.65	.05
1.96	.025
2.33	.01
2.58	.005

Example 6.2

Suppose that we were to draw a random sample of $n = 25$ measurements from a population with mean $\mu = 20$ and standard deviation $\sigma = 5.5$, and find that the sample mean equals 18.8.

a. Find the z-value corresponding to $\bar{x} = 18.8$.

b. Is it likely that the sample mean would be as small as 18.8? Find the probability that $\bar{x}$ will equal 18.8 or less.

Solution

a. To find the value of z corresponding to a value of $\bar{x}$, we substitute the values of $\bar{x}$, μ, and $\text{SE}(\bar{x})$ into the formula for z and solve for z. The first step is to calculate $\text{SE}(\bar{x})$:

$$\text{SE}(\bar{x}) = \frac{\sigma}{\sqrt{n}} = \frac{5.5}{\sqrt{25}} = \frac{5.5}{5} = 1.1$$

Then, substituting $\bar{x} = 18.8$, $\mu = 20$, and $\text{SE}(\bar{x}) = 1.1$ into the formula for z, we obtain

$$z = \frac{\bar{x} - \mu}{\text{SE}(\bar{x})} = \frac{18.8 - 20}{1.1} = -1.09$$

This tells us that $\bar{x} = 18.8$ corresponds to a z-value of -1.09. Or, putting it another way, $\bar{x}$ is located 1.09 standard errors to the left of the mean $\mu = 20$. The location of the sample mean, $\bar{x} = 18.8$, on its sampling distribution is shown in Figure 6.4a. The location of the corresponding $z = -1.09$ is shown on the z distribution, Figure 6.4b. **Note that z will be negative if $\bar{x}$ lies below the mean and positive if it lies above.**

FIGURE 6.4 The probability that $\bar{x}$ is less than or equal to 18.8

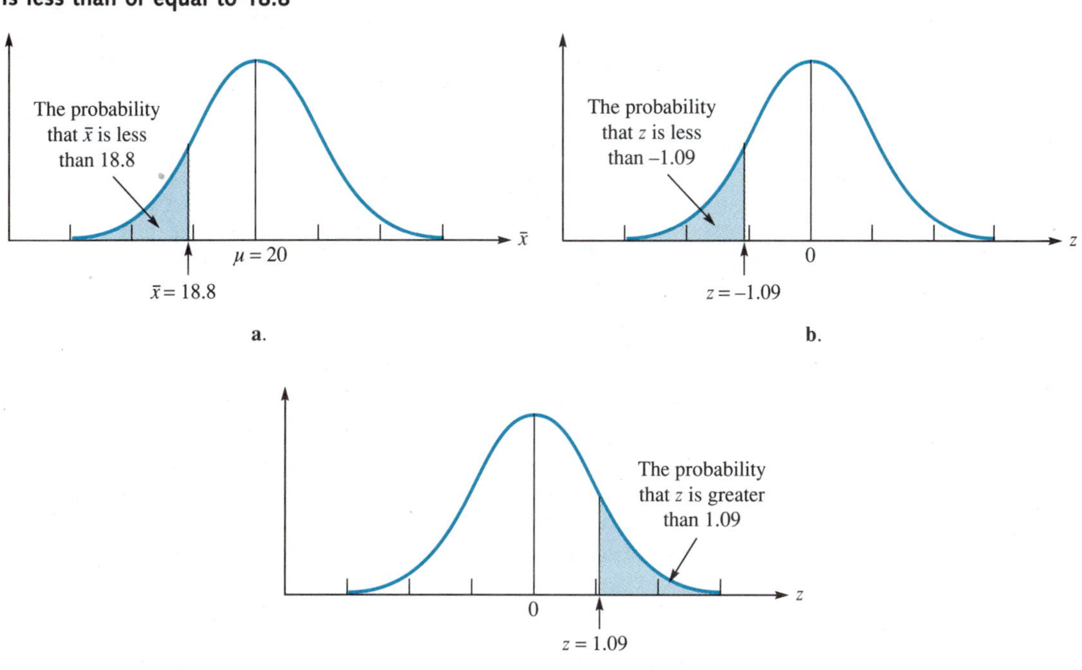

b. The probability that $\bar{x}$ is less than or equal to 18.8 (the shaded area in the lower tail of Figure 6.4a) is equal to the area in the lower tail of the z distribution to the left of $z = -1.09$, Figure 6.4b. Recall that a normal curve is symmetric about its mean. Therefore, the area to the left of $z = -1.09$ in the lower tail of the z distribution, Figure 6.4b, is equal to the area to the right of $z = 1.09$ in the upper tail of the z distribution, Figure 6.4c. Table 6.1 does not give an upper-tail area for $z = 1.09$ but we can approximate the area from Table 6.1. Table 6.1 gives the areas in the upper tail for $z = 1.0$ and $z = 1.28$ as .16 and .10, respectively. Therefore, since $z = 1.09$ falls between $z = 1.0$ and $z = 1.28$, the probability that $\bar{x}$ is less than or equal to 18.8 is between .10 and .16.

If you want an exact value for the probability that $\bar{x}$ is less than or equal to 18.8, see Table 3 of Appendix 2. It gives the area under the z distribution from the mean 0 to a specified value of z. Consulting Table 3 of Appendix 2, we find that the area between $z = 0$ and $z = 1.09$ is .3621. See Figure 6.5 (page 182). Then, since the total area to the right of the mean is .5, the area to the right of $z = 1.09$ is $(.5 - .3621) = .1379$. Therefore, the probability that $\bar{x}$ is less than 18.8 is equal to .1379 (to four decimal places). As expected, this probability falls between .10 and .16.

FIGURE 6.5 Finding the probability that $\bar{x}$ is less than 18.8 using Table 3 of Appendix 2

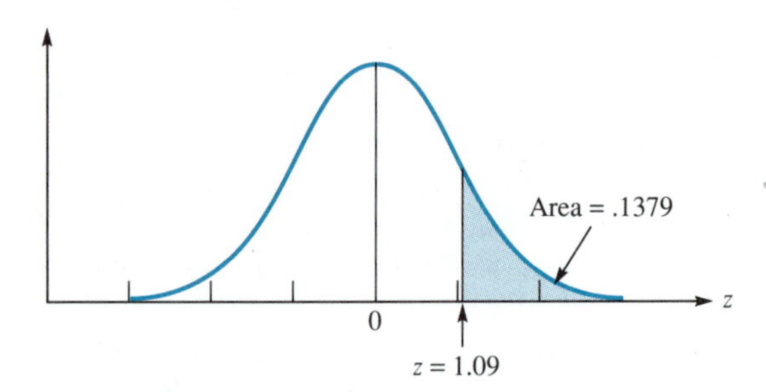

Example 6.3

As a final example, let us return to the problem of estimating the mean percentage iron in the 20,000-ton consignment of iron ore. Use Table 6.1 to find the approximate probability that our estimate, the sample mean $\bar{x}$, will deviate from μ, the true population percentage of iron, by as much as .05%.

Solution

Figure 6.6a shows the sampling distribution of $\bar{x}$ with the deviation, .05% iron, marked off along the horizontal axis. The probability that $\bar{x}$ will deviate from μ by as much as .05% above μ or .05% below μ is equal to the sum of the shaded areas in the tails of the distribution. The first step in finding the area in the upper tail of the distribution is to find how many SE($\bar{x}$)'s the deviation $(\bar{x} - \mu) = .05\%$

FIGURE 6.6 The probability that $\bar{x}$ deviates from μ by .05% iron or more

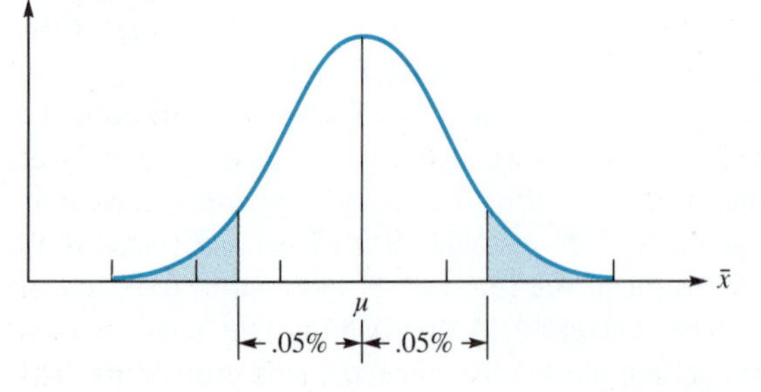

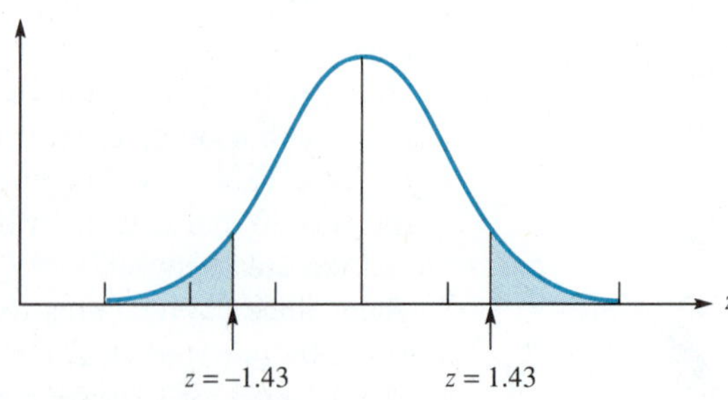

represents. We gave the value of $SE(\bar{x})$ in Example 6.1 as .035. Therefore, a deviation equal to .05% iron corresponds to

$$z = \frac{\bar{x} - \mu}{SE(\bar{x})} = \frac{.05}{.035} = 1.43$$

standard errors above the mean. This value of z is shown on the z distribution, Figure 6.6b. Table 6.1 does not give the upper-tail area for $z = 1.43$ but it does for $z = 1.28$ and $z = 1.65$. The upper-tail areas corresponding to $z = 1.28$ and $z = 1.65$ are .10 and .05, respectively. Therefore, since $z = 1.43$ lies between 1.28 and 1.65, the upper-tail area corresponding to $z = 1.43$ is between .05 and .10. The area to the left of $z = -1.43$ is equal to the area to the right of $z = 1.43$. Therefore, the probability that $\bar{x}$ deviates from μ by as much as .05% is double the area in the upper tail of the z distribution and is between .10 and .20. If we wanted a more accurate value for the probability, we could calculate it using Table 3 of Appendix 2.

Limitations on the Use of the z Statistic

Although Examples 6.1, 6.2, and 6.3 show how to calculate the probability that a sample mean $\bar{x}$ deviates from the population mean μ by more than a specified amount, they ignore a potentially difficult problem: We assumed that we knew the value of the population standard deviation σ.

In most sampling situations neither μ nor σ will be known. Therefore, **the z statistic**

$$z = \frac{\bar{x} - \mu}{SE(\bar{x})}$$

is useful for making inferences about μ only if the population standard deviation σ is known or if the sample size n is large enough (say, $n \geqslant 30$) for the sample standard deviation s to provide a reasonably accurate estimate of σ.
Note: The number (390) of percentage iron measurements in the iron ore data for Examples 6.1 and 6.3 was very large. We were therefore able to use the sample standard deviation s to approximate σ in the formula for $SE(\bar{x})$. In Example 6.2, the sample size was small, but the population standard deviation was assumed known.

Student's t Statistic

What should we do if σ is unknown and we select a small sample—say, $n = 10$ measurements—from a population? The only estimation and tests for a population mean μ available to the statisticians of the early 1900s were based on the z statistic. Faced with small samples, they substituted s for σ in the formula for $SE(\bar{x})$ and hoped for the best. The errors introduced into the methods, as subsequently shown, were sizable for small samples.

This problem was solved by W. S. Gosset, a statistician employed by the Guinness Brewing Company. Gosset discovered the sampling distribution of

$$\frac{\bar{x} - \mu}{s/\sqrt{n}}$$

for samples selected from a **normally distributed population of measurements.** This statistic, which Gosset called a t (instead of a z) is used to make inferences about a population mean when the sample size is small and σ is unknown. Prohibited by company policy from publishing in his own name, Gosset published his findings in 1908 under the pen name "Student." His distribution, which has other applications in statistics, has henceforth been known as a **Student's *t* distribution.**

Properties of a Student's *t* Distribution

A Student's t statistic is very similar to a z statistic. The difference between a t and a z is that a z measures the deviation between $\bar{x}$ and its mean μ in units of the standard error, $\text{SE}(\bar{x}) = \sigma/\sqrt{n}$. In contrast, a Student's t,

$$t = \frac{\bar{x} - \mu}{s/\sqrt{n}} = \frac{\bar{x} - \mu}{\widehat{\text{SE}}(\bar{x})}$$

measures the deviation between $\bar{x}$ and its mean μ in units of the **estimated standard error**, $\widehat{\text{SE}}(\bar{x}) = s/\sqrt{n}$.* Its sampling distribution, like the standard normal distribution, is symmetric about a mean equal to 0 but its spread, which is greater than the spread for a z distribution, depends on a quantity known as the number of degrees of freedom. **The *number of degrees of freedom* (df) for**

$$t = \frac{\bar{x} - \mu}{s/\sqrt{n}}$$

is equal to (n − 1), the divisor of the sum of squares of deviations used in calculating s^2. For example, if you draw a random sample of $n = 10$ measurements from a normal population and calculate $\bar{x}$, s, and t, the t statistic will be based on $(n - 1) = 10 - 1 = 9$ degrees of freedom.

There is no simple intuitive way to describe a "degree of freedom." The explanation is contained in the mathematical theory of statistics and is beyond the level of this course. You will encounter other t statistics in Chapters 7, 8, and 10. The most that we can say is that the number of degrees of freedom associated with a t statistic will always equal the divisor of the sum of squares of deviations used in calculating s^2.

---

*An estimate of a parameter is indicated by placing a "hat" (^) over the symbol for the parameter. For example, $\hat{\mu} = \bar{x}$ and $\hat{\sigma} = s$. A hat placed over $\text{SE}(\bar{x})$, i.e., $\widehat{\text{SE}}(\bar{x})$, indicates that it is an estimate of the standard error of $\bar{x}$.

A Comparison of the z and the t Distributions

Figure 6.7 shows a t distribution with 6 df (degrees of freedom), along with a standard normal distribution. Note that the t distribution, based on 6 df, is more spread out than the standard normal distribution. **The spread of a t distribution decreases as the sample size (and the number of degrees of freedom) increases until, when $n = 30$, it is barely distinguishable from a standard normal z distribution.** Thus, the errors introduced by using the z statistic in place of t when σ is unknown are large for small sample sizes but are negligible for $n = 30$ or more.

FIGURE 6.7 A t distribution and a standard normal distribution

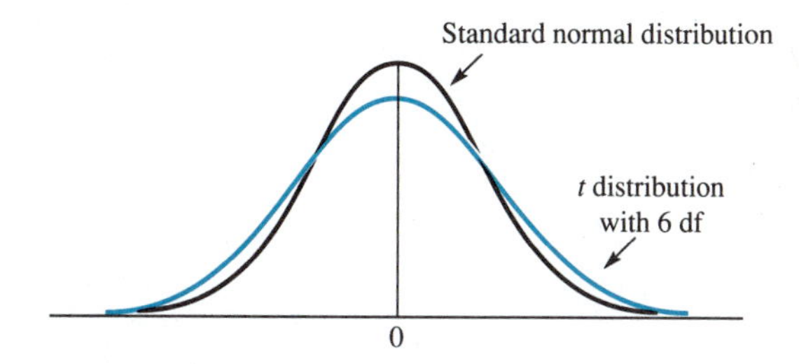

Upper-Tail Areas for the Student's t Distribution

We showed in Examples 6.1–6.3 how to calculate the probability that the error of estimating a population mean was less than some specified amount. These calculations were based on the assumption that the population standard deviation σ was known or that the sample size n was large. Then the probabilities corresponded to tail areas under the sampling distribution for the z statistic.

When the standard deviation σ of the sampled population is unknown and n is small, our methods for making inferences about a population mean are based on a Student's t statistic. Then, as when using the z statistic, we will have to determine the tail areas under the Student's t distribution.

The area in the upper tail of the t distribution depends on the number of degrees of freedom associated with t as well as the upper-tail area a. **The value of t that locates an area a in the upper tail of the t distribution is represented by the symbol t_a. The values of** t_a for different values of a are given in Table 4 of Appendix 2, a portion of which is reproduced in Table 6.2 (page 186). The location of t_a and the area a are shown in Figure 6.8 (page 187).

Reading the t Table, Table 4 of Appendix 2

Values of t_a appear across the top of the table and the degrees of freedom (df) for the Student's t are shown at the left side of the table. To find the value of t_a corresponding to an area a, locate the value of a at the top of the table. Move down that

TABLE 6.2 A Partial Reproduction of Table 4 of Appendix 2: Values of t_a for Values of a

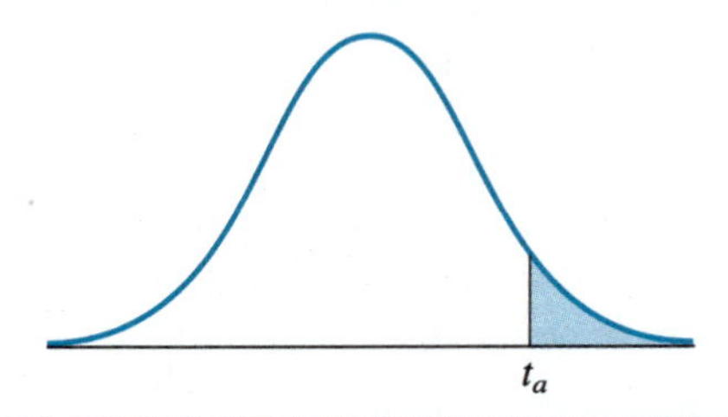

df	$t_{.100}$	$t_{.050}$	$t_{.025}$	$t_{.010}$	$t_{.005}$
1	3.078	6.314	12.706	31.821	63.657
2	1.886	2.920	4.303	6.965	9.925
3	1.638	2.353	3.182	4.541	5.841
4	1.533	2.132	2.776	3.747	4.604
5	1.476	2.015	2.571	3.365	4.032
6	1.440	1.943	2.447	3.143	3.707
7	1.415	1.895	2.365	2.998	3.499
8	1.397	1.860	2.306	2.896	3.355
9	1.383	1.833	2.262	2.821	3.250
10	1.372	1.812	2.228	2.764	3.169
11	1.363	1.796	2.201	2.718	3.106
12	1.356	1.782	2.179	2.681	3.055
13	1.350	1.771	2.160	2.650	3.012
14	1.345	1.761	2.145	2.624	2.977
⋮	⋮	⋮	⋮	⋮	⋮
26	1.315	1.706	2.056	2.479	2.779
27	1.314	1.703	2.052	2.473	2.771
28	1.313	1.701	2.048	2.467	2.763
29	1.311	1.699	2.045	2.462	2.756
30	1.310	1.697	2.042	2.457	2.750
⋮	⋮	⋮	⋮	⋮	⋮
inf.	1.282	1.645	1.960	2.326	2.576

From "Table of Percentage Points of the *t*-Distribution," computed by Maxine Merrington. *Biometrika*, Vol. 32 (1941), p. 300. Reproduced by permission of the Biometrika Trustees.

FIGURE 6.8 A t distribution showing t_a and a

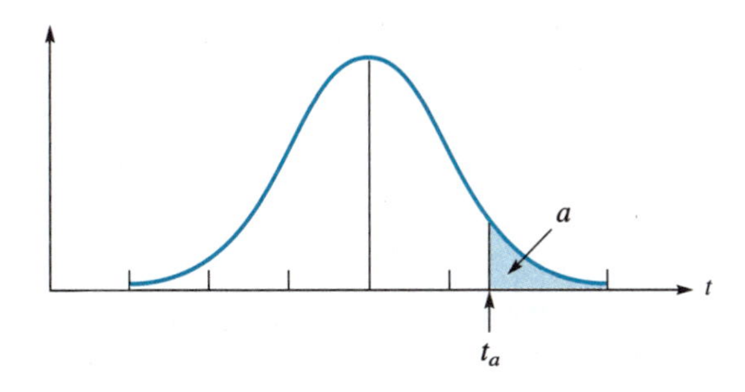

column to the row corresponding to the degrees of freedom for t. The entry will be t_a.

The following examples show how to use the t table, Table 4 of Appendix 2.

Example 6.4

If a t distribution has 13 degrees of freedom, give the value of t that places $a = .05$ in the upper tail of the t distribution.

Solution

The area that we wish to place in the upper tail of the t distribution is $a = .05$. To find the value $t_{.05}$ that places an area $a = .05$ in the upper tail of the t distribution, go to the $t_{.05}$ column in Table 6.2 and move down to the row corresponding to df = 13. The value $t_{.05}$ for df = 13 is 1.771. This t-value is shaded in Table 6.2.

Example 6.5

If a t distribution has 29 degrees of freedom, find the value of t that places $a = .025$ in the upper tail of the t distribution.

Solution

Look in the $t_{.025}$ column of Table 6.2, and move down to the row corresponding to df = 29. The area to the right of this value of t, $t_{.025} = 2.045$, is .025.

Example 6.6

How do the results of Example 6.5, along with the other tabulated values of $t_{.025}$, support the contention that the z and the t distributions become nearly indentical when the sample size n is 30 or larger?

Solution

As you move down the column in Table 6.2 that gives values of $t_{.025}$, the values of

$t_{.025}$ move closer to $z_{.025} = 1.96$, the value of z that places $a = .025$ in the upper tail of the z distribution. When df = 29, i.e., $n = 30$, the value $t_{.025} = 2.045$ is very close to $z_{.025} = 1.96$ and when the number of degrees of freedom is infinitely large, $t_{.025} = z_{.025} = 1.96$. This supports our statement that the curve that outlines the t distribution moves closer and closer to the z distribution until, when the number of degrees of freedom is very large, the two distributions become indistinguishable.

Example 6.7

What is the approximate probability that a t statistic based on 12 degrees of freedom will exceed 1.55?

Solution

Table 4 of Appendix 2 gives values, t_a, that locate an area a in the upper tail of the t distribution. The area a is the probability that t will exceed t_a. Now go to Table 4 of Appendix 2, and move down to the row corresponding to df = 12. The probability that t will exceed $a = .10$ is 1.356 and the probability that it will exceed 1.782 is $a = .05$. Since $t = 1.55$ lies between $t_{.10} = 1.356$ and $t_{.05} = 1.782$, the probability that t will exceed 1.55 is between .05 and .10.

Now that we have learned how to find upper-tail probabilities associated with the z and the t statistics, we will see how they are used to construct confidence intervals for a population mean.

EXERCISES

6.1 When we calculate probabilities associated with $\bar{x}$, why do we divide $(\bar{x} - \mu)$ by SE($\bar{x}$) rather than by σ when forming a standard normal z statistic?

6.2 If we draw a random sample of $n = 36$ measurements from a population with mean $\mu = 100$ and standard deviation $\sigma = 10$, what is the approximate probability that $\bar{x}$ is:

a. Larger than 103?
b. Less than 100?
c. Less than 97?
d. Less than 96?

6.3 If we draw a random sample of $n = 15$ measurements from a population with standard deviation $\sigma = 1.9$, what is the approximate (or exact) probability that the sample mean $\bar{x}$ will:

a. Deviate from its population mean μ by more than 1.96 standard errors of $\bar{x}$?
b. Exceed μ by more than .96?
c. Deviate from μ by more than .96?
d. Deviate from μ by more than .7?

6.4 If we draw a random sample of $n = 40$ measurements from a population with standard deviation $\sigma = 256$, what is the probability that the sample mean $\bar{x}$ will:

a. Lie more than 50 units below the mean μ?

b. Exceed μ by more than 80?
c. Deviate from μ by more than 80?
d. Deviate from μ by more than 100?

6.5 How are the z and the t distributions similar? How do they differ?

6.6 Use Table 4 of Appendix 2 to find the t-value, $t_{.05}$, that locates an area $a = .05$ in the upper tail of a t distribution with 12 df.

6.7 Use Table 4 of Appendix 2 to find the upper-tail value, t_a, for the following combinations of a and df:

a. $a = .025$ and df $= 20$
b. $a = .10$ and df $= 8$
c. $a = .01$ and df $= 10$
d. $a = .025$ and df $= 3$

6.8 Repeat the instructions of Exercise 6.7 for:

a. $a = .05$ and df $= 29$
b. $a = .025$ and df $= 17$
c. $a = .005$ and df $= 18$
d. $a = .05$ and df $= 6$

6.9 What characteristics must a sample and population have in order that $(\bar{x} - \mu)/(s/\sqrt{n})$ will possess a Student's t distribution?

6.10 If a t statistic is based on a sample of n measurements, find the value, t_a, that locates an area a in the upper tail of a t distribution for the following combinations of n and a:

a. $n = 14$ and $a = .025$
b. $n = 5$ and $a = .01$
c. $n = 3$ and $a = .10$
d. $n = 22$ and $a = .05$

6.11 What is the approximate probability that a t statistic based on 5 degrees of freedom will exceed 3.15?

6.3 Confidence Intervals for a Population Mean

How much iron is in the 20,000-ton consignment of iron ore? One way to answer this question is to find a confidence interval for the mean percentage iron in the shipment. Not only does the confidence interval give us an estimate of a population mean μ, but it also tells us how good the estimate is. That is, it tells us how much confidence we can place in the estimate.

In Section 6.2 we talked about the z and the t statistics that we will use to make inferences about a population mean. **The z statistic is used to construct confidence intervals for the rare situations when the population standard deviation σ is known or when the sample size n is large (say, $n \geqslant 30$). When σ is unknown and the population is approximately normally distributed, confidence intervals for μ can be constructed using a Student's t statistic.** This section will review the concepts involved in the construction of a confidence interval, give the formulas for the confidence intervals based on both the z and the t statistics, and demonstrate their use with examples. As you will see, the logic behind the construction of both confidence intervals is the same and the formulas are almost identical.

How a Confidence Interval Is Constructed and Why It Works

We will show you how a confidence interval is constructed (and how and why it works) for the case when σ is known, i.e., when the interval is based on the z statistic. The reasoning for the case when σ is unknown is identical except that the interval is based on the Student's t statistic.

Figure 6.9 shows the sampling distribution for the sample mean $\bar{x}$. Also shown is the interval from $[\mu - 1.96\text{SE}(\bar{x})]$ to $[\mu + 1.96\text{SE}(\bar{x})]$, where $\text{SE}(\bar{x}) = \sigma/\sqrt{n}$ is the standard error of the sampling distribution of $\bar{x}$. If you were to draw a random sample of n measurements from the population and calculate $\bar{x}$, the probability that $\bar{x}$ will fall in the interval from $[\mu - 1.96\text{SE}(\bar{x})]$ to $[\mu + 1.96\text{SE}(\bar{x})]$ is equal to .95.

FIGURE 6.9 The sampling distribution of $\bar{x}$

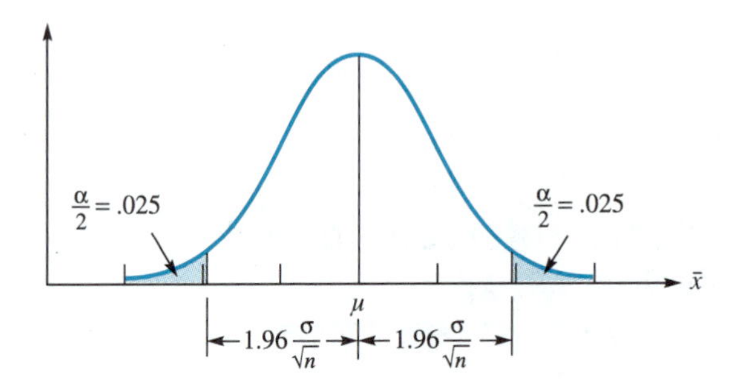

Now look at Figure 6.10. We show a typical value of $\bar{x}$, one that falls within two standard errors of its mean μ, and we show an interval of the same width as the one shown in Figure 6.9, except that this one is formed by moving a distance of 1.96 standard errors from each side of $\bar{x}$. Note that this interval will enclose μ whenever $\bar{x}$ falls within the interval from $[\mu - 1.96\text{SE}(\bar{x})]$ to $[\mu + 1.96\text{SE}(\bar{x})]$ that is shown in Figure 6.9. Since the probability that this will occur is .95, the interval

$$[\bar{x} - 1.96\text{SE}(\bar{x})] \quad \text{to} \quad [\bar{x} + 1.96\text{SE}(\bar{x})]$$

is called a 95% confidence interval for μ. **The lower boundary of the interval is called the *lower confidence limit* (LCL) and the upper boundary is called the *upper confidence limit* (UCL). The *confidence coefficient* for the interval, the probability that an interval will enclose μ, is .95.**

Confidence Intervals for Other Confidence Coefficients

As we have explained, the confidence coefficient measures how confident we are that a particular interval encloses the population mean. If we want to construct an

FIGURE 6.10 A 95% confidence interval for μ when σ is known

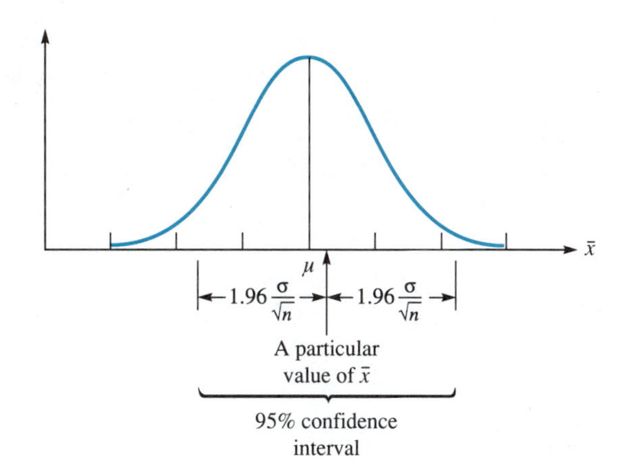

interval that is even more (or less) likely to enclose μ, all that we have to do is to change the confidence coefficient.

To obtain a 95% confidence interval, we used the value $z = 1.96$, a z-value that located points on the distribution (Figure 6.9) that placed .025 in both the lower and upper tails of the sampling distribution of $\bar{x}$. We can construct confidence intervals for any desired confidence coefficient by changing the value of z and the corresponding upper-tail area in Figure 6.9. That is, we replace $z = 1.96$ in the formula for a 95% confidence interval by the appropriate value of z. These values are shown in Table 6.3; the general formula for a population mean when the population variance is known is shown in the box, along with the assumptions upon which the method is based.

TABLE 6.3 Values of z for Different Confidence Intervals

a	Interval Percentage $100(1-a)\%$	Confidence Coefficient $(1-a)$	z
.10	90	.90	1.65
.05	95	.95	1.96
.01	99	.99	2.33

A Confidence Interval for μ When σ Is Known

A 100(1 − a)% Confidence Interval for μ
When the Value of σ Is Known

$$[\bar{x} - z_{a/2}\text{SE}(\bar{x})] \quad \text{to} \quad [\bar{x} + z_{a/2}\text{SE}(\bar{x})]$$

where $\text{SE}(\bar{x}) = \sigma/\sqrt{n}$

Assumptions:

1. **The sample is a random sample.**
2. **The sample mean $\bar{x}$ is approximately normally distributed (i.e., n is large enough so that the Central Limit Theorem applies).**
3. **The population standard deviation σ is known or n is large ($n \geq 30$) so that s provides a good estimate of σ.**

The confidence interval for μ when σ is unknown is identical to the formula used when σ is known, except that you replace $\text{SE}(\bar{x})$ by the estimated standard error, $\widehat{\text{SE}}(\bar{x}) = s/\sqrt{n}$. You also replace the value $z_{a/2}$ by the value $t_{a/2}$, obtained from Table 4 of Appendix 2. For example, suppose that your sample contains $n = 16$ measurements. Then, for a 95% confidence interval (i.e., $a = .05$), you would replace $\text{SE}(\bar{x})$ by its estimate in the formula for the confidence interval and replace $z_{.025} = 1.96$ by the value $t_{.025}$ based on $(n - 1) = 16 - 1 = 15$ degrees of freedom. This value, given in Table 4 of Appendix 2, is $t_{.025} = 2.131$.

A Confidence Interval for μ When σ Is Unknown

A 100(1 − a)% Confidence Interval for μ
When the Value of σ Is Unknown

$$[\bar{x} - t_{a/2}\widehat{\text{SE}}(\bar{x})] \quad \text{to} \quad [\bar{x} + t_{a/2}\widehat{\text{SE}}(\bar{x})]$$

where $\widehat{\text{SE}}(\bar{x}) = s/\sqrt{n}$

Assumptions:

1. **The sample is a random sample.**
2. **The population is normally distributed with unknown mean and standard deviation.**

Finding a Confidence Interval for the Mean Percentage Iron in the Iron Ore, Case 2.1

Example 6.8

Return to the 20,000-ton shipment of iron ore, Case 2.1, and the question, "How much iron does the shipment contain?" Table 1 of Appendix 1 gives the percentage iron measurements for each of 390 1.5-kilogram specimens of iron ore randomly selected from the 20,000-ton consignment and Figure 6.11 shows their relative frequency distribution.

FIGURE 6.11 Relative frequency distribution for the sample of 390 iron ore specimens

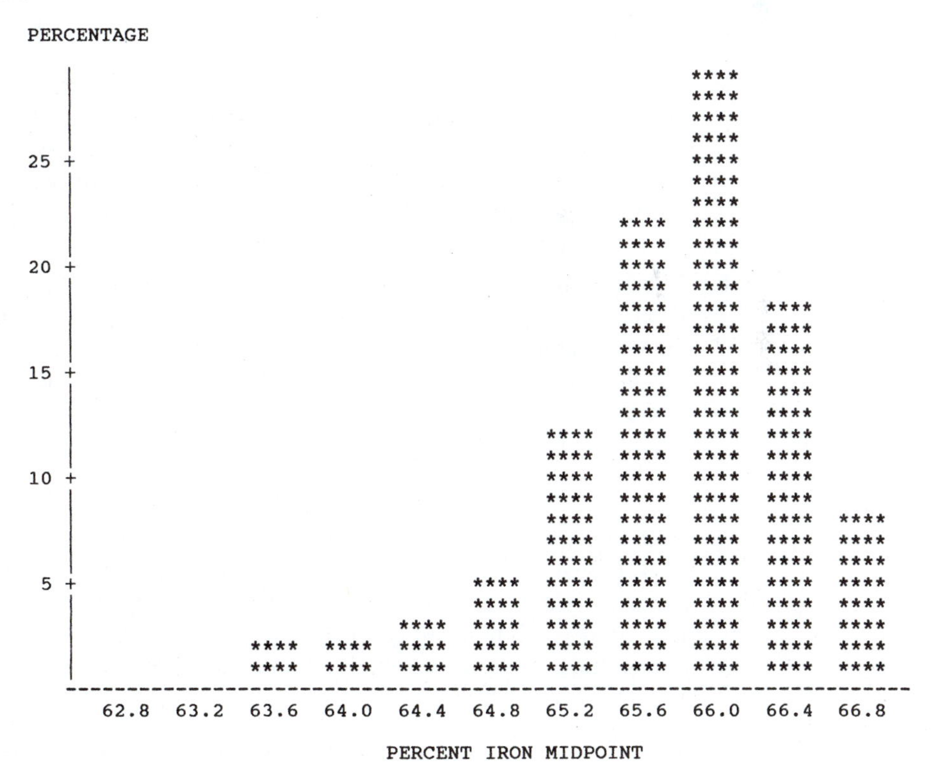

a. If we want to find a 95% confidence interval for the mean percentage of iron in 1.5-kilogram specimens of ore, which confidence interval should we use—the one based on the z or the one based on t?

FIGURE 6.12 A Minitab printout of the confidence interval for the mean percentage of iron in 1.5-kg specimens of ore

```
MTB > tinterval c1

              N       MEAN     STDEV   SE MEAN    95.0 PERCENT C.I.
%iron       390    65.7430    0.6937    0.0351   ( 65.6740, 65.8121)
```

b. Figure 6.12 gives a Minitab computer printout for the confidence interval for the mean percentage iron. Identify the pertinent sample descriptive information and locate the confidence interval.

c. Interpret the confidence interval.

Solution

a. For this problem, a confidence interval based on either the z or the t distribution would be appropriate because the sample size, $n = 390$, is so large. The two methods give almost the same numerical values for the confidence limits when the sample size n is 30 or more. For that reason, **most computer statistical software packages give confidence intervals for population means that are based on the Student's t statistic.**

b. The Minitab computer printout for our data, Figure 6.12, shows (1) sample size n, (2) the sample mean $\bar{x}$, (3) the standard deviation s, (4) the estimated standard error of the mean, and (5) the 95% confidence interval for μ. These quantities are shaded on the printout. Reading from the printout, we see that the lower and upper confidence limits for the mean percentage of iron in a 1.5-kilogram specimen are

$$\text{LCL} = 65.6740 \quad \text{and} \quad \text{UCL} = 65.8121$$

c. Based on these values of LCL and UCL, we estimate that the interval from 65.6740 to 65.8121 contains μ. Or, expressing it another way, we estimate that the mean percentage iron in the 20,000-ton consignment is between 65.67% and 65.81%. How much confidence do we have that this statement is true? Using our procedure, we obtain an interval whose probability of enclosing μ is .95.

Example 6.9

Some software packages may not print out the confidence interval for a population mean. Suppose that your computer printout only gave the values of n, $\bar{x}$, s, and possibly the estimated standard error of the mean. Use the values for these statistics, given in Figure 6.12, to calculate the lower and upper confidence limits for the mean percentage iron and compare your answer with the values of LCL and UCL given on the printout in Figure 6.12.

Solution

From the printout, $n = 390$, $\bar{x} = 65.7430$, and the estimated standard error of the sample mean is $\widehat{\text{SE}}(\bar{x}) = s/\sqrt{n} = .0351$. From Table 6.1, $z_{.025} = t_{.025} = 1.96$. Substituting into the formulas for LCL and UCL, we obtain

$$\text{LCL} = \bar{x} - t_{.025}\widehat{\text{SE}}(\bar{x}) = 65.7430 - 1.96(.0351) = 65.67$$

$$\text{UCL} = \bar{x} + t_{.025}\widehat{\text{SE}}(\bar{x}) = 65.7430 + 1.96(.0351) = 65.81$$

You can see that these confidence limits are the same as those given on the printout, Figure 6.12.

An Example of a Small-Sample Confidence Interval

Example 6.10

Table 6.4 (page 196) gives the carapace lengths in millimeters (mm) of ten *T. orientalis* lobsters that were caught in seas near Singapore. Figure 6.13 shows the Data Desk computer printouts for the numerical descriptive measures and a confidence interval for the mean μ of the sampled population of carapace lengths.

a. Find the sample size n, the sample mean $\bar{x}$, and the sample standard deviation s on the printout.

b. Does the computed value of s appear to agree with the data? Explain.

c. Would it be appropriate to construct a confidence interval for μ based on the Student's t statistic?

d. Find a 95% confidence interval for the mean carapace length of *T. orientalis* lobsters from the seas near Singapore.

e. What does this confidence interval mean?

Solution

a. The sample size $n = 10$, sample mean $\bar{x} = 60.800$, and the sample standard deviation $s = 7.9694$ are shaded on the printout in Figure 6.13a.

b. We learned in Section 4.4 that the range of a moderate sized sample should be no larger than $6s$ and that it could be as small as $3s$ for a sample as small as $n = 10$ (see Table 4.6). Examining the data, we see that the smallest observation is 50 and the largest is 78. Therefore, the range approximation to s is

$$s \approx \frac{\text{Range}}{3} = \frac{78 - 50}{3} = \frac{28}{3} = 9.3$$

TABLE 6.4 Carapace Lengths of $n = 10$ *T. Orientalis* Lobsters

Lobster	1	2	3	4	5	6	7	8	9	10
Carapace Length (mm)	78	66	65	63	60	60	58	56	52	50

Source: Jeffries, Voris, and Yang, "Diversity and Distribution of the Pedunculate Barnacle *Octolasmis* Gray, 1825 Epizoic on the Scyllarid Lobster, *Thenus* orientalis [Lund, 1973]," *Crustaceana* 46, No. 3 (1984).

FIGURE 6.13 A Data Desk printout of a confidence interval for the mean carapace length of *T. Orientalis* lobsters

Summary statistics for **Lobsters**
NumNumeric = 10
Mean = 60.800
Median = 60
Standard Deviation = 7.9694
Interquartile range = 10.250
Range = 28
Variance = 63.511
Minimum = 50
Maximum = 78
25-th %ile = 55
75-th %ile = 65.250

a. Numerical Descriptive Measures

t-intervals
With 95% confidence, $55.099 \le \mu(\text{Lobsters}) \le 66.501$

b. Confidence Interval

This very rough approximation is of the same order as the value $s = 7.9694$ shown on the printout. Therefore, we have no reason to suspect an error in the computed value of s.

c. The confidence interval based on the Student's t statistic requires that the data represent a random sample from a normally distributed population of carapace lengths. The Data Desk sample frequency distribution, shown in Figure 6.14, suggests a mound-shaped population relative frequency with no excessive skewness. If we are willing to accept the assumption that the data represent a random sample from the universe of lobsters in the seas near Singapore, then a confidence interval based on the Student's t statistic would be appropriate.

FIGURE 6.14 The relative frequency distribution of the sample of $n = 10$ carapace lengths

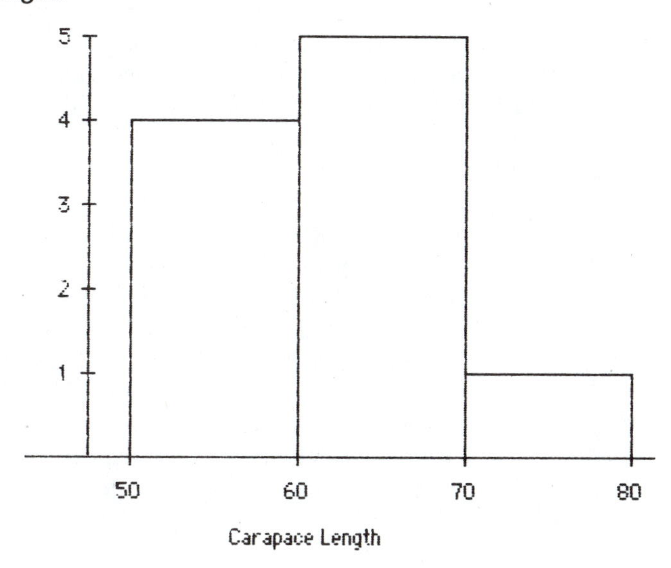

d. The confidence interval, shown on the printout in Figure 6.13b, is 55.099 to 66.501.

e. What does this confidence interval mean? It means that we estimate the mean carapace length of the lobsters to be between 55.099 mm and 66.501 mm. We are 95% confident that this interval encloses the mean carapace length. By this we mean that if we were to draw random samples over and over again, 95% of the intervals would enclose μ.

A Question of Confidence!

Why don't we use 100% confidence intervals and be absolutely certain that our confidence interval encloses μ? Table 6.3, which gives the z-values for various confidence coefficients, shows that we cannot get something for nothing. It shows that **the z-values used to construct a confidence interval increase as the confidence coefficient increases** (the same is true for t-values). Therefore, for a given sample size, the width of a confidence interval increases as the confidence coefficient increases. For example, if we would prefer a 99% to a 95% confidence interval, we pay for the increase in confidence by accepting a wider interval. If we want to be 100% certain that our confidence interval encloses μ, the interval will have to be infinitely wide!

Interpreting Statistical Statements in the News Media and in Scientific Journals

It is important for you to be able not only to perform minor statistical calculations and to read and interpret statistical printouts. You also should be able to interpret statistical results that appear in the newspapers and scientific journals. The following example gives some results from a recent landmark medical study.

Example 6.11

A recently published research paper describes the results of the Framingham Heart Study, a study of the changes in the risk factors and mortality from cardiovascular disease over a 30-year period (Sytkowski, P. A., Kannel, W. B., and D'Agostino, R. B., "Changes in Risk Factors and the Decline in Mortality from Cardiovascular Disease," *The New England Journal of Medicine*, Vol. 322, No. 23, 1990). A large sample of men, aged 50 to 59, was placed under observation beginning on January 1 for each of the years, 1950, 1960, and 1970, and observed over a 10-year period. The men in each sample fell into two groups—those with and those without cardiovascular disease (CVD) at the start (called the baseline) of the observation period. The numbers of men in these two categories for each of the three samples are shown in Table 6.5.

TABLE 6.5 Sample Sizes for the Framingham Study

	Free of CVD at Baseline			CVD Within 10 Years of Baseline		
Baseline	1950	1960	1970	1950	1960	1970
Sample size	485	464	512	64	77	101

Table 6.6 lists seven risk factors with the statistics for each group and time period but shows only the portion of the statistical results that pertain to measurement of the risk factors at the baseline for each time period. We will concern ourselves only with the serum cholesterol statistics. The interval given for each group is the sample mean $\bar{x}$ plus or minus the standard deviation s (written in the paper as "mean $\pm$ SD"). How can we interpret this information?

Solution

The interval "mean $\pm$ SD" (or $\bar{x} \pm s$) has no practical significance as an estimate. The authors are simply giving us the values of $\bar{x}$ and s to do with as we please. Therefore, if we want a measure of confidence associated with an estimate for a particular mean, we need to calculate a confidence interval. To illustrate, we will

TABLE 6.6 Table of Selected Risk Factors from the Paper by Sytkowski et al.

Risk Factor	Free of CVD at Baseline			Incident CVD Within 10 Years of Baseline		
	1950	1960	1970	1950	1960	1970
Serum cholesterol (mg/dl)	228 ± 40	243 ± 37	221 ± 38	239 ± 44	246 ± 35	227 ± 40
Smokers (%)	56	52	34	64	60	57
Definite hypertension (%)	21	23	15	36	41	20
Use of antihypertensive medication (%)	0	11	22	0	11	15
Systolic blood pressure (mm Hg)	139 ± 25	137 ± 21	135 ± 19	152 ± 30	148 ± 22	140 ± 19
Diastolic blood pressure (mm Hg)	85 ± 13	86 ± 12	84 ± 10	91 ± 15	91 ± 12	85 ± 11
Metropolitan relative weight (%)	120 ± 15	120 ± 15	123 ± 17	121 ± 17	123 ± 15	121 ± 18

calculate a 95% confidence interval for the mean cholesterol level for the population of males who were free of cardiovascular disease in 1950.

The sample size, mean, and standard deviation (from Tables 6.5 and 6.6) for the group are $n = 485$, $\bar{x} = 228$, $s = 40$, and, for n as large as 485, $t_{.025}$ is approximately equal to $z_{.025} = 1.96$. Substituting these values into the formulas for $\widehat{\text{SE}}(\bar{x})$ and the lower and upper 95% confidence limits, we obtain

$$\widehat{\text{SE}}(\bar{x}) = \frac{s}{\sqrt{n}} = \frac{40}{\sqrt{485}} = 1.82$$

$$\text{LCL} = \bar{x} - t_{.025}\widehat{\text{SE}}(\bar{x}) = 228 - 1.96(1.82) = 228 - 3.57 = 224.43$$

$$\text{UCL} = \bar{x} + t_{.025}\widehat{\text{SE}}(\bar{x}) = 228 + 1.96(1.82) = 228 + 3.57 = 231.57$$

Therefore the 95% confidence interval for the mean cholesterol level for all men aged 50–59 in 1950 who did not have CVD is 224.43 to 231.57. Intervals constructed using our methodology will enclose the mean μ 95% of the time.

Comments on the Assumptions

It will be a rare situation when we know the standard deviation of a sampled population or know that a population is normally distributed. Therefore, in almost all sampling situations, we will be substituting s for σ in the formula for $\text{SE}(\bar{x})$. Should we use the confidence interval and test of hypothesis based on the z statistic or should we use a t?

Which Method Should We Use?

This question will be academic when the sample size n is large (say, $n \geqslant 30$) because the sampling distribution for a Student's t is almost identical to a standard normal z distribution. Consequently, we can use the methods based on either the z or the t statistic when n is large.*

When the sample size is small, the t distribution is more variable than the z distribution (see Figure 6.7). Then we use the confidence interval and test based on the Student's t statistic. In doing so, we make the assumption that the sampled population is normally distributed.

Will the Methods Still Work When the Assumptions Have Not Been Satisfied?

Will the confidence intervals and tests of hypotheses presented in this chapter still "work" if the assumptions are not satisfied? For example, will a 95% confidence interval still enclose μ 95% of the time? We need not worry too much about the assumption of normality when we use the methods based on Student's t. Experimentation has shown that they are insensitive (statisticians call them **robust**) to moderate departures from the assumption of normality. By that, we mean that the confidence coefficient and value of α employed in a test will be approximately as specified when the population distribution is roughly mound shaped. It is good practice to construct a histogram of the sample data to make sure there are no outliers or extreme skewness.

6.4 Another Way to Report an Estimate

Because the typical newspaper reader gives little thought to estimation error and has no concept of the meaning of a confidence interval, most estimates of a population mean (and other parameters) that appear in newspapers are point estimates. Many articles do not report the reliability of an estimate although some of the

* The number 30 is not magical. The z and the t distributions are very similar for $n = 29$ and $n = 28$. The smaller the sample size, the greater the difference between the two distributions. Many tables give the tail values of t only for df = 1 to df = 30 because the tail values are almost identical to the corresponding z-values for df = 30 and beyond.

larger polling companies give, along with an estimate, a bound (a limit) on the sampling error, i.e., the error of estimation. Typically, a report will give an estimate and say that it is correct to within plus or minus some number, say B. What this means is that the sampling error will be less than B with a high (usually .95) probability.

Figure 6.15 shows an excerpt from an article in *The New York Times* (February 1, 1992) that describes a public opinion poll following President Bush's 1992 State of the Union message. Referring to changes in percentage points from a January 15 poll, the article states that, "Both changes were within the poll's four percentage point margin of error." No probability is attached to the statement.

FIGURE 6.15 A newspaper description of a public opinion poll

A new poll by ABC News and *The Washington Post* suggested that Mr. Bush's State of the Union Message had done little to alter his standing with voters. The poll, conducted on Wednesday, showed that 49 percent of the 769 people surveyed disapproved of Mr. Bush's handling of the Presidency, down three percentage points from a Jan. 15 poll, and that 46 percent approved, up one point. Both changes were within the poll's four percentage point margin of error. Sixty-nine percent said they believed that the President did not understand the economic problems facing Americans.

Commencing in 1990, a few polls (e.g., *The Wall Street Journal*–NBC News Poll and *The New York Times*/CBS News Poll) began to attach a probability of .95 to the bound on the sampling error. Typical is a description of "How the Survey Was Taken" contained in a *New York Times* report (November 26, 1991) for a *New York Times*/CBS News Poll. (This description was shown in Figure 5.2.) Referring to the results of the survey, it states, "In theory, in 19 cases out of 20, the results based on such samples will differ by no more than three percentage points in either direction from what would have been obtained by seeking out [i.e.,sampling] all American adults."

The two ways of reporting an estimate of a population mean, point and interval, are related. If the half-width of a $100(1 - a)\%$ confidence interval for μ is equal to a number B, then the probability that a point estimate will deviate from μ by no more than B is $(1 - a)$, and vice versa. You can see in Figure 6.10 (Section 6.3) that if the half-width of a 95% confidence interval is equal to B, i.e.,

$$B = 1.96\text{SE}(\bar{x})$$

then, whenever the confidence interval encloses μ, $\bar{x}$ will deviate from μ by no more than B. This will occur 95% of the time. Therefore, 95% of the time, the error in estimating μ will be less than B.

EXERCISES

6.12 When is it appropriate to use the confidence interval for μ that is based on the standard normal z statistic?

6.13 When is it appropriate to use the confidence interval for μ that is based on the Student's t statistic?

6.14 A 90% confidence interval for a population mean μ was calculated to be 23.3 to 28.5. What do these numbers mean?

6.15 What is a lower confidence limit? An upper confidence limit?

6.16 What is a confidence coefficient?

6.17 Which is larger, z_a or t_a based on 7 degrees of freedom?

6.18 Suppose that you have a sample and wish to construct a confidence interval for μ. Which will be wider, a 90% or a 95% confidence interval?

6.19 What happens to the width of a confidence interval as the confidence coefficient is increased?

HLTH **6.20** Refer to the Framingham Heart Study statistics given in Table 6.6 (Example 6.11). Calculate a 90% confidence interval for the mean serum cholesterol level for persons aged 50–59 who were free of cardiovascular disease at baseline 1960.

HLTH **6.21** Refer to the Framingham Heart Study statistics given in Table 6.6 (Example 6.11). Calculate a 95% confidence interval for the mean serum cholesterol level for persons aged 50–59 who were free of cardiovascular disease at baseline 1970.

HLTH **6.22** Table 6.7 reproduces the white blood cell count and the lymphocyte count for $n = 50$ West Indian or African workers, originally presented in Chapter 4. Figure 6.16 gives the Minitab printouts of the confidence intervals for mean white blood cell count (WBC) and mean lymphocyte count (LYMPHO).

a. Locate the 95% confidence interval for the mean white blood cell count on the printout. Interpret it.
b. Locate the 95% confidence interval for the mean lymphocyte count on the printout. Interpret it.

HLTH **6.23** Refer to Exercise 6.22. We found the sample mean and standard deviation of the white blood cell counts in Exercise 4.23 to be $\bar{x} = 5{,}334$ and $s = 1{,}387.8$, respectively. Use these values to calculate a 95% confidence interval for the mean white blood cell count. Compare your answer with the confidence interval shown on the printout in Figure 6.16.

HLTH **6.24** Refer to Exercise 6.22. We found the sample mean and standard deviation of the lymphocyte counts in Exercise 4.24 to be $\bar{x} = 22.68$ and $s = 9.54$, respectively. Use these values to calculate a 95% confidence interval for the mean lymphocyte count. Compare your answer with the confidence interval shown on the printout in Figure 6.16.

BIO **6.25** Figure 6.17 (page 204) gives the Minitab printouts of the 90% confidence interval for the mean DDT content for each of the fish species given in Table 2 of Appendix 1.

a. Find the 90% confidence interval for the mean DDT content of channel catfish on the printout. Interpret this estimate.
b. Find the 90% confidence interval for the mean DDT content of small-mouth buffalo. Interpret this estimate.
c. Find the 90% confidence interval for the mean DDT content of large-mouth bass. Interpret this estimate.

BIO **6.26** Find the mean and standard deviation for the DDT content in channel catfish shown on the printout in Figure 6.17. Use these to calculate the confidence interval required in Exercise 6.25(a). Compare with the confidence interval shown on the printout in Exercise 6.25.

BIO **6.27** Find the mean and standard deviation for the DDT content in small-mouth buffalo shown on the printout in Figure 6.17. Use these to calculate the confidence interval required in Exercise 6.25(b). Compare with the confidence interval shown on the printout in Exercise 6.25.

BIO **6.28** Find the mean and standard deviation for the DDT content in large-mouth bass shown on the printout in Figure 6.17. Use these to calculate the confidence interval required in Exercise 6.25(c). Compare with the confidence interval shown on the printout in Exercise 6.25.

6.29 If the width of a 95% confidence interval is equal to 3.6, what is the probability that $\bar{x}$ will lie within 1.8 of μ?

TABLE 6.7 White Blood Cell and Lymphocyte Count Data, Exercise 6.22

Case Number	WBC	LYMPHO	Case Number	WBC	LYMPHO
1	4,100	14	26	4,300	9
2	5,000	15	27	5,200	16
3	4,500	19	28	3,900	18
4	4,600	23	29	6,000	17
5	5,100	17	30	4,700	23
6	4,900	20	31	7,900	43
7	4,300	21	32	3,400	17
8	4,400	16	33	6,000	23
9	4,100	27	34	7,700	31
10	8,400	34	35	3,700	11
11	5,600	26	36	5,200	25
12	5,100	28	37	6,000	30
13	4,700	24	38	8,100	32
14	5,600	26	39	4,900	17
15	4,000	23	40	6,000	22
16	3,400	9	41	4,600	20
17	5,400	18	42	5,500	20
18	6,900	28	43	6,200	20
19	4,600	17	44	4,900	26
20	4,200	14	45	7,200	40
21	5,200	8	46	5,800	22
22	4,700	25	47	8,400	61
23	8,600	37	48	3,100	12
24	5,500	20	49	4,000	20
25	4,200	15	50	6,900	35

Source: Royston, J. P. "Some Techniques for Assessing Multivariate Normality Based on the Shapiro–Wilk *W*." *Applied Statistics*, Vol. 32, No. 2, 1983, pp. 121–133.

FIGURE 6.16 Minitab confidence limits for mean WBC and mean LYMPHO count, Exercise 6.22

```
MTB > tinterval c2

              N      MEAN    STDEV  SE MEAN    95.0 PERCENT C.I.
wbc          50   5334.00  1387.79   196.26   ( 4939.50, 5728.50)

MTB > tinterval c3

              N      MEAN    STDEV  SE MEAN    95.0 PERCENT C.I.
lympho       50     22.68     9.54     1.35   (    19.97,    25.39)
```

FIGURE 6.17 Minitab printouts of confidence intervals for mean DDT content, Exercise 6.25

```
MTB > tinterval 90=k c1

              N        MEAN     STDEV   SE MEAN    90.0 PERCENT C.I.
ddt          96        33.3     119.5      12.2   (     13.0,     53.6)

MTB > describe c1

                  N        MEAN    MEDIAN    TRMEAN     STDEV    SEMEAN
ddt              96        33.3       9.5      14.6     119.5      12.2

                MIN         MAX        Q1        Q3
ddt             0.7      1100.0       5.3      17.0
```

a. Channel catfish

```
MTB > tinterval 90=k c2

              N        MEAN     STDEV   SE MEAN    90.0 PERCENT C.I.
ddt          36        8.16     11.28      1.88   (     4.98,     11.34)

MTB > describe c2

                  N        MEAN    MEDIAN    TRMEAN     STDEV    SEMEAN
ddt              36        8.16      4.65      6.29     11.28      1.88

                MIN         MAX        Q1        Q3
ddt            0.25       48.00      2.35      9.50
```

b. Small-mouth buffalo

```
MTB > tinterval 90=k c3

              N        MEAN     STDEV   SE MEAN    90.0 PERCENT C.I.
ddt          12       1.380     2.043     0.590   (    0.321,     2.439)

MTB > describe c3

                  N        MEAN    MEDIAN    TRMEAN     STDEV    SEMEAN
ddt              12       1.380     0.530     0.905     2.043     0.590

                MIN         MAX        Q1        Q3
              0.110       7.400     0.250     1.975
```

c. Large-mouth bass

EXERCISE FOR YOUR COMPUTER

6.30 As an illustration of estimation, let us regard the 390 percent iron measurements in Table 1 of Appendix 1 as a population.

a. Use the table of random numbers, Table 1 of Appendix 2, to draw a random sample of $n = 5$ measurements from the 390 in our fictitious population.

b. Use a computer to calculate the sample mean and standard deviation.

c. Use a computer to find a 95% confidence interval for μ, the mean percent iron in the set of 390 measurements. (Assume σ is unknown.)

d. What is the probability that your confidence interval will enclose μ?

e. In the usual sampling situation, we do not know μ. For the purpose of this exercise, we chose a population with a known mean because we want you to check whether your interval contains μ. The mean of the 390 measurements in our fictitious population (shown in the Minitab printout, Figure 6.12) is 65.743. Does your confidence interval enclose μ?

f. If possible, have each member of the class work this exercise. What proportion of all intervals calculated by the class enclose μ? Is this proportion reasonably close to your answer to part (d)?

6.5 Why Test Hypotheses About a Population Mean?

Now that we know how to estimate the value of μ, why would we want to test hypotheses about its value? For example, if we estimate that the mean percentage iron in the ore specimens, Example 6.9, is 65.74%, why would we want to know whether the data provide sufficient evidence to indicate that μ is larger than or smaller than some specific value? **The answer depends on the practical question that is posed.**

For example, suppose that it is profitable to buy the 20,000-ton consignment of ore only if the mean percentage of iron in a 1.5-kilogram specimen exceeds 65%. Do the data present sufficient evidence to indicate that the mean percentage μ exceeds 65%? If we say yes, based on the fact that our estimate of μ, $\bar{x} = 65.74\%$, exceeds 65%, we will buy the consignment of ore and begin to worry whether we have made the wrong decision. **In particular, we would like to know the probability that our decision-making procedure would lead us to an incorrect decision. Decision making based on estimation does not provide an answer to this important question. A test of an hypothesis does.**

To illustrate, suppose that we make our decision to buy the shipment based on a test of an hypothesis. The thing we wish to show (since we want to buy the consignment) is that $\mu > 65\%$. Therefore, we will test the null hypothesis, H_0: $\mu = 65\%$, against the alternative hypothesis, H_a: $\mu > 65\%$. If we reject H_0: $\mu = 65\%$ when the p-value for the test is less than α (say, $\alpha = .05$) and conclude that H_a: $\mu > 65\%$ is true (i.e., we decide to buy the consignment), we will make the wrong decision only

5% of the time when H_0 is true. In other words, you will buy consignments of ore that contain less than the specified amount of iron only $\alpha(100)\% = 5\%$ of the time. If we decide to reject H_0 only when the p-value of the test is less than $\alpha = .01$, we will make the wrong decision only $\alpha(100)\% = 1\%$ of the time when H_0 is true. Thus, **a test of an hypothesis allows us to make a decision about μ with a known measure α of risk.**

6.6 Tests of Hypotheses About a Population Mean

Evidence to support the alternative hypothesis that μ is larger than some value, say μ_0, is measured by the difference, $\bar{x} - \mu_0$, in units of its standard error, $\text{SE}(\bar{x}) = \sigma/\sqrt{n}$. **Therefore, we test hypotheses about μ using either the standard normal statistic z or the Student's t statistic.**

When σ is known or when the sample size n is large ($n \geqslant 30$), we use

$$z = \frac{\bar{x} - \mu_0}{\text{SE}(\bar{x})} \quad \text{as the test statistic.}$$

When σ is unknown, we use

$$t = \frac{\bar{x} - \mu_0}{\widehat{\text{SE}}(\bar{x})} \quad \text{as the test statistic.}$$

One- and Two-Tailed Tests for μ: What Are They?

A test designed to detect a value of μ *larger than* some value μ_0 rejects H_0: $\mu = \mu_0$ when the test statistic falls in the *upper tail* of its sampling distribution. Because the rejection region falls in only one tail of the sampling distribution of the test statistic, it is called a *one-tailed statistical test*. For example, suppose that we wanted to detect values of μ greater than $\mu = 65\%$. Then we would reject the null hypothesis, H_0: $\mu = 65$, and accept the alternative hypothesis, H_a: $\mu > 65$, when z (or t) is large and positive. The rejection region for the test would fall in the upper tail of the sampling distribution of z (or t). See Figure 6.18.

A test designed to detect a value of μ that is *less than* a value μ_0 rejects H_0: $\mu = \mu_0$ when the test statistic falls in the *lower tail* of its sampling distribution. This test is also a *one-tailed statistical test*. For example, suppose that we wanted to detect values of μ less than $\mu_0 = 65\%$. Then we would reject the null hypothesis, H_0: $\mu = 65$, and accept the alternative hypothesis, H_a: $\mu < 65$, when z (or t) is large and negative. The rejection region for the test would fall in the lower tail of the sampling distribution of z (or t). See Figure 6.19.

FIGURE 6.18 Rejection region for an upper one-tailed statistical test

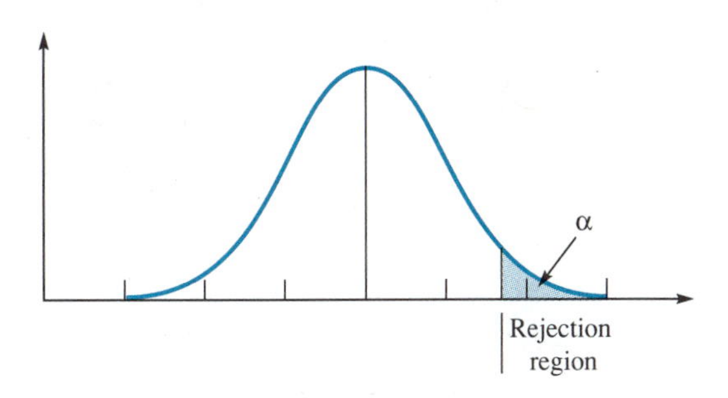

FIGURE 6.19 Rejection region for a lower one-tailed statistical test

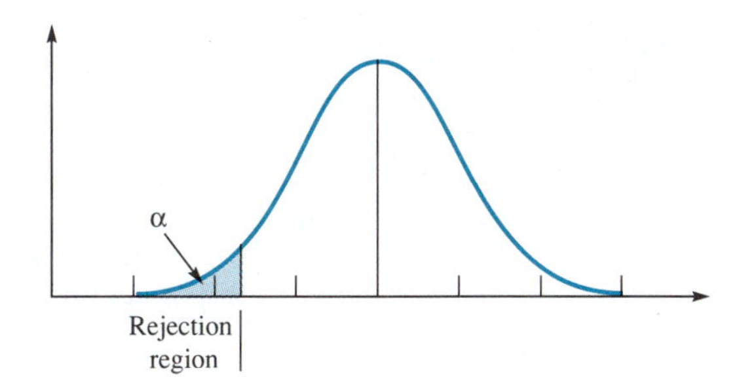

A test designed to detect a value of μ that is *either smaller than or larger than* a value μ_0 rejects $H_0: \mu = \mu_0$ when the test statistic falls in *either the lower or the upper tail* of its sampling distribution. Usually, α is divided with half in each tail. This test is called a *two-tailed statistical test*. For example, suppose that we wanted to detect values of μ either less than or greater than $\mu_0 = 65\%$. Then we would reject the null hypothesis, H_0: $\mu = 65$, and accept the alternative hypothesis, H_a: $\mu \neq 65$, when z (or t) is either large and positive or large and negative. Half of the rejection region for the test would fall in the lower tail and half would fall in the upper tail of the sampling distribution of z (or t). See Figure 6.20, page 208.

Large- and Small-Sample Tests for a Population Mean μ

The z and the t tests are summarized in the accompanying boxes. The examples that follow illustrate their use.

FIGURE 6.20 Rejection region for a two-tailed statistical test

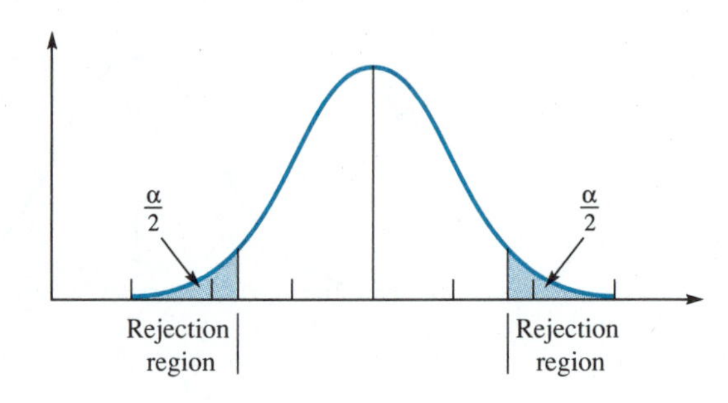

A Standard Normal z Test for a Population Mean

Null hypothesis: $H_0: \mu = \mu_0$

Alternative hypothesis:

1. $H_a: \mu > \mu_0$, an upper-tailed test
2. $H_a: \mu < \mu_0$, a lower-tailed test
3. $H_a: \mu \neq \mu_0$, a two-tailed test

$$\textit{Test statistic:} \quad z = \frac{\bar{x} - \mu_0}{\mathrm{SE}(\bar{x})} \quad \text{where} \quad \mathrm{SE}(\bar{x}) = \frac{\sigma}{\sqrt{n}}$$

Rejection region:

Choose the value of α that is acceptable to you. Then reject H_0 and accept H_a when the p-value for the test is less than or equal to α.

α is in the upper tail of the z distribution when the alternative hypothesis is $H_a: \mu > \mu_0$.

α is in the lower tail of the z distribution when the alternative hypothesis is $H_a: \mu < \mu_0$.

α is divided between the lower and upper tails of the z distribution when the alternative hypothesis is $H_a: \mu \neq \mu_0$.

Assumptions:

1. **The sample is a random sample.**
2. **The sample mean $\bar{x}$ is approximately normally distributed (i.e., n is large enough so that the Central Limit Theorem applies).**
3. **The population standard deviation σ is known or n is large ($n \geq 30$) so that s provides a good estimate of σ.**

A Student's t Test for a Population Mean

The test is identical to the standard normal z test summarized in the previous box, except that the test statistic is

$$t = \frac{\bar{x} - \mu_0}{\widehat{\mathrm{SE}}(\bar{x})} \quad \text{where} \quad \widehat{\mathrm{SE}}(\bar{x}) = \frac{s}{\sqrt{n}}$$

Assumptions:

1. **The sample is a random sample.**
2. **The population is normally distributed with unknown mean and standard deviation. See comments on the assumptions, Section 6.3.**

Example 6.12

In Section 6.5, we explained why a test of an hypothesis is better than estimation when trying to decide whether to buy the 20,000-ton shipment of iron ore. Using a test, we know the probability α of making an error, i.e., buying the shipment when it contains only 65% (or less) iron. Do the data on percentage iron in the 390 ore specimens, Table 1 of Appendix 1, provide sufficient information to conclude that the mean percentage iron, for all 1.5-kilogram specimens in the entire 20,000-ton consignment, exceeds 65%?

a. State the null and alternative hypotheses for the test.
b. Figure 6.21 shows the Minitab printout for the test. Identify the essential information on the printout.
c. Locate the rejection region for the test.
d. Interpret the p-value for the test.
e. State your conclusions.

FIGURE 6.21 A Minitab printout of a test of an hypothesis for the mean percentage of iron in 1.5-kg specimens of ore

```
MTB > ttest mu=65 alternative=+1 c1

TEST OF MU = 65.0000 VS MU G.T. 65.0000

              N      MEAN     STDEV   SE MEAN        T    P VALUE
%iron       390   65.7430    0.6937    0.0351    21.15     0.0000
```

Solution

a. Because we only want to detect values of μ larger than 65%, we want to conduct an **upper one-tailed test** of the null hypothesis H_0: $\mu = 65$ against the alternative hypothesis H_a: $\mu > 65$.

b. Because the sample size is so large, $n = 390$, the mechanics of the z and the t tests are identical. Therefore, we can use the test results produced by most of the standard computer statistical packages. Examine the Minitab printout in Figure 6.21. The null hypothesis, written as TEST OF MU = 65.0000, the alternative hypothesis, written as MU G.T. (greater than) 65.0000, the computed t-value, $t = 21.15$, and the p-value = 0.0000 are shaded on the printout.

c. Large values of t tend to disagree with the null hypothesis that $\mu = 65$ and support the alternative hypothesis that μ is larger than 65%. Therefore, the rejection region for an upper one-tailed test falls in the upper tail of the t distribution (see Figure 6.22).

d. **The p-value for a test measures how unlikely the value of the observed test statistic is, assuming the null hypothesis, H_0: $\mu = \mu_0$, is true.** It tells us whether the observed value of the test statistic represents a rare (improbable) event. The p-value for our test is the probability of observing a value of the test statistic at least as contradictory to the null hypothesis (i.e., that $\mu = 65\%$ or less) as the observed value, $t = 21.15$. It is the area under the t distribution, Figure 6.22, to the right of 21.15. **The smaller the p-value, the greater is the weight of evidence favoring rejection of the null hypothesis and acceptance of the alternative hypothesis.**

e. One look at the computer output, Figure 6.21, and we can see that there is strong evidence to indicate that H_0: $\mu = 65$ is false and that $\mu > 65\%$. The sample mean lies more than 21 $\widehat{\text{SE}}(\bar{x})$ above $\mu = 65$. The probability of this happening, assuming H_0 is true, is (the p-value) only .0000.

FIGURE 6.22 The rejection region for the upper one-tailed test, Example 6.12

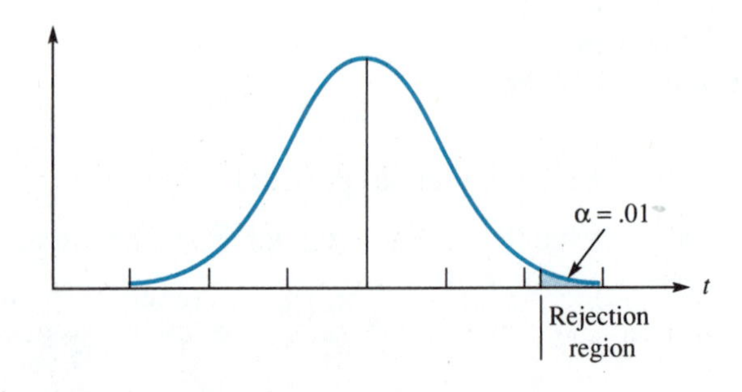

Suppose that we decide to reject H_0: $\mu = 65\%$ and accept H_a: $\mu > 65\%$ if the observed p-value is less than $\alpha = .01$. Since the p-value, 0.0000, shown on the printout is less than $\alpha = .01$, we will reject the null hypothesis and conclude that μ exceeds 65% (and, based on this decision, will buy the consignment of ore). What is our risk of making a wrong decision? Using our decision-making procedure, for $\alpha = .01$, we will wrongly reject the null hypothesis only $\alpha(100)\% = (.01)(100)\% = 1\%$ of the time.

The test in Example 6.12 was an upper one-tailed test. We rejected the null hypothesis only for large values of the test statistic (see Figure 6.22). If we had wanted to detect values of μ smaller than $\mu_0 = 65$, we would have conducted a lower one-tailed test and rejected H_0 only for large negative values of t. If we had wanted to detect values of μ either larger than or smaller than $\mu_0 = 65$, we would have conducted a two-tailed test and rejected H_0 for either large positive or large negative values of t.

Example 6.13

Many scientific measuring instruments tend, over time, to develop a bias. Although the random error in each measurement fluctuates, the mean value of the error tends to become positive or negative, thus tending to give, on the average, readings that are above or below the true reading for the variable being measured. To correct this problem, the instruments are calibrated periodically. This is done by repeatedly measuring a standard for which the exact value is known, estimating the bias, and then adjusting the instrument to correct for it.

Suppose that you want to determine whether your bathroom scale gives biased readings. To establish a standard, you weigh yourself at your doctor's office (the standard) and find that you weigh 150 pounds. You then weigh yourself ten times on your bathroom scale, separating the weighings by 5 or 10 minutes, and record the weight readings shown in Table 6.8. Do the data provide sufficient evidence to indicate that your scale is biased? Use $\alpha = .10$.

TABLE 6.8 Bathroom Scale Weight Readings in Pounds

149	152	153	150	151	151	152	153	150	149

Solution

Does the scale give biased readings—readings that tend to be, on the average, higher than or lower than 150? To answer this question we will want to detect values of the mean scale reading, μ, that are either above 150 or below 150. Thus, we want to conduct a **two-tailed test** of the null hypothesis, H_0: $\mu = 150$, against the alternative hypothesis, H_a: $\mu \neq 150$, i.e., that μ is either greater than 150 or less than 150.

The rejection region for the test locates α, half in the lower tail and half in the upper tail of the t distribution.

The printout for the test, using a Data Desk software package, is shown in Figure 6.23. Note that Data Desk prints the null and alternative hypotheses, gives the calculated value of $\bar{x}$, $\bar{x} = 151$, and the observed value of t, $t = 2.121$ (with 9 degrees of freedom). The p-value for the test, shaded on the printout, is written as Prob $< = 0.0629$. This tells us that the probability of observing a value of the test statistic t that is at least as contradictory to H_0 as the value $t = 2.121$ is 0.0629. Therefore, we would reject H_0 and conclude that the scales are biased for $\alpha = .10$. In fact, it appears that the scale tends to read slightly higher than 150.

FIGURE 6.23 A Data Desk printout of a test for bias of bath scales

t-Tests
Bath Scale: Test H_0: $\mu = 150$ vs H_a: $\mu \neq 150$
Sample mean = 151 t-statistic = 2.121 with 9 d.f.
Fail to reject H_0 at alpha = 0.05 Prob < = 0.0629

Remember that the choice of α is up to you. There is nothing magical about $\alpha = .10$ or $\alpha = .05$. The value that you choose depends on the risk that you are willing to take of rejecting H_0 and concluding that the scales are biased when, in fact, they are not. Since the p-value for the test is quite small and we have little to lose (the sky won't fall down on us) if we incorrectly conclude that the scales are biased, we chose a relatively large value for α, $\alpha = .10$.

Example 6.14

Refer to Example 6.13. Suppose that we had used a computer printout that failed to give the p-value for the test. Use the t-value shown on the Data Desk printout and Table 4 of Appendix 2 to find the approximate p-value for the test. Compare your approximation with the exact value shown on the Data Desk printout in Figure 6.23.

Solution

The observed value of t, obtained from Figure 6.23, is 2.121 with $(n - 1) = (10 - 1) = 9$ degrees of freedom. The p-value for the test is the probability of observing a t value as large as or larger than 2.121 or as small as or smaller than -2.121, the shaded tail areas under the t distribution in Figure 6.24. Now examine Table 4 of Appendix 2. The observed value, $t = 2.121$, falls between $t_{.025} = 2.262$ and $t_{.05} = 1.833$, values of t that locate $a = .025$ and $a = .05$, respectively, in the upper tail of the t distribution. Therefore, half of the p-value, the area in the upper

FIGURE 6.24 ***p*-Value for the test, Example 6.14**

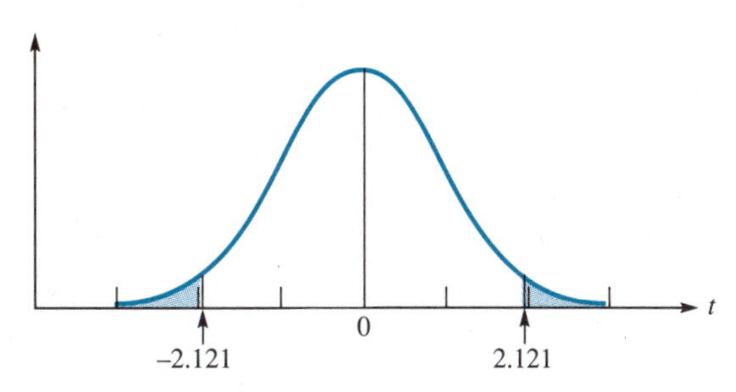

tail of the t-distribution, is between .025 and .05. The total p-value, which includes the area in both tails of the t distribution (Figure 6.24) would fall between .05 and .10. This agrees with the p-value, 0.0629, shown on the printout in Figure 6.23.

EXERCISES

6.31 What are the assumptions required to conduct a standard normal z test for μ? To what extent can these assumptions be relaxed?

6.32 What are the assumptions required to conduct a Student's t test for μ? To what extent can the assumptions be relaxed?

6.33 What is an upper one-sided test of an hypothesis for μ?

6.34 What is a lower one-sided test of an hypothesis for μ?

6.35 What is a two-tailed test of an hypothesis for μ?

MFG **6.36** A manufacturer of a small motor has found that the mean time to assemble a motor is 8.3 minutes. The manufacturer has developed a new assembly procedure that is intended to reduce the mean assembly time. Twenty motors are assembled and the time to assemble is measured for each motor.

a. Which test procedure would you use to determine whether the mean assembly time has been reduced? Why?

b. State the null and alternative hypotheses for the test.

c. Where would the rejection region for the test be located?

d. Is this a one- or a two-tailed test? Why?

6.37 What is the p-value for a statistical test?

6.38 How do you use the p-value for a test to decide whether to reject the null hypothesis?

6.39 A research article gives the p-value for a test of H_0: $\mu = 103$ against the alternative hypothesis, H_a: $\mu < 103$, as .022. If you have chosen $\alpha = .05$ for your test, what is your decision? What is the relevance of α to your decision?

6.40 A research article gives the p-value for a test of H_0: $\mu = 14.2$ against the alternative hypothesis, H_a: $\mu \neq 14.2$ as .09.

a. If you have chosen $\alpha = .05$ for your test, what is your decision? What is the relevance of α to your decision?

b. If you have chosen $\alpha = .10$ for your test, what is your decision? What is the relevance of α to your decision?

c. Explain the conditions under which you might prefer $\alpha = .05$ to $\alpha = .10$.

6.41 Suppose that you were to test H_0: $\mu = 3.12$

against $H_a: \mu > 3.12$ and that the value of your t test statistic, based on 22 degrees of freedom, is $t = 4.78$.

a. How many observations were in your sample?
b. What is the approximate p-value for the test?
c. What is your decision if you have chosen $\alpha = .10$?

6.42 Suppose that the value of the t test statistic for testing $H_0: \mu = 27.3$ against the alternative hypothesis, $H_a: \mu < 27.3$, is $t = -1.75$, where t is based on 14 degrees of freedom.

a. How many observations were in your sample?
b. What is the approximate p-value for the test?
c. What is your decision if you have chosen $\alpha = .05$? What is the relevance of the value $\alpha = .05$ for this decision?
d. What is your decision if you have chosen $\alpha = .10$? What is the relevance of the value $\alpha = .10$ for this decision?

6.43 Suppose that the value of the t test statistic for testing $H_0: \mu = -1.0$ against the alternative hypothesis, $H_a: \mu \neq -1.0$, is $t = 2.55$, based on a sample size of $n = 7$.

a. How many degrees of freedom are associated with t?
b. What is the approximate p-value for the test?
c. What is your decision if you have chosen $\alpha = .10$? What is the relevance of the value $\alpha = .10$ for this decision?
d. What is your decision if you have chosen $\alpha = .01$? What is the relevance of the value $\alpha = .01$ to your decision?

6.44 Refer to Example 6.13 and the test for bias in the measurements of a bathroom scale. Suppose that the test of $H_0: \mu = 150$ against $H_a: \mu \neq 150$ gave the computed value of the test statistic as $t = -1.973$. Find the approximate p-value for the test. If you have chosen $\alpha = .10$, what do you conclude?

6.45 Refer to Exercise 6..44. Explain why it would be reasonable or unreasonable to choose $\alpha = .01$ for this test.

MFG **6.46** The mean amount of cereal discharged per box by a filling machine is checked each hour by sampling the contents of $n = 8$ boxes. The cereal weights are recorded and a test is conducted to determine whether the data provide sufficient evidence to indicate that the mean load per box differs from 12 ounces. The weights of the eight boxes for one test period were 12.1, 11.8, 11.7, 12.0, 11.7, 12.0, 11.7, and 11.9.

a. Examine the data and give an approximate value for the standard deviation of the sample. Justify your answer.
b. State the null and alternative hypotheses that you would use to determine whether the data provide sufficient evidence to indicate that the mean load per box differs from 12 ounces.
c. The mean and standard deviation for the sample are 11.86 and .1598, respectively. How does your check on the calculation of s in part (a) compare with the computed value? Calculate the value of the test statistic and compare with the computed value, $t = -2.434$.
d. Find the approximate p-value for the test.
e. Explain how you would use the p-value from part (d) to decide whether the data indicate that the mean load per box differs from 12 ounces.

HLTH **6.47** Table 6.9 gives measurements of the mean amounts (in milligrams per liter) of lead, copper, and iron in samples of water taken each day for 23 days from the Boston water system (Karalekas, P. C., Jr., Ryan, C. R., and Taylor, F. B., "Control of Lead, Copper, and Iron Pipe Corrosion in Boston," *Journal of the American Water Works Association*, Vol. 75, No. 2, February 1983). Figure 6.25 gives the Minitab printout of the mean, standard deviation, and standard error of the mean for each of the three data sets, along with the 90% confidence interval for each population mean.

a. Interpret the confidence interval for the mean concentration of lead in the water.
b. Use information in Figure 6.25 and Table 4 of Appendix 2 to calculate a 90% confidence interval for the mean concentration of lead. Compare your answer with the confidence limits shown on the printout in Figure 6.25a.

HLTH **6.48** Repeat the instructions of Exercise 6.47 except use the copper data of Table 6.9.

HLTH **6.49** Repeat the instructions of Exercise 6.47 except use the iron data of Table 6.9.

TABLE 6.9 Levels (mg/l) of Lead, Copper, and Iron in Boston Water, 1983 (Exercise 6.47)

LEAD		COPPER		IRON	
.035	.015	.12	.04	.20	.14
.060	.015	.18	.04	.33	.12
.055	.022	.10	.05	.22	.12
.035	.043	.07	.07	.17	.16
.031	.030	.08	.10	.15	.17
.039	.019	.09	.04	.19	.13
.038	.021	.16	.08	.17	.15
.049	.036	.14	.05	.17	.13
.073	.016	.07	.05	.23	.14
.047	.010	.07	.04	.18	.11
.031	.020	.08	.04	.25	.11
.016		.07		.14	

Source: Karalekas, P. C., Jr., Ryan, C. R., and Taylor, F. B. "Control of Lead, Copper, and Iron Pipe Corrosion in Boston." *Journal of the American Water Works Association*, Vol. 75, No. 2, February 1983.

FIGURE 6.25 Minitab printouts of confidence intervals for the data of Table 6.9 (Exercise 6.47)

```
MTB > tinterval k=90 c1

               N       MEAN     STDEV   SE MEAN    90.0 PERCENT C.I.
lead          23    0.03287   0.01630   0.00340   ( 0.02703, 0.03871)
```

a. Lead

```
MTB > tinterval k=90 c2

               N       MEAN     STDEV   SE MEAN    90.0 PERCENT C.I.
copper        23    0.07957   0.03914   0.00816   ( 0.06555, 0.09358)
```

b. Copper

```
MTB > tinterval k=90 c3

               N       MEAN     STDEV   SE MEAN    90.0 PERCENT C.I.
iron          23     0.1687    0.0515    0.0107   (  0.1502,  0.1872)
```

c. Iron

6.50 The Environmental Protection Agency (EPA) upper limit on the amount of lead in drinking water is .05 milligram per liter (mg/l). Do the data in Table 6.9 provide sufficient evidence to indicate that the mean level of lead in the sampled portion of the Boston system is less than the EPA limit?

a. What null and alternative hypotheses should you use to detect a mean concentration that is less than .05 mg/l? Will this result in a one- or two-tailed test?
b. What type of statistical test should you use to test the hypothesis in part (a)? Why?
c. The Data Desk printout of the test results for the lead data, Table 6.9, is shown in Figure 6.26a. Find the computed value of t on the printout. Find the p-value for the test.
d. What does the p-value tell you about the test results?
e. If you choose $\alpha = .10$ for the test, what do you conclude? Why?
f. What is the relevance of $\alpha = .10$ to the test conclusions?

HLTH **6.51** Refer to Exercise 6.50. The EPA upper limit on the amount of copper in drinking water is 1.0 mg/l. Do the data, Table 6.9, provide sufficient evidence to indicate that the mean level of copper in the sampled portion of the Boston system is less than the EPA limit?

a. What null and alternative hypotheses should you use to detect a mean concentration that is less than 1.0 mg/l? Will this result in a one- or two-tailed test?
b. What type of statistical test should you use to test the hypothesis in part (a)? Why?
c. The Data Desk printout of the test results for the copper data, Table 6.9, is shown in Figure 6.26b. Find the computed value of t on the printout. Find the p-value for the test.
d. What does the p-value tell you about the test results?
e. If you choose $\alpha = .10$ for the test, what do you conclude? Why?
f. What is the relevance of $\alpha = .10$ to the test conclusions?

FIGURE 6.26 Data Desk printouts of test results for Exercises 6.50, 6.51, and 6.52

t-Tests
lead: Test H_0: $\mu = 0.05000$ vs H_a: $\mu < 0.05000$
Sample mean = 0.03287 t-statistic = −5.040 with 22 d.f.
Reject H_0 at alpha = 0.10 Prob < = 0.0000

a. Lead

t-Tests
copper: Test H_0: $\mu = 1$ vs H_a: $\mu < 1$
Sample mean = 0.07957 t-statistic = −112.793 with 22 d.f.
Reject H_0 at alpha = 0.10 Prob < = 0.0000

b. Copper

t-Tests
iron: Test H_0: $\mu = 0.30000$ vs H_a: $\mu < 0.30000$
Sample mean = 0.16870 t-statistic = −12.216 with 22 d.f.
Reject H_0 at alpha = 0.10 Prob < = 0.0000

c. Iron

HLTH **6.52** Refer to Exercise 6.50. The EPA upper limit on the amount of iron in drinking water is .3 mg/l. Do the data, Table 6.9, provide sufficient evidence to indicate that the mean level of iron in the sampled portion of the Boston system is less than the EPA limit?

a. What null and alternative hypotheses should you use to detect a mean concentration that is less than .3 mg/l? Will this result in a one- or two-tailed test?

b. What type of statistical test should you use to test the hypothesis in part (a)? Why?

c. The Data Desk printout of the test results for the iron data, Table 6.9, is shown in Figure 6.26c. Find the computed value of t on the printout. Find the p-value for the test.

d. What does the p-value tell you about the test results?

e. If you choose $\alpha = .10$ for the test, what do you conclude? Why?

f. What is the relevance of $\alpha = .10$ to the test conclusions?

6.7 Choosing the Sample Size

Most of the time your involvement with statistics will be interpreting statistical results that appear in newspapers, news magazines, scientific journals, or statistical printouts for data collected by other people. But sometime you may have to collect a sample yourself. The first question that you will face is, **"What sample size should I use?"**

To answer this question, you need to decide how accurate you want your estimate of μ to be and how much confidence you wish to place in your estimate. This means that **you must choose the half-width and the confidence coefficient for the confidence interval that you plan to construct.** For example, suppose (once again!) that we want to estimate the mean amount of iron in some *new* shipment of iron ore. How large a sample—i.e., how many 1.5-kilogram specimens of ore—should we choose for our sample? To answer that question, we would have to decide how accurate we want the estimate to be and how much confidence we want to place in it. Do we want the estimate to be correct to within .02% iron, .05% iron, or what? And what confidence do we want to place in our answer? The more accuracy and the greater the confidence that we desire, the more it is going to cost (in sample size).

Statistical Inference and Verbal Communication: An Analogy

Estimating a population parameter is similar to trying to receive a voice message from a friend across a crowded and noisy room. The greater the background noise (i.e., variation and volume of the noise), analogous to the variation in a population, the more difficult it is to receive the message. If your friend, the sender of the message, wants to increase the likelihood that you will receive the message, he or she will have to increase the volume of their signal (yell a little louder), which is equivalent to increasing the sample size. You can penetrate any background noise (variation in the population) if the signal is loud enough (if the sample is large enough).

Factors That Affect the Width of a Confidence Interval

Understanding the factors that affect the transmission of a verbal signal helps us to understand the factors that affect the amount of information in a sample. That, in turn, helps us select a sample size.

If we look at the formula for the half-width of a confidence interval, Section 6.3, we see that the half-width, B, of a $100(1 - a)\%$ confidence interval for a population mean μ is

$$B = z_{a/2}\text{SE}(\bar{x}) = z_{a/2}\frac{\sigma}{\sqrt{n}}$$

The value of B depends on three quantities:

1. **The amount of variation in the sampled population, measured by its standard deviation σ** (i.e., the level of background noise). Note that σ appears in the numerator of the formula for the half-width of the confidence interval. The greater the variation of data in the population (and hence the larger the value of σ), the larger the sample size must be to obtain an interval of a given width.

2. **The sample size n** (i.e., the volume of the signal). Note that n appears in the denominator of the formula for the half-width of the confidence interval. Thus, the half-width of the interval decreases as the sample size n increases. If you increase the sample size by a multiple, say r, you will multiply the width of the confidence interval by $1/\sqrt{r}$. For example, if you increase the sample size n by a multiple of 4, you will reduce the width of the interval to $1/\sqrt{4} = 1/2$ its original size. If you increase the sample size by a multiple of 9, you will reduce the width of the confidence interval to $1/\sqrt{9} = 1/3$ its original size.

3. **The amount of confidence that you wish to place in your estimate, measured by the confidence coefficient $(1 - a)$.** The larger the confidence coefficient that you choose, the larger will be $z_{a/2}$ and the wider will be the confidence interval.

Suppose that we want the half-width of a 95% confidence interval to equal some number, say B; that is,

$$z_{.025}\frac{\sigma}{\sqrt{n}} = B$$

Substituting $z_{.025} = 1.96$ in the above equation and solving for n, we obtain the following formula for the sample size n.

Choosing the Sample Size

Sample Size Required to Estimate μ Correct to Within B with Probability Equal to .95

$$n = \left[\frac{(1.96)(\sigma)}{B}\right]^2$$

The formula in the box will give you the approximate sample size required for a 95% confidence interval for μ with a half-width equal to B. Or, equivalently, this sample size will allow you to estimate μ with an error of estimation less than B with probability equal to .95.

Example 6.15

To measure the readiness of a forest for harvesting, the owner of a pine tree plantation wants to estimate the mean diameter of trees in a particular planting. A quick examination of the trees indicates that none have diameters less than 6 inches nor greater than 10 inches. Since the amount of usable wood in the stand of timber is dependent on the mean diameter of the trees, the owner wants the half-width of the confidence interval for μ to be small—say, equal to .2 inch—with confidence coefficient .95. How many trees must be included in the sample?

Solution

We are given that the desired half-width of the confidence interval is .2 inch but we must approximate the population standard deviation σ. The owner guesses that most of the tree diameters will vary from 6 to 10 inches. Therefore, the range of the data in the population is

$$\begin{aligned}\text{Range} &= \text{Largest measurement} - \text{Smallest measurement}\\ &= 10 - 6 = 4 \text{ inches}\end{aligned}$$

and the range approximation to σ (see Section 4.4) is

$$\text{Range approximation to } \sigma = \frac{\text{Range}}{4} = \frac{4}{4} = 1.0$$

Then, substituting into the formula for n, we obtain

$$n = \left[\frac{(1.96)(\sigma)}{B}\right]^2 = \left[\frac{(1.96)(1.0)}{.2}\right]^2 = 96.04$$

Therefore, the owner would need a random sample of the diameters of 97 trees selected from the population. [Note that we round the value of n upward to guarantee that the confidence coefficient is at least .95.]

If we were certain that the interval from 6 to 10 inches included the entire population of diameters in Example 6.15, we would have used a divisor of 6 in calculating the range approximation to σ (see Table 4.6). Instead, we used a divisor of 4. This produces a sample size that may be larger than necessary but it gives us assurance that we have enough observations in the sample to meet our specified width and confidence coefficient objectives. If we had used a divisor of 6 in calculating the approximation to σ, we would have arrived at a required sample size of $n = 43$.

Finding an Approximation to σ

You have probably noticed that our sample size formula is useless unless we know (or have an approximation to) the value of the population standard deviation σ. In a practical situation, we will rarely know σ. Therefore, to be able to apply the above formula, we need to approximate its value. There are several ways that this can be done:

1. **Sometimes you will have drawn an earlier sample. You can use the standard deviation s calculated for this prior sample to approximate σ.**

2. **Based on your knowledge of the variable that you intend to measure, you may be able to establish reasonable values for the minimum and the maximum values of x that you might observe.** Then you can approximate σ using the range approximation of Section 4.4. For example, suppose that you plan to sample the heights of American men. You know that the heights of most American males fall between 5.5 feet and 6.75 feet. Therefore, the range for the population is approximately equal to $6.75 - 5.5 = 1.25$ feet. The approximation to σ would equal the range divided by 4, 5, or 6, depending on how conservative you wish to be in selecting the sample size.

3. **If all else fails, you may be forced to draw a small sample from the population, calculate s, and use it to approximate σ.**

Does the Sample Size Depend on the Number of Measurements in the Population?

Before concluding a discussion of sample size, we should answer a question that may have occurred to you. Does the number of observations in the population affect the sample size? For example, does a sample of $n = 50$ from a population containing $N = 1{,}000$ observations provide more information about μ than a sample of $n = 50$ from a population containing $N = 10{,}000{,}000$ observations?

The number of measurements in the population affects the amount of information in the sample pertinent to μ only when the sample size n is large relative to the number N of measurements in the population. As the percent-

age, (n/N) 100%, increases above 10%, the width of the confidence intervals will be smaller than specified in Section 6.3. Therefore since 50 represents less than 10% of 1,000, sampling 50 out of a population of 1,000 gives almost the same confidence interval (all other things being equal) as sampling 50 out of 10,000,000.

Most populations in real sampling situations tend to be large relative to the prospective sample size. For that reason, the size of the population can usually be ignored when choosing the sample size.

EXERCISES

6.53 In which of the following two situations is selection of the sample size more important: (1) testing the ability of a sample of n automobiles to withstand damage when subjected to a head-on collision at 30 miles per hour or (2) testing the voltage of n flashlight batteries? Why?

6.54 If population 1 is more variable than population 2, will the sample size for population 1 have to be larger or smaller than for population 2 if you wish to construct confidence intervals of the same width and confidence coefficient? Why?

6.55 If the data in a population are extremely variable, does it follow that you cannot obtain a very accurate estimate of the population mean? Explain.

6.56 Population 1 contains 5,000 measurements and population 2 contains 1,000,000 measurements, and both possess the same variance σ^2. If you draw random samples of 100 measurements from each and construct confidence intervals for the population means, which, if either, will possess the smaller confidence interval?

6.57 Suppose that you wish to estimate the mean of a population and prior information suggests that the population standard deviation σ is approximately equal to 8.8. How many observations would have to be included in a random sample if you wish to estimate μ correct to within 2.8 with probability equal to .95?

6.58 Suppose that you wish to estimate the mean of a population and prior information suggests that the population standard deviation σ is approximately equal to 70. How many observations would have to be included in a random sample if you wish to estimate μ correct to within 10 with probability equal to .95?

6.59 Suppose that you wish to estimate the mean of a population and you have prior information that indicates that the population measurements vary between 30 and 250. How many observations would have to be included in a random sample if you wish to estimate μ correct to within 12 with probability equal to .95?

6.60 In Example 6.10, we gave the carapace lengths (in mm) for ten *T. orientalis* lobsters. Suppose that you wish to estimate the mean carapace length of *T. orientalis* lobsters to within 2 mm with probability equal to .95. Approximately how many lobsters would have to be included in your sample? Would it be easy to draw a random sample of these lobsters? Explain.

6.61 The half-width of the confidence interval for the mean percentage iron in the ore specimens of Example 6.8 is .0691. Suppose that you wish to reduce the half-width of the confidence interval to .0346. Approximately how many measurements (total number) would be required in the sample?

6.62 Refer to Exercise 6.61. Suppose that you wish to cut the half-width of the confidence interval, Example 6.8, to one-quarter of its original size. Approximately how many measurements (total number) would be required in the sample?

6.8 Why Assumptions Are Important

Each statistical method presented in this chapter and those that follow is based on a set of one or more assumptions. If the assumptions are satisfied, the method will yield an inference with the properties that we describe. The trouble is that we rarely, if ever, know whether the assumptions are satisfied.

Statistics, like the sciences, is divided into two parts—theory and application. For example, a civil engineer uses the theory of structural design and the tabulated strengths of concrete and steel to arrive at a maximum weight load that a bridge will sustain. In practice, steel beams of the same size do not all possess the same strength and neither do specimens of concrete. The designer takes this into account, designs the bridge to sustain loads in excess of any loads that might be expected (using a so-called "safety factor"), and thereby builds a bridge that is usable and safe. In this way, an engineer applies theory to solve a problem and obtain a workable result.

The theory of statistics is similar to the theory of structural design (or any other scientific theory). Just as the civil engineer makes assumptions about the strengths of the steel beams and concrete and then calculates a theoretical maximum weight load that a bridge will sustain, a theoretical statistician makes assumptions about the nature of a sampled population and then derives all of the probabilistic quantities associated with statistical inference, sampling distributions, confidence coefficients for confidence intervals, and the p-values associated with tests of hypotheses.

Although we rarely know the exact nature of a sampled population or whether our assumptions are satisfied, we often know how lack of satisfaction of the assumptions affects the confidence coefficients for confidence intervals and the p-values for statistical tests. Therefore, like the civil engineer, we use statistical theory to give us a theoretical result. We then allow for the realities of life (the fact that our assumptions are not exactly satisfied) and come up with a useful statistical inference that, while not exactly correct, is correct for all practical purposes. Methods called **robust statistical methods** give answers that are approximately correct, even when there are substantial departures from the assumptions. Therefore, we need to know the assumptions associated with each method and how robust the method is to departures from the assumptions.

6.9 Key Words and Concepts

▶ This chapter tells you how to make inferences about a population mean. To answer the question "What is μ?", we ***estimate*** its value using a confidence interval. To answer the question "Is μ larger than some number, say 49?", we ***test an hypothesis*** about its value.

- A $100(1 - a)\%$ ***confidence interval*** is composed of two numbers calculated from the data in a sample, the ***lower and upper confidence limits.*** The probability that the interval will enclose μ, called the ***confidence coefficient,*** is equal to $(1 - a)$. We presented two similar confidence intervals for μ: one based on the standard normal z statistic, which is used when σ is known; and one based on the t statistic, which is appropriate when σ is unknown and the sampled population is approximately normally distributed. The two confidence intervals are identical, for all practical purposes, when n is equal to 30 or larger.
- The width of a confidence interval depends on the variation of the data (measured by the population standard deviation σ), the confidence coefficient that we choose, and the sample size n. The width increases as the variation and the confidence coefficient increase and decreases as the sample size increases.
- Tests of hypotheses about μ are also based on the z and the t statistics. An ***upper one-tailed test*** of an hypothesis attempts to detect values of μ larger than a value μ_0 that we specify in the null hypothesis. The rejection region for the test is located entirely in the upper tail of the sampling distribution of the test statistic. Therefore, we reject H_0 only for large positive values of the test statistic.
- A ***lower one-tailed test*** of H_0: $\mu = \mu_0$ attempts to detect values of μ smaller than μ_0. It places the rejection region in the lower tail of the sampling distribution of the test statistic and rejects H_0 when the test statistic is large and negative.
- A ***two-tailed test*** of H_0: $\mu = \mu_0$ attempts to detect values of μ that are either smaller than or larger than μ_0. It divides the rejection region, half in one tail of the sampling distribution of the test statistic and half in the other, and rejects H_0 for either very large negative or positive values of the test statistic.
- The ***p-value*** for a test is the probability of observing a value of the test statistic at least as contradictory to the null hypothesis as the value computed from the sample data. Thus, the p-value for a test measures how unlikely the value of the observed test statistic is, assuming the null hypothesis to be true.
- We reject the null hypothesis for a statistical test when the p-value for the test is less than a value α.
- The quantity α is the probability of rejecting the null hypothesis when it is true. It measures the risk that we have erroneously rejected the null hypothesis when, in fact, it is true.

SUPPLEMENTARY EXERCISES

BIO **6.63** Reedfish, a native of tropical Africa, can breathe in both air and water. As a consequence, they can live in water containing a low oxygen content. Table 6.10 (page 224) gives the weight, the oxygen uptake in air and water, and the percentage of oxygen uptake by air and water, at 25° Celsius, for $n = 11$ reedfish (Pettit, M. J. and Beitinger, T. L., "Oxygen Acquisition of the Reedfish, *Erpetoichthys calabaricus*," *Journal of Experimental Biology*, Vol. 114, 1985). Figure 6.27a gives the Minitab

TABLE 6.10 Weight and Oxygen Uptake Data for 11 Reedfish

Weight (grams)	Oxygen uptake ($ml\ O_2\ g^{-1}h^{-1}$) Air	Water	Percent Air	Water
22.1	.043	.045	49	51
21.1	.034	.066	34	66
13.4	.026	.084	24	76
19.5	.048	.045	52	48
21.9	.009	.056	14	86
19.5	.028	.072	28	72
16.5	.032	.081	28	72
18.5	.045	.051	47	53
18.0	.061	.041	60	40
15.4	.028	.060	32	68
13.4	.024	.113	18	82

FIGURE 6.27 Minitab printouts of the fish weight data of Table 6.10

```
Histogram of wt

Midpoint   Count
      13       2  **
      14       0
      15       1  *
      16       0
      17       1  *
      18       1  *
      19       1  *
      20       2  **
      21       1  *
      22       2  **
```

a. Relative frequency distribution

```
MTB > describe c2

              N      MEAN    MEDIAN    TRMEAN     STDEV    SEMEAN
wt           11    18.118    18.500    18.200     3.123     0.942

            MIN       MAX        Q1        Q3
wt       13.400    22.100    15.400    21.100
```

b. Numerical descriptive measures

printout of the sample relative frequency distribution and Figure 6.27b gives the numerical descriptive measures for the 11 fish weights.

a. Do the mean and standard deviation of the fish weights provide an adequate description of the sample relative frequency distribution? Explain.

b. Find a 95% confidence interval for the weight of the reedfish employed in the experiment and explain what it means.

BIO **6.64** Refer to Exercise 6.63. Figure 6.28a gives the Minitab printout of the sample relative frequency distribution and Figure 6.28b gives the numerical descriptive measures for the oxygen uptake by air data of Table 6.10. Follow the instructions of Exercise 6.63.

BIO **6.65** Refer to Exercise 6.63. Figure 6.29a (page 226) gives the Minitab printout of the sample relative frequency distribution and Figure 6.29b gives the numerical descriptive measures for the oxygen uptake by water data of Table 6.10. Follow the instructions of Exercise 6.63.

BIO **6.66** Refer to Exercise 6.63. Suppose that you had a theory that the maximum uptake of oxygen by air for reedfish in 25°C water is less than 45%. Do the data support this theory?

a. What is your null hypothesis?

b. What is your alternative hypothesis?

c. Is this a one- or a two-tailed test? Explain.

d. What test statistic would you use for the test and where would you locate the rejection region?

e. Figure 6.30 (page 226) gives the Minitab printout for this test. Find the p-value for the test. What does the p-value mean?

f. Suppose that you decide to use $\alpha = .05$ for the test. What do you conclude? What are the practical implications of your conclusions?

g. What is the relevance of your choice of $\alpha = .05$ for the test?

FIGURE 6.28 Minitab printouts for the oxygen uptake by air data of Table 6.10

```
Histogram of air   N = 11

Midpoint   Count
   0.010       1  *
   0.015       0
   0.020       0
   0.025       2  **
   0.030       3  ***
   0.035       1  *
   0.040       0
   0.045       2  **
   0.050       1  *
   0.055       0
   0.060       1  *
```

a. Relative frequency distribution

```
MTB > describe c3

                N      MEAN    MEDIAN    TRMEAN     STDEV    SEMEAN
air            11   0.03436   0.03200   0.03422   0.01411   0.00425

              MIN       MAX        Q1        Q3
air       0.00900   0.06100   0.02600   0.04500
```

b. Numerical descriptive measures

FIGURE 6.29 Minitab printouts for the oxygen uptake by water data of Table 6.10

```
Histogram of water    N = 11

Midpoint   Count
    0.04       1  *
    0.05       3  ***
    0.06       2  **
    0.07       2  **
    0.08       2  **
    0.09       0
    0.10       0
    0.11       1  *
```

a. Relative frequency distribution

```
MTB > describe c4

                  N     MEAN    MEDIAN    TRMEAN     STDEV    SEMEAN
water            11  0.06491   0.06000   0.06222   0.02156   0.00650

                MIN      MAX        Q1        Q3
water       0.04100  0.11300   0.04500   0.08100
```

b. Numerical descriptive measures

FIGURE 6.30 A Minitab printout for the test, Exercise 6.66

```
TEST OF MU = 45.000 VS MU L.T. 45.000

              N      MEAN     STDEV   SE MEAN          T    P VALUE
air%         11    35.091    14.876     4.485      -2.21      0.026
```

EXERCISES FOR YOUR COMPUTER

GEN **6.67** A study to investigate the relationship between a set of variables and a competitive runner's finish time in a 10-kilometer race gives the finish times for nine runners (Powers, Scott K. et al., "Ventilatory Threshold, Running Economy and Distance Running Performance of Trained Athletes," *Research Quarterly for Exercise and Sport*, Vol. 54, No. 3, 1983). The finish times (in minutes) are shown in Table 6.11. Assume that the nine finish times represent a random sample from the population of finish times for the current crop of competitive 10-km runners.

a. Construct a relative frequency distribution for the data.

b. Find the numerical descriptive measures for the data. Do the mean and standard deviation provide an adequate description of the distribution in part (a)? Explain.

TABLE 6.11 Finish Times in a 10-Kilometer Race

Runner	1	2	3	4	5	6	7	8	9
Finish time	33.15	33.33	33.50	33.55	33.73	33.86	33.90	34.15	34.90

c. Find a 95% confidence interval for the mean finish time. Interpret the interval.

d. Give a point estimate of the mean finish time. How accurate is your point estimate? Explain.

e. Use the values of $\bar{x}$ and s given on the printout to calculate a 95% confidence interval for the mean finish time. Compare your answer with the interval shown on the printout, part (c).

6.68 Select a random sample of $n = 10$ percentage iron ore measurements from the set of 390 measurements given in Table 1 of Appendix 1.

a. Construct a relative frequency distribution for the data.

b. Find the numerical descriptive measures for the data. Do the mean and standard deviation provide an adequate description of the distribution in part (a)? Explain.

c. Find a 95% confidence interval for the mean percentage iron ore. Interpret the interval.

d. Give a point estimate of the mean percentage iron ore. How accurate is your point estimate? Explain.

e. Use the values of $\bar{x}$ and s given on the printout to calculate a 95% confidence interval for the mean percentage iron ore. Compare your answer with the interval shown on the printout, part (c).

f. How does your confidence interval compare with the confidence interval based on all 390 measurements that we calculated in Example 6.8? In what respects should the two confidence intervals be similar? In what respects should they be different? Explain.

6.69 Refer to Exercise 6.68.

a. Do the data provide sufficient information to indicate that the mean percentage iron in the ore exceeds 65%? Give the null and alternative hypotheses that you would use to answer this question.

b. What test statistic should you use to test your hypothesis and where will you locate the rejection region?

c. Use your computer to conduct the test. Find the p-value for the test. If you choose $\alpha = .01$, what do you conclude?

d. How does the p-value for your test result compare with the test based on all 390 measurements, Example 6.12? In what respects should the two tests be similar? In what respects should they be different? Explain.

6.70 Use your computer and the data in Table 3 of Appendix 1 to find a confidence interval for the mean TCHOL level for the universe of women in Florida for which the sample is representative. Interpret the interval and discuss its limitations.

6.71 Use your computer and the data in Table 3 of Appendix 1 to find a confidence interval for the mean TCHOL level for the universe of men in Florida for which the sample is representative. Interpret the interval and discuss its limitations.

References

1. Freedman, D., Pisani, R., Purves, R., and Adhikari, A., *Statistics*, 2nd edition. New York: W. W. Norton & Co., 1991.

2. McClave, J. and Dietrich, F., *Statistics*, 5th edition. San Francisco: Dellen Publishing Co., 1991.
3. Mendenhall, W. and Beaver, R., *Introduction to Probability and Statistics*, 8th edition. Boston: PWS-Kent, 1991.
4. Moore, D. and McCabe, G., *Introduction to the Practice of Statistics,* 2nd edition. New York: W. H. Freeman & Co., 1993.
5. Sincich, T., *Statistics by Example*, 4th edition. San Francisco: Dellen Publishing Co., 1990.

seven

Comparing Two Population Means

In a Nutshell

Ralston Purina wants to know whether cats tend to eat more of catfood A than catfood B. Why? So that they can sell more catfood. This chapter presents another useful application of statistical inference: how to estimate and test hypotheses about the difference between two population means.

7.1 The Problem

7.2 Independent Random Sampling and Matched-Pairs Designs

7.3 Comparing Population Means: Independent Random Samples

7.4 Comparing Population Means: A Matched-Pairs Design

7.5 Choosing the Sample Size

7.6 Key Words and Concepts

7.1 The Problem

Making inferences about a single population mean is an important problem, but not nearly as important as the comparison of two population means. The way to progress is paved with change. We change a set of conditions and then examine how this change affects some variable of interest to us. Very often this leads us to the comparison of the mean value of the variable after the change with the mean value before the change.

For example, a producer of steel bars would want to compare the mean strength of bars produced by a conventional method A with the mean strength of steel produced by a new method B. Similarly, a pharmaceutical firm would want to compare the mean blood pressure of a set of patients before and after receiving treatment with a new drug. Or, we might want to compare the mean time to deterioration for two varieties of apples that have been placed in cold storage.

This chapter is concerned with the comparison of the means of two populations. We will estimate the difference between two population means and will test hypotheses to detect whether a difference really exists. As we proceed through this chapter, you will see that our methods bear a strong resemblance to those of Chapter 6.

7.2 Independent Random Sampling and Matched-Pairs Designs

We explained in Chapter 5 that statistical inferences are based on **probability samples**, the simplest of which is the random sample. Without probability sampling, we would have been unable to calculate the probabilities associated with the confidence intervals and statistical tests of Chapter 6.

The Design of an Experiment: What It Is and Why We Do It

There are many different ways of selecting a probability sample. **A design or plan for selecting a sample is called a *design of an experiment*. Designs for public opinion or consumer preference polls are called *sample survey designs*.** The design of an experiment is important because the design affects the quantity of information in the sample data. For example, suppose that one particular design requires random samples of 1,000 people from each of two cities to estimate the difference in mean family income correct to within \$500. Another design may give the same confidence interval at 1/10 or 1/100 the cost of the random sample. Other experiments, such as the testing of space rockets, are so costly that each single observation (a rocket firing) is a major expense. As a consequence, the design of experiments, buying sample information at minimum cost, is a very broad and important subspecialty in statistics.

This chapter will present methods for comparing two population means for each of two experimental designs. One design, based on independent random samples, gives us a probabilistic sample but little more. The second design also gives a probability sample, but can sometimes also produce substantial savings in cost.

An Independent Random Samples Design

Selecting random samples from each of the two populations in an independent manner, i.e., so that the observations selected in one sample do not influence the selection of observations for the second sample, is called an *independent random samples design.* See Section 5.3 for a definition of independence.

Independent random samples can arise in two ways:

1. The difference between the elements in one sample and the elements in another sample is created by the experimenter. For example, suppose that you wish to determine whether a new drug is useful in lowering serum cholesterol. The universe consists of all people with high cholesterol; the experimenter has a sample of 80 patients from the universe. The 80 patients are randomly divided into two groups. This is done by numbering the patients from 1 to 80 and then using the random number tables (see Section 5.4 and Table 1 of Appendix 2) to select 40 for group 1. The remaining 40 patients would be group 2. The experimenter then applies **treatments** to the elements of each sample. The patients in sample 1 are assigned to receive the new drug—call this treatment 1—over a fixed period of time. Those in sample 2 receive a **placebo**—treatment 2, a sugar-coated pill that possesses no therapeutic value. The two treatments are "receive drug" and "receive placebo." Note that the treatments are applied after the elements (the patients) have been assigned to the samples.

By treating one group of patients differently from the other, the experimenter has created two universes from the one original universe of high-cholesterol patients. We view the 40 patients on the new drug as a sample from the universe of all high-cholesterol patients who could (conceptually) receive the drug. Similarly, the 40 patients receiving the placebo represent a sample from the universe of all high-cholesterol patients who could (conceptually) receive the placebo. The object of the sampling is to determine, after a specified period of time, whether the mean cholesterol level for the universe of patients receiving the drug is less than the mean cholesterol level for the universe of patients receiving the placebo.

2. The elements of the two samples are inherently different. No treatment need be applied because the elements in the two universes are, by their very nature, different. For example, we might wish to compare the strengths of two different types of steel alloys—say, carbon and nickel. The universes consist of steel bars, thousands of bars of each type, collected over a period of time. Although nothing is actually done to "treat" the elements in the samples after they have been collected, statisticians often refer to the conditions that make the two samples (and populations) different as **treatments**. For example, they would refer to the two "treatments" in this experiment as "carbon" and "nickel." To select random samples in this type of situation, number the elements in each universe and use the table of random numbers, Section 5.4, to draw the samples, first from one universe and then from the other.

A Matched-Pairs Design

In Section 6.7, we drew an analogy between (1) sampling to detect the value of a population parameter and (2) trying to detect a voice signal sent across a crowded room. Variation in the sample data (background noise) increases the error in estimating the population parameter (it makes it difficult for us to receive the voice signal). In contrast, increasing the sample size (the volume of the audio message) reduces the error in estimating the population parameter (and makes it easier to receive the voice signal).

The same logic applies when we want to compare the means of two populations. The more variable the differences between two observations, one from each population (i.e., the greater the noise), the more difficult it will be to detect and estimate the difference between the two population means. The greater the number of pairs of differences (i.e., the greater the volume of the signal), the easier it is to estimate and detect differences between two population means.

The preceding analogy identifies one of the strategies employed in the statistical design of experiments: **We try to select the data in such a way that we minimize the variation (noise) contributed by extraneous variables.**

To apply this strategy to the estimation (or test) of the difference between two population means, we need to identify variables that might produce unwanted variation in our data. For example, suppose that we want to compare the mean weight gain of 2-month-old babies subjected to two different treatments, Diet 1 and Diet 2, over a 3-month period of time. We draw samples of weight gain measurements from each of the two populations, calculate the sample means $\bar{x}_1$ and $\bar{x}_2$, and then calculate our estimate $(\bar{x}_1 - \bar{x}_2)$ of the difference in mean weight gains, $(\mu_1 - \mu_2)$, for babies on the two diets. Why is this difference in weight gain not the same for every pair of babies?

One answer is that, in addition to diet, differences in weight gains are affected by genetic differences between babies. To eliminate the effects of this variable, we would want to compare the mean weight gains for Diets 1 and 2 using pairs of babies that are genetically similar.

For example, we would suspect that the variation in the differences in weight gains between pairs of twins on the same diet would be less than the variation in the differences in weight gains between pairs of babies selected at random. If our speculation is correct, we can eliminate the genetic variation between the two observations in a pair by applying the two treatments (Diets 1 and 2), one to each member of a pair of twins. We would expect 20 differences in weight gains between pairs of twins to contain more information about the difference in population means than

20 differences in weight gains contained in two independent random samples, 20 babies per sample.

Consider another example where we would want to compare two treatments within pairs of similar elements. Suppose that we want to compare the mean amounts of material learned when students are subjected to two different methods of teaching, Method 1 and Method 2. We would expect the difference in the amount of material learned between pairs of students with the same intellectual ability to be less variable than for any two students selected at random. Therefore, we would match pairs of students on some measure of intellectual ability—say, IQ. One member of each pair would be randomly assigned to learn with Method 1; the other would be assigned to Method 2. We would expect data provided by this design to give more information on differences in mean learning (measured by achievement test scores) than if we based our inference on the differences in scores between two students selected at random, one from each of the two populations.

These two examples explain how we reduce the effects of extraneous variation: We make comparisons of two treatments within matched (i.e., similar) pairs of elements selected from a universe.

Randomly assigning two treatments, one to each of a number of matched pairs of elements, is called a *matched-pairs design* or a *paired-difference experiment*. The concept of matching pairs of similar elements is illustrated symbolically in Figure 7.1.

FIGURE 7.1 Symbolic representation of matched and unmatched pairs of elements

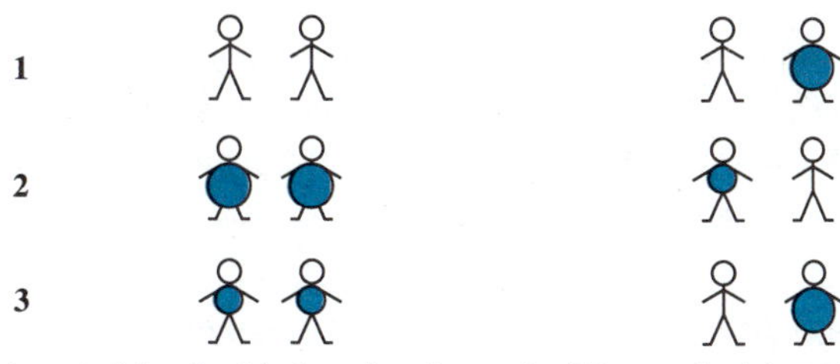

EXERCISES

7.1 Describe the two experimental designs for comparing population means, an independent random samples design and a matched-pairs design.

7.2 Suppose that an experimenter wishes to compare the amount of rise produced in cakes by two different cake recipes.

a. Identify the elements of the two universes.

b. Identify the two treatments.

c. Explain how the observations might be paired to possibly increase the information about the difference in population means.

7.3 Describe a sampling situation where pairing might yield more information than independent random sampling for comparing two population means.

EDUC **7.4** Suppose that you plan to compare two teaching methods by teaching a specific amount of course material to two groups of students, one group taught by Method 1 and the other by Method 2. A standard test will be given to each student at the end of the teaching period to measure student achievement.

a. Explain how the students would be selected if you wish to use an independent samples design.

b. Explain how you might match students and achieve more information about the difference in mean achievement for the two teaching methods by using a matched-pairs design.

BUS **7.5** Suppose that you wish to compare the difference in the mean amount of iron ore excavated and loaded per hour using two different methods. The ore is to be removed from an ore deposit that covers a large area.

a. Explain why the amount of ore removed might vary from day to day even though an excavator used the same method of removal and loading.

b. Explain how you would conduct the experiment using an independent random samples design.

c. Explain how you might pair observations, one for each method, to reduce the variability of differences between pairs.

7.3 Comparing Population Means: Independent Random Samples

Do women tend to have higher (or lower) cholesterol levels than men? If a difference does exist between the levels of cholesterol in women and men, is it caused by differences in diet, genetic differences, or what? Can answers to these questions help researchers improve the treatment of high cholesterol?

Table 3 of Appendix 1 may help us answer the first question. It gives the total cholesterol readings (as well as other data) for 49 women and 58 men. A summary of the sample statistics for the data, including the standard errors of the sample means, is shown in Table 7.1.

TABLE 7.1 Female and Male TCHOL Sample Statistics for Table 3 of Appendix 1

Population	Sex	Sample Size n	Sample Mean $\bar{x}$	Standard Deviation s	Standard Error SE($\bar{x}$)
1	Female	49	219.8	60.7	8.7
2	Male	58	216.1	46.5	6.1

What do these two samples tell us about the difference in mean TCHOL levels for the populations of TCHOL measurements for men and women? First we would estimate that the difference in the mean TCHOL level for women versus the mean

TCHOL level for men is equal to the difference in their respective sample means, $(\bar{x}_1 - \bar{x}_2) = 219.8 - 216.1 = 3.7$ milligrams per deciliter. From a practical point of view, this does not appear to be a very large difference. But, what is the error of estimation? How far from the bull's-eye (the difference in population means) is our estimate 3.7 likely to be? To answer this question, we need to know the characteristics of the sampling distribution of the difference $(\bar{x}_1 - \bar{x}_2)$ in the sample means.

The Sampling Distribution of $(\bar{x}_1 - \bar{x}_2)$

We will assume that the two samples of cholesterol data were collected according to an independent random samples design from populations with means and variances (μ_1, σ_1^2) and (μ_2, σ_2^2), respectively, and that the sample means are approximately normally distributed (which will usually be the case because of the Central Limit Theorem, Section 5.8). Then it can be shown (proof omitted) that **the sampling distribution of the difference $(\bar{x}_1 - \bar{x}_2)$ in the sample means will be approximately normally distributed (see Figure 7.2) with a mean and standard error as shown in the box.**

Mean and Standard Error of the Difference $(\bar{x}_1 - \bar{x}_2)$ in Sample Means

$$\text{Mean} = (\mu_1 - \mu_2)$$

and

$$SE(\bar{x}_1 - \bar{x}_2) = \sqrt{\frac{\sigma_1^2}{n_1} + \frac{\sigma_2^2}{n_2}} = \sqrt{SE(\bar{x}_1)^2 + SE(\bar{x}_2)^2}$$

where μ_1 and σ_1^2 are the mean and variance of population 1, and μ_2 and σ_2^2 are the mean and variance of population 2.

FIGURE 7.2 The sampling distribution of $(\bar{x}_1 - \bar{x}_2)$

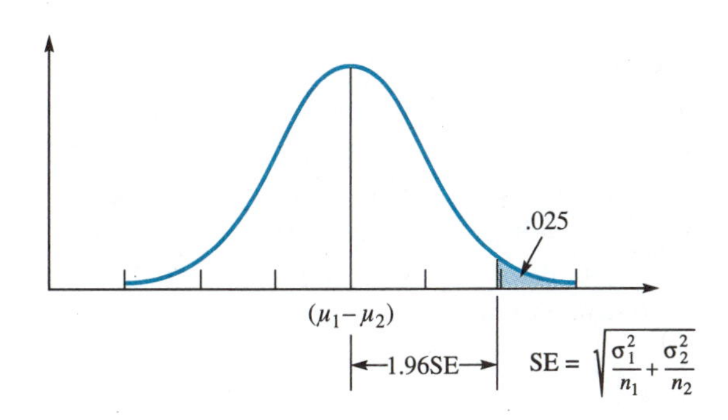

We interpret the sampling distribution of $(\bar{x}_1 - \bar{x}_2)$ in exactly the same way that we interpreted the sampling distribution of $\bar{x}$ in Chapter 6. Examining Figure 7.2, we see that the mean of the sampling distribution is the parameter $(\mu_1 - \mu_2)$ that we want to estimate. Therefore, the difference in sample means is an unbiased estimator of the difference $(\mu_1 - \mu_2)$ in population means. Most (95%) of the time, the difference between a pair of sample means will fall within 1.96 standard errors, i.e., within $1.96\text{SE}(\bar{x}_1 - \bar{x}_2)$, of its mean $(\mu_1 - \mu_2)$, a statement that agrees with the Empirical Rule. The probability that the difference $(\bar{x}_1 - \bar{x}_2)$ in a pair of sample means falls more than 1.96 standard errors away from its mean $(\mu_1 - \mu_2)$ is .05.

The Standard Normal *z* Statistic

The probability that a particular estimate $(\bar{x}_1 - \bar{x}_2)$ will deviate from $(\mu_1 - \mu_2)$ by more than some specified amount depends on how many standard errors this deviation represents. This number is the familiar standard normal z statistic of Chapters 5 and 6,

$$z = \frac{\text{Normally distributed point estimate} - \text{Mean of the point estimate}}{\text{Standard error of the point estimate}}$$

or

$$z = \frac{(\bar{x}_1 - \bar{x}_2) - (\mu_1 - \mu_2)}{\text{SE}(\bar{x}_1 - \bar{x}_2)}$$

The following example shows how to find the probability that the error of estimating $(\mu_1 - \mu_2)$ will exceed some specified amount.

Example 7.1

Refer to the sampling of the TCHOL levels of women and men, Table 7.1. Find the approximate probability that a difference $(\bar{x}_1 - \bar{x}_2)$ in a pair of sample means will deviate from the difference $(\mu_1 - \mu_2)$ in population means by as much as 10 milligrams per deciliter.

Solution

To answer this question, we must calculate the standard error of $(\bar{x}_1 - \bar{x}_2)$. We do not know the values of σ_1^2 and σ_2^2 that appear in its formula (shown in the box) but we do have estimates of σ_1 and σ_2 based on large samples. These estimates, which provide good approximations to σ_1 and σ_2, are given in Table 7.1 as $s_1 = 60.7$ and $s_2 = 46.5$. Corresponding estimates of the standard errors of the individual sample means are $\text{SE}(\bar{x}_1) = 8.7$ and $\text{SE}(\bar{x}_2) = 6.1$. Substituting these values into the formula for $SE(\bar{x}_1 - \bar{x}_2)$, we obtain

$$\begin{aligned} SE(\bar{x}_1 - \bar{x}_2) &= \sqrt{SE(\bar{x}_1)^2 + SE(\bar{x}_2)^2} \\ &= \sqrt{(8.7)^2 + (6.1)^2} \\ &= \sqrt{112.9} = 10.6 \end{aligned}$$

The next step is to determine how many standard errors the deviation 10 mg/dl represents. Substituting into the formula for z gives

$$z = \frac{(\bar{x}_1 - \bar{x}_2) - (\mu_1 - \mu_2)}{SE(\bar{x}_1 - \bar{x}_2)} = \frac{10}{10.6} = .94$$

Figure 7.3a shows the 10 mg/dl deviation marked off to the left and to the right of $(\mu_1 - \mu_2)$ on the sampling distribution of $(\bar{x}_1 - \bar{x}_2)$. Figure 7.3b locates the corresponding points on the z distribution. The probability that $(\bar{x}_1 - \bar{x}_2)$ will deviate

FIGURE 7.3 Sampling distributions of $(\bar{x}_1 - \bar{x}_2)$ and z for Example 7.1

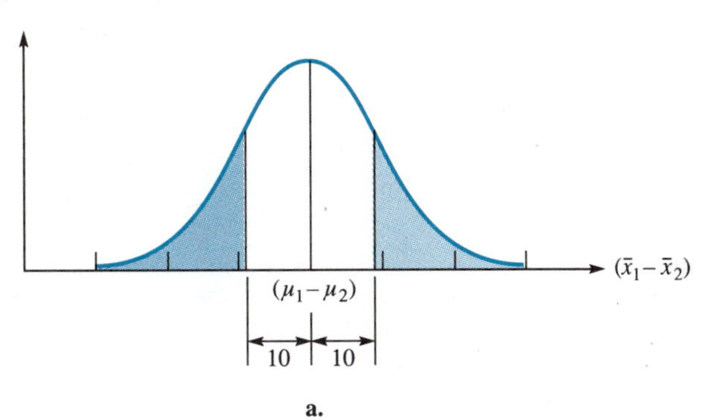

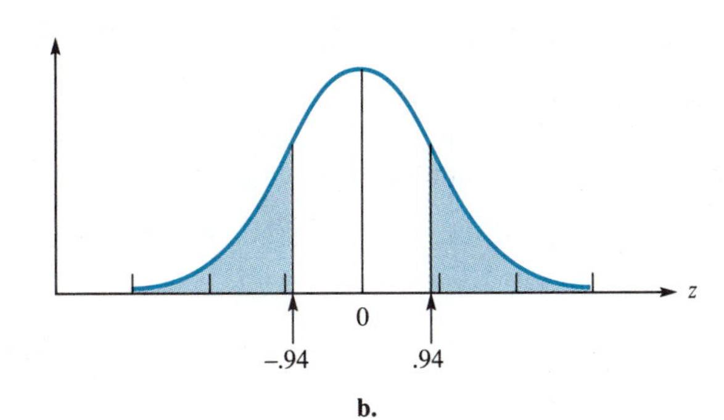

from the difference $(\mu_1 - \mu_2)$ in population means by more than 10 mg/dl is the probability that z will deviate from 0 by more than .94. This probability is equal to the sum of the shaded areas in the lower and upper tails of the z distribution, Figure 7.3b.

We can find the approximate area in the upper tail of the z distribution by using Table 7.2, or we can calculate the area correct to four decimal places using Table 3 of Appendix 2. Table 7.2 shows that the upper-tail area to the right of $z = 1.0$ is .16. The calculated value, $z = .94$, is close to 1.0. Therefore, the area in the upper tail of the z distribution is close to (slightly larger than) .16 and the sum of the areas in the upper and lower tails of the z distribution is approximately $2(.16) = .32$. This tells us that an error as large as or larger than 10 mg/dl in estimating the difference in the means of the female and male TCHOL populations is quite probable. It is approximately 1 chance in 3.

TABLE 7.2 Some Upper-Tail Areas for the Standard Normal Distribution

z_a	Upper-Tail Area a
.84	.20
1.00	.16
1.28	.10
1.65	.05
1.96	.025
2.33	.01
2.58	.005

A Large-Sample Confidence Interval for $(\mu_1 - \mu_2)$

The $100(1 - a)\%$ confidence interval for $(\mu_1 - \mu_2)$ when σ_1^2 and σ_2^2 are known or when n_1 and n_2 are large ($n_1 \geqslant 30$ and $n_2 \geqslant 30$) is very similar to the confidence interval for a single population mean discussed in Chapter 6:

$$\text{LCL} = \text{Point estimate} - z_{a/2}\text{SE(point estimate)}$$

and

$$\text{UCL} = \text{Point estimate} + z_{a/2}\text{SE(point estimate)}$$

The formula for the confidence interval for $(\mu_1 - \mu_2)$ is shown in the box and the following example shows how we can use it to find a confidence interval for the difference between the mean cholesterol levels for women and men.

A 100(1 − a)% Confidence Interval for ($\mu_1 - \mu_2$)
When σ_1^2 and σ_2^2 Are Known

$$\text{LCL} = (\bar{x}_1 - \bar{x}_2) - z_{a/2}\text{SE}(\bar{x}_1 - \bar{x}_2)$$
$$\text{UCL} = (\bar{x}_1 - \bar{x}_2) + z_{a/2}\text{SE}(\bar{x}_1 - \bar{x}_2)$$

where

$$\text{SE}(\bar{x}_1 - \bar{x}_2) = \sqrt{\frac{\sigma_1^2}{n_1} + \frac{\sigma_2^2}{n_2}} = \sqrt{\text{SE}(\bar{x}_1)^2 + \text{SE}(\bar{x}_2)^2}$$

Assumptions:

1. The samples are independently and randomly drawn from the two populations.
2. The sample means, $\bar{x}_1$ and $\bar{x}_2$, are approximately normally distributed.
3. The population variances, σ_1^2 and σ_2^2, are known or n_1 and n_2 are large enough ($n_1 \geqslant 30$ and $n_2 \geqslant 30$) so that s_1^2 and s_2^2 can be used to approximate σ_1^2 and σ_2^2 in the formula for $\text{SE}(\bar{x}_1 - \bar{x}_2)$.

Example 7.2

Use the data in Table 3 of Appendix 1 and the Minitab printout shown in Figure 7.4.

FIGURE 7.4 A Minitab printout of the confidence interval for the difference in mean female and male total cholesterol count

```
TWOSAMPLE T FOR tchol
sex    N       MEAN        STDEV    SE MEAN
1      49      219.8        60.7        8.7
2      58      216.1        46.5        6.1

95 PCT CI FOR MU 1 - MU 2: (-17.3, 24.8)

TTEST MU 1 = MU 2 (VS NE): T= 0.35    P=0.73    DF=  89
```

a. Find the sample mean, standard deviation, and standard error of the mean for each sample.

b. Find a 95% confidence interval for the difference in mean cholesterol levels of the sampled female and male TCHOL populations and interpret it.

c. Suppose that you used a computer printout that gave the sample statistics but did not give the confidence interval for the difference in population means. Use the relevant information in Example 7.1 to verify the interval shown on the Minitab printout. Assume that the samples can be regarded as random and independent.

Solution

a. The Minitab printout in Figure 7.4 gives the sample mean, standard deviation, and standard error of the sample mean for each sample. Sex is coded 1 for female and 2 for male. The first row, opposite 1, gives the sample size and statistics for females. Reading from the printout,

$$n_1 = 49 \qquad \bar{x}_1 = 219.8 \qquad s_1 = 60.7$$

The corresponding statistics for males are

$$n_2 = 58 \qquad \bar{x}_2 = 216.1 \qquad s_2 = 46.5$$

b. The 95% confidence interval for $(\mu_1 - \mu_2)$ is shaded on the printout: LCL $= -17.3$ and UCL $= 24.8$. Therefore, we estimate that the difference in mean cholesterol levels between women and men is somewhere between -17.3 and 24.8 mg/dl. Note that the confidence interval includes 0 (that is, no difference) as a possible value for $(\mu_1 - \mu_2)$. This suggests that there is little evidence of a difference in the population means. A test for the difference in population means, which we will subsequently present, will confirm this conclusion.

c. Recall that $z_{.025}$, required for a 95% confidence interval, is 1.96. From Example 7.1, $(\bar{x}_1 - \bar{x}_2) = 3.7$ and $\mathrm{SE}(\bar{x}_1 - \bar{x}_2) = 10.6$. Substituting these values into the formula shown in the box, we obtain

$$\begin{aligned} \mathrm{LCL} &= (\bar{x}_1 - \bar{x}_2) - z_{.025}\mathrm{SE}(\bar{x}_1 - \bar{x}_2) \\ &= 3.7 - 1.96(10.6) = 3.7 - 20.8 = -17.1 \end{aligned}$$

and

$$\begin{aligned} \mathrm{UCL} &= (\bar{x}_1 - \bar{x}_2) + z_{.025}\mathrm{SE}(\bar{x}_1 - \bar{x}_2) \\ &= 3.7 + 20.8 = 24.5 \end{aligned}$$

These values of LCL and UCL differ slightly from the values given on the printout, Figure 7.4, due to rounding errors in computation.

A Small-Sample Confidence Interval for $(\mu_1 - \mu_2)$

Asphalt concrete is used as the surface for many of our highways. The more permeable the surface, the more likely it is that water will seep into and erode the road foundation. The data shown in Table 7.3 are measurements on the permeability of two types of asphalt: one a mix of 3% asphalt, the other 7% asphalt by weight (Woelfl, G., Wei, I., Faulstich, C., and Litwack, H., "Laboratory Testing of Asphalt Concrete for Porous Pavements," *Journal of Testing and Evaluation*, Vol. 9, No. 4,

TABLE 7.3 Permeability Data for Asphalt Concrete

Asphalt Content	
3%	7%
1,189	853
840	900
1,020	733
980	785

July 1981). Permeability of the concrete specimens was measured by passing de-aired water over the specimens and measuring (in inches per hour) the water loss. The greater the water loss through the pavement, the greater is the permeability of the concrete. One object of the experiment was to estimate the difference in mean permeability for the two types of concrete. Since the population variances are unknown and the sample sizes are small, we cannot use the confidence interval based on the z statistic. Once again, we must rely on a Student's t.

When the population variances, σ_1^2 and σ_2^2, are unknown and the sample sizes are small, we can use a t statistic to construct confidence intervals and test hypotheses about $(\mu_1 - \mu_2)$ if we make the following assumptions:

1. **Both populations are normally distributed.**
2. **The population variances, σ_1^2 and σ_2^2, are equal, i.e., they equal a common variance σ^2.**

If these assumptions are satisfied, it can be shown (proof omitted) that

$$t = \frac{\text{Normally distributed point estimate} - \text{Mean of the point estimate}}{\textbf{Estimated}\text{ standard error of the point estimate}}$$

or

$$t = \frac{(\bar{x}_1 - \bar{x}_2) - (\mu_1 - \mu_2)}{\widehat{\text{SE}}(\bar{x}_1 - \bar{x}_2)}$$

is a Student's t statistic with $(n_1 + n_2 - 2)$ degrees of freedom.

The estimated standard error of $(\bar{x}_1 - \bar{x}_2)$ is

$$\widehat{\text{SE}}(\bar{x}_1 - \bar{x}_2) = s\sqrt{\frac{1}{n_1} + \frac{1}{n_2}}$$

where

$$s^2 = \frac{(n_1 - 1)s_1^2 + (n_2 - 1)s_2^2}{n_1 + n_2 - 2}$$

The quantity s^2 is called a *pooled estimator* of the population variance. Theoretically, we could use either s_1^2 or s_2^2 to estimate the common population variance σ^2. However, the pooled estimator is better. It combines (or pools) the information in both samples to obtain an estimate of σ^2 that has a smaller error of estimation than either s_1^2 or s_2^2. Note that the degrees of freedom for the pooled estimator is equal to the sum of the degrees of freedom for s_1^2 and s_2^2, i.e., $(n_1 + n_2 - 2) = (n_1 - 1) + (n_2 - 1)$.

The confidence interval for the difference $(\mu_1 - \mu_2)$ in population means based on the t statistic is similar to the confidence interval for a single population mean (Chapter 6) when the population variance σ^2 is unknown. The lower and upper confidence limits for a $100(1 - a)\%$ confidence interval (as in Chapter 6) are equal to

$$\text{LCL} = \text{Point estimate} - t_{a/2}\{\textbf{Estimated} \text{ standard error of } (\bar{x}_1 - \bar{x}_2)\}$$
$$\text{UCL} = \text{Point estimate} + t_{a/2}\{\textbf{Estimated} \text{ standard error of } (\bar{x}_1 - \bar{x}_2)\}$$

The formula for the confidence interval for $(\mu_1 - \mu_2)$ when σ_1^2 and σ_2^2 are unknown is shown in the box.

A $100(1 - a)\%$ Confidence Interval for $(\mu_1 - \mu_2)$ When σ_1^2 and σ_2^2 Are Unknown

$$\text{LCL} = (\bar{x}_1 - \bar{x}_2) - t_{a/2}\widehat{\text{SE}}(\bar{x}_1 - \bar{x}_2)$$
$$\text{UCL} = (\bar{x}_1 - \bar{x}_2) + t_{a/2}\widehat{\text{SE}}(\bar{x}_1 - \bar{x}_2)$$

where

$$\widehat{\text{SE}}(\bar{x}_1 - \bar{x}_2) = s\sqrt{\frac{1}{n_1} + \frac{1}{n_2}}$$

and s^2, the pooled estimate of the common population variance, σ^2, is

$$s^2 = \frac{(n_1 - 1)s_1^2 + (n_2 - 1)s_2^2}{(n_1 + n_2 - 2)}$$

where s_1^2 and s_2^2 are the variances of samples 1 and 2, respectively, and $t_{a/2}$ is based on $(n_1 + n_2 - 2)$ degrees of freedom.

Assumptions:

1. **Independent random samples are selected from the two populations.**
2. **The two populations are normally distributed.**
3. **The population variances, σ_1^2 and σ_2^2 are equal, i.e., $\sigma_1^2 = \sigma_2^2 = \sigma^2$, but the exact value of σ^2 is unknown.**

Example 7.3

Refer to the asphalt concrete permeability data of Table 7.3.

a. Figure 7.5a shows a Data Desk computer printout of the means and standard deviations for the data in Table 7.3. Find the mean and standard deviation for each sample.

b. Find the 90% confidence interval for the difference in mean permeability for the two mixes. This is shown on the Data Desk printout in Figure 7.5b.

c. Interpret the confidence interval.

d. Suppose that you were unable to obtain a computer printout of the confidence interval in part (b). Use the sample means and standard deviations from part (a) to calculate the lower and upper confidence limits for a 90% confidence interval for the difference in mean permeability for the two mixes. Compare your answers with those found on the printout in part (b).

FIGURE 7.5 A Data Desk computer printout for a comparison of concrete permeability

```
Summary statistics for   Asp 3
NumNumeric = 4
Mean = 1007.2
Standard Deviation = 143.66

Summary statistics for   Asp 7
NumNumeric = 4
Mean = 817.75
Standard Deviation = 73.627
```

a. Means and standard deviations

t-intervals
pooled estimate of σ^2

With 90% confidence, $32.662 \leqslant \mu(\text{Asp 3}) - \mu(\text{Asp 7}) \leqslant 346.34$

b. Confidence interval

Solution

a. The means and standard deviations for samples 1 and 2, 3% asphalt and 7% asphalt, respectively, are shaded on the Data Desk printout in Figure 7.5a: $\bar{x}_1 = 1{,}007.2$, $\bar{x}_2 = 817.75$, $s_1 = 143.66$, and $s_2 = 73.627$.

b. The 90% confidence interval for $(\mu_1 - \mu_2)$, from Figure 7.5b, is 32.662 to 346.34

c. We estimate that the interval from 32.662 to 346.34 encloses the difference in the mean permeability for the two concrete mixes. We cannot be certain that this particular interval encloses $(\mu_1 - \mu_2)$ but we know that our procedure will produce intervals that will enclose $(\mu_1 - \mu_2)$ 90% of the time.

d. The first two steps in finding LCL and UCL are to calculate s^2, the pooled estimate of the variance, and the estimated standard error of $(\bar{x}_1 - \bar{x}_2)$. Substituting first into the formula for s^2, we obtain

$$\begin{aligned} s^2 &= \frac{(n_1 - 1)s_1^2 + (n_2 - 1)s_2^2}{(n_1 + n_2 - 2)} \\ &= \frac{(4 - 1)(143.66)^2 + (4 - 1)(73.627)^2}{(4 + 4 - 2)} \\ &= \frac{61{,}914.587 + 16{,}262.805}{6} \\ &= 13{,}029.565 \end{aligned}$$

or

$$s = \sqrt{s^2} = \sqrt{13{,}029.565} = 114.15$$

Then

$$\begin{aligned} \widehat{\text{SE}}(\bar{x}_1 - \bar{x}_2) &= s\sqrt{\frac{1}{n_1} + \frac{1}{n_2}} \\ &= 114.15\sqrt{\frac{1}{4} + \frac{1}{4}} \\ &= 80.71 \end{aligned}$$

For a 90% confidence interval, $a = .10$ and $a/2 = .05$. Therefore we want to find $t_{.05}$ based on $(n_1 + n_2 - 2) = (4 + 4 - 2) = 6$ degrees of freedom. This value, given in Table 4 of Appendix 2, is $t_{.05} = 1.943$. Then we substitute $\bar{x}_1$, $\bar{x}_2$, $\text{SE}(\bar{x}_1 - \bar{x}_2)$, and $t_{.05}$ into the formulas for LCL and UCL:

$$\begin{aligned} \text{LCL} &= (\bar{x}_1 - \bar{x}_2) - t_{.05}\widehat{\text{SE}}(\bar{x}_1 - \bar{x}_2) \\ &= (1{,}007.2 - 817.75) - 1.943(80.71) \\ &= 189.45 - 156.82 = 32.63 \end{aligned}$$

and

$$\begin{aligned} \text{UCL} &= (\bar{x}_1 - \bar{x}_2) + t_{.05}\widehat{\text{SE}}(\bar{x}_1 - \bar{x}_2) \\ &= 189.45 + 156.82 \\ &= 346.27 \end{aligned}$$

You can see that these values for LCL and UCL agree (except for rounding error) with the computed values shown on the printout in Figure 7.5b.

Tests of the Difference Between Population Means Based on Independent Random Samples

The large- and small-sample tests of hypotheses for the difference in population means are, with minor exceptions, the same as those for a single population mean (Chapter 6). The differences are that $\bar{x}$ is replaced by $(\bar{x}_1 - \bar{x}_2)$ and $SE(\bar{x})$ is replaced by $SE(\bar{x}_1 - \bar{x}_2)$. The two tests are summarized in the boxes and demonstrated by the examples that follow.

A Standard Normal z Test for a Difference Between Two Population Means Based on Independent Random Samples

Null hypothesis: $H_0: \mu_1 = \mu_2$

Alternative hypothesis:

1. $H_a: \mu_1 > \mu_2$, an upper-tailed test
2. $H_a: \mu_1 < \mu_2$, a lower-tailed test
3. $H_a: \mu_1 \neq \mu_2$, a two-tailed test

Test statistic: $z = \dfrac{(\bar{x}_1 - \bar{x}_2)}{SE(\bar{x}_1 - \bar{x}_2)}$

where $SE(\bar{x}_1 - \bar{x}_2) = \sqrt{\dfrac{\sigma_1^2}{n_1} + \dfrac{\sigma_2^2}{n_2}} = \sqrt{SE(\bar{x}_1)^2 + SE(\bar{x}_2)^2}$

Rejection region:

Choose the value of α that is acceptable to you.

Then reject H_0 and accept H_a when the p-value for the test is less than or equal to α.

Assumptions:

1. Independent random samples are selected from the two populations.
2. The sampling distributions of the sample means are approximately normally distributed.
3. The population variances, σ_1^2 and σ_2^2, are known or n_1 and n_2 are large enough ($n_1 \geq 30$ and $n_2 \geq 30$) so that s_1^2 and s_2^2 can be used to approximate σ_1^2 and σ_2^2 in the formula for $SE(\bar{x}_1 - \bar{x}_2)$.

A Student's t Test for a Difference Between Two Population Means Based on Independent Random Samples

Null hypothesis: $H_0: \mu_1 = \mu_2$

Alternative hypothesis:

1. $H_a: \mu_1 > \mu_2$, **an upper-tailed test**
2. $H_a: \mu_1 < \mu_2$, **a lower-tailed test**
3. $H_a: \mu_1 \neq \mu_2$, **a two-tailed test**

Test statistic: $t = \dfrac{(\bar{x}_1 - \bar{x}_2)}{\widehat{SE}(\bar{x}_1 - \bar{x}_2)}$

where $\widehat{SE}(\bar{x}_1 - \bar{x}_2) = s\sqrt{\dfrac{1}{n_1} + \dfrac{1}{n_2}}$

s^2, the pooled estimate of the population variance σ^2, is

$$s^2 = \frac{(n_1 - 1)s_1^2 + (n_2 - 1)s_2^2}{(n_1 + n_2 - 2)}$$

where s_1^2 and s_2^2 are the variances of samples 1 and 2, respectively, and $t_{\alpha/2}$ is based on $(n_1 + n_2 - 2)$ degrees of freedom.

Rejection region:

Choose the value of α that is acceptable to you.

Then reject H_0 and accept H_a when the *p*-value for the test is less than or equal to α.

Assumptions:

1. **Independent random samples are selected from the two populations.**
2. **The two populations are normally distributed.**
3. **The population variances, σ_1^2 and σ_2^2, are equal, i.e., $\sigma_1^2 = \sigma_2^2 = \sigma^2$, but the exact value of σ^2 is unknown.**

Example 7.4

Suppose we believe that the mean cholesterol levels of men and women may differ. Do the total cholesterol data on 49 women and 58 men given in Table 3 of Appendix 1 support this theory?

a. State the appropriate null and alternative hypotheses for testing this theory.

b. Which test statistic would you use for the test and why?

c. Where would you locate the rejection region for the test?

d. Use a computer printout to find the computed p-value for the test. What is the interpretation of this p-value?

e. Do the data support the theory that the mean cholesterol levels for men and women differ? Explain.

f. Suppose that you decide to use $\alpha = .05$ for the test. What are the implications of this choice?

Solution

a. Let μ_1 and μ_2 be the mean cholesterol levels for women and men, respectively. Since we want to detect a difference in population means, either $\mu_1 > \mu_2$ or vice versa, we will want to test the null hypothesis H_0: $\mu_1 = \mu_2$ against the alternative hypothesis H_a: $\mu_1 \neq \mu_2$.

b. We would use the standard normal z statistic because both sample sizes are large—both are equal to 30 or larger.

c. The alternative hypothesis in part (a) implies a two-tailed test. Therefore, we will reject H_0 and conclude that the means differ if z falls in either the upper or the lower tail of the z distribution (see Figure 7.6).

FIGURE 7.6 The rejection region for the test in Example 7.4

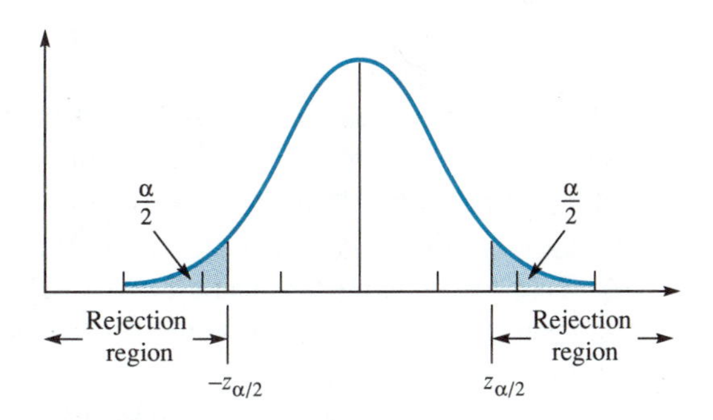

d. The SAS printout shown in Figure 7.7 (page 248) gives the mean, standard deviation, and standard error of the mean for females and males, respectively, in the top two rows of the printout.

FIGURE 7.7 A SAS printout of a *t* test for a difference in mean TCHOL

```
Variable: TCHOL

SEX        N          Mean         Std Dev       Std Error        Minimum         Maximum
-----------------------------------------------------------------------------------------
F         49   219.8163265   60.65052675    8.66436096    123.0000000    414.0000000
M         58   216.0844828   46.48648097    6.10397399    120.0000000    399.0000000

Variances          T          DF      Prob>|T|
----------------------------------------------
Unequal       0.3521        89.0        0.7256
Equal         0.3600       105.0        0.7196

For H0: Variances are equal, F' = 1.70    DF = (48,57)    Prob>F' = 0.0546
```

SAS gives the value of t (rather than z) for the test of population means. This creates no problem. **When the sample sizes are large, the computed values of t and z, along with their p-values, will be almost identical.** In fact, SAS prints the results for two t tests, the standard t test and a more complicated version that is sometimes used when the population variances are unequal. The first, in the "Equal" row of the printout, is the standard t test based on the assumption that the population variances are equal. We will not use the second test, given in the row identified as "Unequal."

See the results for the t test shown in the "Equal" row (which is shaded) on the printout in Figure 7.7. The observed value of t is $t = .3600$. **SAS prints the p-value for a two-tailed test**; in this case it prints the probability that t is either greater than $t = .3600$ or less than $t = -.3600$. This quantity is shown on the printout as Prob $> |T| = .7196$. Therefore, the p-value for our test is .7196. **(The p-value for a one-tailed test would be half this value.)** This p-value tells us that the probability of observing a value of the test statistic at least as contradictory to H_0 as the value we observed is .7196.

e. Because the p-value for the test, $p = .7196$, is larger than .05, we do not reject H_0. This large p-value tells us that there is little or no evidence to indicate that the mean total cholesterol reading for women differs from the mean reading for men.

f. $\alpha = .05$ is the probability of rejecting the null hypothesis, $H_0: \mu_1 = \mu_2$, when it is true. Since we did not reject H_0, α is irrelevant to our decision.

Are we surprised that this test did not show evidence of a difference in the mean TCHOL levels for women and men? Not really! The confidence interval for the difference in means, computed in Example 7.2, included 0 as a very possible value for the difference in means.

Example 7.5

Suppose that we wish to test the hypothesis that there is no difference in mean permeability for the two types of asphalt concrete discussed in Example 7.3.

a. State the appropriate null and alternative hypotheses for the test.

b. Which test statistic should be used for the test and why?

c. Suppose that you decided to use $\alpha = .10$ for the test. In which tail (or tails) of the t distribution is α located?

d. Suppose that your source of information on the concrete experiment, a computer printout or a research paper, did not give either the p-value for the test or the computed value of t. Use the sample means shown on the computer printout in Figure 7.5 and the estimated standard error of $(\bar{x}_1 - \bar{x}_2)$ to calculate the value of the test statistic.

e. Use the value of the t statistic in part (d) and Table 4 of Appendix 2 to find the approximate p-value for the test.

f. Do the data provide sufficient evidence to indicate that the mean permeability of the concrete differs for the two mixes?

Solution

a. Since we are asking whether a difference exists between μ_1 and μ_2, i.e., either $\mu_1 > \mu_2$ or $\mu_1 < \mu_2$, we will test the null hypothesis $H_0\colon \mu_1 = \mu_2$ against the alternative hypothesis $H_a\colon \mu_1 \neq \mu_2$.

b. Because both n_1 and n_2 are small, we use a Student's t as the test statistic with $(n_1 + n_2 - 2) = 4 + 4 - 2 = 6$ degrees of freedom.

c. Evidence contradictory to the null hypothesis will occur when $(\bar{x}_1 - \bar{x}_2)$ is either large and positive (suggesting that $\mu_1 > \mu_2$) or large and negative (suggesting that $\mu_1 < \mu_2$). Therefore, we will conduct a two-tailed test and split α evenly, with half ($\alpha/2 = .10/2 = .05$) in each tail of the t distribution. See Figure 7.8.

FIGURE 7.8 Rejection region for the test in Example 7.5

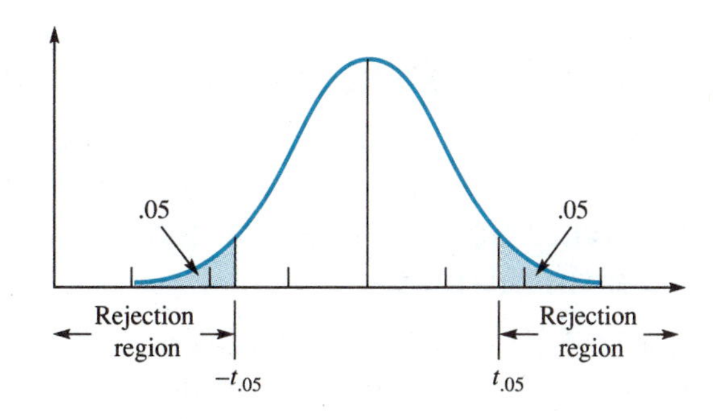

d. The values of the sample means given in the printout in Figure 7.5(a) are $\bar{x}_1 = 1{,}007.2$ and $\bar{x}_2 = 817.75$. The estimated standard error of $(\bar{x}_1 - \bar{x}_2)$, computed in Example 7.3, is $\widehat{SE}(\bar{x}_1 - \bar{x}_2) = 80.71$. Substituting these values into the formula for the test statistic, we find

$$t = \frac{(\bar{x}_1 - \bar{x}_2)}{\widehat{SE}(\bar{x}_1 - \bar{x}_2)} = \frac{1{,}007.2 - 817.75}{80.71} = 2.347$$

You can see that this is the same value (except for rounding error) as the value, $t = 2.348$, given on the Data Desk printout shown in Figure 7.9.

FIGURE 7.9 Data Desk printout of test results for Example 7.5

t-Tests
pooled estimate of σ^2

Test Ho: μ(Asp 3) − μ(Asp 7) = 0
vs Ha: μ(Asp 3) − μ(Asp 7) ≠ 0
Sample mean(Asp 3) = 1007.2 Sample mean(Asp 7) = 817.75
t-statistic = 2.348 with 6 d.f.
Reject Ho at alpha = 0.10

e. Because this is a two-tailed test, the p-value is the probability of observing a value of t as large as or larger than 2.347 or as small as or smaller than -2.347. The values of t_a that locate an area a in the upper tail of the t distribution are given in Table 4 of Appendix 2. The upper-tail areas corresponding to df = 6 degrees of freedom and to $a = .05$ and $a = .025$ are $t_{.05} = 1.943$ and $t_{.025} = 2.447$, respectively. The observed value, $t = 2.347$, falls between these two values. Therefore, the probability of observing a value of t as large as or larger than 2.347 is between .025 and .05. Since the p-value includes an equal area in the lower tail of the t distribution, the p-value for the test falls between .05 and .10.

f. We have chosen $\alpha = .10$. Since the p-value for the test is less than .10, we reject H_0 and conclude that the two types of asphalt concrete differ in mean permeability. Note that this agrees with the Data Desk printout (Figure 7.9), which states "Reject H_0 at alpha = 0.10." The 3% asphalt concrete appears to be more porous than the 7%.

Verifying Statistical Statements Contained in Scientific Articles

It is often difficult to verify claims made in scientific articles because the articles rarely contain the sample data and they often fail to give the sample variances (or standard deviations) that are so necessary for statistical inference. The following example illustrates that, despite these omissions, we can sometimes check the article's conclusions.

Example 7.6

In Section 3.2, we referred to the Minnesota Heart Survey, a study designed to see whether trends in public awareness and treatment of high levels of blood cholesterol produced a change in the mean cholesterol levels of men and women from the period 1980–1982 to the period 1985–1987. Independent random samples of men and women were selected during each time period and the total cholesterol level was measured and recorded for each person. Figure 7.10 shows the relative frequency distributions of total cholesterol (in millimoles per liter),* one for 1,777 women for the time period 1980–1982 and the other for 2,353 women for the time period 1985–1987.

FIGURE 7.10 Relative frequency distributions of total cholesterol for women, 1980–1982 and 1985–1987

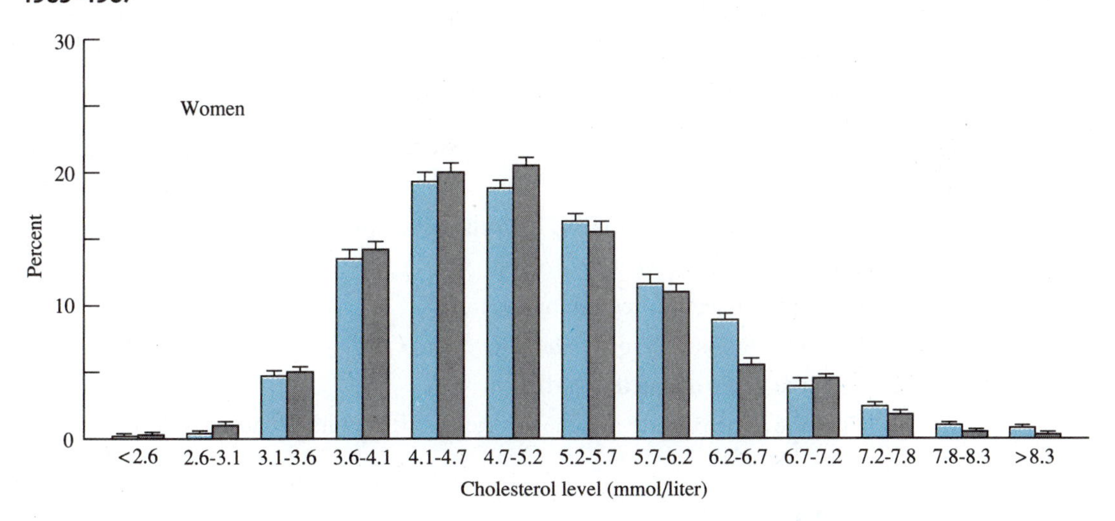

Color bars indicate levels in 1980-1982, and gray bars levels in 1985-1987. T bars indicate the standard error of the mean.

Source: Burke, Gregory L. et al., "Trends in Serum Cholesterol Levels from 1980 to 1987," *The New England Journal of Medicine,* Vol. 324, No. 14, 1991 (Figure 1).

* The total cholesterol readings discussed in Case 2.3 and presented in Table 3 of Appendix 1 are given in milligrams per deciliter. One millimole per liter equals 38.46 milligrams per deciliter.

Table 7.4 gives the number of women measured in each of the two time periods, the mean cholesterol level (in millimoles/liter) for each of the two groups, and the 95% confidence interval for the mean for each of the two groups.

TABLE 7.4 Confidence Intervals for the Mean Cholesterol Levels of Women, 1980–1982 to 1985–1987: The Minnesota Heart Survey

	1980–1982	1985–1987
Sample size	1,777	2,353
Mean	5.19	5.04
95% confidence interval	5.14 to 5.23	5.00 to 5.08

These values were abstracted from Tables 1 and 2 of the article by Burke et al.

Examine Figure 7.10. If current thinking is correct, trends in public awareness of the harmful effects of high cholesterol should have produced an overall decrease in cholesterol levels from 1980–1982 to 1985–1987; that is, the mean μ_1 of the 1980–1982 distribution should be located to the right of the mean μ_2 for the 1985–1987 distribution. Do the data provide sufficient evidence to indicate that the mean μ_1 of the population of total cholesterol measurements for women for the 1980–1982 period is larger than the mean μ_2 for the 1985–1987 period?

a. State the null and alternative hypotheses for the test.

b. Which test should we use?

c. Locate the rejection region for the test.

d. Find the standard error of the difference $(\bar{x}_1 - \bar{x}_2)$ in sample means.

e. Find the value of the test statistic.

f. Find the approximate p-value for the test. If you have chosen $\alpha = .05$, what do you conclude?

Solution

a. We want to show that the mean μ_1 for the 1980–1982 period is larger than the mean μ_2 for the 1985–1987 period. Therefore, the alternative hypothesis is $H_a\colon \mu_1 > \mu_2$. The null hypothesis is that there is no difference in the population means, i.e., $H_0\colon \mu_1 = \mu_2$.

b. Because the sample sizes are so large, we will use the z test and eliminate any need to make an assumption about the equality of the population variances.

c. Large values of $(\bar{x}_1 - \bar{x}_2)$ would provide evidence to indicate that $\mu_1 > \mu_2$.

Therefore, the rejection region for the test is located in the upper tail of the z distribution (see Figure 7.11).

FIGURE 7.11 The rejection region for a test of the difference in mean total cholesterol

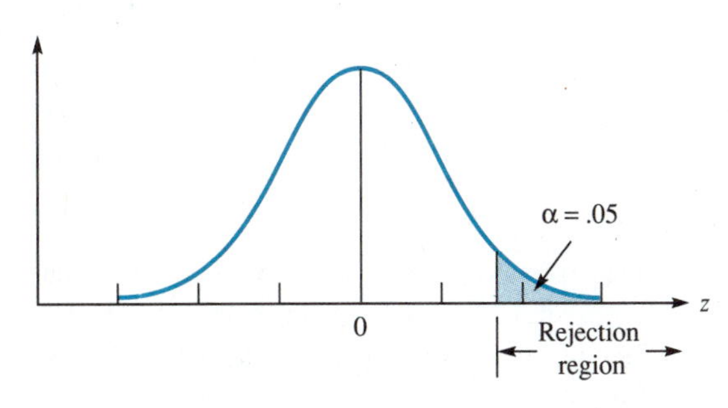

d. The standard error of the sampling distribution of the difference $(\bar{x}_1 - \bar{x}_2)$ between two sample means is

$$\text{SE}(\bar{x}_1 - \bar{x}_2) = \sqrt{\frac{\sigma_1^2}{n_1} + \frac{\sigma_2^2}{n_2}} = \sqrt{\text{SE}(\bar{x}_1)^2 + \text{SE}(\bar{x}_2)^2}$$

At first glance, it would seem impossible to calculate $\text{SE}(\bar{x}_1 - \bar{x}_2)$ because the article by Burke et al. does not give the sample standard deviations. We can, however, calculate $\text{SE}(\bar{x}_1)$ and $\text{SE}(\bar{x}_2)$ from the confidence intervals for the population means. For example, the 95% confidence interval for the mean of population 1 is

$$\bar{x}_1 \pm z_{.025}\text{SE}(\bar{x}_1)$$

and the half-width of the interval is $1.96\text{SE}(\bar{x}_1)$. Table 7.4 shows that the confidence interval for μ_1 is 5.14 to 5.23 millimoles per liter (mmol/liter). Therefore, the width of the interval is .09 mmol/liter and the half-width is .045. Equating the half-width .045 to $1.96\text{SE}(\bar{x}_1)$ gives

$$1.96\text{SE}(\bar{x}_1) = .045 \text{ mmol/liter}$$

and solving for $\text{SE}(\bar{x}_1)$,

$$\text{SE}(\bar{x}_1) = \frac{.045}{1.96} = .0230$$

Using a similar procedure,

$$\text{SE}(\bar{x}_2) = \frac{.04}{1.96} = .0204$$

Substitute these quantities into the formula for $SE(\bar{x}_1 - \bar{x}_2)$ to obtain

$$SE(\bar{x}_1 - \bar{x}_2) = \sqrt{SE(\bar{x}_1)^2 + SE(\bar{x}_2)^2} = \sqrt{(.0230)^2 + (.0204)^2}$$
$$= \sqrt{.0005290 + .0004162} = \sqrt{.0009452} = .0307$$

e. The value of the test statistic is

$$z = \frac{(\bar{x}_1 - \bar{x}_2)}{SE(\bar{x}_1 - \bar{x}_2)}$$

The sample means $\bar{x}_1 = 5.19$ and $\bar{x}_2 = 5.04$ are given in Table 7.4 and we found $SE(\bar{x}_1 - \bar{x}_2) = .0307$ in part (d). Therefore,

$$z = \frac{(\bar{x}_1 - \bar{x}_2)}{SE(\bar{x}_1 - \bar{x}_2)} = \frac{(5.19 - 5.04)}{.0307} = \frac{.15}{.0307} = 4.89$$

f. The approximate p-value for the test is near 0. The probability of observing a value of $z = 1.96$ or larger is only .025. The probability that z will be as large as or larger than the observed value, $z = 4.89$, is too small to calculate. Therefore, if we choose $\alpha = .05$, we reject H_0 and conclude that the mean cholesterol level for the 1980–1982 population is larger than the mean for the 1985–1987 population. It appears that something has happened over the period 1980–1982 to 1985–1989 to produce a reduction in the mean cholesterol level of women sampled in the Minnesota Study.

Statisticians would say that the test results in part (f) of Example 7.6 are **statistically significant**—i.e., the test results provide sufficient evidence to reject the null hypothesis and accept the alternative hypothesis. Statistical significance does not imply that the difference in means is large enough to be of **practical significance.**

Comments on the Assumptions

1. Like the confidence interval and test for a single population mean based on a Student's t statistic (Chapter 6), the small-sample interval and test for the difference $(\mu_1 - \mu_2)$ in a pair of population means are fairly robust to the assumption of normality. If the population relative frequency distributions are mound shaped, the methods "work." The confidence coefficient for a confidence interval and the value of α for a test will be approximately the values that we specify.

2. The confidence interval and test are moderately robust (i.e., insensitive) to departures from the assumption that the population variances are equal when the sample sizes n_1 and n_2 are equal. Unequal population variances can affect the values of the confidence coefficients and values of α if the sample sizes are substantially different.

Which Method Should We Use?

Most statistical software packages compute confidence intervals and tests based on the Student's t statistic. The reason is that methods based on the z and the t statistics give approximately the same results when σ_1^2 is approximately equal to σ_2^2 and n_1 and n_2 are large, say $n_1 \geqslant 30$ and $n_2 \geqslant 30$. If the sample sizes are small, only the methods based on the t statistic are appropriate. *Remember:* The validity of the results depends on the assumptions.

EXERCISES

7.6 Suppose that you drew independent random samples of n_1 and n_2 observations, respectively, from populations with means and variances (μ_1, σ_1^2) and (μ_2, σ_2^2).

a. Give the mean and standard error of the sampling distribution of $(\bar{x}_1 - \bar{x}_2)$.
b. When will the sampling distribution of $(\bar{x}_1 - \bar{x}_2)$ be (approximately) a normal distribution?
c. Give the formula for the z statistic used to test hypotheses about $(\mu_1 - \mu_2)$ when σ_1^2 and σ_2^2 are known.

7.7 What assumptions must be satisfied to assure the validity of a confidence interval or test based on the z statistic?

7.8 What assumptions must be satisfied to assure the validity of a confidence interval or test for $(\mu_1 - \mu_2)$ based on the t statistic? How many degrees of freedom will t possess?

7.9 Give the formula for the t statistic used to test hypotheses about $(\mu_1 - \mu_2)$ when σ_1^2 and σ_2^2 are unknown.

7.10 Why do we use a pooled estimate of σ^2, instead of s_1^2 or s_2^2, in the formula for $\widehat{\text{SE}}(\bar{x}_1 - \bar{x}_2)$?

7.11 Give the formula for the pooled estimate of σ^2.

HLTH **7.12** Refer to Table 3 of Appendix 1, which gives cholesterol readings on 49 women and 58 men. As noted earlier, many physicians regard the ratio, RATIO = LDL/HDL, as a better measure of the risk of heart disease than the total cholesterol or TCHOL measurement. Figure 7.12 shows the

FIGURE 7.12 A Minitab printout of the confidence interval for the difference in mean male and female ratios

```
MTB > twot 95=k c6 c4

TWOSAMPLE T FOR ratio
sex     N        MEAN      STDEV    SE MEAN
0      58        3.39       1.29       0.17
1      49        2.58       1.56       0.22

95 PCT CI FOR MU 0 - MU 1: (0.25, 1.36)

TTEST MU 0 = MU 1 (VS NE): T= 2.89  P=0.0048  DF=  93
```

Minitab printout of the means, standard deviations, and standard errors of the means of the ratios, LDL/HDL, for the two samples as well as the 95% confidence interval for the difference $(\mu_1 - \mu_2)$ in mean ratios. A coding of 0 denotes males; 1 denotes females.

a. Find the 95% confidence interval for $(\mu_1 - \mu_2)$, the difference in mean LDL/HDL ratios for men and women, on the printout.

b. Give the formula used to calculate the confidence interval in part (a). State the assumptions required for the confidence interval and explain why they are or are not satisfied.

c. Use the numerical descriptive measures on the printout in Figure 7.12 to calculate $SE(\bar{x}_1 - \bar{x}_2)$, LCL, and UCL. Compare your answers for LCL and UCL with the computed values in part (a).

d. Interpret the confidence interval in part (a).

BIO **7.13** Bees and other flying insects expend large amounts of energy as they hover in flight. One measure of this energy expenditure is the wing stroke frequency of the insect. Researchers conducted an experiment to compare the wing stroke frequencies for two species of bees. The data in Table 7.5 show the wing frequencies (in hertz) for $n = 4$ *Euglossa mandibularis Friese* bees and $n = 6$ *Euglossa imperialis Cockerell* bees (Casey, T. M., May, M. L., and Morgan, K. R., "Flight Energetics of Euglossine Bees in Relation to Morphology and Wing Stroke Velocity," *Journal of Experimental Biology*, Vol. 116, 1985). Figure 7.13a gives the Data Desk printout for the numerical descriptive measures for the samples. Figure 7.13b gives a 90% confidence interval for $(\mu_1 - \mu_2)$, the difference in mean stroke velocity for the two species, and Figure 7.13c gives the results for a statistical test that we will use in Exercise 7.15.

a. Locate the confidence interval on the printout and interpret it.

b. Why would we choose a confidence coefficient as low as .90 in part (a)? Why did we not construct a 99%, or even a 99.9%, confidence interval for $(\mu_1 - \mu_2)$ so that we could be very confident of our results?

c. Find the mean and standard deviation for each of the two samples on the printout. Use them to calculate the pooled estimate of the common variance and the estimate of the standard error of $(\bar{x}_1 - \bar{x}_2)$.

d. Use the results of part (c) to calculate LCL and UCL for the confidence interval. Compare your answers with the values given on the printout in part (a).

HLTH **7.14** Refer to the comparison of the mean of the LDL/HDL ratios for men and women, Exercise 7.12. Suppose that we suspected that the mean RATIO for men is greater than the mean RATIO for women. Do the data support this theory? The information needed to answer this question is contained on the SAS printout in Figure 7.14 (page 258).

a. If you wish to answer the question "Do the data provide sufficient evidence to indicate that the mean RATIO for men is greater than the mean RATIO for women?", what would you choose for the null and alternative hypotheses?

TABLE 7.5 Wing Stroke Frequencies (in hertz)

Sample 1 (*Euglossa mandibularis Friese*)	Sample 2 (*Euglossa imperialis Cockerell*)
235	180
225	169
190	180
188	185
	178
	182

FIGURE 7.13 Data Desk printout for a comparison of wing stroke frequencies, Exercise 7.13

Summary statistics for **Bees 1**
NumNumeric = 4
Mean = 209.50
Median = 207.50
Standard Deviation = 24.035
Interquartile range = 44
Range = 47
Variance = 577.67
Minimum = 188
Maximum = 235
25-th %ile = 188.50
75-th %ile = 232.50

Summary statistics for **Bees 2**
NumNumeric = 6
Mean = 179
Median = 180
Standard Deviation = 5.4406
Interquartile range = 7
Range = 16
Variance = 29.600
Minimum = 169
Maximum = 185
25-th %ile = 175.75
75-th %ile = 182.75

a. Numerical descriptive measures

t-intervals
pooled estimate of σ^2

With 90% confidence, $12.094 \leqslant \mu(\text{Bees 1}) - \mu(\text{Bees 2}) \leqslant 48.906$

b. Confidence interval

t-Tests
pooled estimate of σ^2

Test Ho: $\mu(\text{Bees 1}) - \mu(\text{Bees 2}) = 0$
vs Ha: $\mu(\text{Bees 1}) - \mu(\text{Bees 2}) \neq 0$
Sample mean(Bees 1) = 209.50 Sample mean(Bees 2) = 179
t-statistic = 3.081 with 8 d.f.
Reject Ho at alpha = 0.10

c. Statistical test

FIGURE 7.14 SAS printout for a comparison of mean LDL/HDL ratios, men versus women (Exercise 7.14)

```
Variable: RATIO

SEX      N          Mean        Std Dev      Std Error       Minimum        Maximum
---------------------------------------------------------------------------------------
F       49    2.57189847    1.54190706    0.22027244    0.46391753    6.48712121
M       58    3.37786322    1.27842775    0.16786579    1.31818182    8.57352941

Variances        T          DF      Prob>|T|
---------------------------------------------
Unequal     -2.9102        93.4      0.0045
Equal       -2.9563       105.0      0.0038

For H0: Variances are equal, F' = 1.45    DF = (48,57)    Prob>F' = 0.1746
```

b. Find the value of the test statistic on the printout.

c. How many degrees of freedom are shown on the printout?

d. Use the printout to find the computed *p*-value for the test.

e. Suppose that you decide to use $\alpha = .05$ for the probability of a Type I error. What is the implication of this decision?

f. What is your conclusion? Do the data provide sufficient evidence to indicate that the mean RATIO for men is greater than the mean RATIO for women?

BIO **7.15** Refer to Exercise 7.13 and the wing stroke frequencies for the two species of bees.

a. Suppose that we wish to determine whether the data support the theory that the mean wing stroke frequencies for the two species differ. Which test statistic should we use for the test and why?

b. State the null and alternative hypotheses for the test.

c. Where is the rejection region located for the test in part (b)?

d. Find the *p*-value for the test on the printout in Figure 7.13c. What does it mean?

e. If you have chosen $\alpha = .10$ for the test, what are your conclusions?

f. In what sense does $\alpha = .10$ measure the risk of a faulty conclusion for this test?

HLTH **7.16** Recall that the Framingham Heart Study, Example 6.11, was conducted to detect changes in the risk factors and mortality from cardiac disease. Samples of men between the ages of 50 and 59 were collected at the beginning of each of the baseline years, 1950, 1960, and 1970. Each of these groups was divided into two subgroups, those with cardiovascular disease (CVD) and those without. The six sample sizes are shown in Table 7.6 and statistics from the paper by Sytkowski et al. are reproduced in Table 7.7. See the line in Table 7.7 that gives the sample statistics for serum cholesterol. The interval, for each of the six samples, is the sample mean $\bar{x}$ plus or minus the sample standard deviation s,

TABLE 7.6 Sample Sizes for the Framingham Study

	Free of CVD at Baseline			CVD Within 10 Years of Baseline		
Baseline	1950	1960	1970	1950	1960	1970
Sample size	485	464	512	64	77	101

TABLE 7.7 Table of Selected Risk Factors from the Paper by Sytkowski et al.

Risk Factor	Free of CVD at Baseline 1950	1960	1970	Incident CVD Within 10 years of Baseline 1950	1960	1970
Serum cholesterol (mg/dl)	228±40	243±37	221±38	239±44	246±35	227±40
Smokers (%)	56	52	34	64	60	57
Definite hypertension (%)	21	23	15	36	41	20
Use of antihypertensive medication (%)	0	11	22	0	11	15
Systolic blood pressure (mm Hg)	139±25	137±21	135±19	152±30	148±22	140±19
Diastolic blood pressure (mm Hg)	85±13	86±12	84±10	91±15	91±12	85±11
Metropolitan relative weight (%)	120±15	120±15	123±17	121±17	123±15	121±18

Source: Sytkowski, P. A., Kannel, W. B., and D'Agostino, R. B., "Changes in Risk Factors and the Decline in Mortality from Cardiovascular Disease," *The New England Journal of Medicine*, Vol. 322, No. 23, 1990 (Table 6).

i.e., $(\bar{x} \pm s)$. For example, the mean and standard deviation for the "Free of CVD" group in 1950 are $\bar{x} = 228$ and $s = 40$.

Suppose that we theorize that the mean level of cholesterol in the population of "Free of CVD" men, ages 50 to 59, decreased from 1960 to 1970 (perhaps due to the publicity about the harmful effects of high cholesterol). If our theory is correct, then the mean total cholesterol μ_1 in 1960 would be larger than the mean μ_2 for 1970.

a. State the null and alternative hypotheses that you should use to test your theory.

b. Which test statistic should you use for the test and why?

c. Is this a one- or two-tailed test? Explain.

d. Find the value of the test statistic.

e. Find the approximate p-value for the test.

f. Suppose that you choose $\alpha = .10$ for the test. What are the implications of this choice?

g. Do the data present sufficient evidence to indicate that the mean cholesterol level in men, ages 50 to 59, decreased from 1960 to 1970?

h. What is the probability that your answer in part (g) will be incorrect?

HLTH **7.17** Repeat the instructions of Exercise 7.16, but test the theory for the group of subjects who had CVD within 10 years of the baseline.

HLTH **7.18** Find a 95% confidence interval for the difference in mean cholesterol levels for non-CVD men in 1960 versus non-CVD men in 1970.

7.4 Comparing Population Means: A Matched-Pairs Design

Recall from Section 7.2 that a matched-pairs design makes a comparison of two treatment means within matched pairs of elements. The objective of the design is to buy a specified amount of information—say, a confidence interval of a specified width—at a **lower cost** than would be required using independent random sampling.

The comparison of two different teaching methods, call them Method 1 and Method 2, is typical of experiments for which the paired-difference design would be appropriate. Twenty 6-year-old children were matched on IQ scores to form $n = 10$ matched pairs. One member of each pair was randomly selected and assigned to learn with Method 1; the remainder were assigned to learn with Method 2. After a 4-month period, each child took a standardized achievement test. The scores are shown in columns 2 and 3 of Table 7.8. The difference d in the scores for each matched pair is shown in the last column of Table 7.8.

TABLE 7.8 Achievement Test Scores for a Matched-Pairs Design

Pair	Method 1	Method 2	Difference d
1	78	69	9
2	63	56	7
3	95	77	18
4	75	62	13
5	65	60	5
6	79	59	20
7	82	85	−3
8	85	72	13
9	67	51	16
10	72	71	1
Sample means	76.10	66.20	9.90

The last row of Table 7.8 gives the sample mean for each method and the mean of the sample differences. Note that the difference in sample means $(\bar{x}_1 - \bar{x}_2) = 76.10 - 66.20 = 9.90$ is equal to the mean $d = 9.90$ of the paired differences. Similarly, the difference $(\mu_1 - \mu_2)$ in population means is equal to the mean μ_d of the population of paired differences. Therefore, we can use the mean $\bar{d}$ of the n paired differences to estimate and test hypotheses about the difference μ_d in population means using the methods of Chapter 6.

Confidence Intervals for the Difference in Population Means Based on Matched Pairs

The formulas for confidence intervals and tests of hypotheses about μ_d are exactly the same as those given in Chapter 6 for a single population mean μ,

except that the x's in the formulas are replaced by d's. The sample mean $\bar{d}$ and standard deviation s_d of the n sample differences are calculated in the same way that we calculated $\bar{x}$ and s in Chapter 4:

$$\bar{d} = \frac{\sum d}{n} \quad \text{and} \quad s_d = \sqrt{\frac{\sum(d - \bar{d})^2}{n - 1}}$$

The confidence interval for the difference in two population means when the variance σ_d^2 of the differences is known or when the number n of differences is large ($n \geqslant 30$) is shown in the box.

A 100(1 − a)% Paired-Difference Confidence Interval for $\mu_d = \mu_1 - \mu_2$ When the Value of σ_d^2 Is Known

$$[\bar{d} - z_{a/2}\text{SE}(\bar{d})] \quad \text{to} \quad [\bar{d} + z_{a/2}\text{SE}(\bar{d})]$$

where the standard error of $\bar{d}$ is

$$\text{SE}(\bar{d}) = \frac{\sigma_d}{\sqrt{n}}$$

and $\bar{d}$ is the mean of the n differences.

Assumptions:

1. **The n differences represent a random sample from the population of all paired differences.**
2. **The sample mean $\bar{d}$ is approximately normally distributed (i.e., n is large enough so that the Central Limit Theorem applies).**
3. **The standard deviation σ_d of the population of differences is known or n is large enough ($n \geqslant 30$) so that s_d provides a good estimate of σ_d in the formula for SE($\bar{d}$).**

A corresponding confidence interval for μ_d when σ_d^2 is unknown is based on a Student's t statistic with $(n - 1)$ degrees of freedom. It is given in the accompanying box.

A 100(1 − a)% Confidence Interval for μ_d When the Value of σ_d^2 Is Unknown

$$[\bar{d} - t_{a/2}\widehat{\text{SE}}(\bar{d})] \quad \text{to} \quad [\bar{d} + t_{a/2}\widehat{\text{SE}}(\bar{d})]$$

where the estimated standard error of $\bar{d}$ is

$$\widehat{\text{SE}}(\bar{d}) = \frac{s_d}{\sqrt{n}}$$

(continued)

$\bar{d}$ and s_d are the mean and standard deviation of the n differences, and t is based on df $= n - 1$ degrees of freedom.

Assumptions:

1. **The n differences represent a random sample from the population of all paired differences.**
2. **The population of differences is normally distributed with unknown mean μ_d and standard deviation σ_d.**

Example 7.7

Refer to the paired-difference experiment summarized in Table 7.8.

a. Use a computer to find $\bar{d}$, s_d, and a 90% confidence interval for the difference in the mean achievement test scores for the two teaching methods.

b. Interpret the interval.

c. Use the range of the differences to check on the computed value of s_d.

d. Suppose that your computer printout gave only the values of $\bar{d}$ and s_d. Use the values of n, $\bar{d}$, and s_d, found in part (a), to calculate $\widehat{\text{SE}}(\bar{d})$ and the lower and upper 90% confidence limits for $(\mu_1 - \mu_2)$. Compare your answers with the values given for LCL and UCL in Figure 7.15b.

FIGURE 7.15 Data Desk printouts of descriptive statistics and a confidence interval based on a matched-pairs design

Summary statistics for **Difference**
NumNumeric = 10
Mean = 9.9000
Standard Deviation = 7.4751

a. Description of differences

t-interval, paired samples

With 90% confidence, $5.5668 \leqslant \mu(\text{Method 1}) - \mu(\text{Method 2}) \leqslant 14.233$

b. Confidence interval

Solution

a. Figure 7.15 shows the Data Desk printouts for the numerical descriptive measures of the sample of $n = 10$ differences and the 90% confidence interval for μ_d or, equivalently, for $(\mu_1 - \mu_2)$. The values of $\bar{d} = 9.9$, $s_d = 7.4751$, and the confidence interval, LCL = 5.5668 and UCL = 14.233, are shaded on the printout.

b. We estimate that the difference $(\mu_1 - \mu_2)$ in mean achievement test scores is contained in the interval from 5.5668 to 14.233. That is, we estimate that the mean achievement test scores for students taught by Method 1 exceed those for students taught by Method 2 by as little as 5.5668 to as much as 14.233. How confident are we of the truth of this statement? The answer is that confidence intervals constructed using this methodology will enclose $(\mu_1 - \mu_2)$ with probability equal to the confidence coefficient, .90.

c. The range of the $n = 10$ differences is

$$\text{Range} = \text{Largest} - \text{Smallest} = 20 - (-3) = 23$$

Consulting Table 4.6, we find that the range will equal approximately 3 standard deviations (instead of the usual 4 or 6) for a sample of only 10 measurements. Therefore, the range approximation to s_d is

$$s_d \approx \frac{\text{Range}}{3} = \frac{23}{3} = 7.67$$

You can see that this very rough approximation agrees reasonably well with the computed value, $s_d = 7.4751$, found in part (a).

d. Since $n = 10$ and the printout, Figure 7.15a, shows $s_d = 7.4751$, the estimated standard error of $\bar{d}$ is

$$\widehat{\text{SE}}(\bar{d}) = \frac{s_d}{\sqrt{n}} = \frac{7.4751}{\sqrt{10}} = 2.3638$$

To calculate LCL and UCL for a 90% confidence interval, we need to find the value of $t_{.05}$ based on $(n - 1) = (10 - 1) = 9$ degrees of freedom. This value, given in Table 4 of Appendix 2, is $t_{.05} = 1.833$. Substituting $t_{.05}$, $\bar{d}$, and s_d into the formulas for LCL and UCL, we obtain

$$\text{LCL} = \bar{d} - t_{.05}\widehat{\text{SE}}(\bar{d}) = 9.90 - 1.833(2.3638) = 5.567$$

$$\text{UCL} = \bar{d} + t_{.05}\widehat{\text{SE}}(\bar{d}) = 9.90 + 1.833(2.3638) = 14.233$$

You can see that these calculated values of LCL and UCL agree with those given by the Data Desk printout in Figure 7.15b.

Matched-Pairs Tests for a Difference in Population Means

The tests of hypotheses based on an analysis of the differences from a matched-pairs experiment are identical to those of Section 6.6 except that

the x's in the formulas are replaced by d's. The formulas are shown in the boxes and the small-sample test (based on Student's t) is illustrated with an example.

A Standard Normal z Test for a Difference Between Two Population Means Based on Matched Pairs

Null hypothesis: $H_0: \mu_d = 0$ (i.e., $\mu_1 = \mu_2$)

Alternative hypothesis:

1. $H_a: \mu_d > 0$ (i.e., $\mu_1 > \mu_2$), an upper-tailed test
2. $H_a: \mu_d < 0$ (i.e., $\mu_1 < \mu_2$), a lower-tailed test
3. $H_a: \mu_d \neq 0$ (i.e., $\mu_1 \neq \mu_2$), a two-tailed test

Test statistic: $z = \dfrac{\bar{d}}{SE(\bar{d})}$ where $SE(\bar{d}) = \dfrac{\sigma_d}{\sqrt{n}}$

and $\bar{d}$ is the mean of the n differences.

Rejection region:

Choose the value of α that is acceptable to you.

Then reject H_0 and accept H_a when the p-value for the test is less than or equal to α.

Assumptions:

1. The n differences represent a random sample from the population of all paired differences.
2. The sample mean $\bar{d}$ is approximately normally distributed (i.e., n is large enough so that the Central Limit Theorem applies).
3. The standard deviation σ_d of the population of differences is known or n is large enough ($n \geqslant 30$) so that s_d provides a good estimate of σ_d in the formula for $SE(\bar{d})$.

A Student's t Test for a Difference Between Two Population Means Based on Matched Pairs

The test is identical to the standard normal z test except that the test statistic is

$$t = \frac{\bar{d}}{\widehat{SE}(\bar{d})} \quad \text{where} \quad \widehat{SE}(\bar{d}) = \frac{s_d}{\sqrt{n}}$$

and t is based on df $= n - 1$ degrees of freedom.

Assumptions:

1. **The n differences represent a random sample from the population of all paired differences.**
2. **The population of differences is normally distributed with unknown mean μ_d and standard deviation σ_d.**

Example 7.8

Refer to the data on the two teaching methods given in Table 7.8. Suppose that Method 2 is the method currently employed for teaching 6th graders in a particular school system and the school board is trying to decide whether Method 2 should be replaced with a new teaching method, Method 1. Do the data in Table 7.8 provide sufficient evidence to indicate that Method 1 produces a higher mean score on the achievement test than Method 2?

a. State the null and alternative hypotheses for the test.

b. Identify the test statistic that you will use for the test. Explain why it was chosen.

c. Obtain a computer printout of the test results. Find the p-value for the test. What is the interpretation of the p-value?

d. If you have chosen $\alpha = .10$, state your test conclusions.

e. What is the relevance of your choice of $\alpha = .10$ to the test?

Solution

a. Since we suspect that Method 1 is better than Method 2 (which is why we ran the experiment), we want to show that $\mu_1 > \mu_2$ or, equivalently, that $\mu_d > 0$. Therefore, we will conduct an upper one-tailed test of the null hypothesis that there is no difference in the mean achievement of students taught by the two methods [H_0: $\mu_d = (\mu_1 - \mu_2) = 0$] against the alternative hypothesis that the mean achievement test score for Method 1 is larger than the mean score for Method 2 [H_a: $\mu_d > 0$].

b. Because σ_d is unknown and the sample size $n = 10$ is small, we will conduct the test using the Student's t as the test statistic.

c. A Data Desk printout of the test results is shown in Figure 7.16 (page 266). The p-value for the test, written as "Prob < = 0.0023," is .0023. This means that the probability of observing a test result at least as contradictory as H_0 as that which we have observed, assuming that H_0 is true, is only .0023.

d. The small p-value tells us that we have observed a very rare (improbable) event if, in fact, H_0 is true. We have decided to reject H_0 if the p-value for the test is less than $\alpha = .10$. Since p-value $= .0023$ is less than $\alpha = .10$, we reject H_0 and conclude that $\mu_d > 0$, i.e., the mean achievement test score for Method 1 is larger than the mean score for Method 2.

FIGURE 7.16 Data Desk printout of matched-pairs test for teaching methods

t-Test, paired samples

Method 1 − Method 2: Test H0: $\mu = 0$ vs Ha: $\mu > 0$
Sample mean = 9.9000 t-statistic = 4.188 with 9 d.f.
Reject H0 at alpha = 0.10 Prob < = 0.0023

e. Since we rejected H_0, we have a measure of our risk of being wrong. We decided to reject H_0 when the p-value for the test was less than or equal to $\alpha = .10$. Using this procedure, we will wrongly reject H_0, when it is true, 10% of the time.

Use the Method of Analysis Appropriate for Your Design

The properties of a confidence interval or a test of an hypothesis are derived assuming that the probability sample has been selected in some specified way. For example, suppose that you calculate a 95% confidence interval for the difference in a pair of population means using the formula appropriate for an independent random samples design (Section 7.3). Then we know that our confidence interval will enclose $(\mu_1 - \mu_2)$ 95% of the time. If you used the same (the wrong) formula for data collected according to a matched-pairs (or some other) design, we do not know what the confidence coefficient will be. It may be .90, .80, .5, or whatever.

The point is that you must analyze your data using the method derived for your sampling procedure. If you used an independent random samples design, then you should use the confidence intervals and tests of hypotheses appropriate for that design—namely, those of Section 7.3. If a matched-pairs design was employed, you should use the confidence intervals and tests of Section 7.4.

You often cannot tell from looking at a data set how the data were collected (and, therefore, how they should be analyzed). For example, suppose that columns 1 and 4 were missing from Table 7.8, i.e., that the table contained only the 20 achievement test scores, 10 in a column for Method 1 and 10 in a column for Method 2. You would not know whether the data were collected in independent random samples or in matched pairs. If you do not know how the data were collected, ask! Otherwise, there is no way to tell whether the data have been properly analyzed.

How Much Money Can We Save by Using a Matched-Pairs Design?

To answer this question, we must have a way of measuring the amount of information in a sample relevant to a particular parameter. One good method is to use the width of a 95% confidence interval for the parameter. We can then compare the approximate "cost" of two different sampling designs by comparing the sample sizes required to obtain a confidence interval of the same width.

We learned in Section 6.7 that the half-width of a confidence interval for a population mean is inversely proportional to the square root of the sample size. This relationship holds, approximately, for most confidence intervals encountered in statistics. For example, we would have to increase the sample size by a multiple of 4 to cut the width of a confidence interval in half, and by a multiple of 9 to cut it to one-third its original size.

Example 7.9

Let's use these ideas to find the savings obtained by using a matched-pairs design instead of an independent random samples design for the comparison of student achievement scores, Example 7.7.

Solution

The confidence interval for the difference in the mean achievement test scores between Methods 1 and 2 using a matched-pairs design (Example 7.7) is 5.567 to 14.233 points. We cannot calculate the corresponding confidence interval for an independent random samples design because we did not collect the data that way. But we can approximate the confidence interval by treating the matched-pairs data as if they had been collected using an independent samples design.

The Data Desk printout of the 90% confidence interval for $(\mu_1 - \mu_2)$ using the unpaired method of analysis, Section 7.3, is shown in Figure 7.17. The width of this interval is $(17.776 - 2.024) = 15.752$. The width of the correct confidence interval, the interval based on the paired-data analysis, Section 7.4 (shown in Figure 7.15), is $(14.233 - 5.567) = 8.666$. Therefore, the confidence interval based on

FIGURE 7.17 Data Desk printout of a confidence interval based on an independent random samples design

t-intervals
pooled estimate of σ^2

With 90% confidence, $2.0243 \leqslant \mu(\text{Method 1}) - \mu(\text{Method 2}) \leqslant 17.776$

the matched-pairs design is a little more than one-half as wide as the confidence interval (as it might have been) had we run an independent random samples design. How do we compare the relative costs of the two designs? If we base the costs solely on the number of elements included in the sample, the matched-pairs design buys approximately the same amount of information as an independent random samples design at a little more than one-fourth the cost (sample size).

EXERCISES

7.19 Why can we use the confidence intervals and tests of hypotheses appropriate for making inferences about a single mean (Sections 6.3 and 6.6) to make inferences about $(\mu_1 - \mu_2)$ when the data have been collected according to a matched-pairs design?

7.20 If you are presented with two samples and you wish to test an hypothesis about the difference in the population means, how can you decide whether to use a paired or an unpaired test?

GEN **7.21** Suppose that you want to compare the mean time to type a document using two different word-processing packages. Explain how you might pair the "length of time to type the document" for times recorded using the two word processors. Remember that the objective in pairing is to reduce the variation in differences between matched pairs.

MFG **7.22** The strength and other properties of concrete vary from one batch of concrete to another. Can you suggest a way to increase the information in the experiment described in Example 7.3?

MFG **7.23** Suppose that you wish to estimate the difference in an electrical characteristic of some electronic tubes using two different pieces of test equipment. You think that one piece of test equipment may be reading higher (or lower) than the other for the same electronic tube. Can you suggest a way to match the observations to increase the information in the experiment? A number of tubes are available for testing and the tubes can be retested on either piece of test equipment.

7.24 Suppose that the variation of the differences in matched pairs of observations, one from each of two populations, is the same as the variation in the differences between two observations selected at random, one from each population. Will a matched-pairs design provide more information about $(\mu_1 - \mu_2)$ than would an independent random samples design? Assume that the number of differences is the same for both methods. Explain.

7.25 Suppose that a matched-pairs design involves 16 matched pairs.

a. How many degrees of freedom will be associated with the t test statistic?

b. Suppose that the test is two-tailed and that the observed value of t is 2.10. What is the approximate p-value for the test?

c. If you have chosen $\alpha = .05$, what is your test conclusion?

d. What is the relevance of α to your decision?

7.26 Suppose that a matched-pairs design involves 8 matched pairs.

a. How many degrees of freedom will be associated with the t test statistic?

b. Suppose that the test is two-tailed and that the observed value of t is 2.65. What is the approximate p-value for the test?

c. If you have chosen $\alpha = .05$, what is your test conclusion?

d. What is the relevance of α to your decision?

BUS **7.27** Two procedures for packing china in shipping containers were compared in the following manner. Each of 20 container packers was selected to pack a box using each of the two procedures and the difference $d = (x_1 - x_2)$ in packing time was calculated for each pair.

a. If the objective of the experiment is to determine whether the data provide sufficient evidence to indicate that the mean packing time for a new method, Method 2, is less than the mean packing time for the existing method, Method 1, what null and alternative hypotheses would you use for a test?

b. What test statistic would you use for the test? Why?

c. If $\bar{d} = 37$ seconds and $s_d = 5.6$ seconds, find the value of the test statistic. What is the approximate p-value for the test?

d. If you have chosen $\alpha = .05$, what are your test conclusions? Do the data indicate that the mean time to pack a container using Method 2 is less than the mean time for Method 1?

e. What is the likelihood that your decision in part (d) is incorrect?

f. Find a 95% confidence interval for $(\mu_1 - \mu_2)$ and interpret it.

HLTH **7.28** Weight was recorded before and after 3 weeks of participation in a weight reduction program for each of 50 participants.

a. If the objective of the experiment is to determine whether the program produces a mean weight loss for people participating in the program, what null and alternative hypotheses would you use for a test?

b. What test statistic would you use for the test? Why?

c. If $\bar{d} = 9.3$ pounds and $s_d = 4.1$ pounds, find the value of the test statistic. What is the approximate p-value for the test?

d. If you have chosen $\alpha = .05$, what are your test conclusions? Do the data indicate that the weight reduction program produces a mean loss over a 3-week period?

e. What is the likelihood that your decision in part (d) is incorrect?

f. Find a 95% confidence interval for $(\mu_1 - \mu_2)$ and interpret it.

MFG **7.29** The data in Table 7.9 were obtained from an experiment that was conducted to compare the mean power level readings (in watts) on a type of military electronic tube by two identical pieces of test equipment. The power output for each of 10 military electronic tubes, randomly selected from production, was measured by both pieces of test equipment.

TABLE 7.9 Power Output Measured by Two Pieces of Test Equipment

Tube Number	Tester 1	Tester 2
1	2,563	2,556
2	2,665	2,479
3	2,460	2,426
4	2,650	2,619
5	2,610	2,617
6	2,657	2,491
7	2,529	2,590
8	2,427	2,466
9	2,448	2,516
10	2,480	2,428

Source: Unpublished report by Burnett Tyson, Williamsport, Pennsylvania.

a. What type of experimental design was used for this experiment? Justify your answer.

b. The objective of the experiment was to detect a difference in mean power level readings for the two pieces of test equipment if, in fact, a difference exists. State the null and alternative hypotheses you would use for a test.

c. What test statistic would you use for the test? Why?

d. The means and standard deviations of samples 1 and 2 are shown in rows 1 and 2 of the SAS printout in Figure 7.18a (page 270). The mean and standard deviation of the sample differences appear in the third row (opposite T1-T2). The SAS printout of the results of a paired-difference test is shown in Figure 7.18b. Find the observed value of the test statistic. Find the p-value for the test. (Remember that SAS prints out the p-value appropriate for a two-tailed test.)

e. If you have chosen $\alpha = .10$, what are your test conclusions?

f. What is the relevance of $\alpha = .10$ to your conclusion?

7.30 Refer to Exercise 7.29. Use the information in Figure 7.18a to calculate the value of the test statistic for the test. Does your answer agree with the value shown on the printout in Figure 7.18b?

FIGURE 7.18 SAS printout of test results for Exercise 7.29

N Obs	Variable	N	Minimum	Maximum	Mean	Std Dev
10	TESTER1	10	2427.00	2665.00	2548.90	92.7391204
	TESTER2	10	2426.00	2619.00	2518.80	73.1403521
	T1_T2	10	-68.0000000	186.0000000	30.1000000	86.8439853

a. Means and standard deviations

Analysis Variable : T1_T2

N Obs	N	T	Prob>|T|
10	10	1.0960409	0.3015

b. Test results

7.5 Choosing the Sample Size

When we sample and estimate a population mean or the difference between a pair of means, we have some practical objective in mind. We know that our estimate will not be on the bull's-eye but the error of estimation cannot be too large or the estimate will be useless.

The upper limit on the error of estimation—the largest value that we are willing to tolerate (with reasonable probability)—is specified by the researchers collecting the sample. For example, the educational researchers attempting to estimate the difference in mean academic achievement scores for the two teaching methods (Example 7.7) must decide (from a practical point of view) how large an error of estimation they are willing to tolerate. Are they willing to tolerate an error in their estimate as large as 5 test score points? Ten? If the error is too large, they will be unable to tell whether teaching Method 2 produces an increase in mean score over Method 1 that is large enough to be of practical value.

The larger the sample size, the smaller will be the error of estimation. Choosing an upper limit, call it B, on the error of estimation that we are willing to tolerate is equivalent to specifying how much information that we want to buy. This, in turn, determines the sample size.

The method for choosing the sample size to estimate the difference between two population means is exactly the same as the method used when estimating a single population mean. We decide how large an error of estimation that we are willing to tolerate. Suppose that we want the half-width of the confidence interval for $(\mu_1 - \mu_2)$ to equal B or, equivalently, that we want to estimate $(\mu_1 - \mu_2)$ with a probability equal to .95 that the error of estimation is less than B. The sample sizes, for the independent random samples and the matched-pairs designs, are given in the accompanying boxes.

Sample Sizes Required to Estimate $(\mu_1 - \mu_2)$ Correct to Within B with Probability Equal to .95: Independent Random Samples Design

$$n_1 = n_2 = 2\left[\frac{(1.96)(\sigma)}{B}\right]^2$$

Assumptions:

1. **The sample sizes are equal, i.e., $n_1 = n_2$.**
2. **The standard deviation of each population is approximately equal to σ.**

Sample Size Required to Estimate $(\mu_1 - \mu_2)$ Correct to Within B with Probability Equal to .95: A Matched-Pairs Design

$$n = \left[\frac{(1.96)(\sigma_d)}{B}\right]^2$$

where n is the number of pairs and σ_d is the standard deviation of the population of differences.

We cannot solve for n in the sample size equations unless we have approximations to the unknown population variances. This is not too difficult a problem for the independent random samples design. We use the same techniques as described in Section 6.7. We approximate σ with the sample standard deviation from an earlier sample, use a range approximation to σ, or, as a last resort, draw a small sample from the population, calculate s, and use it to estimate σ.

Finding an approximation to σ_d for a matched-pairs design is more difficult. The best procedure is to acquire an estimate s_d based on a small sample, and use it to approximate σ_d.

Example 7.10

Suppose that we want to estimate the difference between the mean permeability of 3% and 7% asphalt concrete specimens, Example 7.3, correct to within 30 inches per hour with probability equal to .95. Assume that the data will be collected using an independent random samples design. How many specimens would be required for each sample (assume equal sample sizes)?

Solution

We can use the pooled estimate of the common population standard deviation found in Example 7.3, $s = 114.15$, to approximate the population standard deviation σ. Substituting this value, along with the bound on the error, $B = 30$, into the formula for the sample size gives

$$n_1 = n_2 = 2\left[\frac{(1.96)(\sigma)}{B}\right]^2 = 2\left[\frac{(1.96)(114.15)}{30}\right]^2 = 111.2$$

or

$$n_1 = n_2 = 112$$

Therefore, we will need approximately 112 specimens for each type of asphalt in order to estimate the difference in mean permeability correct to within 30 inches per hour.

EXERCISES

7.31 Suppose that you wish to estimate the difference in two population means using an independent random samples design. What do you need to know in order to calculate the approximate sample sizes? How will you get this information?

7.32 Suppose that you wish to estimate the difference in two population means using a matched-pairs design. What do you need to know in order to calculate the approximate sample sizes? How will you get this information?

7.33 Suppose that you want to estimate the difference in two population means using an independent random samples design and that you want the half-width of the confidence interval to equal 3. If you suspect that the standard deviation for each of the populations is approximately equal to 8.5, how many measurements would you have to include in each sample?

7.34 Suppose that you want to estimate the difference in two population means using an independent random samples design and that you want your estimate to lie within 12 units of $(\mu_1 - \mu_2)$ with probability equal to .95. If you suspect that the standard deviation for each of the populations is approximately equal to 56, how many measurements would you have to include in each sample?

7.35 Suppose that you want to estimate the difference in two population means using a matched-pairs design and that you want your estimate to lie within 2 units of $(\mu_1 - \mu_2)$ with probability equal to .95. If prior sampling suggests that the standard deviation of the population of differences is approximately equal to 5, how many measurements would you have to include in each sample?

GEN **7.36** Suppose that you want to estimate the difference in mean weight gains for chickens on two different diets and that you want your estimate to be correct to within $B = 2$ ounces with probability equal to .95. You plan to collect the data using an independent random samples design and you have prior information that suggests that the population standard deviation of weight gains for each diet is equal to 6 ounces. Approximately how many chickens would you have to include in each sample?

BUS **7.37** Suppose that you want to estimate the difference in mean percentage iron in ore specimens collected from two different locations and that you want your estimate to be correct to within $B = .1$ percent with probability equal to .95. You plan to sample the two locations using an independent random samples design and you have prior information that suggests that the population standard deviation of the percentage measurements at each location is approximately equal to .7 percent. Approximately how many ore specimens would you have to include in each sample?

MFG **7.38** Refer to Exercise 7.29. Suppose that you want to estimate the difference in mean power output readings for the two electronic tube testers, correct to within 25 watts of power output with probability equal to .95. How many tubes would have to be included in your experiment? (Use the estimate s_d found in Exercise 7.29 to approximate σ_d.)

7.6 Key Words and Concepts

- The concepts introduced in this chapter are, with few exceptions, identical to those of Chapter 6. The major difference between the two chapters is that they have different practical objectives. Chapter 6 was concerned with making inferences about a single population mean. This chapter is concerned with the procedures for making (and the practical implications to be derived from) inferences about the difference between a pair of population means.
- The plan to be followed in collecting a sample is called the ***design of an experiment***. A design for sampling opinions or other characteristics of our society is called a ***sample survey design***.
- The objective of an experimental design is to buy a specified amount of information about a population parameter—say, a confidence interval of a specified width—at minimum cost.
- This chapter presents two designs for comparing two population means: an independent random samples design and a matched-pairs design.
- An ***independent random samples design*** involves the selection of random samples from each of the two populations, with the selection of one sample completely independent of the other.
- A ***matched-pairs design*** involves the selection of a random sample of matched pairs of observations, with one member of a pair from each population.
- Pairs are matched when we believe that the variation between matched pairs is less than the variation between any two measurements, one from each population, selected at random. If this belief is correct, data from a matched-pairs design will contain more information about $(\mu_1 - \mu_2)$ than an independent random samples design of the same size (i.e., one containing the same number of pairs of observations).
- Confidence intervals and tests of hypotheses about $(\mu_1 - \mu_2)$ are based on the familiar z and t statistics of Chapter 6.
- The formulas for z and t are different from those used in Chapter 6, but the procedures used for constructing the confidence intervals and performing the tests of hypotheses for $(\mu_1 - \mu_2)$ are identical to those used to make inferences about a single population mean.

▶ The procedure for selecting the sample size(s) is similar to the method employed in Chapter 6 but the formulas are slightly different.

SUPPLEMENTARY EXERCISES

MFG **7.39** The experiment described in Exercise 7.29 was repeated for 10 different electronic tubes and the following statistics were recorded: $\bar{x}_1 = 2{,}647.9$, $s_1 = 101.2$, $\bar{x}_2 = 2{,}550.4$, $s_2 = 82.3$, $\bar{d} = 97.5$, and $s_d = 123.6$ (unpublished report by Burnett Tyson, Williamsport, Pennsylvania).

a. What type of experimental design was used for this experiment? Justify your answer.

b. The objective of the experiment was to detect a difference in mean power level readings for the two pieces of test equipment if, in fact, a difference exists. State the null and alternative hypotheses you would use for a test.

c. What test statistic would you use for the test? Why?

d. Are any of the statistics given above irrelevant to the test? Explain.

e. Calculate the test statistic for the test.

f. Calculate the approximate p-value for the test.

g. If you have chosen $\alpha = .10$, what is your test conclusion?

h. What is the relevance of $\alpha = .10$ to your conclusion?

BIO **7.40** J. S. Diana conducted an experiment to investigate the effect of water temperature on the growth of large-mouth bass ("The Growth of Large-Mouth Bass, *Micropterus salmoides* (Lacepede), *Journal of Fish Biology*, Vol. 24, 1984). Sixty fish were randomly separated into six groups of ten fish each. The groups were then randomly assigned, one to each of six different water temperature conditions. The mean of the growth rates of the ten fish after 15 days of feeding and the standard error of the mean for two conditions were as follows: For "constant warm," $\bar{x}_1 = 40.8$ and $\widehat{SE}(\bar{x}_1) = 12.5$; for "start warm," $\bar{x}_2 = 46.3$ and $\widehat{SE}(\bar{x}_2) = 8.6$.

a. Based on the difference between the sample means, do you think that the mean growth rate differs for the two water temperature conditions?

b. Find the sample variance for each sample and then find the pooled estimate of σ^2.

c. Do the data provide sufficient evidence to indicate a difference in mean growth rates for the two conditions? Find the approximate p-value for the test. If you have chosen $\alpha = .05$, what do you conclude? What do the data tell you about feeding bass?

HLTH **7.41** Refer to Example 7.6 and our discussion of the Minnesota Heart Survey, a study designed to see whether trends in public awareness and treatment of high levels of blood cholesterol produced a change in the mean cholesterol levels of men and women from the period 1980–1982 to the period 1985–1987. Example 7.6 compared the mean total cholesterol for women for the time period 1980–1982 with the corresponding mean for the time period 1985–1987. Figure 7.19 shows the relative frequency distributions of total cholesterol (in millimoles per liter) for a sample of 1,588 men assessed during the time period 1980–1982 and the other for 2,192 men for the time period 1985–1987.

Table 7.10 gives the number of men measured in each of the two time periods, the mean cholesterol level (in millimoles/liter) for each of the two groups, and the 95% confidence interval for the mean for each of the two groups.

Examine Figure 7.19. If current thinking is correct, trends in public awareness of the harmful effects of high cholesterol should have produced an overall decrease in cholesterol levels from 1980–1982 to 1985–1987; that is, the mean μ_1 of the 1980–1982 distribution should be located to the right of the mean μ_2 for the 1985–1987 distribution. Do the data in Table 7.10 provide sufficient evidence to indicate that the mean μ_1 of the population of total cholesterol measurements for

FIGURE 7.19 Relative frequency distributions of total cholesterol for men, 1980–1982 and 1985–1987

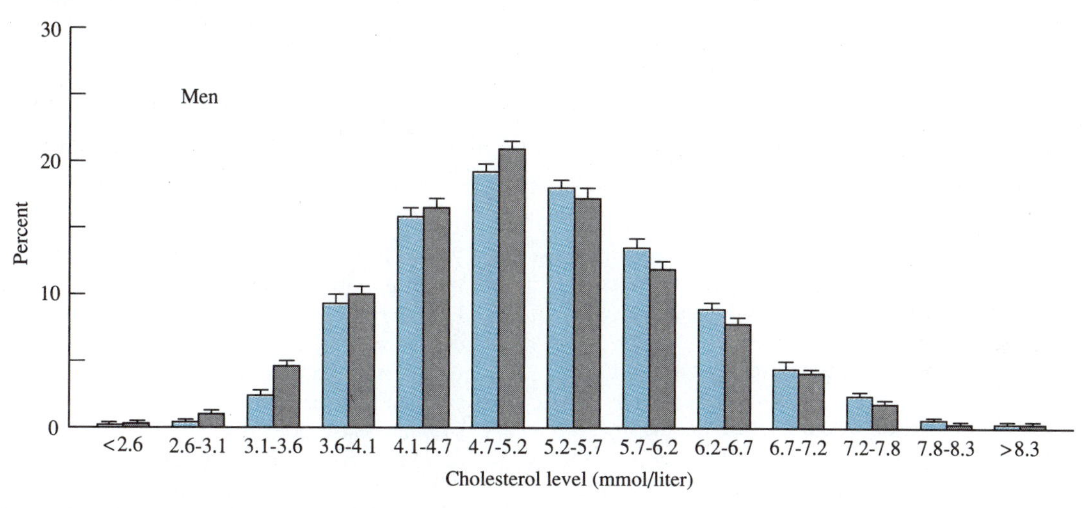

Color bars indicate levels in 1980-1982, and gray bars levels in 1985-1987. T bars indicate the standard error of the mean.

Source: Burke, Gregory L. et al., "Trends in Serum Cholesterol Levels from 1980 to 1987," *The New England Journal of Medicine,* Vol. 324, No. 14, 1991 (Figure 1).

TABLE 7.10 Confidence Intervals for the Mean Cholesterol Levels of Men, 1980–1982 to 1985–1987: The Minnesota Heart Survey

	1980–1982	1985–1987
Sample size	1,588 mmol/liter	2,192 mmol/liter
Mean	5.19	5.04
Confidence interval	5.25 to 5.35	5.12 to 5.20

These values were abstracted from Tables 1 and 2 of the article by Burke et al.

men for the 1980–1982 period is larger than the mean μ_2 for the 1985–1987 period?

a. State the null and alternative hypotheses for the test.

b. Which test should we use?

c. Locate the rejection region for the test.

d. Find the standard error of the difference $(\bar{x}_1 - \bar{x}_2)$ in sample means.

e. Find the value of the test statistic.

f. Find the approximate p-value for the test. If you have chosen $\alpha = .05$, what do you conclude?

BIO

7.42 Rats, like dogs, mark their territory using urine or other secretions. Linda I. A. Birke and Dawn Saddler conducted a study of the marking habits of rats ("Scent Marking Behavior to Conspecific Odors by the Rat, *Rattus norvegicus*," *Animal Behavior*, Vol. 32, 1984). Thirty male rats were placed, ten each in three different environments: one odor-free, one containing female scents, and one containing male scents. The number of markings made by each rat was recorded. The mean and standard deviation (in parentheses) for each of the three environments are shown in Table 7.11.

Do the data present sufficient evidence to indicate a difference in the mean number of markings for those exposed to the female scent versus those exposed to the male scent?

a. State the null and alternative hypotheses for the test.

b. What test statistic would you use for the test?

TABLE 7.11 The Mean and Standard Deviation of the Number of Rat Markings Recorded for Three Environments

Odor Source		
None (control)	Female	Male
9.0 (1.1)	19.4 (2.85)	24.2 (3.4)

c. Where is the rejection region located for the test of part (b)?

d. Find the approximate p-value for the test. What does it mean?

e. If you have chosen $\alpha = .10$ for the test, what are your conclusions?

f. In what sense does $\alpha = .10$ measure the risk of a faulty conclusion for this test?

EXERCISES FOR YOUR COMPUTER

GEN

7.43 An experiment was conducted to compare the density of cakes prepared from two different cake mixes, A and B. Six cake pans received batter A and six received batter B. Expecting a variation in oven temperature from one spot in the oven to another, the experimenter placed an A cake and a B cake side-by-side at six different locations. The six paired cake density readings are shown in Table 7.12.

Do the data present sufficient evidence to indicate a difference in the mean density for cakes prepared by the two types of batter?

a. State the null and alternative hypotheses for the test.

b. What test statistic would you use for the test?

c. Where is the rejection region located for the test of part (b)?

d. Find the value of the test statistic. Find the p-value for the test. What does it mean?

e. If you have chosen $\alpha = .05$ for the test, what are your conclusions?

f. In what sense does $\alpha = .05$ measure the risk of a faulty conclusion for this test?

TABLE 7.12 Paired Cake Density Measurements for Two Cake Mixes

Mix	Density (oz/in.^{3})					
A	.135	.102	.098	.141	.131	.144
B	.129	.120	.112	.152	.135	.163

BIO **7.44** Refer to the study of the oxygen uptake by reedfish, Exercise 6.63. Table 7.13 gives the oxygen uptake readings for reedfish exposed to two temperature environments, one group of $n_1 = 11$ reedfish in water at 25°C and the other group of $n_2 = 12$ reedfish in water at 33°C (Pettit, M. J. and Beitinger, T. L., "Oxygen Acquisition of the Reedfish, *Erpetoichthys calabaricus*," *Journal of Experimental Biology*, Vol. 114, 1985).

a. Do the data present sufficient evidence to indicate a difference in the percentage of oxygen uptake by air between reedfish exposed to water at 25°C and those at 33°C? State the null and alternative hypotheses for the test.

b. Describe the test and the test statistic appropriate for the test.

c. Where is the rejection region for the test?

d. Find the value of the test statistic.

e. Find the *p*-value for the test.

f. If you have chosen $\alpha = .05$ for the test, what is your conclusion? Does the percentage oxygen uptake by air differ from one water temperature environment to another?

TABLE 7.13 Oxygen Uptake Data for Some Reedfish

		Oxygen Uptake (ml O_2 $g^{-1}h^{-1}$)		Percent	
Temperature	Weight (grams)	Air	Water	Air	Water
25°C	22.1	.043	.045	49	51
	21.1	.034	.066	34	66
	13.4	.026	.084	24	76
	19.5	.048	.045	52	48
	21.9	.009	.056	14	86
	19.5	.028	.072	28	72
	16.5	.032	.081	28	72
	18.5	.045	.051	47	53
	18.0	.061	.041	60	40
	15.4	.028	.060	32	68
	13.4	.024	.113	18	82
33°C	22.1	.020	.051	28	72
	21.1	.044	.036	55	45
	19.5	.035	.043	45	55
	21.9	.051	.049	51	49
	19.5	.035	.051	41	59
	16.5	.027	.073	27	73
	18.5	.043	.054	44	56
	18.9	.053	.057	48	52
	15.4	.055	.046	54	46
	21.8	.075	.037	67	33
	22.1	.042	.049	46	54
	13.2	.055	.039	59	41

References

1. Freedman, D., Pisani, R., Purves, R., and Adhikari, A., *Statistics,* 2nd edition. New York: W. W. Norton & Co., 1991.
2. McClave, J. and Dietrich, F., *Statistics,* 5th edition. San Francisco: Dellen Publishing Co., 1991.
3. Mendenhall, W. and Beaver, R., *Introduction to Probability and Statistics.* 8th edition. Boston: PWS-Kent, 1991.
4. Moore, D. and McCabe, G., *Introduction to the Practice of Statistics,* 2nd edition. New York: W. H. Freeman & Co., 1993.
5. Sincich, T., *Statistics by Example,* 4th edition. San Francisco: Dellen Publishing Co., 1990.

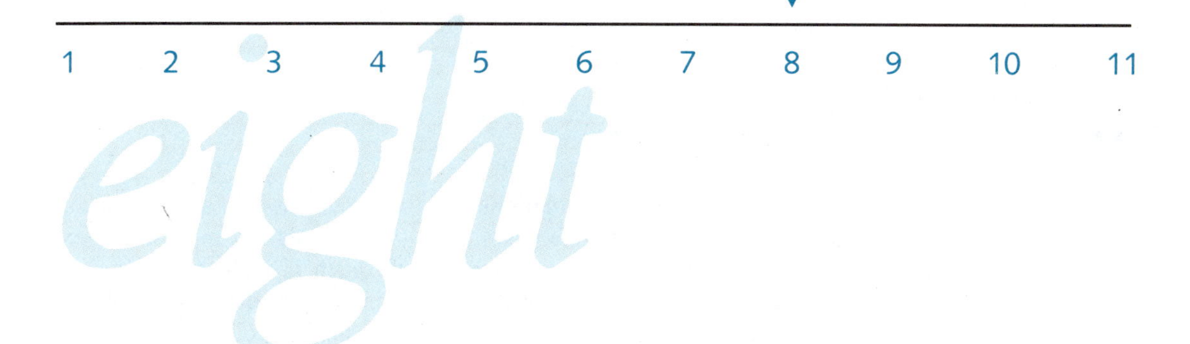

Correlation and Linear Regression Analysis

In a Nutshell

An application of statistical inference: Can your aptitude test score be used to help predict the grade you will get in this course? A linear regression analysis will provide an answer to this question.

8.1 The Problem

8.2 Scattergrams

8.3 A Linear Relation Between x and y

8.4 The Method of Least Squares

8.5 The Pearson Coefficient of Correlation

8.6 A Simple Linear Regression Analysis: Questions It Will Answer

8.7 A Simple Linear Regression Analysis: Assumptions

8.8 A Simple Linear Regression Analysis: Typical Computer Outputs

8.9 A Multiple Linear Regression Analysis

8.10 Key Words and Concepts

8.1 The Problem

In Chapters 3–7, we described populations of data associated with a **single quantitative variable** and we made inferences about the population means. For example, although we recorded the measurements for each of eight variables (sex, age, total cholesterol, triglycerides, etc.) on each person in the sample of 107 "healthy adults" in Table 3 of Appendix 1, we described (in Chapters 3 and 4) the samples associated with each of these variables separately. We then used the sample data associated with the quantitative variables to make inferences about their respective population means (Chapters 6 and 7). Now consider a different problem.

Suppose that we observe the value for each of **two quantitative variables** on each of the elements in a sample and **we want to know whether the two variables are related**. For example, suppose that we want to study the relationship between level of cigarette smoking and a person's diastolic blood pressure. To perform this study, we would draw a sample from a population of smokers and would record the average number x of cigarettes smoked per day and the blood pressure y for each of the smokers in the sample. From these data, we would like to determine whether x and y are related. Does the level of diastolic blood pressure y tend to rise as the level x of smoking increases? That is, does y increase as x increases? Or perhaps y decreases as x increases, or maybe x and y are completely unrelated.

Pairs of measurements (x, y) made on a set of elements are called *bivariate data.* Table 8.1 shows a sample of bivariate data. Twelve students—elements of the sample—were selected from a universe of students and the quantitative aptitude test score and achievement test score were recorded for each student. Why? The admissions officer of the college wants to know whether the quantitative aptitude test score is a good predictor of a student's mathematics achievement test score. Ultimately, he or she wants to know whether you will survive the mathematics courses of the college.

TABLE 8.1 Quantitative Aptitude and Achievement Test Scores

Student	1	2	3	4	5	6	7	8	9	10	11	12
x	88	57	76	97	71	90	66	58	92	85	51	85
y	620	495	549	635	480	568	570	437	655	547	395	662

This chapter is concerned with describing bivariate data and then using the data to determine whether a relationship exists between two variables, x and y.

Ultimately, we would like to obtain a mathematical equation that relates y to x and to be able to use it to estimate the mean value of y or to predict y for a given value of x.

8.2 Scattergrams

One way to see whether a relationship exists between two variables x and y in a set of bivariate data is to plot the data, one point for each pair of measurements, on a graph. Pairs of measurements (x, y) are often called *data points* and the resulting plot is called a *scattergram*. A scattergram for the data in Table 8.1 is shown in Figure 8.1. The scattergram shows 12 points, one for each pair of grades. For example, student 1 is represented by the point (circled) that is located at $x = 88$ and $y = 620$. Note that the dots tend to rise as you move from the left side of the plot to the right. This suggests (as we might suspect) that a student's achievement test score is related to the student's quantitative aptitude score. Specifically, it appears that the achievement test score increases as the aptitude score increases.

FIGURE 8.1 Scattergram for the quantitative aptitude and achievement test scores, Table 8.1

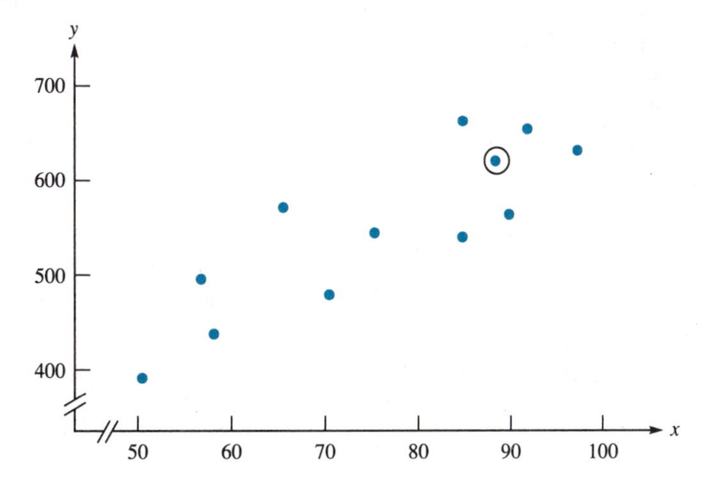

Other sets of data might produce scattergrams similar to those shown in Figure 8.2 (page 282). For example, Figure 8.2a shows a scattergram for a set of data where the value of y tends to decrease as the value of x increases, whereas in Figure 8.2b, y appears to rise and then fall as x increases.

FIGURE 8.2 Typical scattergrams

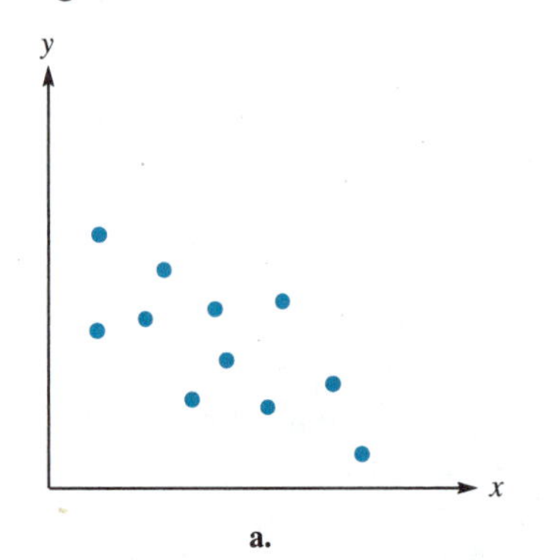

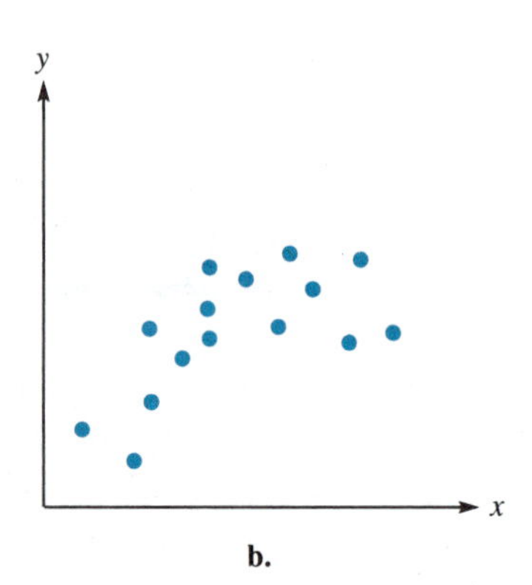

EXERCISES

8.1 What do we mean by the term *bivariate data*?

8.2 Describe a practical situation where bivariate data might be collected on the elements of a universe.

8.3 Describe a practical situation where we would be interested in the association between two variables. Describe the associated universe, elements of the universe, and the bivariate data that you would expect to sample.

8.4 What is a data point?

8.5 What is a scattergram for a bivariate set of data?

8.6 List two quantitative variables that might be observed for each of the following elements. Describe the universe and the population of bivariate data associated with each.

a. A heart in a live human
b. A 6″ × 6″ × 24″ concrete test specimen
c. A work day at a manufacturing plant
d. A mature golden retriever dog

ECON **8.7** Does it pay to save? There is a theory that heavy taxation of investment income, such as savings, reduces a country's national savings rate. Table 8.2 gives the personal savings rate and the taxation rate, i.e., the investment income tax liability, for a sample of eight countries (Blotnick, S., "Psychology and Investing." *Forbes*, May 25, 1981).

a. Identify the elements in this experiment.
b. Describe the universe from which the sample was drawn.
c. Describe the variable(s) observed on each element of the universe.
d. Describe the data collected on each element.
e. Describe the sampled population. Is it univariate or bivariate?
f. Construct a scattergram for the data.
g. What does the scattergram suggest (if anything) concerning the relationship between the rate of personal savings and investment income tax liability?

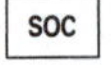

8.8 An article in *The New York Times* (March 4, 1985) reports that 84% of arriving inmates at state prisons in 1979 were repeat offenders. Table 8.3 shows the percentage y of repeat offenders returning within 1 year for each of four age groups.

a. Identify the elements in this experiment.
b. Describe the data collected on each element.
c. Construct a scattergram for the data.
d. What does the scattergram suggest (if anything) concerning the relationship between the tendency for an offender to repeat within 1 year and the offender's age?

TABLE 8.2 Personal Savings Rate and Investment Income Tax Liability for Eight Countries

Country	Personal Savings Rate (%)	Investment Income Tax Liability (%)
Italy	23.1	6.4
Japan	21.5	14.4
France	17.2	7.3
W. Germany	14.5	11.8
United Kingdom	12.2	32.5
Canada	10.3	30.0
Sweden	9.1	52.7
United States	6.3	33.5

Source: New York Stock Exchange with assistance of Price Waterhouse.

TABLE 8.3 Percentage of Prisoners Returning Within 1 Year in Four Age Brackets

Age Group Interval (years)	Midpoint x of Age Interval	Percentage y Returning Within 1 Year
18 to 24	21.5	22
25 to 34	30	12
35 to 44	40	7
45 and older	Approx. 55	2

8.3 A Linear Relation Between x and y

Describing the Relationship Between Two Variables

How can we characterize the relationship between two variables, x and y? One way to do it—a graphical method—is to draw a straight line through the points so as to minimize, according to some criterion, the deviations of the points from the line. For example, the line drawn through the data points of Figure 8.1, shown in Figure 8.3 on page 284, provides a description of the relationship between the aptitude test and the achievement test scores of Table 8.1.

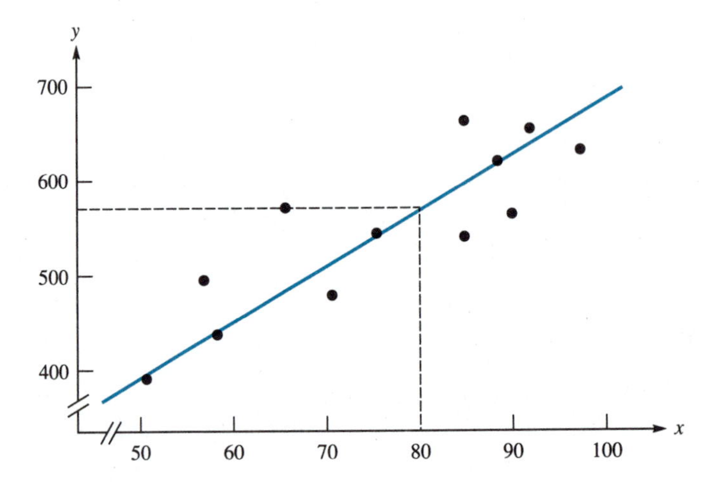

FIGURE 8.3 Line drawn through the points on the scattergram, Figure 8.1

It suggests that y increases in value as x increases and we can use the line to predict values of y for given values of x. For example, if we want to predict the value of y for $x = 80$, we look for the point on the line corresponding to $x = 80$ and find that $y = 570$.

The Equation of a Line

Each line, such as the one shown in Figure 8.3, can be represented by a mathematical equation of the type

$$y = \beta_0 + \beta_1 x$$

That is, for every line that we draw on a piece of graph paper, there corresponds one and only one equation of the type $y = \beta_0 + \beta_1 x$. The line consists of the set of all points (x, y) that satisfy the equation.

The symbols β_0 (beta-naught) and β_1 (beta-one) are constants. Each pair of values corresponds to a particular line. For example, consider the equation $y = 1 + 2x$. For this line, $\beta_0 = 1$ and $\beta_1 = 2$. If $x = 0$, $y = 1 + 2(0) = 1$. Then the point (0, 1) falls on the line $y = 1 + 2x$. Similarly, if $x = 1$, $y = 1 + 2(1) = 3$ and if $x = 2$, $y = 1 + 2(2) = 5$. Therefore, the points (1, 3) and (2, 5) also fall on the line. The three points and the line $y = 1 + 2x$ are shown in Figure 8.4.

Interpreting β_0 and β_1

The constants β_0 and β_1 have special meanings. If $x = 0$, $y = \beta_0$. Therefore, β_0, called the *y-intercept* for the line, is the y-value for the point where the line inter-

FIGURE 8.4 Graph of the line $y = 1 + 2x$

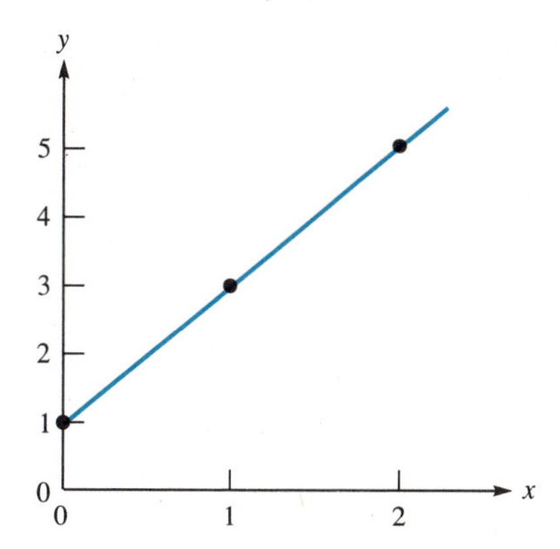

FIGURE 8.5 y-intercept and slope for the line $y = 1 + 2x$

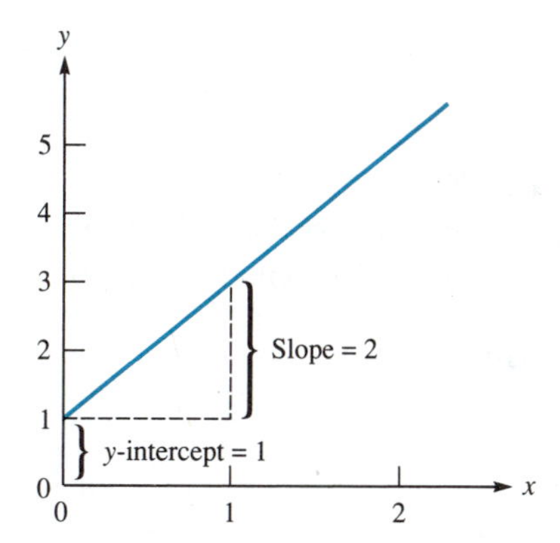

cepts the y-axis of the graph. For example, the y-intercept for the line $y = 1 + 2x$ is $\beta_0 = 1$. This intercept is located on the graph in Figure 8.5. You can see that when $x = 0$, $y = 1$.

The constant β_1, called the *slope* of the line, is the amount that y increases (or decreases) for each 1-unit increase in x. For example, the slope of the line $y = 1 + 2x$ is $\beta_1 = 2$ (see Figure 8.5). If x increases by 1, the value of y increases by 2.

The Relationship Between the Slope and the Sign of β_1

If the slope of a line is positive, the line slopes upward to the right. If negative, it slopes downward to the right. If the slope is 0, the line is horizontal and has no slope (see Figure 8.6 on page 286).

FIGURE 8.6 Lines with positive, negative, and zero slopes

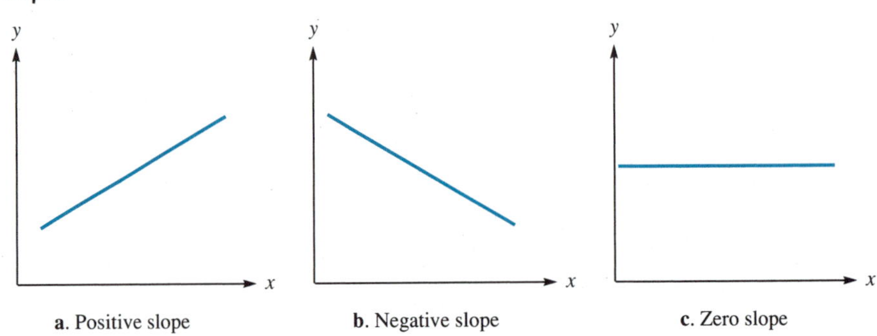

Any equation in x and y of the form $Ax + By = C$, where A, B, and C are constants, is also the equation of a line because we can always change the equation into its *slope–intercept form*, $y = \beta_0 + \beta_1 x$, by solving the equation for y. For example, the equations $-2x + 4y = 7$ and $5x = 2y + 4$ are equations of straight lines. If we solve for y in either of these equations, we transform the equation into its slope–intercept form.

Example 8.1

Find the slope–intercept form for the equation $-2x + 4y = 7$. Then find the y-intercept and the slope and interpret them.

Solution

Solving the equation $-2x + 4y = 7$ for y, we find that $4y = 7 + 2x$ or

$$y = \frac{7}{4} + \frac{2}{4}x = \frac{7}{4} + \frac{1}{2}x$$

This is the slope–intercept form of the equation. Therefore we know that the constant $\beta_0 = \frac{7}{4}$ is the y-intercept and the coefficient of x, $\beta_1 = \frac{1}{2}$, is the slope of the line. Since the slope of the line is positive and equal to $\frac{1}{2}$, we know that the line slopes upward to the right with y increasing $\frac{1}{2}$ unit for every 1-unit increase in x. Since $\beta_0 = \frac{7}{4}$, the line intersects the y-axis at $\frac{7}{4}$.

Example 8.2

Draw a graph for the line of Example 8.1.

Solution

We need two points to graph a line. We already know that the line intersects the

y-axis at $y = \frac{7}{4} = 1.75$. Therefore, the line passes through the point (0, 1.75). To find a second point on the line, we choose a value of x, say $x = 2$, substitute it into the equation for the line, and solve for y:

$$y = \frac{7}{4} + \frac{1}{2}x = \frac{7}{4} + \frac{1}{2}(2) = \frac{7}{4} + 1 = 2.75$$

Therefore the second point is (2, 2.75). The two points, (0, 1.75) and (2, 2.75), are then plotted on a graph (see Figure 8.7) and the line represented by the equation $y = \frac{7}{4} + \frac{1}{2}x$ is drawn through the points.

FIGURE 8.7 Graph for the line $y = \frac{7}{4} + \frac{1}{2}x$

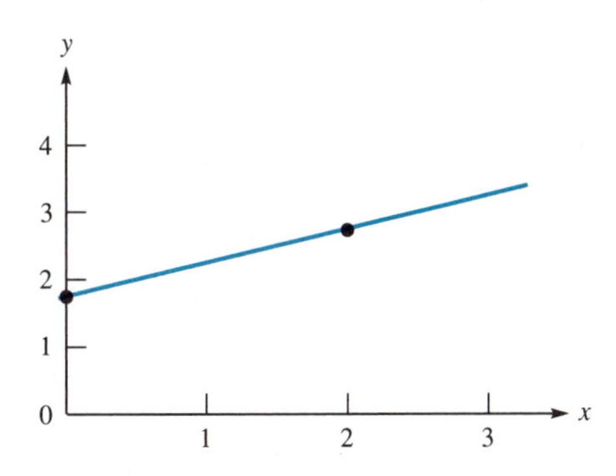

EXERCISES

8.9 Give the slope–intercept form for the equation of a straight line. Identify the y-intercept and the slope in the equation. What is the practical interpretation of the y-intercept? The slope?

8.10 Suppose that a line has the equation $y = -2 + 3x$. Describe in words how the line would appear graphically.

8.11 Suppose that a line has the equation $y = \frac{1}{2} - x$. Describe in words how the line would appear graphically.

8.12 Suppose that a line has the equation $2 = x - 3y$. Find the slope-intercept form for the equation of the line. Then describe in words how the line would appear graphically.

8.13 Suppose that a line has the equation $0 = 4x + 2y - 3$. Find the slope-intercept form for the equation of the line. Then describe in words how the line would appear graphically.

8.14 Graph the line in Exercise 8.10.

8.15 Graph the line in Exercise 8.11.

8.16 Graph the line in Exercise 8.12.

8.17 Graph the line in Exercise 8.13.

8.4 The Method of Least Squares

Finding the Best-Fitting Straight Line for a Set of Data

Now that we know how to write the equation for a line, we will find the equation of the line that, under certain circumstances, best describes the linear relation between two variables x and y.

Look at the line passing through the scattergram of points for the mathematics aptitude–achievement test scores (see Figure 8.8). Intuitively, the best-fitting line would minimize, in some sense, the deviations of the points from the fitted line. We will define the deviation of a point from a fitted line in the following way.

FIGURE 8.8 A fitted line for the aptitude–achievement data of Table 8.1

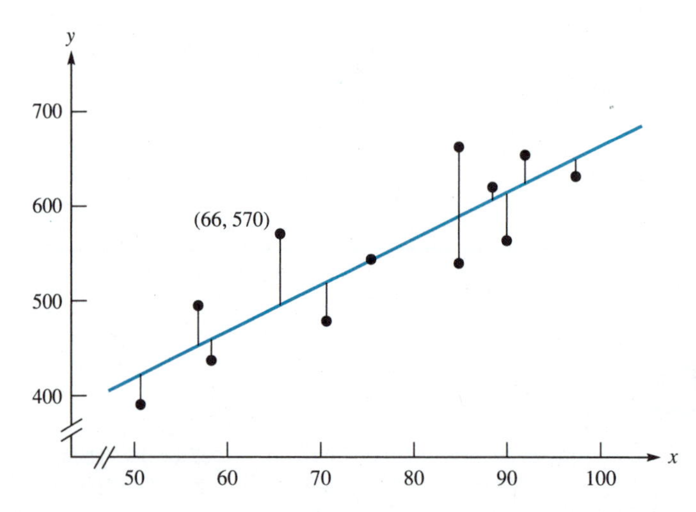

The *deviation of a point from a line* is defined to be the difference between the observed value of y and the predicted value, i.e., the value of y predicted by the line. For example, the deviation for the point (66, 570) from the line in Figure 8.8 is the difference between $y = 570$ and the value of y predicted by the fitted line when $x = 66$. This deviation is represented by the vertical line segment connecting the point and the line. The deviations of the other 11 points from the fitted line are also represented as line segments in Figure 8.8.

We could specify many different criteria for selecting the line that best fits a set of data. For reasons that we will explain in Section 8.7, we will use the method of least squares.

The Method of Least Squares

The *method of least squares* chooses the "best"-fitting straight line as the one that minimizes the sum of the squares of the deviations of the points from the fitted line. This line is called *the least squares line*. The sum of squares of deviations of the y-values about the least squares line is called *the sum of squares for error* and is represented by the symbol *SSE*. To be sure that you understand what we mean when we speak of the method of least squares, pass a line (any line) through the points on the scattergram in Figure 8.8. Imagine that you calculate the deviation of each point from the line, square each deviation, and then sum the squares. If you did this for every possible line and chose the line with the smallest sum of squares of deviations, this would be the least squares line. The sum of squares of deviations for this least squares line would equal SSE.

The Least Squares Line

The least squares line is called the *prediction equation* because we will use it to predict values of y for given values of x. It is given by the equation shown in the box.

The Least Squares Line

$$y = b_0 + b_1 x$$

where

$$b_0 = y\text{-intercept}$$
$$b_1 = \text{slope}$$

The formulas for finding b_0, b_1, and SSE (and, therefore, the least squares line) can be derived mathematically. We will omit the formulas and accept the output from one of the many reputable computer software packages that are available. We will show the output for the mathematics aptitude–achievement test score data in Section 8.8.

While scattergrams and least squares lines are useful in indicating a possible linear relationship between a pair of variables, we need some way to measure the strength of this relationship. How close do the points fall to the line? Is it a close fit or are the points widely scattered? The most common measures of the strength of the relationship between two variables, x and y, for a set of bivariate data are **the Pearson coefficient of correlation** and **the coefficient of determination**.

8.5 The Pearson Coefficient of Correlation

A Coefficient of Correlation: What It Is

In the early 1900s the English statistician Karl Pearson (1857–1936) proposed a measure of the strength of the linear relationship between two variables. We can better understand the logic behind Pearson's coefficient of correlation by examining Figure 8.9, the scattergram for our quantitative aptitude–achievement test scores of Table 8.1.

FIGURE 8.9 Scattergram for aptitude–achievement test scores, Table 8.1

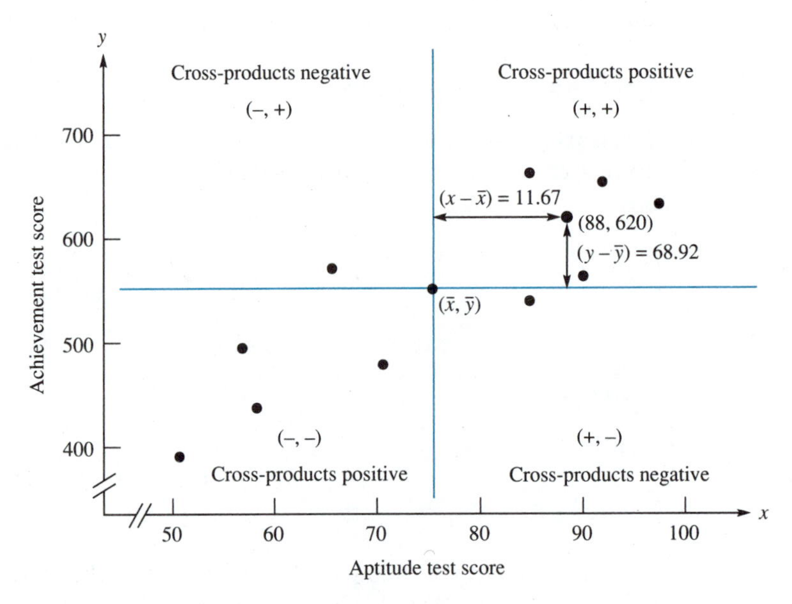

First, locate a point "in the middle" of the scatter of points. A natural choice would be the point located at $\bar{x}$ and $\bar{y}$—that is, the point $(\bar{x}, \bar{y})$. The mean of the 12 aptitude test scores is $\bar{x} = 76.33$ and the corresponding mean of the achievement test scores is $\bar{y} = 551.08$. This point is plotted on the scattergram and you can see that it is located "in the middle" of the points.

Now draw a horizontal line and a vertical line through the point $(\bar{x}, \bar{y})$ in Figure 8.9. This divides the area in the figure into four quadrants. Pick a point in the data set—say, the point for student 1: (88, 620). Note (in Figure 8.9) that this point lies in the upper right-hand quadrant and deviates $(x - \bar{x}) = (88 - 76.33) = 11.67$

test points to the right of the point $(\bar{x}, \bar{y})$ and $(y - \bar{y}) = (620 - 551.08) = 68.92$ above it. Both deviations are positive and the cross-product $(x - \bar{x})(y - \bar{y}) = (11.67)(68.92)$ is positive. In fact, **for any point in the upper right-hand quadrant, both $(x - \bar{x})$ and $(y - \bar{y})$ will be positive (indicated by the symbol (+, +)) and the cross-product $(x - \bar{x})(y - \bar{y})$ will be positive.**

What about the cross-products of the deviations in the other quadrants? For all points in the lower right-hand quadrant, the $(x - \bar{x})$ deviations will be positive and the deviations $(y - \bar{y})$ will be negative. This situation is indicated in the lower right-hand quadrant of Figure 8.9 by the symbol (+, −). Similarly, the deviations of the points in the upper and lower left-hand quadrants are (−, +) and (−, −), respectively. Therefore, the cross-product for every point in the (−, −) quadrant is positive. All cross-products in the (−, +) and the (+, −) quadrants are negative.

Pearson used the cross-product of the deviations of the data points from $(\bar{x}, \bar{y})$ to form a measure of the linear relationship between two variables, x and y. If the points slope upward to the right (see Figure 8.10a on page 292), most of the points will fall in the lower left (−, −) and the upper right (+, +) quadrants, and the sum of the cross-products will be *a large positive number.* If the points tend to slope downward to the right, most of the cross-products will be negative and the sum of the cross-products will be a *large negative number* (see Figure 8.10b). If the points tend to be splattered, neither rising nor falling as x increases (see Figure 8.10c), some of the cross-products will be positive, some will be negative, and they will tend to cancel each other. Therefore, their sum will be a *small positive or negative number.*

The *Pearson coefficient of correlation* r is given by the formula

$$r = \frac{SS_{xy}}{\sqrt{SS_x SS_y}}$$

where

$$SS_x = \sum(x - \bar{x})^2$$

is the sum of squares of deviations of the x-values about their mean $\bar{x}$, the same quantity that we calculated in order to compute the sample variance s^2 in Chapter 4. Similarly,

$$SS_y = \sum(y - \bar{y})^2$$

is the sum of squares of the deviations of the y-values about their mean $\bar{y}$.

The numerator of r is the sum of the cross-products of the deviations of the points from $(\bar{x}, \bar{y})$,

$$SS_{xy} = \sum(x - \bar{x})(y - \bar{y})$$

It will be positive or negative depending on whether the points rise or fall as x increases.

The divisor, $\sqrt{SS_x SS_y}$, is always a positive number. It adjusts SS_{xy} for the difference in the units of measurement of x and y so that r always assumes a value

FIGURE 8.10 Sum of cross-products of deviations for different scattergrams

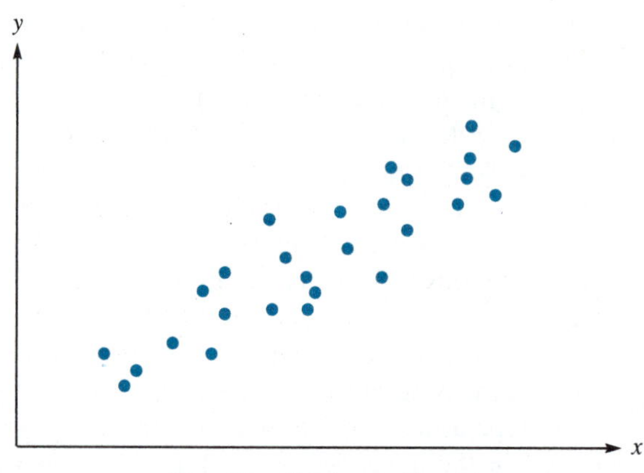

a. $\Sigma(x - \bar{x})(y - \bar{y})$ is large and positive.

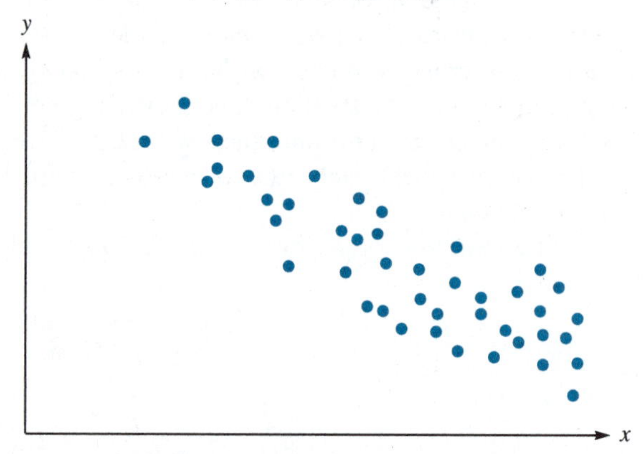

b. $\Sigma(x - \bar{x})(y - \bar{y})$ is large and negative.

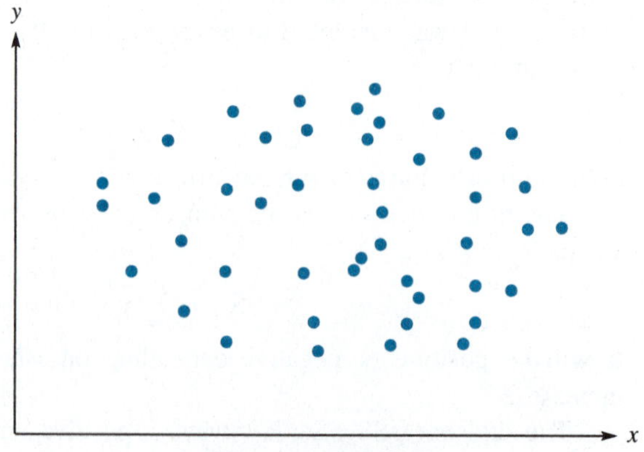

c. $\Sigma(x - \bar{x})(y - \bar{y})$ is near 0.

in the interval from -1 to $+1$. All three of the quantities, SS_x, SS_y, and SS_{xy}, as well as r, are calculated on a computer using one of many computer programs that are available.

How a Correlation Coefficient Measures the Strength of the Linear Relationship Between Two Variables, *x* and *y*

Figure 8.11 shows what a Pearson coefficient of correlation r tells us about the relationship between two variables, x and y. **The value of r will always be in the interval from -1 to $+1$.** The sign of r will always be the same as the sign of SS_{xy} because the denominator of r is always a positive number. **If the data points rise as x increases, as shown in Figure 8.11a, r will be positive and will assume a value between 0 and $+1$.** The closer the points fall to the line, the nearer r will be to 1. If all of the points fall on the line, r will equal $+1$.

If the data points fall as x increases, as shown in Figure 8.11b, r will be negative and will assume a value between 0 and -1. The closer the points fall to the line, the nearer r will be to -1. If all of the points fall on a line, $r = -1$.

FIGURE 8.11 Scattergrams showing different types of correlation

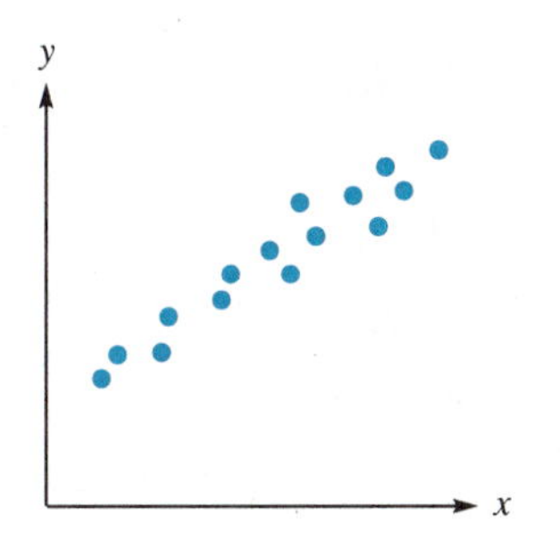

a. Strong positive linear correlation: r is near 1

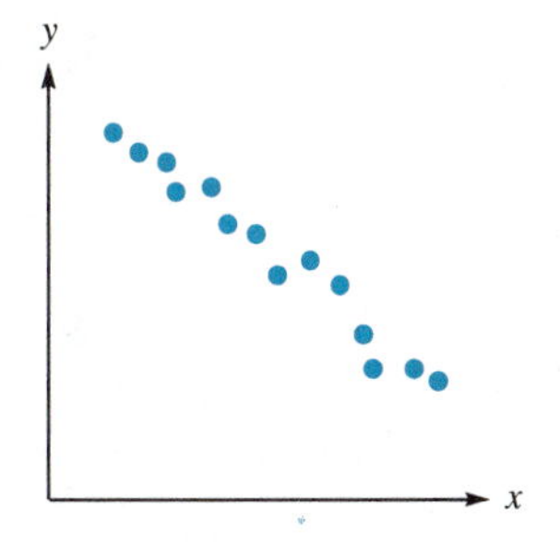

b. Strong negative linear correlation: r is near -1

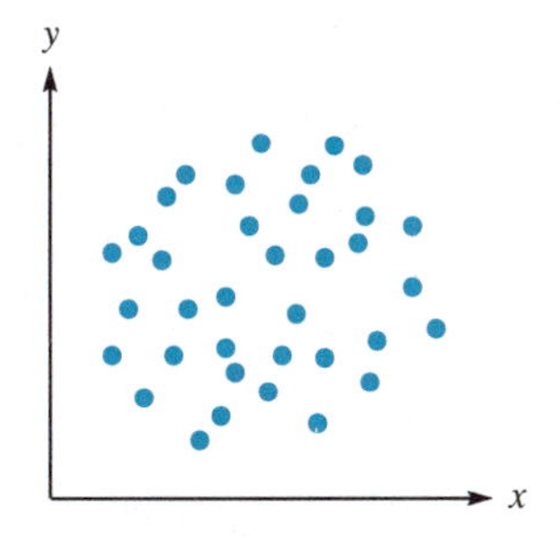

c. No apparent linear correlation: r is near 0

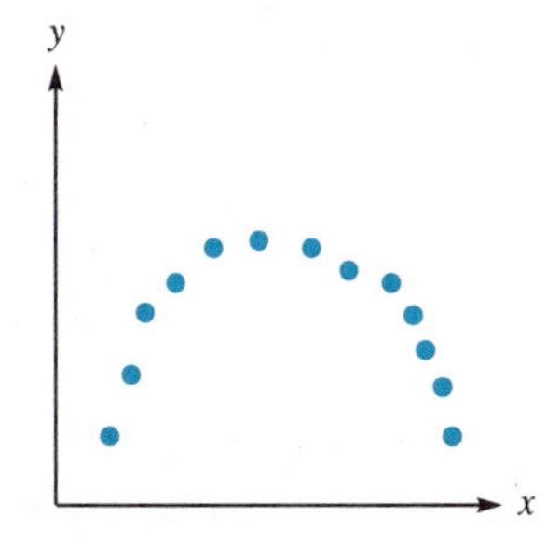

d. Curvilinear, but not linear correlation: r is near 0

If y tends neither to increase nor to decrease as x increases (i.e., the data appear as shown in Figure 8.11c), r will be near 0. The coefficient of correlation r may also be very small or near 0 if the plot of the data is as shown in Figure 8.11d, in which the data show a very close curvilinear relationship but little linear relationship.

Example 8.3

Look once again at the scattergram for the aptitude–achievement test scores shown in Figure 8.1. We will subsequently learn that the coefficient of correlation r for the data is approximately equal to .865. Does this value of r agree with what you see on the scattergram?

Solution

The value $r = .865$ suggests a linear relationship between x and y with the points sloping upward as x increases in value. Not all of the points fall on the line. Since $r = .865$ is close to 1, we might expect the points to fall more closely to the line than they do. The more scattergrams that you see with their associated correlation coefficients, the more you will realize that $r = .5$ is not "halfway" between no correlation and perfect correlation. In that respect, r^2 (our next topic) provides a better measure of the strength of the linear relationship between two variables x and y.

Another Measure of the Strength of the Linear Relationship Between *x* and *y*

The sign of r tells us whether y increases or decreases as x increases, but its numerical value is difficult to interpret. As explained in Example 8.3, a set of data points with $r = .5$ does not represent "halfway between perfect correlation and no correlation." In contrast, the quantity r^2, called ***the coefficient of determination***, has some meaning. Consider the following explanation.

A Practical Interpretation of r^2

The best way to determine whether x and y are related is to see whether x allows us to predict y with a smaller error of prediction than we would have if we did not use x at all. If we did not use x (or any other variable) to predict y, our best predictor of some new student's grade would be the sample mean. For example, if we were to predict the achievement test score for a student, based on only the 12 achievement test scores of Table 8.1 (in other words, the aptitude test scores were unavailable), our best guess for the student's test score would be the sample mean, $\bar{y} = 551.08$.

The variation of the y-values about their predicted values for this "no-other-information" predictor would be measured by the sum of squares of deviations SS_y of the y-values about their mean. The 12 deviations used in calculating SS_y

are shown in Figure 8.12a. **Since SS_y measures the variation in the prediction errors when no x variable is used to aid in prediction, SS_y is called *the total sum of squares of deviations*.**

In contrast, after we fit a least squares line to the aptitude–achievement test score data, the sum of squares of deviations of the observed values of y about their predicted values (i.e., values predicted by the least squares line) is reduced to SSE. The 12 deviations used in calculating SSE are shown in Figure 8.12b.

FIGURE 8.12 Graphical representation of two sources of variation

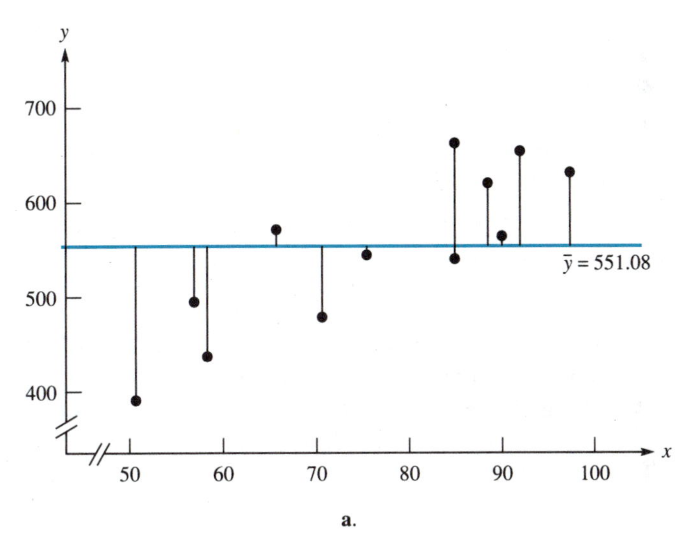

a.

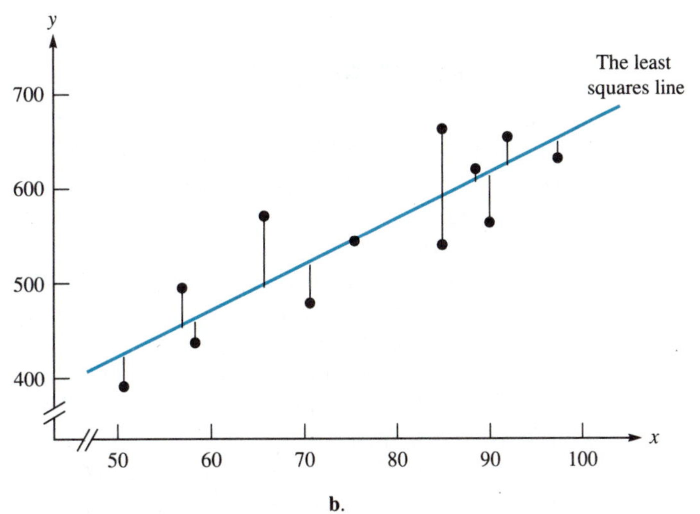

b.

You can see graphically that the deviations of the points from the least squares line in Figure 8.12b, measured by SSE, are much smaller than the deviations of the points from $\bar{y}$, measured by SS_y, in Figure 8.12a. The reduction in the sum of squares of the errors in prediction explained by the variable x is $(SS_y - SSE) = (\text{Total SS} - SSE)$. This reduction, expressed as a proportion of the Total SS, is (proof omitted) equal to the coefficient of determination, r^2. Therefore, we say that **r^2 is the proportion of the total variability of the y-values that can be explained by the variable x.**

Coefficient of Determination

$$r^2 = \frac{\text{Total SS} - \text{SSE}}{\text{Total SS}} = \frac{\text{SS}_y - \text{SSE}}{\text{SS}_y}$$

Example 8.4

The coefficient of determination for the aptitude–achievement test score data of Table 8.1 is equal to $r^2 = (.865)^2 = .748$. What does this mean?

Solution

It means that 74.8% of the total variation of the y-values, measured by Total $SS = SS_y$, is explained by x. The remaining 25.2% of the variation in y is explained by other variables or by the fact that the relationship between x and y is not exactly linear. Therefore, if we are attempting to find an equation that will enable us to predict y with a small error of prediction, we have ample room for improvement.

More on the Interpretation of r^2

Suppose that, in attempting to obtain a good predictor of a student's achievement in mathematics, we calculated the correlation coefficient for two or more different x variables. For example, suppose that we found the correlation between a student's aptitude test score x_1 and mathematics achievement test score y to be .6, and the correlation between the student's high school mathematics grade point average x_2 and y to be .7. These two correlation coefficients tell us that x_1 explains $(.6)^2 = .36$ or 36% of the variability of the y-values; x_2 explains $(.7)^2 = .49$ or 49% of the variability of the y-values. Variable x_2 gives better predictions of y than x_1, but a linear prediction equation based solely on x_2 leaves much room for improvement. Fifty-one percent of the variation of the y-values about the prediction equation is unexplained.

Would a prediction equation based on both x_1 and x_2 (the topic of Section 8.9) give better predictions than the straight-line prediction equation involving only x_2? Maybe. The fact that $r_1^2 = .36$ and $r_2^2 = .49$ **does not imply** that a prediction equation using both x_1 and x_2 would have an r^2 equal to $r_1^2 + r_2^2 = .36 + .49 = .85$. It is likely that x_1 and x_2 contribute overlapping information—i.e., much of the information in x_1 is already contained in x_2.

EXERCISES

8.18 What is meant by "the method of least squares"?

8.19 What is SSE?

8.20 What do the symbols SS_x, SS_y, and SS_{xy} represent?

8.21 Express r in terms of SS_x, SS_y, and SS_{xy}.

8.22 What is a coefficient of determination and what does it mean?

8.23 If the coefficient of correlation for a set of bivariate data is equal to .8, what can you say about the linear relationship between the two variables?

8.24 If the coefficient of correlation for a set of bivariate data is equal to $-.9$, what can you say about the linear relationship between the two variables?

8.25 If the coefficient of correlation for a set of bivariate data is equal to $-.2$, what can you say about the linear relationship between the two variables?

8.26 If the coefficient of correlation for a set of bivariate data is equal to 0, what can you say about the linear relationship between the two variables? If it equals 1? If it equals -1?

8.27 How does the coefficient of determination for a set of data relate to the total sum of squares of deviations (Total SS)?

BUS **8.28** In a study of the relationship between strike activity and productivity growth, data were collected on productivity and man-days lost per employee due to strikes for 20 OECD (Organization for Economic Development) countries during the period from 1967 to 1975 (Maki, Dennis R., "Strike Activity and Productivity Growth: Evidence from Twenty Countries," *Columbia Journal of World Business*, Summer 1983). A plot of the data is shown in Figure 8.13 on page 298. The coordinates (x, y) for a point give the calculated rate y of productivity growth for a particular country and the corresponding calculated man-days lost due to strikes per employee. For example, Japan had the highest rate of productivity growth, $y = .029$ with $x = .15$ man-day lost per employee. The least squares line is also shown on the graph in Figure 8.13.

a. Examine the graph and find the approximate values for b_0 and b_1, the constants in the equation of the least squares line.

b. Use your answer to part **b** to give the equation of the least squares line.

c. Using the information in your answers to parts **a** and **b**, what can you say about the correlation coefficient for the data?

8.29 This exercise will give you a better understanding of the coefficient of correlation r. Suppose that a set of data consists of the four points (pairs of x and y values) shown in Table 8.4 (page 298). Figure 8.14 shows a scattergram for the data.

a. Calculate $\bar{x}$ and $\bar{y}$.

b. Calculate the deviations, $(x - \bar{x})$ and $(y - \bar{y})$, for each data point.

c. Calculate the square of each of the deviations and find the cross-product of the deviations for each data point. Insert these quantities in the appropriate row and column of the computing table (Table 8.5 on page 299).

d. Find SS_x, SS_y, and SS_{xy}.*

e. Find r.

f. Is your value of r consistent with the scattergram in Figure 8.14? Explain.

* A computer is programmed to follow a slightly different procedure (one with smaller computational rounding errors) than the procedure employed in parts **a** and **b**.

FIGURE 8.13 A scattergram and least squares line for rate of productivity growth versus man-days lost per employee

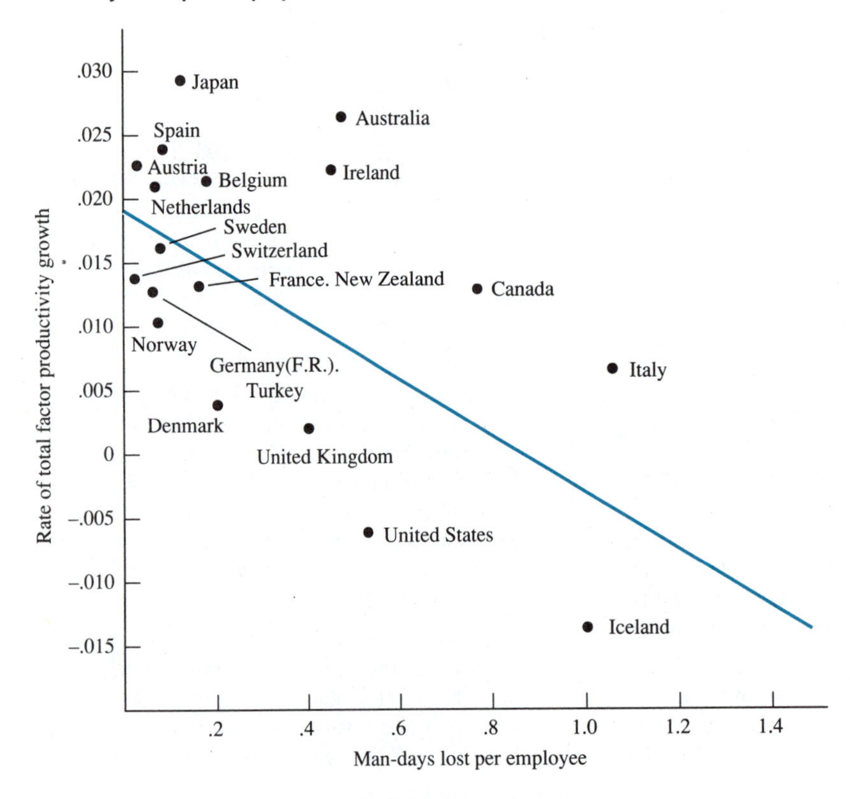

TABLE 8.4
A Data Set, $n = 4$

x	y
1	0
3	2
4	4
2	2

FIGURE 8.14 Scattergram for data of Table 8.4

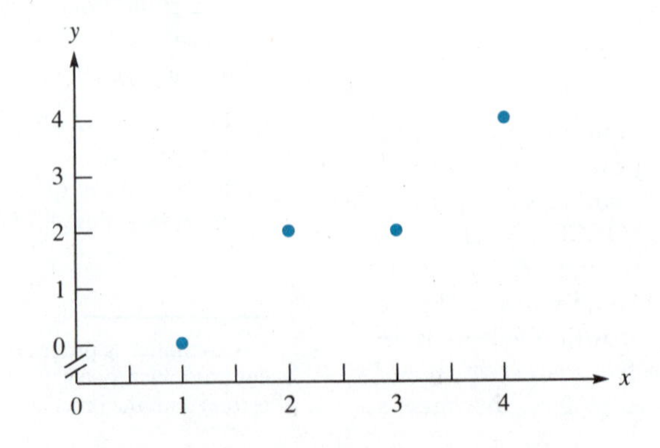

TABLE 8.5 Procedure for Calculating SS_x, SS_y, SS_{xy} for a Small Set of Data

	x	y	$(x-\bar{x})$	$(x-\bar{x})^2$	$(y-\bar{y})$	$(y-\bar{y})^2$	$(x-\bar{x})(y-\bar{y})$
	1	0	−1.5	2.25	−2	4	3
	3	2					
	4	4					
	2	2					
Sums	10	8	0	5.0	0	8	6

8.6 A Simple Linear Regression Analysis: Questions It Will Answer

What It Is

Wouldn't it be wonderful if we had an equation that would enable us to predict the value of some variable y—say, your grade in this course, the shooting proficiency of a basketball player, or the price of Walt Disney stock on January 1 of next year? **A *simple linear regression analysis* is a catch-all term for the process of fitting a straight-line model to a set of bivariate data and, among other things, using the prediction equation to predict a value of y to be observed in the future.** It represents a special case of **multiple regression analysis**, a powerful tool for developing a prediction equation for y based on any number of x variables. We will show you and will interpret the regression analysis for the academic achievement data of Table 8.1. But first we will list the questions that a regression analysis will answer and will list the assumptions on which the validity of the analysis is based.

Questions a Simple Linear Regression Analysis Will Answer

1. **What are the values of b_0 and b_1 and, therefore, what is the least squares prediction equation?** A regression analysis of the data in Table 8.1 fits a least squares line to the data and gives us estimates of the y-intercept b_0 and the slope b_1. With these values, we can write the equation of the least squares line—i.e., the prediction equation. The prediction equation relates the mathematics achievement test score y to the quantitative aptitude score x. We can use this equation to predict the mathematics achievement test score for a student who has received a score x on the quantitative aptitude test.

2. **Do the data provide sufficient evidence to indicate that x contributes information for the prediction of y when the mean value of y and the value of x are linearly related?** Or is the upward (or downward) slope of the points that we think we see in a scattergram due simply to chance or sampling error?

3. **How well does the prediction line fit the data?** To answer this question, we would like to know the value of the simple coefficient of correlation r and, more important, the value of the coefficient of determination r^2.

4. **What is the confidence interval for the mean value of y when x is equal to some particular value?** For example, we might want to estimate the mean achievement test score for all students who scored 85 on the mathematics aptitude test.

5. **What is the prediction interval for y when x is equal to some particular value?** For example, we might want to predict the achievement test score y for one particular student who scored 85 on his or her mathematics aptitude test.

What Is a Prediction Interval?

Prediction intervals are to random variables what confidence intervals are to population parameters. **A $(1 - a)100\%$ *prediction interval* will enclose the random variable y with a probability equal to $(1 - a)$.** This probability, a confidence coefficient, measures how confident we are that the interval encloses the value of y that we are attempting to predict. A 95% prediction interval for a student's achievement score y, given that the student's aptitude score is $x = 85$, is an interval that encloses y with probability equal to .95. Ninety-five percent of the prediction intervals constructed under similar circumstances will enclose y.

8.7 A Simple Linear Regression Analysis: Assumptions

Before we use a regression analysis to make some practical inferences, we need to know the rules of the game—the assumptions on which our methods are based. Otherwise, we might use a method and place more confidence in our results than they warrant.

The inferences that are derived from a simple linear regression analysis are valid if the following assumptions are satisfied (see Section 6.8):

1. **The mean value of y for a given value of x, represented by the symbol $E(y)$, is given by the equation***

$$E(y) = \beta_0 + \beta_1 x$$

* The mean value of y is also called the "expected value of y."

Remember that this is just an assumption. We assume that each point on the line $E(y) = \beta_0 + \beta_1 x$ is the mean of a population of y-values for that particular value of x and, therefore, that the line is a "**line of means.**" We do not know the values of β_0 and β_1 but we estimate their values when we fit a least squares line to a set of data and calculate b_0 and b_1.

2. For any given value of x, y is a normally distributed random variable with mean equal to $E(y)$ and with a common variance equal to some value, call it σ^2. In fact, we envision a population of y-values corresponding to each value of x that could be collected if we were inclined and able to do so (which we are not). Individual values of y observed for a particular value of x will vary in a random manner about the line $E(y) = \beta_0 + \beta_1 x$. The distribution of the y-values for a given value of x is a normal distribution with a mean $E(y)$ and with a variance equal to σ^2 (the standard deviation of the distribution would equal σ).

The graphical implications of the first two assumptions are depicted in Figure 8.15. The graph shows distributions of y-values for several values of x. Note that each of the normal distributions is centered on the line of means, $E(y) = \beta_0 + \beta_1 x$, and all have the same spread, measured by the standard deviation σ.

FIGURE 8.15 A graphical representation of assumptions 1 and 2

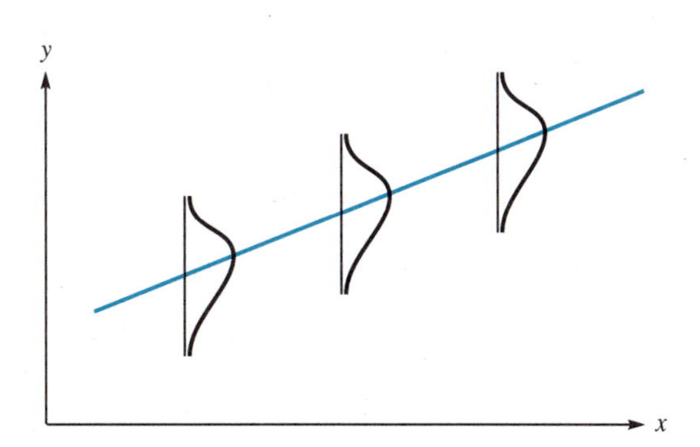

3. The third assumption is that any pair of y-values are independent in a probabilistic sense. That is, an observed value of y will be unaffected by any other value of y. For example, the fact that one value of y fell above the line of means would not affect the value of the next y-value to be observed.

Comments on the Assumptions

If the assumptions are satisfied, least squares estimates of β_0 and β_1 have smaller errors of estimation than any other estimators. Tests of hypotheses based on a regression analysis are also best in the sense that they are less likely to lead to incorrect decisions.

Note that the assumptions for a regression analysis bear a similarity to the assumptions required for the methods of Chapters 6 and 7. As with the methods of Chapters 6 and 7, the results of regression analyses do not appear to be greatly disturbed by moderate departures from the assumptions. We do not need to have exact normality; a mound-shaped distribution of deviations from the line of means is satisfactory. We would like the variance of y to be approximately the same for all values of x, but moderate differences in variation for different values of x do not appear to greatly disturb the results of a regression analysis. If disagreement with the assumptions is extreme, it is sometimes possible to transform the y-values (for example, use the square roots of the y-values) to make the data better fit our assumptions. Consult the references for more detail.

8.8 A Simple Linear Regression Analysis: Typical Computer Outputs

Figure 8.16 gives the Data Desk, Minitab, and SAS regression analyses for the mathematics aptitude–achievement test scores of Table 8.1. We will examine the printouts and look for answers to the five questions that can be answered by a regression analysis.

Finding the Equation of the Least Squares Line

1. **What are the values of b_0 and b_1 and, therefore, what is the prediction equation?**

The estimates b_0 and b_1, shaded and numbered 1 and 2, respectively, on each of the printouts, are

$$b_0 = 184.658 \quad \text{and} \quad b_1 = 4.800$$

Therefore, the least squares prediction equation is

$$y = 184.66 + 4.80x$$

An Estimator of σ^2

The estimator of the variance σ^2 of the y-values about the line of means (see assumption 2 in Section 8.7) is called the *mean square for error* and is designated by the symbol MSE or s^2. This quantity is important because it measures the variation of the data points about the fitted line and it appears in the formula for every confidence interval and statistical test in a regression analysis. It is equal to the sum of squares of the deviations* of the y-values about the least squares line,

* The sum of squares of deviations about the least squares line is also called **sum of squares of residuals** and **sum of squares for error**.

FIGURE 8.16 Data Desk, Minitab, and SAS regression analyses for the aptitude–achievement data of Table 8.1

```
  Dependent variable is: y
  R² = 74.8%      R²(adjusted) = 72.2%
5 s = 45.24 with 12 - 2 = 10 degrees of freedom

  Source        Sum of Squares   df         Mean Square   F-ratio
  Regression    60665.0           1               60665      29.6
  Residual      20468.0          10             2046.80 4

  Variable     Coefficient    s.e. of Coeff    t-ratio
  Constant     184.658 1           68.56         2.69
  x            4.80033 2          0.8817 8       5.44 6
```

a. Data Desk

```
 ROW      C1      C2

   1     620      88
   2     495      57
   3     549      76
   4     635      97
  .   .   .

MTB > regress c1 on 1 predictor c2

The regression equation is
y = 185 + 4.80 x

Predictor           Coef         Stdev      t-ratio         p
Constant          184.66 1       68.56         2.69     0.023
x                 4.8003 2      0.8817 8       5.44 6   0.000 7

s = 45.24 5        R-sq = 74.8%       R-sq(adj) = 72.2%

Analysis of Variance

SOURCE          DF          SS            MS          F          p
Regression       1       60665         60665      29.64      0.000
Error           10       20468 3        2047 4
Total           11       81133
```

b. Minitab

(continued)

FIGURE 8.16 ***Continued***

```
Model: MODEL1
Dependent Variable: Y

                          Analysis of Variance

                              Sum of           Mean
Source            DF         Squares         Square      F Value       Prob>F

Model              1     60664.96029    60664.96029       29.639       0.0003
Error             10     20467.95638 3   2046.79564 4
C Total           11     81132.91667

     Root MSE       45.24153 5     R-square        0.7477
     Dep Mean      551.08333       Adj R-sq        0.7225
     C.V.            8.20956

                          Parameter Estimates

                    Parameter        Standard    T for H0:
Variable  DF         Estimate           Error   Parameter=0    Prob > |T|

INTERCEP   1     184.658205 1  68.56135015          2.693        0.0226
X          1       4.800329 2   0.88173748 8        5.444 6      0.0003 7
```

c. SAS

represented by the symbol SSE, divided by $(n - 2)$, the number of degees of freedom* for the estimator s^2.

Least Squares Estimator of σ^2

$$s^2 = \frac{\text{Sum of squares of deviations}}{n-2} = \frac{\text{SSE}}{n-2}$$

Degrees of freedom = $n - 2$

The quantities SSE = 20468.0 and $s^2 = 2046.8$ are shaded and numbered 3 and 4, respectively, on the printouts. The value of s^2 appears in the Data Desk printout in the "Mean Square" column and the "Residual" row, in the Minitab printout in the "MS" column and the "Error" row, and in the SAS printout in the "Mean Square" column and the "Error" row.

The square root of s^2, $s = \sqrt{s^2} = \sqrt{2{,}046.8} = 45.24$, is the sample standard deviation of the residuals, i.e., the deviations of the *y*-values about the least squares line. The value of s is shaded and numbered 5 on all three of the

*The number of degrees of freedom associated with a t statistic in Chapters 6, 7, and 8 is always the divisor of the sum of squares of deviations in the formula for s^2. For example, the number of degrees of freedom for the unpaired comparison of means, Chapter 7, is the divisor in the formula for the pooled estimator of σ^2—namely, $(n_1 + n_2 - 2)$.

printouts of Figure 8.16. It measures the spread of the data points about the least squares line. The Empirical Rule suggests that all or almost all of the data points should fall within $3s$ and most should fall within $2s$ of the line. For example, Figure 8.17 shows a scattergram of the data of Table 8.1, with a graph of the least squares line superimposed. We have drawn lines on the graph located a vertical distance of $2s = 2(45.24) = 90.48$ above and below the least squares line. For this example, all of the data points fall within a $2s$-band of the least squares line. This gives us a practical check on the calculation of s. If many of the points deviate from the least squares line by more than $2s$, something is wrong with our analysis.

FIGURE 8.17 A scattergram of the data points and a graph of the least squares line

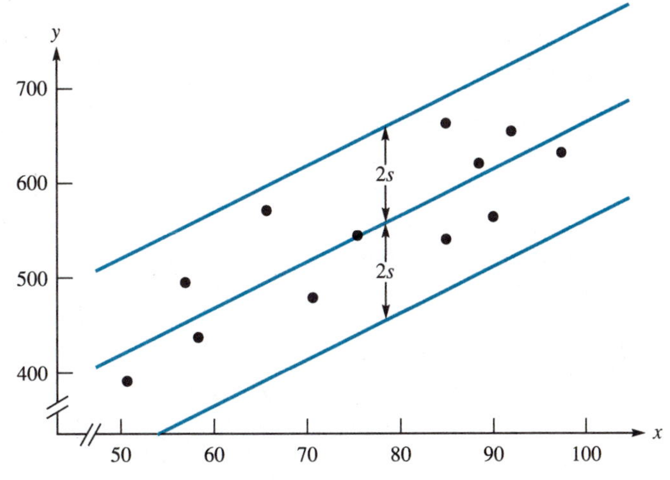

2. **Do the data provide sufficient evidence to indicate that x contributes information for the prediction of y when $E(y)$ and x are linearly related?**

If the slope of the line of means

$$E(y) = \beta_0 + \beta_1 x$$

equals 0 (that is, the line is parallel to the x-axis), the predicted value of y for a given value of x will always equal β_0, regardless of the value of x (see Figure 8.18 on page 306). Therefore, to answer our question, **we will want to test the null hypothesis that the slope β_1 of the line of means equals 0 against the alternative hypothesis that $\beta_1 \neq 0$. If we reject the null hypothesis, we conclude that the line slopes either upward or downward and, therefore, that x contributes information for the prediction of y.**

FIGURE 8.18 Graph of line when slope equals 0 and *x* contributes no information for the prediction of *y*

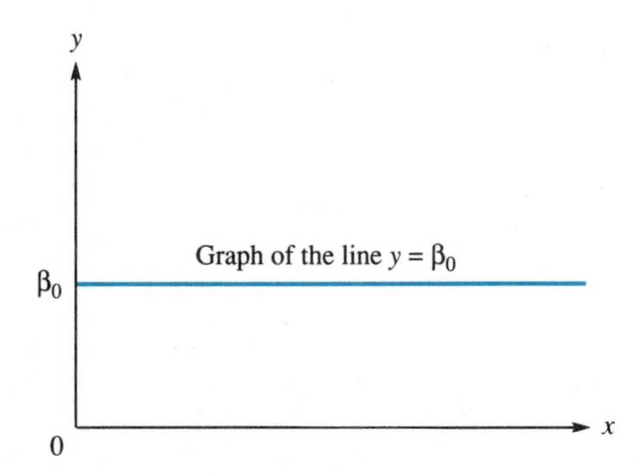

A Test and Confidence Interval for β_1

The confidence interval and test of hypothesis for β_1 are based on a Student's t statistic with $(n - 2)$ degrees of freedom. The formulas for the test and confidence interval for β_1 are identical to those for a single mean given in Chapter 6, except that β_1 replaces μ, the estimate b_1 replaces $\bar{x}$, and $\widehat{\text{SE}}(b_1)$ replaces $\widehat{\text{SE}}(\bar{x})$. They are summarized in the boxes.

A $100(1 - a)\%$ Confidence Interval for β_1

$$\text{LCL} = b_1 - t_{a/2}\widehat{\text{SE}}(b_1)$$
$$\text{UCL} = b_1 + t_{a/2}\widehat{\text{SE}}(b_1)$$

where $t_{a/2}$ is based on $(n - 2)$ degrees of freedom.

Assumptions: **See Section 8.7.**

A Test to Determine Whether x Contributes Information for the Prediction of y (i.e., Whether the Slope β_1 Differs from 0)

Null hypothesis: $H_0: \beta_1 = 0$

Alternative hypothesis:

1. $H_a: \beta_1 > 0$, **an upper-tailed test**
2. $H_a: \beta_1 < 0$, **a lower-tailed test**

3. H_a: $\beta_1 \neq 0$, a two-tailed test

Test statistic: $t = \dfrac{b_1}{\widehat{\text{SE}}(b_1)}$

where t is based on $(n - 2)$ degrees of freedom.

Rejection region: **Choose the value of α that is acceptable to you. Then reject H_0 and accept H_a when the p-value for the test is less than or equal to α.**

Assumptions: **See Section 8.7.**

Testing the Adequacy of the Prediction Equation

Example 8.5

The regression analysis on the aptitude–achievement data of Table 8.1 produced the prediction equation $y = 184.66 + 4.80x$. Does the admissions officer have a prediction equation that really works or is the apparent relationship between y and x a result of chance? Does the aptitude score x contribute information for the prediction of achievement score y, or do we need to find a better prediction equation? Use the test for β_1 to decide whether the test score data provide sufficient evidence to indicate that the aptitude test score x contributes information for the prediction of the achievement test score y.

Solution

We want to test

$$H_0: \quad \beta_1 = 0$$

against the alternative hypothesis

$$H_a: \quad \beta_1 > 0$$

i.e., that β_1 is larger than 0. We have chosen this one-sided alternative because we believe that if there is any relationship between the aptitude and the achievement test scores, it must be positive—that is, if the slope differs from 0, it must be positive. Therefore, we will conduct an upper one-tailed test and will reject H_0 for very large positive values of t. The rejection region is shown in Figure 8.19 on page 308.

The computed value of the t statistic and the p-value (if given) for the test of H_0: $\beta_1 = 0$ are shaded on the printouts in Figure 8.16 and are numbered 6 and 7, respectively. The Data Desk printout shows that the "t ratio" for testing β_1, the coefficient of x, is $t = 5.44$, but it does not give the p-value for the test. The Minitab and SAS printouts give both the t and the p-values for the test. Minitab gives the p-value correct to three decimal places as 0.000. SAS shows the p-value correct

FIGURE 8.19 Rejection region for detecting positive values of β_1

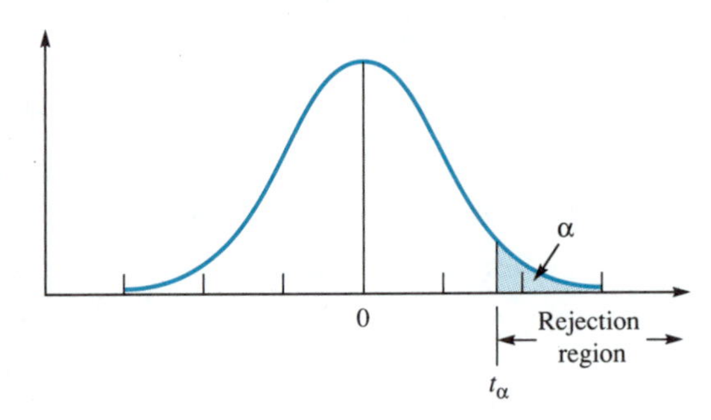

to four decimal places as 0.0003. **Both Minitab and SAS give the *p*-value for a two-tailed test**; i.e., they give PROB > |T|, the probability of observing a value of t as large as or larger than 5.44 or as small as or smaller than -5.44, if the null hypothesis is true. Since we want only the probability that t falls in the *upper* tail tail of the t distribution, the p-value for our test is half the printed value, $.0003/2 = .00015$, or only 15 chances in 100,000. Consequently, we would reject H_0 for values of α as small as .00015. There is overwhelming evidence to indicate that the aptitude score x contributes information for prediction of the mathematics achievement test score y. The aptitude test score is useful in predicting academic achievement—just what we would expect!

Example 8.6

Under what circumstances would we employ a two-tailed test for β_1?

Solution

We would use a two-tailed test if it was conceivable that β_1 could be *either positive or negative* if, in fact, it differed from 0. In choosing a one-tailed test for Example 8.5, we assumed that the people who constructed the aptitude test intended high aptitude test scores to correspond to high achievement test scores; i.e., they intended to create a positive correlation between aptitude and achievement test scores. If for some reason it was possible that the aptitude test could have been designed to create *either* a positive or a negative correlation between the aptitude and achievement test scores (i.e., β_1 could have been either positive or negative if it differed from 0), then we would have chosen H_a: $\beta_1 \neq 0$ and employed a two-tailed test.

Calculating the p-Value When Testing the Slope of a Line

Example 8.7

Suppose that the printout for your software package does not give the p-value of the test (Data Desk does not, for example). Find the approximate p-value for the test of Example 8.5, using the procedure employed in Chapters 6 and 7.

Solution

The data set consists of $n = 12$ data points. Therefore, the number of degrees of freedom for the computed t is $(n - 2) = (12 - 2) = 10$. Looking in the row corresponding to df = 10 in Table 4 of Appendix 2 (reproduced in Table 8.6), we find that the value of t that locates an area equal to .005 in the upper tail of the t distribution is $t = 3.169$. Therefore, the probability of observing a t as large as or larger than 3.169 is .005, if the null hypothesis is true. Since our observed value of t, $t = 5.44$, is larger than 3.169, the p-value for our test is less than .005. (We know from Example 8.5 that the p-value is .00015.)

TABLE 8.6 A Partial Reproduction of Table 4 of Appendix 2

df	$t_{.100}$	$t_{.050}$	$t_{.025}$	$t_{.010}$	$t_{.005}$
1	3.078	6.314	12.706	31.821	63.657
2	1.886	2.920	4.303	6.965	9.925
3	1.638	2.353	3.182	4.541	5.841
4	1.533	2.132	2.776	3.747	4.604
5	1.476	2.015	2.571	3.365	4.032
6	1.440	1.943	2.447	3.143	3.707
7	1.415	1.895	2.365	2.998	3.499
8	1.397	1.860	2.306	2.896	3.355
9	1.383	1.833	2.262	2.821	3.250
10	1.372	1.812	2.228	2.764	3.169
11	1.363	1.796	2.201	2.718	3.106
12	1.356	1.782	2.179	2.681	3.055

Estimating the Slope of the Least Squares Line

Recall that the slope β_1 of a line is the change in y for a 1-unit increase in x. Suppose that we have a regression prediction equation that predicts the profit y of a dress factory as a function of the number x of sewing machines. Then the mean increase in profit obtained by adding one sewing machine is the slope β_1 of the line of means. We can estimate β_1 using the point estimate b_1, the slope in the least squares line, or we can find a confidence interval for β_1.

Example 8.8

Find a 95% confidence interval for the mean increase in achievement test score for a 1-point increase in aptitude test score.

Solution

None of the printouts in Figure 8.16 gives a confidence interval for β_1 but all give the parameter estimates and the standard errors of the estimates. The estimated slope is $b_1 = 4.800$, the estimated standard error of b_1 (shaded and numbered as 8 on all three printouts) is

$$\widehat{\text{SE}}(b_1) = .8817$$

and the value of $t_{.025}$ based on $(n - 2) = (12 - 2) = 10$ degrees of freedom (from Table 4 of Appendix 2) is $t = 2.228$. Substituting these values into the formulas for the confidence limits (see the box on page 306) gives the 95% limits:

$$\text{LCL} = b_1 - t_{.025}\widehat{\text{SE}}(b_1) = 4.800 - 2.228(.8817) = 2.84$$
$$\text{UCL} = b_1 + t_{.025}\widehat{\text{SE}}(b_1) = 4.800 + 2.228(.8817) = 6.76$$

Estimating the Mean Value of *y* and Predicting a Particular Value of *y* for a Value of *x*

Question 3—"How well does the prediction line fit the data?"—was answered in Section 8.5. Questions 4 and 5 are restated below. We will comment on the two questions jointly.

4. **What is the confidence interval for the mean value of y when x is equal to some particular value?** For example, we might want to estimate the mean achievement test score for all students who scored 85 on the mathematics aptitude test. Or . . .
5. **What is the prediction interval for y when x is equal to some particular value? . . .** we might want to predict the achievement test score y for one particular student who scored 85 on his or her mathematics aptitude test.

Notice the difference between these two intervals. One is a confidence interval for the mean value $E(y)$ of a population of y scores for students with an aptitude test score of $x = 85$. The other is a prediction interval for some new value of y to be observed in the future when $x = 85$. **Both the point estimate of $E(y)$ and the predicted value of y for $x = 85$ will be the same; they will be the value of y predicted by the least squares line for $x = 85$:**

$$y = 184.66 + 4.80x = 184.66 + 4.80(85) = 592.66$$

The difference between the two intervals is that the prediction interval is wider than the confidence interval; the error of prediction is much larger than the error of estimation. You can see this in Figure 8.20. The error in estimating the mean value of y when $x = 85$ is the vertical distance between the line of means and the least squares line. In contrast, the error of predicting some new value of y when $x = 85$ is that error *plus* the distance between y and the line of means.

FIGURE 8.20 Comparing the error of estimation and the error of prediction

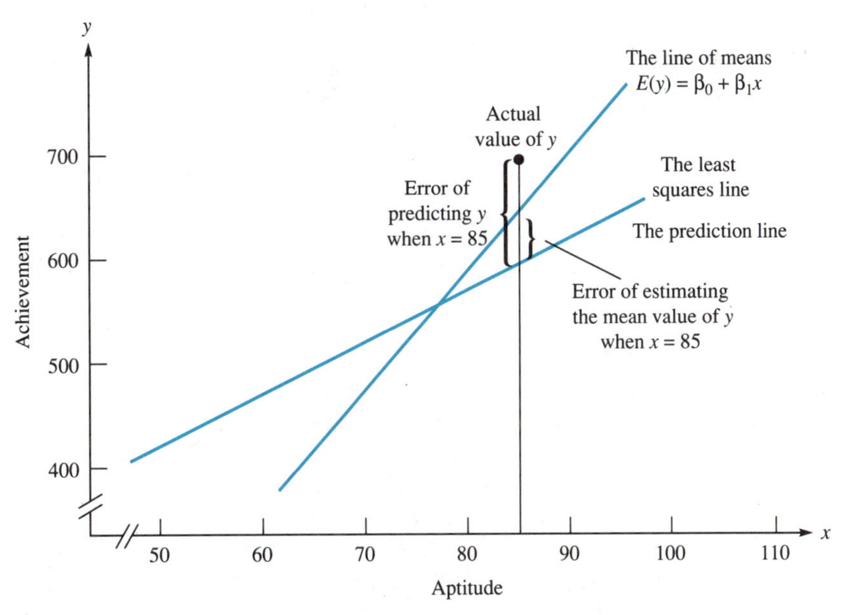

Some (but not all) regression software packages will, upon request, print confidence intervals for the mean and prediction intervals for a particular value of x.* Figure 8.21 (page 312) gives the Minitab and SAS printouts for the confidence intervals

* The half-width of the confidence interval for the mean value of y when $x = x_0$ is

$$t_{a/2}s\sqrt{\frac{1}{n} + \frac{(x_0 - \bar{x})^2}{SS_x}}$$

The corresponding half-width of the prediction interval is

$$t_{a/2}s\sqrt{1 + \frac{1}{n} + \frac{(x_0 - \bar{x})^2}{SS_x}}$$

FIGURE 8.21 Minitab and SAS printouts of confidence and prediction intervals for the aptitude–achievement test score data, Table 8.1

```
     Fit  Stdev.Fit          95% C.I.              95% P.I.
   592.7       15.1    ( 559.0,   626.4) 1 (  486.4,   699.0) 2
```

a. Minitab

Obs	Dep Var Y	Predict Value	Std Err Predict	Lower95% Mean	Upper95% Mean	Lower95% Predict	Upper95% Predict
1	620.0	607.1	16.625	570.0	644.1	499.7	714.5
2	495.0	458.3	21.475	410.4	506.1	346.7	569.9
3	549.0	549.5	13.063	520.4	578.6	444.6	654.4
4	635.0	650.3	22.419	600.3	700.2	537.8	762.8
5	480.0	525.5	13.881	494.6	556.4	420.0	630.9
6	568.0	616.7	17.770	577.1	656.3	508.4	725.0
7	570.0	501.5	15.924	466.0	537.0	394.6	608.3
8	437.0	463.1	20.782	416.8	509.4	352.1	574.0
9	655.0	626.3	19.010	583.9	668.6	516.9	735.6
10	547.0	592.7	15.131	559.0	626.4	486.4	699.0
11	395.0	429.5	25.875	371.8	487.1	313.3	545.6
12	662.0	592.7	15.131	559.0	626.4	486.4	699.0
13	.	592.7	15.131	559.0	626.4 1	486.4	699.0 2

b. SAS

and prediction intervals for $x = 85$. The Minitab printout, Figure 8.21a, gives only the intervals for the specified value of x. For $x = 85$, the confidence interval on the mean value of y is 559.0 to 626.4. The prediction interval for some new value of y when $x = 85$ is 486.4 to 699.0.

The SAS printout, Figure 8.21b, will give the intervals for every value of x used in the original data set. For example, the x value for the first observation in the data set of Table 8.1 is $x = 88$. The observed and predicted values of y, 620.0 and 607.1, the standard error of the predicted value, and the intervals appear in the row corresponding to Obs 1. The next eleven rows give the intervals corresponding to the remaining eleven observations. The confidence and prediction intervals for the requested value, $x = 85$, always appear in the last row—in this case, in the row corresponding to Obs 13. The intervals are shaded and marked 1 and 2, respectively, on the printouts.

A Word of Caution

Confidence intervals and prediction intervals can be grossly incorrect if the assumption that $E(y) = \beta_0 + \beta_1 x$ is not satisfied (see Section 8.7), i.e., if the relationship between the mean value of y and the value of x does not graph as a straight line. Suppose that the relationship between x and y is curvilinear, as shown in Figure 8.22. The error of estimating $E(y)$ and predicting y will be larger than expected

FIGURE 8.22 Errors that can occur when a straight line is fit to data that are curvilinearly related

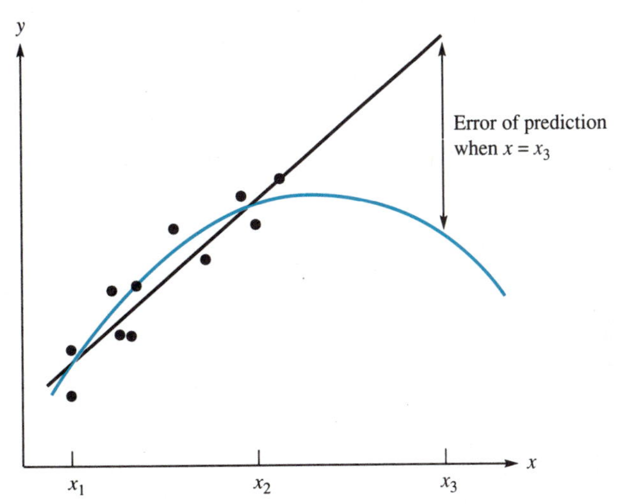

if the value of x that you choose falls outside the set of x-values used in your regression analysis. For example, if you collected your data over the interval from x_1 to x_2, the errors of prediction would be quite small. If you used this fitted line to predict y for $x = x_3$, the error of prediction would be very large. Most relationships between two variables, x and y, in the real world are curvilinear. Nevertheless, the relationship over a small interval on the x-axis will often be approximately linear—close enough so that our least squares line gives good estimates of $E(y)$ and good predictions of y as long as we choose values of x within the interval covered by the data set.

EXERCISES

GEN **8.30** A study was conducted to investigate the relationship between "running economy" and distance running performance for trained, competitive male runners (Powers, Scott K., et al., "Ventilatory Threshold, Running Economy and Distance Running Performance of Trained Athletes," *Research Quarterly for Exercise and Sport,* Vol. 54, No. 3, 1983). "Running economy" for a runner is defined to be the runner's steady-state oxygen consumption for a standardized running speed. The study involved nine trained runners. Each runner competed in two 10-kilometer races and the better of the two finish times was recorded. In addition, a measure of each runner's running economy was recorded. These data, along with the number of years of competitive running experience, are shown for each runner in Table 8.7 (page 314). Let us examine the relationship between a runner's running economy x and finish time y. The SAS output for the regression analysis for the data is shown in Figure 8.23.

TABLE 8.7 Years of Experience, Running Economy, and Finish Times for Nine Runners

Runner	Years Competitive Running	Running Economy	10 Kilometers Finish Time (min.)
1	9	46.2	33.15
2	13	43.2	33.33
3	5	46.7	33.50
4	7	48.9	33.55
5	12	47.1	33.73
6	6	50.5	33.86
7	4	53.5	33.90
8	5	46.3	34.15
9	3	50.5	34.90

FIGURE 8.23 A SAS printout of the regression analysis for the finish time–running economy data of Table 8.7

```
Model: MODEL1
Dependent Variable: Y

                              Analysis of Variance

                                Sum of          Mean
        Source          DF     Squares        Square       F Value         Prob>F

        Model            1     0.56610       0.56610         2.510         0.1572
        Error            7     1.57893       0.22556
        C Total          8     2.14502

            Root MSE       0.47493     R-square       0.2639
            Dep Mean      33.78556     Adj R-sq       0.1588
            C.V.           1.40573

                              Parameter Estimates

                         Parameter       Standard      T for H0:
      Variable  DF        Estimate          Error    Parameter=0     Prob > |T|

      INTERCEP   1       29.610573     2.64011905         11.216         0.0001
      X          1        0.086798     0.05478936          1.584         0.1572
```

a. Construct a scattergram for the data.

b. Find the estimates of β_0 and β_1. Give the equation of the least squares line for the data. Graph the least squares line on the scattergram of part **a**.

c. Do the data provide sufficient evidence to indicate that x contributes information for the prediction of y? State the null and alternative hypotheses for the test. Explain the reasons for the choice of your alternative hypothesis. Give the value of the test statistic and the p-value for the test. If you were to choose $\alpha = .01$, what would you conclude?

d. Find a 95% confidence interval for the mean increase (or decrease) in finish time for a 1-unit increase in running economy.

GEN **8.31** Refer to Exercise 8.30. Estimate the mean finish time y for all runners who have a running economy equal to $x = 50$. Predict the actual finish time y for a runner who had a running economy equal to $x = 50$. The estimate and the prediction are numerically identical. In what sense do they differ?

GEN **8.32** Refer to Exercise 8.30. Write a brief essay that explains the practical implications of the regression analysis. (Take your time. Explain in detail and justify your statements.)

8.33 Do football coaches get better with age? If Coach Bear Bryant, the legendary coach of the University of Alabama, is typical of other football coaches, universities may want to hire older coaches for their teams. In his 30s, Coach Bryant had 59 wins, 23 losses, and 5 ties. In his 40s, his record improved to 73 wins, 24 losses, and 8 ties. In his 50s, it was 88, 22, and 3; and in his 60s, his record was 95, 12, and 1. The percentage wins y in a 10-year interval and the midpoint x of a 10-year age interval for Coach Bryant (*Sports Illustrated*), Vol. 57, No. 11, 1982) are given in Table 8.8. Figure 8.24 gives the Minitab printout for these data.

TABLE 8.8 Coach Bear Bryant's Percentage Wins for Four Age Brackets

Midpoint Age Interval	35	45	55	65
Percentage Wins	67.8	69.5	77.9	88.0

FIGURE 8.24 Minitab printout of the data of Table 8.8

```
 ROW       y      x

   1    67.8     35
   2    69.5     45
   3    77.9     55
   4    88.0     65

MTB > regress c1 on 1 predictor c2

The regression equation is
y = 41.3 + 0.690 x

Predictor       Coef       Stdev    t-ratio        p
Constant      41.300       7.042       5.87    0.028
x             0.6900      0.1374       5.02    0.037

s = 3.073       R-sq = 92.6%      R-sq(adj) = 89.0%

Analysis of Variance

SOURCE         DF        SS          MS          F          p
Regression      1    238.05      238.05      25.20      0.037
Error           2     18.89        9.45
Total           3    256.94
```

Explain why you think there is or is not a relationship between Coach Bryant's win record and his age. What can you say about the strength of the relationship if it exists?

BIO **8.34** Table 8.9 gives the diameters and heights of ten fossil specimens of a species of small shellfish, *Rotularia (Annelida) fallax*, that were unearthed in a mapping expedition near the Antarctic Peninsula (Macellari, Carlos E, "Revision of Serpulids of the Genus *Rotularia (Annelida)* at Seymour Island (Antarctic Peninsula) and Their Value in Stratigraphy," *Journal of Paleontology*, Vol. 58, No. 4, July 1984). Column 1 of the table gives the identification symbol for the fossil specimen; column 2 gives the diameter of the fossil, in millimeters (mm); and column 3 gives its height (in mm). Figure 8.25 gives the SAS printout of the regression analysis for the fossil data.

a. Construct a scattergram for the data. Let $x =$ height and $y =$ diameter of a fossil.

b. Find the estimates of β_0 and β_1. Give the equation of the least squares line for the data. Graph the least squares line on the scattergram of part **a**.

c. Do the data provide sufficient evidence to indicate that x contributes information for the prediction of y? State the null and alternative hypotheses for the text. Explain the reasons for the choice of your alternative hypothesis. Give the value of the test statistic and the p-value for the test. If you were to choose $\alpha = .01$, what would you conclude?

d. Find a 95% confidence interval for the mean change in fossil diameter for a 1-millimeter increase in fossil height. Interpret the interval.

e. Find a 95% confidence interval for the mean fossil diameter when $x = 74$ mm. Interpret the interval. (*Note:* The requested interval appears in the row opposite Obs 11.)

f. Notice the printout of the residuals—the deviations of the y-values from the least squares line. The sum of squares of the residuals is given as 4,073.2098. Where else does this appear on the printout and in what context?

8.35 Refer to Exercise 8.34. Write a brief essay that explains the practical implications of the regression analysis. (Take your time. Explain in detail and justify your statements.)

TABLE 8.9 Diameters and Heights for Ten Fossil Specimens of *Rotularia (Annelida) fallax*

Specimen	Diameter	Height	D/H
OSU 36651	185	78	2.37
OSU 36652	194	65	2.98
OSU 36653	173	77	2.25
OSU 36654	200	76	2.63
OSU 36655	179	72	2.49
OSU 36656	213	76	2.80
OSU 36657	134	75	1.79
OSU 36658	191	77	2.48
OSU 36659	177	69	2.57
OSU 36660	199	65	3.06
Mean:	184.5	73	2.54
s:	21.5	5	

FIGURE 8.25 SAS regression analysis for ten fossil specimens of *Rotularia (Annelida) fallax*

```
Model: MODEL1
Dependent Variable: Y

                         Analysis of Variance

                              Sum of         Mean
       Source         DF     Squares       Square      F Value       Prob>F

       Model           1    91.29018     91.29018        0.179       0.6831
       Error           8  4073.20982    509.15123
       C Total         9  4164.50000

           Root MSE      22.56438     R-square       0.0219
           Dep Mean     184.50000     Adj R-sq      -0.1003
           C.V.          12.23002

                         Parameter Estimates

                      Parameter      Standard    T for H0:
      Variable  DF     Estimate         Error   Parameter=0     Prob > |T|

      INTERCEP   1   231.102679  110.28922771         2.095         0.0694
      X          1    -0.638393    1.50764603        -0.423         0.6831

            Dep Var   Predict   Std Err  Lower95%  Upper95%  Lower95%  Upper95%
     Obs       Y       Value    Predict     Mean      Mean   Predict   Predict

       1     185.0     181.3     10.380    157.4     205.2     124.0     238.6
       2     194.0     189.6     14.014    157.3     221.9     128.4     250.9
       3     173.0     181.9      9.343    160.4     203.5     125.6     238.3
       4     200.0     182.6      8.448    163.1     202.1     127.0     238.1
       5     179.0     185.1      7.293    168.3     202.0     130.5     239.8
       6     213.0     182.6      8.448    163.1     202.1     127.0     238.1
       7     134.0     183.2      7.746    165.4     201.1     128.2     238.2
       8     191.0     181.9      9.343    160.4     203.5     125.6     238.3
       9     177.0     187.1      9.343    165.5     208.6     130.7     243.4
      10     199.0     189.6     14.014    157.3     221.9     128.4     250.9
      11         .     183.9      7.293    167.0     200.7     129.2     238.5

     Obs  Residual

       1    3.6920
       2    4.3929
       3   -8.9464
       4   17.4152
       5   -6.1384
       6   30.4152
       7  -49.2232
       8    9.0536
       9  -10.0536
      10    9.3929
      11         .

Sum of Residuals                1.705303E-13
Sum of Squared Residuals           4073.2098
Predicted Resid SS (Press)         5512.2037

                               OBS      Y       X

                                 1    185      78
                                 2    194      65
                                 3    173      77
                                 4    200      76
                                 5    179      72
                                 6    213      76
                                 7    134      75
                                 8    191      77
                                 9    177      69
                                10    199      65
                                11      .      74
```

GEN **8.36** Daily readings of air temperature and soil moisture percentage were collected at four elevations—670, 825, 1,145, and 1,379 meters—in the White Mountains of New Hampshire (Reiners, William A., et al., "Temperature and Evapotranspiration Gradients of the White Mountains, New Hampshire, U.S.A.," *Arctic and Alpine Research*, Vol. 16, No. 1, 1984). Among the analyses of the data are simple linear regressions relating the air temperature y (in degrees Centigrade) to altitude x in meters. For example, the regression analysis for July produced the following statistics: $b_0 = 21.4$, $b_1 = -.0063$, $r^2 = .99$, and $\widehat{\text{SE}}(b_1) = .0007$.

a. Give the least squares prediction equation relating air temperature y to altitude x for the month of July.

b. How many degrees of freedom are associated with s^2 and the t statistic that is used to make inferences for this regression analysis? (Assume that readings were taken each day during the month of July at all four elevations.)

c. Interpret the value of r^2.

d. Estimate the mean temperature for an altitude of 1,000 meters.

e. Do the data provide sufficient evidence to indicate that x contributes information for the prediction of y? State the null and alternative hypotheses for the test. Explain the reasons for the choice of your alternative hypothesis. Give the value of the test statistic and the approximate p-value for the test. If you were to choose $\alpha = .05$, what would you conclude?

f. Find a 95% confidence interval for the mean increase or decrease in temperature for a 1-meter increase in altitude. Interpret the interval.

8.37 Refer to Exercise 8.36. Write a brief essay that explains the practical implications of the regression analysis. (Take your time. Explain in detail and justify your statements.)

EDUC **8.38** Most sophomore physics students are required to conduct an experiment verifying Hooke's Law. Hooke's Law states that when a pulling force is applied to a body that is long relative to its cross-sectional area, the change y (i.e., the stretch) in its length is proportional to the force x; that is,

$$y = \beta_1 x$$

where β_1 is the constant of proportionality. The results of an actual student's laboratory experiment (real data!) are shown in Table 8.10 (Mendenhall, C. M., physics laboratory notes, University of Florida, 1972–1973). Six lengths of steel wire, .34 millimeter (mm) in diameter and 2 meters (m) long, were used to obtain the six force–length change measurements. A Minitab printout for a simple linear regression analysis of the data is shown in Figure 8.26.

a. Find the equation of the least squares line.

b. Plot the data points and graph the least squares line.

c. Find the estimated standard deviation of the deviations of the points about the least squares line. How many degrees of freedom is it based on?

d. How many data points would you expect to deviate more than $2s$ from the least squares line? Why? How many of the data points in Table 8.10 actually deviate more than $2s$ from the least squares line? To answer this question, construct lines parallel to and a vertical distance of $2s$ from the least squares line.

e. Find r^2 and interpret its value.

f. Do the data provide sufficient evidence to indicate that x contributes information for the prediction of y? State the null and alternative hypotheses for the test. Explain the reasons for the choice of your alternative hypothesis. Give the value of the test statistic and the p-value for the test. If you were to choose $\alpha = .05$, what do you conclude?

g. If Hooke's Law is correct, the line of means must pass through the point (0, 0); that is, β_0 must equal 0. The test of H_0: $\beta_0 = 0$ is a t test exactly like the test for the slope β_1 except that the test statistic is

$$t = \frac{b_0}{\widehat{\text{SE}}(b_0)}$$

The value of b_0, the estimated standard error of b_0, the value of the test statistic t, and the p-value for the test are shaded on the Minitab printout in Figure 8.26. Conduct the test. State the null and alternative hypotheses for the test. Explain the reasons for the choice of your alternative hypothesis. Give the value of the test

TABLE 8.10 Physics Laboratory Data to Verify Hooke's Law

Force x (kg)	Change in Length y (mm)
29.4	4.25
39.2	5.25
49.0	6.50
58.8	7.85
68.6	8.75
78.4	10.00

FIGURE 8.26 Minitab printout of a simple linear regression analysis for the Hooke's Law experiment

```
 ROW        y       x

   1     4.25    29.4
   2     5.25    39.2
   3     6.50    49.0
   4     7.85    58.8
  .   .   .

MTB > regress c1 on 1 predictor c2

The regression equation is
y = 0.720 + 0.118 x

Predictor         Coef        Stdev     t-ratio        p
Constant        0.7200       0.1570        4.59    0.010
x             0.118367     0.002781       42.56    0.000

s = 0.1140        R-sq = 99.8%      R-sq(adj) = 99.7%

Analysis of Variance

SOURCE        DF           SS          MS          F          p
Regression     1       23.548      23.548    1811.39      0.000
Error          4        0.052       0.013
Total          5       23.600
```

statistic and the p-value for the test. If you were to choose $\alpha = .01$, what do you conclude? What are the implications of the choice of $\alpha = .01$?

GEN **8.39** It is generally thought that the temperature on football fields with synthetic turf is higher than the temperature on a natural grass surface. Eighty-one observations were made of the surface temperature of the synthetic turf at the Texas Tech University football stadium and also at a nearby natural grass surface (Ramsey, Jerry D., "Environmental Heat from Synthetic and Natural

Turf," *Research Quarterly for Exercise and Sport,* Vol. 53, No. 1, 1982). The computer printout for a simple linear regression analysis of the data is shown in Figure 8.27.

a. Give the least squares prediction equation relating the dry bulb synthetic surface temperature (DBS) y to the dry bulb natural surface temperature (DBN) x.

b. Do the data present sufficient evidence to indicate that the dry bulb natural grass temperature (DBN) provides information for the prediction of the dry bulb synthetic turf temperature (DBS)? Explain.

FIGURE 8.27 A regression computer printout relating natural grass dry bulb temperature x to the temperature y on a synthetic surface

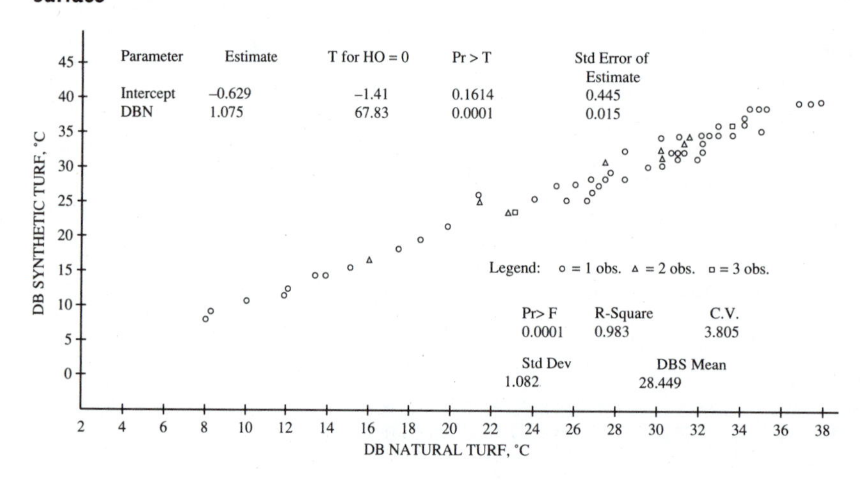

8.9 A Multiple Linear Regression Analysis

Sometimes the points on a scattergram suggest a curvilinear (rather than a straight-line) relationship between x and y. If the curvature is pronounced, we may want to replace the straight-line model with a model that allows for curvature.

We can construct many different models that allow for curvature in the graphical relationship between x and y. One type that will often provide a good fit to the data points is called a *general linear model.* As you will subsequently see, these models will contain more than two β parameters.

The General Linear Model

A *general linear model* is one that is linear in the parameters, $\beta_0, \beta_1, \beta_2, \ldots, \beta_k$. For example,

$$E(y) = 2\beta_0 + 3\beta_1 - 5\beta_2$$
$$E(y) = \beta_0 + \beta_1 x + \beta_2 x^2$$
$$E(y) = \beta_0 + \beta_1 x + \beta_2 x^2 + \beta_3 x^3$$

are examples of linear functions of the β parameters. The linear function involves the sum (or difference) of a number of terms. Each term includes a parameter multiplied by a constant or by some function of x. For example, in the linear function

$$E(y) = \beta_0 + \beta_1 x + \beta_2 x^2$$

the coefficient of β_0 is 1, the coefficient of β_1 is x, and the coefficient of β_2 is x^2. The coefficient of a β can be any function of x as long as the function does not involve any unknown parameters.

The model

$$E(y) = \beta_0 + \beta_1 x$$

(the equation of a straight line) is called a **first-order linear model in x** (see Figure 8.28a). A **second-order model in x,**

$$E(y) = \beta_0 + \beta_1 x + \beta_2 x^2$$

includes an x^2 term and graphs as a segment of a parabola (Figure 8.28b). A **third-order model in x,**

$$E(y) = \beta_0 + \beta_1 x + \beta_2 x^2 + \beta_3 x^3$$

includes an x^3 term. It graphs (see Figure 8.28c) as a curve that rises, falls, and then rises again (or falls, rises, and then falls, depending on the values of the

FIGURE 8.28 Graphs of first-, second-, and third-order linear models

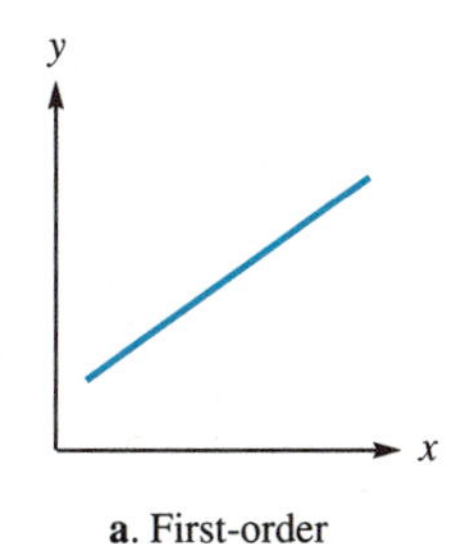

a. First-order

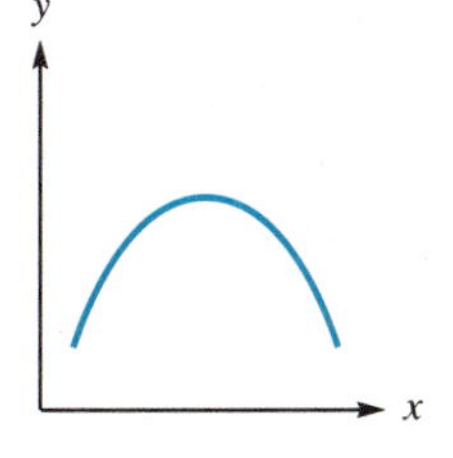

b. Second-order

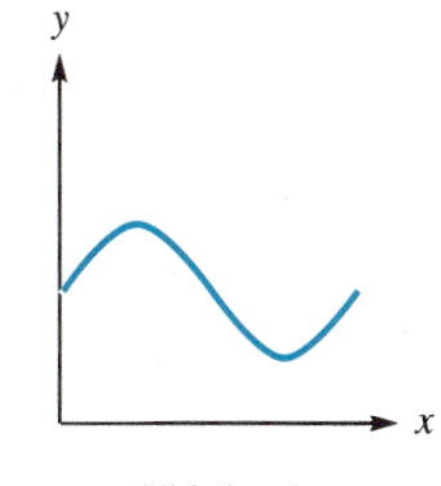

c. Third-order

parameters). Because all of these models are linear functions of the β parameters, they satisfy the definition of the general linear model. In most practical situations, the relationship between x and y can be approximated by either a first- or a second-order linear model.

The Computer Output for a Multiple Regression Analysis: What It Contains

A multiple regression analysis fits a general linear model to a set of data using the method of least squares. A computer output for a multiple regression analysis is quite similar to the output for a simple linear regression analysis; the major difference is that the model contains more parameters.

1. **A multiple regression analysis output gives estimates of the β parameters in the general linear model.** These estimates enable you to write the corresponding least squares prediction equation.

2. **For each parameter β in the model, the output gives a test of the null hypothesis that the particular parameter equals 0.** For example, if $\beta_2 = 0$, the term involving β_2 drops out of the model. Therefore, a test to decide whether the term involving β_2 should be retained in the model is a test of $H_0\colon \beta_2 = 0$. As in the case of the simple linear regression, the output gives estimates b_0, b_1, b_2, etc., of the β parameters, estimates of their standard errors, the values of the test statistics, and the p-values for the tests. The formula for the test statistic is identical to that used in a simple linear regression analysis. For example, the test statistic used to test $H_0\colon \beta_2 = 0$ is $t = b_2/\widehat{\text{SE}}(b_2)$. **The number of degrees of freedom (df) for t is always equal to the number n of data points minus the number of β parameters in the linear model.** For example, if you fit the two-variable model, $E(y) = \beta_0 + \beta_1 x_1 + \beta_2 x_2$, to the data, the model contains three parameters and the number of degrees of freedom associated with t is $(n - 3)$. The larger the number of degrees of freedom, the better. If possible, have enough data points so that the number of degrees of freedom for t is 4 or more.

3. **The output for a multiple regression analysis gives the value of s, the standard deviation of the deviations of the points about the least squares "curve."** Approximately 95% of the data points should lie within $2s$ of the values predicted by the multiple regression model. The printout also gives R^2, the multiple coefficient of determination, which gives the percentage of the Total SS that can be explained by the fitted multiple regression model.

4. **The output gives a confidence interval for $E(y)$ for a specified value of x (or, in the case of several independent variables, for specified values of x_1, x_2, x_3, etc.).** The number of degrees of freedom for t is n minus the number of β parameters in the linear model.

5. **The output gives a prediction interval for some new value of y for a specified value of x (or, in the case of several independent variables, for specified values of x_1, x_2, x_3, etc.).** The number of degrees of freedom for t is n minus the number of β parameters in the linear model.
6. **The computer printout permits us to test an hypothesis about a set of model parameters.** We will omit discussion of this test and the practical situation that would motivate it. See Mendenhall and Sincich (1993).

A Multiple Regression Analysis: An Example

Example 8.9

Exercise 8.7 gives data on the personal savings rate and the investment income tax liability for eight countries. The purpose of the study was to see whether a country's savings rate is related to its tax level on invested income (i.e., on savings). Table 8.11 gives the personal savings rate y and the taxation rate x (i.e., the investment income tax liability) for a sample of eight countries (Blotnick, S., "Psychology and Investing," *Forbes*, May 25, 1981). Suppose we think that the relationship between savings rate and taxation rate is curvilinear and fit to the data the second-order model

$$E(y) = \beta_0 + \beta_1 x + \beta_2 x^2$$

A SAS multiple regression analysis for the data of Table 8.11 is shown in Figure 8.29 on page 324.

TABLE 8.11 Personal Savings Rate and Investment Income Tax Liability for Eight Countries

Country	Personal Savings Rate (%)	Investment Income Tax Liability (%)
Italy	23.1	6.4
Japan	21.5	14.4
France	17.2	7.3
W. Germany	14.5	11.8
United Kingdom	12.2	32.5
Canada	10.3	30.0
Sweden	9.1	52.7
United States	6.3	33.5

Source: New York Stock Exchange with assistance of Price Waterhouse.

FIGURE 8.29 A SAS multiple regression analysis for the data of Table 8.11

Model: MODEL1
Dependent Variable: Y

Analysis of Variance

Source	DF	Sum of Squares		Mean Square		F Value	Prob>F
Model	2	180.86774		90.43387		6.620	0.0393
Error	5	68.30726	5	13.66145	6		
C Total	7	249.17500					

Root MSE	3.69614	7	R-square	0.7259	8
Dep Mean	14.27500		Adj R-sq	0.6162	
C.V.	25.89240				

Parameter Estimates

Variable	DF	Parameter Estimate		Standard Error		T for H0: Parameter=0		Prob > |T|	
INTERCEP	1	25.135210		3.93495137		6.388		0.0014	
X	1	-0.717497	1	0.34480494	2	-2.081	3	0.0920	4
X2	1	0.007693		0.00605049		1.271		0.2595	

Obs	Dep Var Y	Predict Value	Std Err Predict	Lower95% Mean	Upper95% Mean		Lower95% Predict	Upper95% Predict
1	23.1000	20.8583	2.319	14.8983	26.8184		9.6426	32.0741
2	21.5000	16.3984	1.589	12.3128	20.4840		6.0561	26.7407
3	17.2000	20.3074	2.157	14.7625	25.8523		9.3067	31.3082
4	14.5000	17.7399	1.653	13.4915	21.9883		7.3322	28.1475
5	12.2000	9.9420	2.030	4.7225	15.1614		-0.8984	20.7823
6	10.3000	10.5337	2.037	5.2967	15.7708		-0.3151	21.3826
7	9.1000	8.6880	3.652	-0.6988	18.0748		-4.6680	22.0440
8	6.3000	9.7322	2.022	4.5339	14.9305		-1.0980	20.5624
9	.	13.8624	1.766	9.3215	18.4032	9	3.3319	24.3928

Obs	Residual
1	2.2417
2	5.1016
3	-3.1074
4	-3.2399
5	2.2580
6	-0.2337
7	0.4120
8	-3.4322
9	.

Sum of Residuals	1.332268E-14
Sum of Squared Residuals	68.3073
Predicted Resid SS (Press)	422.8373

OBS	Y	X	X2
1	23.1	6.4	40.96
2	21.5	14.4	207.36
3	17.2	7.3	53.29
4	14.5	11.8	139.24
5	12.2	32.5	1056.25
6	10.3	30.0	900.00
7	9.1	52.7	2777.29
8	6.3	33.5	1122.25
9	.	20.0	400.00

a. Find b_0, b_1, and b_2. Then give the equation of the fitted second-order model.

b. Find the estimated standard errors of b_0, b_1, and b_2.

c. Find the computed values of the t statistics for testing hypotheses that the individual β parameters equal 0. Find the p-value for each test.

d. Do the data provide sufficient evidence to indicate that the relationship between x and y is curvilinear? State the null and alternative hypotheses for the test. Give the value of the test statistic and the p-value for the test. If you were to choose $\alpha = .05$, what would you conclude?

Solution

a. The estimates of the three β parameters of the second-order model, shaded and designated as 1 on the printout in Figure 8.29, are $b_0 = 25.135210$,

FIGURE 8.30 **A scattergram and graph of the least squares curve for the data of Table 8.11**

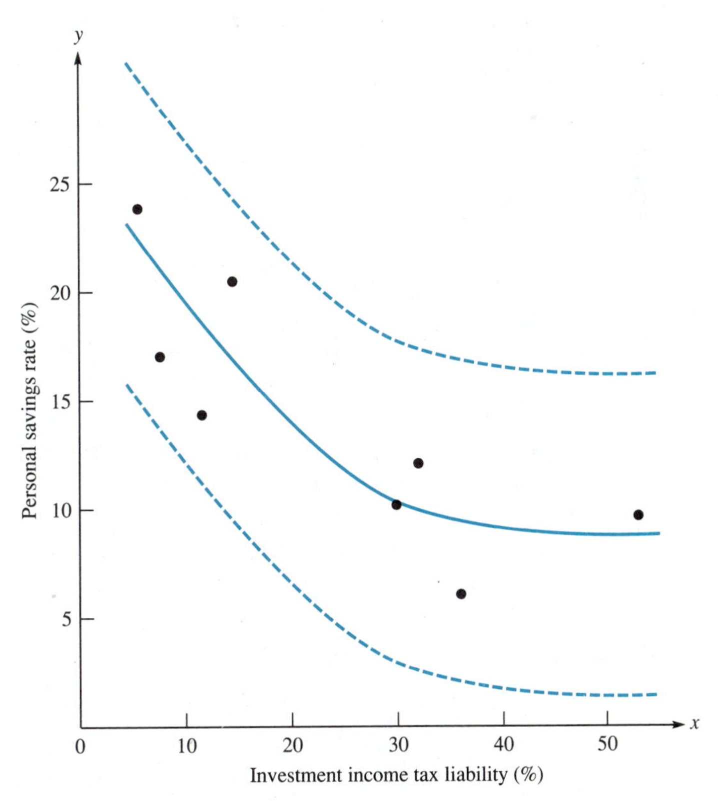

$b_1 = -.717497$, and $b_2 = .007693$. The equation of the fitted model is

$$y = 25.1352 - .7175x + .0077x^2$$

A plot of the data points and a graph of the fitted curve are shown in Figure 8.30.

b. The estimated standard errors of b_0, b_1, and b_2 are shaded and marked 2 on the printout. The standard error for each estimate appears in the same row as the estimate. For example, $\widehat{\text{SE}}(b_2) = .00605049$.

c. The computed values of the t statistics and the p-values for the tests are shaded and numbered 3 and 4, respectively, on the printout.

d. The term that produces curvilinearity in the second-order model is the term involving x^2. Therefore, to determine whether the relationship between x and y is curvilinear, we test the null hypothesis H_0: $\beta_2 = 0$ against the two-sided alternative H_a: $\beta_2 \neq 0$. This implies a two-tailed test with α split equally between the lower and upper tails of the t distribution. The value of the t statistic for testing H_0: $\beta_2 = 0$ is $t = 1.271$ and the corresponding p-value is .2595. This large p-value tells us that there is approximately 1 chance in 4 of observing a t-value as large as or larger than 1.271, if $\beta_2 = 0$. Since $p = .2595$ is larger than $\alpha = .05$, we do not reject H_0: $\beta_2 = 0$. There is insufficient evidence to indicate that β_2 differs from 0 or, equivalently, there is insufficient evidence to indicate curvilinearity in the relationship between x and y.

Example 8.10

Refer to the savings and investment tax analysis of Example 8.9.

a. Find SSE, s^2, and the estimated standard deviation s of the deviations of the points about the least squares line on the printout in Figure 8.29. How many degrees of freedom are s^2 and s based on?

b. How many data points would you expect to deviate more than $2s$ from the least squares line? Why? How many of the data points in Table 8.11 actually deviate more than $2s$ from the least squares line? To answer this question, construct curves a vertical distance of $2s$ above and below the least squares curve.

c. Find the multiple correlation coefficient R. Find the multiple coefficient of determination R^2 and interpret its value.

Solution

a. $\widehat{\text{SE}} = 68.30726$ and $s^2 = 13.66145$ are shaded and marked 5 and 6, respectively, on the printout in Figure 8.29. The standard deviation $s = 3.69614$ is shaded and marked 7. The number of degrees of freedom associated with s^2

is equal to n minus the number of β parameters in the model, or $(8 - 3) = 5$. This number is shown in the "DF" column in the "Error" row.

b. Based on our knowledge of the Empirical Rule (Chapter 4), we would expect almost all of the points to fall within $2s$ of the least squares curve. Figure 8.30 shows a scattergram of the data points and dashed curves located a vertical distance of $2s$ above and below the least squares curve. You can see from Figure 8.30 that all of the points fall within this interval.

c. The multiple coefficient of determination, $R^2 = .7259$, is shaded and numbered 8 on the printout in Figure 8.29. It tells us that approximately 73% of the total sum of squares of deviations in the y-values, SS_y, can be explained by a second-order model in x. The unexplained 27% is caused either by other variables that are related to y and have not been included in the model, or by errors in the equation assumed for the model, or both.

Example 8.11

Refer to the savings and investment tax analysis of Example 8.9. Find a 95% confidence interval for the mean savings rate for a country with an investment income tax liability of 20%.

Solution

The confidence interval for the mean value of y and the prediction interval for y for each of the eight observations in the data set are shown in the SAS printout in Figure 8.29. The corresponding intervals for $x = 20\%$ appear in the row for Obs 9. The confidence interval on the mean savings rate when the investment income tax liability is $x = 20\%$, shaded and numbered 9 on the printout, is 9.3215% to 18.4032%.

EXERCISES

GEN **8.40** Cucumbers are usually preserved by fermenting them in a low-salt brine (6% to 9% sodium chloride) and then storing them in a high-salt brine until they are used by processors to produce various types of pickles. The high-salt brine is needed to retard softening of the pickles and to prevent freezing when stored outside in northern climates. Data showing the reduction in firmness of pickles stored over time in a high-salt brine are shown in Table 8.12 on page 328 (Buescher, R. W., Hudson, J. M., Adams, J. R., and Wallace, D. H., "Calcium Makes It Possible to Store Cucumber Pickles in Low-Salt Brine," *Arkansas Farm Research,* Vol. 30, No. 4, July/August 1981).

a. Construct a scattergram for the data and discuss the type of model that you think is most appropriate for the data.

b. Write a second-order model for the data.

c. Figure 8.31 (page 328) gives a SAS multiple regression analysis for the data of Table 8.12. Find the parameter estimates and the least squares prediction equation.

d. Graph the least squares curve on the scattergram in part **a**.

TABLE 8.12 Pickle Firmness (in pounds) *y* Versus Weeks *x* in Storage at 72°F

Weeks in Storage x	Firmness y
0	19.8
4	16.5
14	12.8
32	8.1
52	7.5

FIGURE 8.31 SAS multiple regression analysis: Pickle firmness versus storage time

```
                        Analysis of Variance

                          Sum of         Mean
Source          DF       Squares        Square      F Value       Prob>F

Model            2     112.05359      56.02680      155.975       0.0064
Error            2       0.71841       0.35920
C Total          4     112.77200

    Root MSE        0.59934     R-square        0.9936
    Dep Mean       12.94000     Adj R-sq        0.9873
    C.V.            4.63166

                        Parameter Estimates

                      Parameter      Standard     T for H0:
Variable   DF          Estimate         Error   Parameter=0    Prob > |T|

INTERCEP    1         19.296817    0.47114379        40.957        0.0006
X           1         -0.559332    0.05352823       -10.449        0.0090
XSQ         1          0.006413    0.00100311         6.393        0.0236

        Dep Var    Predict    Std Err   Lower95%   Upper95%
  Obs         Y      Value    Predict       Mean       Mean   Residual

    1   19.8000    19.2968      0.471    17.2696    21.3240     0.5032
    2   16.5000    17.1621      0.359    15.6156    18.7086    -0.6621
    3   12.8000    12.7231      0.386    11.0627    14.3836     0.0769
    4    8.1000     7.9653      0.479     5.9039    10.0266     0.1347
    5    7.5000     7.5527      0.590     5.0145    10.0909    -0.0527

Sum of Residuals                 -7.99361E-15
Sum of Squared Residuals               0.7184
Predicted Resid SS (Press)             5.8087
```

e. Find SSE, s^2, and s. How many degrees of freedom are associated with these quantities?
f. Find R^2 and interpret its value.
g. Test to determine whether the data provide sufficient evidence to indicate curvature in the relationship between y and x. Describe each step and explain your conclusion.
h. Suppose that you want to compute a 95% confidence interval for the mean firmness of pickles stored for 10 weeks and you also want to find a prediction interval for y for pickles stored 10 weeks. Describe the difference between the two intervals.

8.10 Key Words and Concepts

- This chapter is about ***bivariate data***. Two quantitative measurements are observed on each element–one measurement on a variable x and one on a variable y. A random sample of n elements is selected from the universe and the values of quantitative variables x and y are recorded for each element. These bivariate data are then used to determine whether x and y are related. Ultimately, we would like to find a prediction equation that will enable us to predict values of y for given values of x.
- The relationship between two variables x and y is best described by a mathematical equation, the simplest of which is the equation of a straight line. Each mathematical equation graphs as a unique line or curve.
- The relationship between the paired observations of the data points can be explored by plotting the pairs of (x, y) values on graph paper. Each point, corresponding to a single pair of values of x and y, is called a ***data point*** and the resulting scatter of points is called a ***scattergram***. A quick glance at a scattergram will tell you whether the y-values of the points tend to rise or fall as x increases, or whether the points appear to vary in a random pattern. A pattern, if apparent, can then be summarized by drawing a straight line through the data points in a manner that "minimizes" the deviations of the points from the line.
- The ***least squares line*** is the one that minimizes the sum of squares of the deviations of the y-values from the fitted line.
- A ***simple linear regression analysis*** (i.e., a regression analysis for the straight-line model) does the following:

  1. Fits a least squares line to the data
  2. Tests to determine whether x contributes information for the prediction of y
  3. Gives us a measure of how well the model fits the data points. It calculates the ***simple linear coefficient of correlation*** r and the ***coefficient of determination*** r^2.
  4. Enables us to calculate a confidence interval for the mean value of y for a given value of x
  5. Enables us to find a prediction interval for some value of y to be observed in the future for a given value of x

- The inferences drawn from a simple linear regression analysis are based on the assumption that the mean value of y is linearly related to x:

$$E(y) = \beta_0 + \beta_1 x$$

That is, the relationship between $E(y)$ and x graphs as a straight line. In addition, we assume that y is normally distributed about the line of means with a common variance σ^2 and with pairs of values of y independent of each other.

- The sign of the coefficient of correlation r tells us whether the line is sloping upward or downward and indicates how well the line fits the data, with -1 and $+1$ representing perfect fits and 0 representing no fit at all. The value of r^2 gives us the proportion of the total sum of squares of deviations SS_y explained by the variable x.

- A ***general linear model*** relates the mean value of y to more than two parameters in an equation of the form

$$E(y) = \beta_0 + \beta_1 x_1 + \beta_2 x_2 + \cdots + \beta_k x_k$$

The x variables that appear in the model may represent different variables or they may represent functions, e.g., x^2, x^3, or $x_1 x_2$, of one or more variables. These variables are assumed to be measured without error.

- The process of fitting a general linear model and drawing inferences from it is called a ***multiple regression analysis***. A multiple regression analysis answers essentially the same questions as a simple linear regression except that it pertains to a more complex model.

SUPPLEMENTARY EXERCISES

GEN **8.41** Table 8.13 gives the number x of robberies and the number y of robbery arrests per month in New York City during the first half of 1988. A Data Desk regression analysis for the data is shown in Figure 8.32.

a. Without looking at the data or your printout, explain why you think that robbery arrrests per month tend to increase, decrease, or be unrelated to the level x of robberies per month.

b. Give the least squares equation that relates the number y of robbery arrests per month to the number x of robberies per month.

c. What do the coefficient of correlation r and the coefficient of determination r^2 tell you?

d. Suppose that you want to test your theory in part **a**. What hypothesis would you want to test? Use your regression output to conduct the test.

TABLE 8.13 Robberies and Robbery Arrests in New York City

Month	Robberies	Robbery Arrests
January	475	180
February	465	155
March	470	160
April	500	190
May	550	225
June	600	220
July	602	223

Source: Data estimated from graph in *The New York Times*, October 2, 1988.

FIGURE 8.32 A Data Desk regression analysis for the New York robbery data

```
Dependent variable is:    Arrests
R² = 84.4%    R²(adjusted) = 81.3%
s = 12.93 with 7 - 2 = 5 degrees of freedom

Source        Sum of Squares   df   Mean Square   F-ratio
Regression    4527.58           1          4528     27.1
Residual      835.849           5       167.170

Variable     Coefficient    s.e. of Coeff    t-ratio
Constant     -44.5533               45.96     -0.969
Robberies    0.454635              0.0874       5.20
```

EXERCISES FOR YOUR COMPUTER

BUS **8.42** Geothermal energy is heat contained in hot water drawn from deep wells drilled into the earth's surface. Since the energy derived from a well is a function of the water temperature, we might wonder whether deeper wells produce hotter water (and, therefore, more energy) than shallower wells. Table 8.14 gives the average drill hole depth x in meters and the average maximum water temperature (Centigrade) y for nine thermal wells (Ellis, A. J., "Geothermal Systems," *American Scientist*, September–October 1975). Use your computer to perform a linear regression analysis on the data. Is hole depth useful in predicting water temperature? Write a brief essay describing the results of the regression analysis.

TABLE 8.14 Geothermal Well Depths and Water Temperature

Location of Well	Average Drill Hole Depth (meters)	Average Temperature (degrees Centigrade)
El Tateo, Chile	650	230
Ahuachapan, El Salvador	1,000	230
Namafjall, Iceland	1,000	250
Larderello (region), Italy	600	200
Matsukawa, Japan	1,000	220
Cerro Prieto, Mexico	800	300
Walrakei, New Zealand	800	230
Kizildere, Turkey	700	190
The Geysers, United States	1,500	250

ECON **8.43** Table 8.15 shows the amount of money paid by each state to the federal government in income taxes (on a per-capita basis) and the amount of money poured back into the state in federal aid for the fiscal year 1986. Use your computer to perform a simple linear regression analysis relating aid y to taxes x. Then answer the following questions.

a. Without looking at the data or your printout, explain why you think that aid y to a state will tend to increase, decrease, or be unrelated to the level x of a state's income taxes.

b. What do the coefficient of correlation r and the coefficient of determination r^2 tell you?

c. Suppose that you want to test your theory in part **a**. What hypothesis would you want to

TABLE 8.15 Federal Income Tax Paid and Aid Received per State

State	Taxes	U.S. Aid	State	Taxes	U.S. Aid
AL	$ 740	$ 434	CO	$ 718	$374
AK	3490	1244	CT	1202	471
AZ	975	367	DE	1343	495
AR	770	474	FL	780	278
CA	1144	419	GA	806	448
HI	1400	446	NM	989	579
ID	743	434	NY	1278	697
IL	848	434	NC	881	360
IN	810	363	ND	907	638
IA	863	434	OH	843	443
KS	777	359	OK	895	423
KY	863	479	OR	715	497
LA	807	453	PA	898	481
ME	940	573	RI	908	585
MD	1047	439	SC	863	391
MA	1314	528	SD	570	646
MI	1019	476	TN	682	443
MN	1163	501	TX	667	313
MS	731	512	UT	820	485
MO	712	391	VT	923	617
MT	755	723	VA	836	345
NE	700	413	WA	1169	427
NV	1084	434	WV	964	554
NH	472	394	WI	1148	483
NJ	1096	440	WY	1569	929

Source: Data from U.S. Commerce Department, Bureau of the Census, *World Almanac & Book of Facts*, 1989, p. 141.

test? Use your regression output to conduct the test. Explain each step and state your conclusion.

d. Find a 95% confidence interval for the change in the amount a state will receive in aid for each additional dollar paid (per capita) in income tax. Interpret the interval.

References

1. Freedman, D., Pisani, R., Purves, R., and Adhikari, A., *Statistics*, 2nd edition. New York: W. W. Norton & Co., 1991.
2. McClave, J. and Dietrich, F., *Statistics*, 5th edition. San Francisco: Dellen Publishing Co., 1991.
3. Mendenhall, W. and Sincich, T., *A Second Course in Business Statistics: Regression Analysis*, 4th edition. San Francisco: Dellen Publishing Co., 1993.
4. Mendenhall, W. and Beaver, R., *Introduction to Probability and Statistics*, 8th edition. Boston: PWS-Kent, 1991.
5. Moore, D. and McCabe, G., *Introduction to the Practice of Statistics,* 2nd edition. New York: W. H. Freeman & Co., 1993.
6. Neter, J. and Wasserman, W., *Applied Linear Statistical Models.* Homewood, Ill: Richard D. Irwin, Inc., 1974.
7. Ott, L., *An Introduction to Statistical Methods and Data Analysis*, 4th edition. Boston: Duxbury Press, 1991.
8. Sincich, T., *Statistics by Example*, 4th edition. San Francisco: Dellen Publishing Co., 1990.

1 2 3 4 5 6 7 8 9 10 11

nine

Inferences from Qualitative Data

In a Nutshell

Another application: Using statistical inference to make inferences from public opinion and other polls.

9.1 The Problem

9.2 Assumptions

9.3 Inferences About a Population Proportion

9.4 Comparing Two Population Proportions

9.5 Choosing the Sample Size

9.6 A Contingency Table Analysis

9.7 Key Words and Concepts

9.1 The Problem

This chapter explains how to make inferences based on a sample of qualitative data. The data are usually collected by observing a random sample of elements from a universe, but more complex sampling designs, involving random selections from subsets of the universe, are used as well. Typical of these are the Gallup, Harris, and news media public opinion polls that have become a familiar part of our daily lives. We want to know (and therefore the news media want to know) what the rest of us think about various social and political issues. Although public opinion polls are the most familiar to us, the collection of qualitative data also occurs in business and scientific research.

Recall from Chapter 2 that qualitative data arise when we classify the elements of a universe according to one or more criteria. For example, Table 9.1 presents partial data from a telephone poll of 1,486 persons living in the small towns, rural areas, and farms of Tennessee. The poll was conducted by Jerry R. Lynn ("Newspaper Ad Impact in Nonmetropolitan Markets," *Journal of Advertising Research,* Vol. 21, No. 4, 1981) to obtain information on the readership of a particular newspaper, the characteristics of the readership, and the impact of advertising on the market that they represent. The universe that Lynn sampled is the collection of all readers and potential readers in the newspaper's market area and an element of the universe is a single person.

Table 9.1 shows that two qualitative observations were made on each of the 1,486 persons in the sample. Each person was asked whether they live in an urban, a rural, or on a farm community, and whether they read the newpaper. Therefore, each person was classified according to two qualitative variables, call them "community" and "reading." The community variable can assume one of three "values" (urban, rural, or farm) and the reading variable can assume one of two "values" or **levels** (reader or nonreader). Table 9.1 shows the number of persons falling in each of the six categories or "cells" of the table. For example, 237 persons read the

TABLE 9.1 Characteristics of Potential Readers in a Newspaper's Market Area

		Reading	
		Reader	Nonreader
	Urban	529	121
Community	Rural	373	137
	Farm	237	89

paper and live on a farm. The sum of the six cell counts is equal to the total number, $n = 1{,}486$, of people in the sample.

The Types of Inferences We Want to Make

1. **We will want to estimate or test hypotheses about the proportion of elements in the universe that fall into one or more of the categories.** For example, we might want to estimate the proportion p of all people in the marketing area who read the paper and live on farms. Or we might want to know whether the data provide sufficient evidence to indicate that the readership in the marketing area is less than 30%.

2. **If two qualitative variables have been observed on each element, we will (as in the case of quantitative variables in Chapter 8) want to determine whether the variables are associated.** For example, we might want to know whether reading (i.e., whether a person reads the paper) is dependent on community (i.e., the type of community in which the person lives).

Before we attempt to answer these questions, we will list the assumptions on which our methods will be based. Then we will present methods for estimating and testing hypotheses about population proportions and for determining whether two qualitative variables are dependent.

9.2 Assumptions

The Characteristics of a Multinomial Experiment

The statistical methods presented in this chapter assume that the data collection satisfies the characteristics of a *multinomial experiment*. These characteristics are as follows:

1. **The experiment consists of n identical trials.** Observing a single element drawn from a universe would represent a trial. For example, the random selection of a single person in the telephone survey, Table 9.1, is a trial. Observing a sample of $n = 1{,}486$ people would represent $n = 1{,}486$ trials.

2. **The outcome of each trial will fall into one and only one, of k categories or cells.** For example, each person questioned in the telephone survey was classified and assigned to one and only one of the $k = 6$ categories of Table 9.1.

3. **The probability that the outcome for a single trial will fall in a particular category, say category i, is p_i.** If there are k categories, the sum of all the category probabilities is equal to 1:

$$p_1 + p_2 + p_3 + \cdots + p_k = 1$$

The cell probabilities are the population parameters about which we will want to make inferences. For example, if we were to designate the farm–reader cell of Table 9.1 as cell 1, then p_1 is the probability of selecting a person in the marketing area who reads the paper and lives on a farm.

4. **The trials are independent in a probabilistic sense, i.e., the outcome of any one trial does not affect the probability of the outcome for any other trial.** For example, if the probability of drawing a farm–reader on the first draw from the universe is .1, then we assume that the probability of drawing a farm–reader on the second draw is also .1, and so on.

5. **We are interested in observing $n_1, n_2, n_3, \ldots, n_k$, the numbers of outcomes falling in the k cells.** For example, in the telephone survey of people in the marketing area, n is the number 1,486 of people in the sample and $n_1, n_2, \ldots, n_6$ are the numbers of subscribers in the sample who fall into response categories $1, 2, \ldots, 6$. Since every element in a sample must fall into one, and only one, of the k cells, it follows that the sum of the cell counts must equal n:

$$n_1 + n_2 + n_3 + \cdots + n_k = n$$

You can verify that the sum of the six cell counts in Table 9.1 is $n = 1{,}486$.

The Relationship Between a Binomial and a Multinomial Experiment

A binomial experiment (Chapter 5) is a multinomial experiment with $k = 2$ cells. In fact, any multinomial experiment can be reduced to a binomial experiment by combining the cells in a multinomial table into $k = 2$ cells. For example, if we are interested only in making inferences about the proportion p of readers in the marketing area, we could combine all of the reader cells (and cell counts) into one cell and all of the nonreader cells into another. The collapsed table would appear as shown in Table 9.2.

TABLE 9.2 Reducing Table 9.1 to Output for a Binomial Experiment

Reading		
Readers	Nonreaders	Total
1,139	347	1,486

A Comment on the Assumptions

Public opinion and consumer preference polls, employed to investigate the attitudes of the public on specific issues, yield qualitative data. None

satisfy exactly the requirements of a multinomial experiment but, if the number of elements in the universe is large (which is almost always the case), the requirements, for all practical purposes, will be satisfied.

If the number of elements in the universe is small, outcomes of successive draws will be dependent events and the sampling will not represent a multinomial experiment. For example, suppose that there were only 20 (instead of 1,486) people in the newspaper survey's marketing area and, of these, only 3 were readers who lived on farms. If you were to draw a sample of $n = 5$ people from this universe of 20 people, the probability of selecting a farm–reader on the first draw is 3/20. The probability of selecting a farm–reader on the second draw is either 2/19 or 3/19, depending on whether you drew or did not draw a farm–reader on the first draw. Therefore, outcomes for successive draws are dependent events and the sampling does not satisfy the requirements of a multinomial experiment. In contrast, if the number of elements in the universe is large, as was the case for the telephone market survey, the probability that an element will fall in a particular category changes so little from draw to draw that the change is negligible. Therefore, Lynn's sampling satisfies, for all practical purposes, the requirements of a multinomial experiment.

EXERCISES

9.1 Explain the difference between qualitative and quantitative variables.

9.2 Chapters 1–4 and 6–8 presented ways to describe or make inferences about qualitative variables, quantitative variables, or both. Explain, for each of these chapters, whether the topic was descriptive or inferential and whether the chapter dealt with qualitative variables, quantitative variables, or both.

9.3 What types of problems do we want to solve in this chapter?

9.4 Describe the five characteristics of a multinomial experiment.

9.5 How is a multinomial experiment related to a binomial experiment?

9.6 Explain why a consumer preference poll satisfies, approximately, the characteristics of a multinomial experiment.

9.3 Inferences About a Population Proportion

Sampling Procedure, Symbols, and Objectives

Suppose we were to select a random sample of n elements from a large universe and that we want to estimate or test hypotheses about the proportion p of elements that possess a particular characteristic. If x of the n elements in the sample possess the characteristic, then **the best estimate of p is the *sample proportion*,**

***x/n*. It is represented by the symbol $\hat{p}$, i.e., $\hat{p} = x/n$.** For example, refer to Lynn's telephone survey and suppose that we want to estimate the proportion p of people in the market area who read the newspaper. Tables 9.1 and 9.2 show that $x = 1{,}139$ of the $n = 1{,}486$ people in the sample read the newspaper. Therefore, our best estimate of the proportion p of people in the universe who read the newspaper (i.e., in the market area) is $\hat{p} = x/n = 1{,}139/1{,}486 = .77$.

Symbols

Population proportion: p

Sample proportion: $\hat{p} = \dfrac{x}{n}$

The Sampling Distribution of the Sample Proportion

In the following sections, we will use the sample proportion $\hat{p}$ both to estimate and to test hypotheses about a multinomial (or binomial) population parameter p. How far from a population proportion p is the sample proportion likely to deviate? For example, is it likely that a sample proportion of people in Lynn's telephone survey deviates from the population proportion p by as much as .02? To answer this question, we need to know the properties of the sampling distribution of $\hat{p}$.*

We know from Chapter 5 that the number x of successes in a binomial experiment is approximately normally distributed when the number n of trials is large. Therefore, it is no surprise that the sample proportion, x divided by the constant n, is also normally distributed when n is large. The mean of the sampling distribution (proof omitted) of $\hat{p}$ is p and the standard error is $SE(\hat{p}) = \sqrt{p(1-p)/n}$. The sampling distribution of $\hat{p}$ is shown in Figure 9.1 (page 340) and the mean and standard error for the distribution are shown in the box.

Mean and Standard Error of $\hat{p}$

Mean of $\hat{p}$: p

Standard error of $\hat{p}$: $SE(\hat{p}) = \sqrt{\dfrac{p(1-p)}{n}}$

*Does this sound familiar? We introduce the sampling distribution of the sample proportion $\hat{p}$ in the same manner as we introduced the sampling distributions of $\bar{x}$ in Chapter 6 and $(\bar{x}_1 - \bar{x}_2)$ in Chapter 7. The reasoning is the same; only the symbols differ.

FIGURE 9.1 The sampling distribution of a sample proportion $\hat{p}$

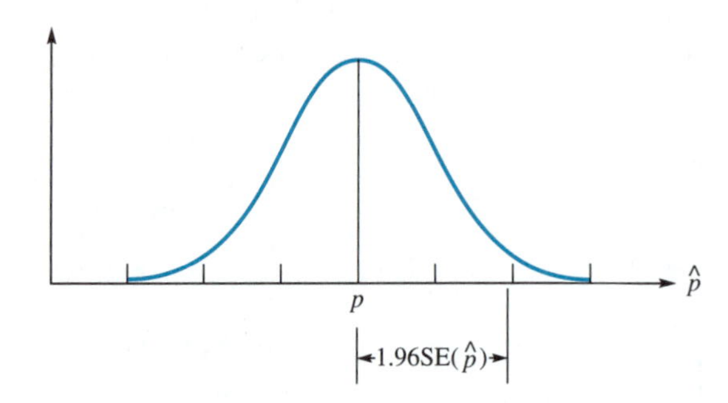

Except for a difference in the formulas for standard errors, the sampling distribution of $\hat{p}$ has the same properties as the sampling distribution of $\bar{x}$. The probability that a sample proportion will fall within 1.96 standard errors of p [i.e., within $1.96\text{SE}(\hat{p})$ of p] is .95. The probability that it will deviate from p by $1.96\text{SE}(\hat{p})$ or more is only .05. See Figure 9.1.

To find the probability that a sample proportion will deviate from p by some specified amount, we need to express the deviation in units of $\text{SE}(\hat{p})$—that is, in units of the standard normal variable—as follows:

$$z = \frac{\hat{p} - p}{\text{SE}(\hat{p})}$$

The following example will refresh your memory.

Example 9.1

Consider the problem of estimating the proportion of readers in Lynn's survey who read the newspaper. Find the approximate probability that a sample proportion, based on a random sample of 1,468 people, will deviate from the population proportion by as much as .02 (that is, by .02 or more).

Solution

We want to find the approximate probability that $\hat{p}$ will deviate from p by .02 or more. The first step is to find the standard error of $\hat{p}$:

$$\text{SE}(\hat{p}) = \sqrt{\frac{p(1-p)}{n}}$$

You can see that we need to know both n and p in order to calculate $\text{SE}(\hat{p})$. We know the sample size is $n = 1{,}486$, but the value of p is unknown. (That is what we are trying to estimate!) No problem! We will approximate p using the sample proportion $\hat{p} = x/n = 1{,}139/1{,}486 = .77$. Then

$$\text{SE}(\hat{p}) = \sqrt{\frac{p(1-p)}{n}} = \sqrt{\frac{.77(1-.77)}{1{,}486}} = .011$$

and the deviation, $(\hat{p} - p) = .02$, expressed in units of $\text{SE}(\hat{p})$, is

$$z = \frac{\hat{p} - p}{\text{SE}(\hat{p})} = \frac{.02}{.011} = 1.82$$

Figure 9.2a shows the sampling distribution of the sample proportion with the deviation .02 marked off to the left and to the right of p. Figure 9.2b shows the deviation, in units of $\text{SE}(\hat{p})$, marked off to the left and to the right of the mean 0 of the z distribution. The probability that $\hat{p}$ deviates from p by as much as .02 is the shaded area in the tails of the sampling distribution of $\hat{p}$, Figure 9.2a. It is also equal to the shaded area in the tails of the z distribution.

FIGURE 9.2 The probability that $\hat{p}$ deviates from p by .02 or more

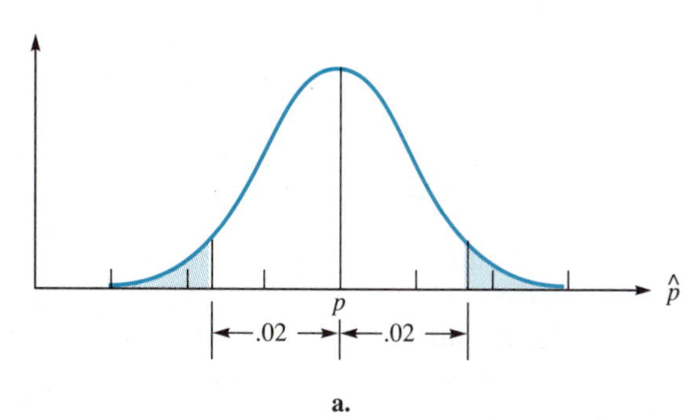

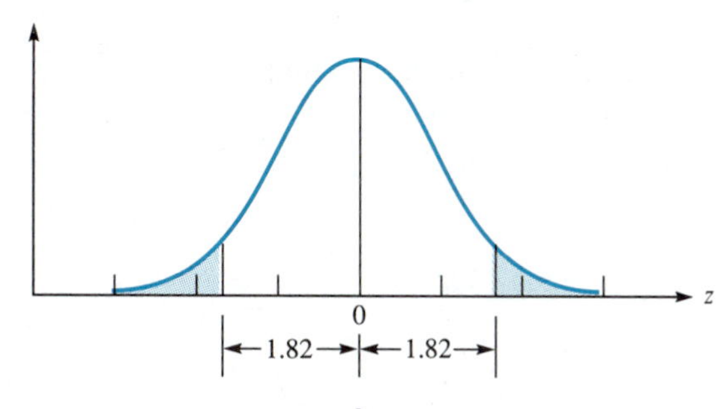

The table of some upper-tail areas for the z distribution is reproduced in Table 9.3. You can see that the area in the upper tail of the z distribution to the right of $z = 1.65$ is $a = .05$ and the area to the right of $z = 1.96$ is $a = .025$. Since $z = 1.82$ falls between 1.65 and 1.96, the area to the right of $z = 1.82$ is between .025 and .05. The area in the lower tail of the z distribution, Figure 9.2b, is equal to the area in the upper tail. Therefore, the probability that $\hat{p}$ will deviate from p by .02 or more is double the area in the upper tail. This probability will be between .05 and .10.

TABLE 9.3 Some Upper-Tail Areas for the Standard Normal Distribution

z_a	Upper-Tail Area a
.84	.20
1.00	.16
1.28	.10
1.65	.05
1.96	.025
2.33	.01
2.58	.005

Using $\hat{p}$ to Approximate p in the Formula for SE($\hat{p}$)

Example 9.1 shows how to calculate the probability that $\hat{p}$ will deviate from p by some specified amount—the kind of calculation that we will use later to find the p-value for a statistical test. But are the calculations really valid? Or are we introducing large errors into our calculations when we use the sample proportion $\hat{p}$ to approximate p in the formula for SE($\hat{p}$)? Not really! You can see that p is used to calculate the numerator, $\sqrt{p(1-p)}$, in the formula for SE($\hat{p}$). Table 9.4 gives values of $\sqrt{p(1-p)}$ for different values of p. You can see that small changes in p produce even smaller changes in $\sqrt{p(1-p)}$. For example, if we were in error by as much as .02 in approximating p when p is actually .7, the error in approximating $\sqrt{p(1-p)}$, SE($\hat{p}$), z, and the tail areas under the z distribution would be very small.

TABLE 9.4 Table of $\sqrt{p(1-p)}$ for Various Values of p

p	.9	.8	.7	.6	.5	.4	.3	.2	.1
$\sqrt{p(1-p)}$	.30	.40	.46	.49	.50	.49	.46	.40	.30

A Confidence Interval and Test for a Population Proportion p

The z statistic is used to find a confidence interval and a test of an hypothesis for a population proportion p. A $100(1 - a)\%$ confidence interval for p, based on the standard normal z statistic

$$z = \frac{\hat{p} - p}{\text{SE}(\hat{p})}$$

is shown in the box.

A $100(1 - a)\%$ Confidence Interval for a Population Proportion p

$$\text{LCL} = \hat{p} - z_{a/2}\text{SE}(\hat{p})$$
$$\text{UCL} = \hat{p} + z_{a/2}\text{SE}(\hat{p})$$

where

$$\text{SE}(\hat{p}) = \sqrt{\frac{p(1 - p)}{n}}$$

Note: **Use $\hat{p}$ to approximate p in the formula for $\text{SE}(\hat{p})$.**

The computation of a confidence interval for a population proportion p is an option for some statistical software packages but most, because the calculations are so easy, leave the calculations to you.

Example 9.2

Refer to Lynn's telephone survey, Table 9.1, and find a 95% confidence interval for the proportion p of people in the market area who read the newspaper.

Solution

The value of z for a 95% confidence interval is $z_{.025} = 1.96$ and, from Example 9.1, $\hat{p} = .77$ and $\text{SE}(\hat{p}) = .011$. Substituting these values into the formula for a 95% confidence interval for p gives

$$\text{LCL} = \hat{p} - z_{.025}\text{SE}(\hat{p}) = .77 - 1.96(.011) = .75$$
$$\text{UCL} = \hat{p} + z_{.025}\text{SE}(\hat{p}) = .77 + 1.96(.011) = .79$$

Therefore, we estimate that the interval from .75 to .79 encloses the proportion of readers of the newspaper in the market area. We are reasonably confident of this estimate. Ninety-five percent of the intervals constructed using our methodology will enclose the population proportion p.

A test of the null hypothesis that p equals some value, say $p = p_0$, is conducted in the same manner as the z tests of Chapters 6 and 7. The test is summarized in the box and illustrated in Example 9.3.

A Test Concerning a Population Proportion p

Null hypothesis: $H_0 : p = p_0$

Alternative hypothesis:

1. $H_a : p > p_0$, **an upper-tailed test**
2. $H_a : p < p_0$, **a lower-tailed test**
3. $H_a : p \neq p_0$, **a two-tailed test**

Test statistic: $z = \dfrac{\hat{p} - p_0}{SE(\hat{p})}$ **where** $SE(\hat{p}) = \sqrt{\dfrac{p(1-p)}{n}}$

Rejection region: **Choose the value of α that is acceptable to you. Then reject H_0 and accept H_a when the p-value* for the test is less than or equal to α.**

Assumptions:

1. **The sampling satisfies the requirements of a binomial experiment.**
2. **The sampling distribution of x (and, therefore, of $\hat{p}$) is approximately normally distributed. (That is, n is large enough so that the interval $np \pm 3\sqrt{np(1-p)}$ falls within the interval from 0 to n. See Section 5.6.)**

Example 9.3

An article in *The Wall Street Journal* (August 27, 1985) states that the National Union of Hospital and Health Care Workers won 56 of 80 union representation elections in 1984 compared with a success rate of 55% for all health care unions during the same time period. Do these data provide sufficient evidence to indicate that the probability p of an election success for the National Union of Hospital and Health Care Workers in 1984 was higher than .55, the probability for all health unions taken as a whole?

a. Is n large enough for us to use the z test shown in the box?

b. State the null and alternative hypotheses for a test designed to answer this question.

c. Locate the rejection region for the test.

d. Find $\hat{p}$, $SE(\hat{p})$, and the value of the test statistic.

* Do not confuse the p in the "p-value" with a population proportion p.

e. Find the approximate p-value for the test.

f. State your conclusion.

Solution

a. In Section 5.6, we learned that x will be approximately normally distributed if the interval $np \pm 3\sqrt{np(1-p)}$ falls within the interval from 0 to n. We are hypothesizing that $p = .55$. Therefore,

$$\begin{aligned} np \pm 3\sqrt{np(1-p)} &= 80(.55) \pm 3\sqrt{80(.55)(1-.55)} \\ &= 44 \pm 3(4.45) \\ \text{or} \quad & 30.65 \text{ to } 57.35 \end{aligned}$$

Since this interval falls within the interval from 0 to $n = 80$, the distribution of x (and, therefore, $\hat{p}$) will be approximately normal, and we can use the z test given in the box.

b. We want to know whether the data provide sufficient evidence to indicate that p exceeds .55. Therefore, we will test the null hypothesis $H_0\colon p = .55$ (i.e., $p_0 = .55$) against the alternative hypothesis $H_a\colon p > .55$.

c. The larger the value of $\hat{p}$ relative to .55, the greater is the evidence to indicate that p is larger than .55. Therefore, we will reject H_0 and accept H_a for large values of $\hat{p}$ or, equivalently, for large positive values of the test statistic z.

d. The observed success rate for the union is $\hat{p} = x/n = 56/80 = .7$. The standard error of $\hat{p}$, assuming H_0 is true (i.e., $p = .55$), is

$$\text{SE}(\hat{p}) = \sqrt{\frac{p(1-p)}{n}} = \sqrt{\frac{.55(1-.55)}{80}} = .056$$

and

$$z = \frac{\hat{p} - p_0}{\text{SE}(\hat{p})} = \frac{.7 - .55}{.056} = 2.68$$

e. Since this is an upper one-tailed test, the p-value for the test is the probability of observing a value of z as large as or larger than $z = 2.68$ (see Figure 9.3 on page 346). We know from Table 9.3 (or see Table 3 of Appendix 2) that z-values that locate an area a in the upper tail of the normal distribution are

$z = 1.96$ for $a = .025$

$z = 2.33$ for $a = .010$

$z = 2.58$ for $a = .005$

Since the observed value of z is slightly larger than 2.58, the area in the upper tail of the normal distribution (and the approximate p-value for the test) is smaller than .005.

f. Suppose we decided that we were willing to risk rejecting H_0 when it is true with probability $\alpha = .01$. Since the test tells us to reject H_0 when the p-value

FIGURE 9.3 The p-value for the test of Example 9.3

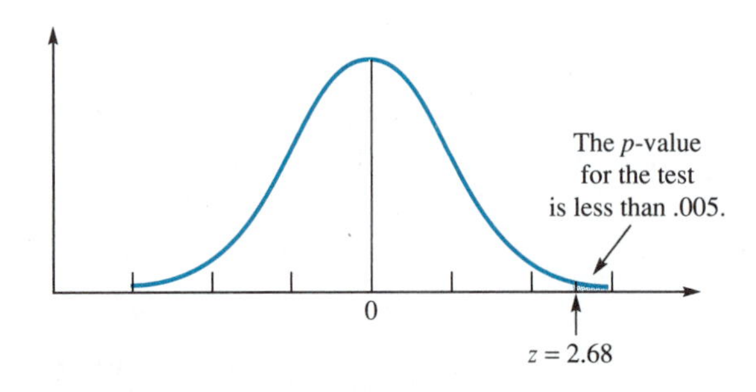
The p-value
for the test
is less than .005.
0
z = 2.68

for the test is smaller than α, we reject $H_0 : p = .55$, and conclude that the probability of a successful election for the National Union of Hospital and Health Care Workers is larger than .55, the probability of success for all health care unions.

EXERCISES

SOC **9.7** A report by the Bureau of Justice Department of Statistics, that 24% of state inmates in prison for murder killed a close relative or an intimate friend, is based on a "representative sample of 13,711 inmates" (*The Williamsport Sun Gazette*, July 30, 1990).

a. Find a 95% confidence interval for the proportion of convicted murderers who murdered a relative or a close associate.

b. Have the assumptions required for your method been satisfied? Explain.

HLTH **9.8** A survey sponsored by the Robert Wood Johnson Foundation sampled 1,300 people from among the U.S. population and found that "nearly 70% of Americans believe that mental illness is on the rise, but only one in three of those people would welcome a mental health facility in their neighborhood" (*The Tampa Tribune-Times*, April 8, 1990).

a. How accurate is the estimate of the proportion of Americans who believe that mental illness is on the rise?

b. What assumptions must be satisfied to assure the validity of your inference in part (a)?

9.9 Refer to Exercise 9.8.

a. Estimate the proportion of Americans who believe that mental illness is on the rise but are unwilling to have a mental facility in their neighborhood. How accurate is the estimate?

b. What assumptions must be satisfied to assure the validity of your inference in part (a)?

HLTH **9.10** *The Atlanta Journal* (May 29, 1990) reports on some research conducted by Dr. Russell P. Sherwin, a pathologist at the University of Southern California. Sherwin conducted autopsies on 100 Los Angeles youths, ages 14 to 25, who had died in accidents or homicides, and found that "27% had suffered severe lung damage, and 80% had lung tissue abnormalities."

a. Find a 90% confidence interval for the proportion of all youths in Los Angeles who will die as the result of accidents or murder, who will be found to have severe lung damage.

b. Is this sample large enough for your method to yield valid results? Explain.

c. Does this sampling method satisfy the assumptions required for your method? Explain.

MED **9.11** Human skin can be removed from a dead person, stored in a skin bank, and then grafted onto the body of a burn victim. Because only skin taken from flat areas (the chest, back, etc.) is acceptable and because some donors with infections or disease must be excluded, only 7% of potential donors are acceptable. A *New York Times* article (June 14, 1981) notes that a burn research center at a large city hospital screens approximately 8,000 deaths annually and, from these, gets only 80 acceptable donors. Do these data present sufficient evidence to indicate that the burn center is selecting from a universe of donors that contains a lower proportion of acceptables than the proportion in the universe at large (i.e., less than .07)?

a. What value of α are you going to choose for the test? Why?

b. Conduct the test, step by step, and state your conclusions.

c. Can you say anything about the probability that you have arrived at an incorrect conclusion?

d. Do you think that the assumptions required for the test were adequately satisfied? Explain.

PSY **9.12** Many people believe that "to spare the rod is to spoil the child." A public opinion poll in Britain, conducted by Marketing and Opinion Research International for *The Times of London*, found that of 604 parents questioned, 63% were in favor of corporal punishment in the schools (*The New York Times*, August 18, 1985). Suppose that the 604 parents represent a random sample of all parents in Britain. Do the data present sufficient evidence to indicate that a majority of parents in Britain favor the retention of corporal punishment in the schools? Conduct the test. Then state your conclusions and write a brief explanation to support your conclusions.

9.4 Comparing Two Population Proportions

Sampling Procedure, Symbols, and Objectives

The problem of comparing proportions from two different populations is identical to the problem of comparing two population means except that the populations consist of qualitative data. We select independent random samples from two binomial populations, n_1 elements from population 1 and n_2 elements from population 2. The objective of the sampling is to estimate the difference between the proportions, p_1 and p_2, of elements in the two populations that fall into a particular category or to test hypotheses concerning the difference between p_1 and p_2.

For example, William B. Waegel ("The Use of Lethal Force by Police: The Effect of Statutory Change," *Crime and Delinquency*, Vol. 30, No. 1, 1984) presents data on 459 police shootings in Philadelphia, Pennsylvania, over the period 1970–1978. The data are shown in Table 9.5 (page 348). The objective of the study was to see whether restrictive laws governing the use of lethal force, introduced in 1973, produced a change (presumably a reduction) in the proportion of cases where lethal force was unjustified.

Suppose that we want to compare the probability p_1 that a case will involve the unjustified use of lethal force by police in the time period 1970–1972 with the corresponding probability p_2 for the time period 1974–1978. Universe 1 consists of all cases that occurred, *or conceptually could have occurred*, during the time

TABLE 9.5 Data on the Use of Lethal Force over Three Time Periods

	Time Period		
Case Type	Total for Period 1970–1972	1973	Total for Period 1974–1978
Justified	72	30	173
Not Justified	11	9	59
Unable to Determine	18	7	53
Accidental	10	5	12
Totals	111	51	297

period 1970–1972 in which police used lethal force. The sample selected from this universe consists of $n_1 = 111$ cases, of which $x_1 = 11$ were considered to have involved the unjustified use of lethal force. Universe 2 is the similar collection of cases for the period 1974–1978. The sample from this universe, *selected independently of the elements selected in sample 1*, contained $n_2 = 297$ cases, of which $x_2 = 59$ were judged to have involved the unjustified use of lethal force. Did the probability of the use of lethal force by police in Philadelphia decrease, increase, or remain unchanged after passage of the restrictive law? Although the intent of the law was to reduce the unjustified use of lethal force by police, the effect of the law could have been nullified by any one or more other factors (e.g., an increase in violence toward police by suspected criminals might motivate police to use lethal force). Therefore, we want to know whether a change in p occurred (either up or down) from 1970–1972 to 1974–1978; i.e., we want to know whether p_1 differs from p_2, and we want to know the direction of change, and how much.

Problems of the type that we have just described imply the need for a test of an hypothesis concerning the equality of p_1 and p_2 (i.e., did a change occur?) as well as a confidence interval for the difference between p_1 and p_2 (i.e., what is the estimated change?).

The Sampling Distribution of the Difference Between Two Sample Proportions

Since, for large samples, both $\hat{p}_1$ and $\hat{p}_2$ are approximately normally distributed, it follows (proof omitted) that the difference $(\hat{p}_1 - \hat{p}_2)$ in the sample proportions will be approximately normally distributed with mean and standard error shown in the box.

Mean and Standard Error of the Sampling Distribution of $(\hat{p}_1 - \hat{p}_2)$

Mean: $(p_1 - p_2)$

Standard error: $$\mathrm{SE}(\hat{p}_1 - \hat{p}_2) = \sqrt{\frac{p_1(1 - p_1)}{n_1} + \frac{p_2(1 - p_2)}{n_2}}$$

A Confidence Interval and Test for the Difference Between Two Population Proportions

The confidence interval and the test of an hypothesis for $(p_1 - p_2)$ are based on the standard normal z statistic,

$$z = \frac{(\hat{p}_1 - \hat{p}_2) - (p_1 - p_2)}{\mathrm{SE}(\hat{p}_1 - \hat{p}_2)}$$

They are shown in the accompanying boxes.

A $100(1 - a)\%$ Confidence Interval for $(p_1 - p_2)$

$$\mathrm{LCL} = (\hat{p}_1 - \hat{p}_2) - z_{a/2}\mathrm{SE}(\hat{p}_1 - \hat{p}_2)$$
$$\mathrm{UCL} = (\hat{p}_1 - \hat{p}_2) + z_{a/2}\mathrm{SE}(\hat{p}_1 - \hat{p}_2)$$

where

$$\mathrm{SE}(\hat{p}_1 - \hat{p}_2) - \sqrt{\frac{p_1(1 - p_1)}{n_1} + \frac{p_2(1 - p_2)}{n_2}}$$

Assumptions:

1. **The samples are independently and randomly drawn from the two populations.**
2. **The sample sizes, n_1 and n_2, are large enough so that the numbers of successes, x_1 and x_2, are approximately normally distributed.**
3. **The values of p_1 and p_2 in the formula for $\mathrm{SE}(\hat{p}_1 - \hat{p}_2)$ can be approximated by $\hat{p}_1$ and $\hat{p}_2$.**

A Standard Normal z Test for a Difference Between Two Population Proportions Based on Independent Samples

Null hypothesis: $H_0: p_1 = p_2 = p$

Alternative hypothesis:

1. $H_a: p_1 > p_2$, an upper-tailed test
2. $H_a: p_1 < p_2$, a lower-tailed test
3. $H_a: p_1 \neq p_2$, a two-tailed test

Test statistic: $z = \dfrac{\hat{p}_1 - \hat{p}_2}{\mathrm{SE}(\hat{p}_1 - \hat{p}_2)}$

where

$$\mathrm{SE}(\hat{p}_1 - \hat{p}_2) = \sqrt{p(1-p)\left(\frac{1}{n_1} + \frac{1}{n_2}\right)}$$

and the value of p (which is unknown) in the formula for SE $(\hat{p}_1 - \hat{p}_2)$ is approximated by

$$\hat{p} = \frac{x_1 + x_2}{n_1 + n_2}$$

Rejection region: Choose the value of α that is acceptable to you. Then reject H_0 and accept H_a when the p-value* for the test is less than or equal to α.

Assumptions:

1. The samples are independently and randomly drawn from the two populations.
2. The sample sizes, n_1 and n_2, are large enough so that the numbers of successes, x_1 and x_2, are approximately normally distributed.

Example 9.4

Do the data in Table 9.5 provide sufficient evidence to indicate that the probability of the unjustifiable use of lethal force by police changed from the time period 1970–1972 to the period 1974–1978?

a. State the null and alternative hypotheses for a test designed to answer this question.

b. Locate the rejection region for the test.

c. Find $(\hat{p}_1 - \hat{p}_2)$, $\mathrm{SE}(\hat{p}_1 - \hat{p}_2)$, and the value of the test statistic.

* Do not confuse the p in "p-value" with the population proportion p.

d. Find the approximate p-value for the test.

e. State your conclusion.

Solution

a. We want to know whether the data provide sufficient evidence to indicate that p_1 differs from p_2. Therefore, we will test the null hypothesis $H_0: p_1 = p_2$ against the two-sided alternative hypothesis $H_a: p_1 \neq p_2$.

b. The greater the difference between $\hat{p}_1$ and $\hat{p}_2$, the greater is the evidence to indicate that p_1 differs from p_2. Therefore, we will conduct a two-tailed test and will reject H_0 and accept H_a for either large positive or large negative values of the test statistic z.

c. The sample proportions of cases in which lethal force was unjustifiably used during time periods 1970–1972 and 1974–1978 are $\hat{p}_1 = x_1/n_1 = 11/111 = .099$ and $\hat{p}_2 = x_2/n_2 = 59/297 = .199$, respectively, and the difference is $(\hat{p}_1 - \hat{p}_2) = (.099 - .199) = -.100$. The standard error of $(\hat{p}_1 - \hat{p}_2)$, assuming H_0 is true (i.e., $p_1 = p_2 = p$), is

$$\mathrm{SE}(\hat{p}_1 - \hat{p}_2) = \sqrt{p(1-p)\left(\frac{1}{n_1} + \frac{1}{n_2}\right)}$$

where

$$\hat{p} = \frac{x_1 + x_2}{n_1 + n_2} = \frac{11 + 59}{111 + 297} = .172$$

Then

$$\mathrm{SE}(\hat{p}_1 - \hat{p}_2) = \sqrt{.172(1 - .172)\left(\frac{1}{111} + \frac{1}{297}\right)} = .042$$

and

$$z = \frac{\hat{p}_1 - \hat{p}_2}{\mathrm{SE}(\hat{p}_1 - \hat{p}_2)} = \frac{-.100}{.042} = -2.38$$

d. Since this is a two-tailed test, the p-value for the test is the probability of observing a value of z as large as or larger than $z = 2.38$ or as small as or smaller than -2.38 (see Figure 9.4, page 352). We know from Table 9.3 (or see Table 3 of Appendix 2) that z-values that locate an area a in the upper tail of the normal distribution are

$z = 1.96$ for $a = .025$

$z = 2.33$ for $a = .010$

$z = 2.58$ for $a = .005$

Since the magnitude of the observed value of z is slightly larger than 2.33, the area in one tail of the normal distribution is smaller than .01. Therefore, the p-value for the test, the area above $z = 2.38$ and the area below $z = -2.38$, will be slightly less than $2(.01) = .02$.

FIGURE 9.4 The *p*-value for the test in Example 9.4

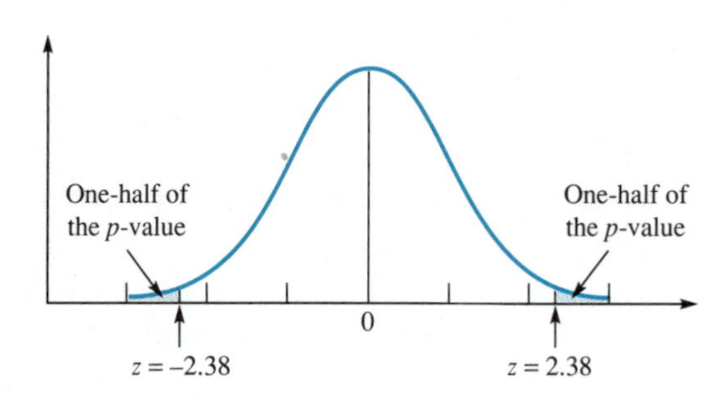

e. Suppose we decided that we were willing to risk rejecting H_0 when it is true with probability $\alpha = .05$. Since the *p*-value for the test is smaller than $\alpha = .05$, we reject $H_0: p_1 = p_2$, and conclude that the probability of the unjustified use of force by police during the period 1970–1972 differs from the probability during the period 1974–1978.

Example 9.5

Refer to Example 9.4 and to the lethal force data given in Table 9.5. Find a 95% confidence interval for the difference in the probability of the unjustified use of lethal force between the time periods 1970–1972 and 1974–1978.

Solution

The confidence limits for the confidence interval for the difference in two population proportions are

$$\text{LCL} = (\hat{p}_1 - \hat{p}_2) - z_{a/2}\text{SE}(\hat{p}_1 - \hat{p}_2)$$
$$\text{UCL} = (\hat{p}_1 - \hat{p}_2) + z_{a/2}\text{SE}(\hat{p}_1 - \hat{p}_2)$$

where $a = .05$, $z_{.025} = 1.96$, and, from Example 9.4, $(\hat{p}_1 - \hat{p}_2) = -.100$ and

$$\text{SE}(\hat{p}_1 - \hat{p}_2) = \sqrt{\frac{p_1(1 - p_1)}{n_1} + \frac{p_2(1 - p_2)}{n_2}}$$
$$= \sqrt{\frac{.099(1 - .099)}{111} + \frac{.199(1 - .199)}{297}} = .037$$

Substituting into the formulas for LCL and UCL, we obtain

$$\text{LCL} = -.100 - 1.96(.037) = -.173$$
$$\text{UCL} = -.100 + 1.96(.037) = -.027$$

Therefore, we estimate that the probability of the unjustified use of lethal force increased (rather than decreased) by an amount between .027 and .173 after passage of the law. We are reasonably confident of this estimate. Ninety-five percent of the intervals constructed using our methodology will enclose the difference $(p_1 - p_2)$.

EXERCISES

HLTH **9.13** Most AIDS deaths are caused by opportunistic infections—those that attack a patient already weakened by AIDS. An old drug, Imuthiol, made by Pasteur Merieux Serums and Vaccines, was tested for effectiveness against these secondary infections. In a study of 389 patients with AIDS or AIDS-related complex, half (say $n_1 = 195$) were randomly selected and assigned to receive Imuthiol and the remainder ($n_2 = 194$) were given a placebo, i.e., an inactive substance. Of those receiving Imuthiol, 10 contracted opportunistic infections as compared with 21 in the untreated (those receiving placebos) group (*The Wall Street Journal*, April 1, 1991). Do the data provide sufficient evidence to indicate a difference between treated and untreated patients in the proportions contracting an opportunistic infection?

a. State the null and alternative hypotheses for a test designed to answer this question.
b. Locate the rejection region for the test.
c. Find $(\hat{p}_1 - \hat{p}_2)$, SE $(\hat{p}_1 - \hat{p}_2)$, and the value of the test statistic.
d. Find the approximate p-value for the test.
e. State your conclusion.

9.14 Refer to Exercise 9.13. Find a 95% confidence interval for the difference between treated and untreated patients in the probability of contracting an opportunistic infection. Interpret the interval.

SOC **9.15** Examples 9.4 and 9.5 used the data of Table 9.5 to show that, despite the restrictive law of 1973, the probability of the unjustified use of lethal force by police increased from 1970–1972 to 1974–1978. Is it possible that it became more difficult to classify the cases after passage of the law in 1973?

a. To answer this question, test to see whether the probability that a case would be classified as "unable to determine" increased from the 1970–1972 period to the 1974–1978 period. State the null and alternative hypotheses for the test.
b. Suppose that you were to choose $\alpha = .05$. Conduct the test and state your conclusion.
c. Of what relevance is the value $\alpha = .05$ to your conclusions?

9.16 Refer to Exercise 9.15. Find a 95% confidence interval for the change in the probability that a case will fall in the "unable to determine" category from 1970–1972 to 1974–1978. Interpret the interval.

HLTH **9.17** An article in *The Wall Street Journal* (April 16, 1990), titled "Study Finds Doctors Tend to Postpone Heart Surgery for Women, Raising Risk," attempts to explain why a higher proportion of women (4.6% of 482) than men (2.6% of more than 1,800) died after coronary bypass surgery. We will avoid an explanation of why this might be true and ask whether the data support the contention that the two population proportions differ. Do the data provide sufficient evidence to indicate that the mortality rates after bypass surgery differ for men and women? (Assume that the sample size for men is 1,800.)

9.18 Refer to Exercise 9.17. Find a 95% confidence interval for the difference in mortality rates between men and women after bypass surgery.

9.5 Choosing the Sample Size

The method for choosing the sample size to estimate a population proportion or the difference between two population proportions is exactly the same as the method used to estimate population means. We decide how large an error of estimation we are willing to tolerate. For example, we may specify that we want the half-width of a confidence interval to equal some number B or, equivalently, that we want to estimate p or $(p_1 - p_2)$ with a probability equal to .95 that the error of estimation is less than B. The sample sizes, based on independent random sampling, are given in the boxes.

Sample Size Required to Estimate a Population Proportion p Correct to Within Some Number B with Probability Equal to .95

$$n = \left[\frac{1.96}{B}\right]^2 (p)(1 - p)$$

The value of p required in the formula is unknown. You can approximate p using a guessed value or an estimate based on a prior sample.

Sample Sizes Required to Estimate $(p_1 - p_2)$ Correct to Within Some Number B with Probability Equal to .95

$$n_1 = n_2 = \left[\frac{1.96}{B}\right]^2 [p_1(1 - p_1) + p_2(1 - p_2)]$$

The values of p_1 and p_2 required in the formula are unknown. You can approximate p_1 and p_2 using guessed values or estimates obtained from prior sampling.

The fact that we must substitute an approximate value for p (or p_1 and p_2) into the sample size equations means that the calculated sample sizes will be approximate. Does that bother you? Take comfort. Your answer will not be off by too much because $p(1 - p)$ does not change much for small changes in p as long as p is not near 0 or 1. If you want to be conservative—i.e., you want to be sure that your sample size is large enough–substitute $p = .5$ into the formulas. The solution for $p = .5$ always gives the largest possible value for n.

Example 9.6

Suppose that we want to conduct a public opinion poll of all adults in the state of Illinois. Like most of the major polls, we want to estimate the percentage of people in that universe who possess a particular opinion correct to within 3%; that is, we want to estimate the population proportion correct to within .03. How many adults should we include in a random sample of the universe?

Solution

The closer p is to .5, the larger will be the required sample size. Presumably, the poll will consist of a number of questions, and p will vary from one question to another. To be sure that our sample size is large enough for all questions, we will substitute $p = .5$ into the formula for n. Since we have chosen $B = .03$,

$$n = \left[\frac{1.96}{B}\right]^2 (p)(1 - p) = \left[\frac{1.96}{.03}\right]^2 (.5)(1 - .5) = 1{,}067.1$$

or, rounding upward,

$$n = 1{,}068$$

Therefore, we will need to select a random sample of 1,068 people in order to estimate the population proportion correct to within .03.

Example 9.7

A newspaper that distributes large numbers of grocery coupons with its Friday edition wants to estimate the difference between the proportion of rural newspaper subscribers who use grocery coupons and the corresponding proportion of urban subscribers. If the paper wants to estimate the difference in proportions correct to within .04, how many subscribers would have to be included in each sample?

Solution

Suppose we think that the proportions in the two categories are in the neighborhood of $p = .20$. We will use this value to approximate the values of both p_1 and p_2 in the formula for n_1 and n_2. Then, for $B = .04$,

$$n_1 = n_2 = \left[\frac{1.96}{B}\right]^2 [p_1(1 - p_1) + p_2(1 - p_2)]$$

$$= \left[\frac{1.96}{.04}\right]^2 [(.2)(1 - .2) + (.2)(1 - .2)] = 768.3$$

Therefore, we would need independent random samples of approximately 769 people from each of the two populations to estimate the difference in the proportions of rural versus urban subscribers who use grocery coupons.

EXERCISES

PSY **9.19** A psychologist is planning an experiment to estimate the probability that a subject will give a response, call it R, when subjected to a particular stimulus. The psychologist plans to subject a random sample of n individuals to the stimulus and record the number x of times that R occurs. How large should the sample be if it is desired to estimate the probability correct to within .05? Assume that prior information suggests that p is near .8.

9.20 Refer to Exercise 9.19. What value should be used for the approximation to p to be certain that the sample size is adequate? With this condition, what sample size would be required?

GEN **9.21** Suppose that you want to conduct an opinion poll to estimate the proportion of voters in a city who will vote for candidate Smith and you want your estimate to lie within .06 of the true proportion. Approximately how many voters would you have to include in the sample? Assume that you have no prior information on the value of p.

MED **9.22** An article in *The Wall Street Journal* (January 24, 1985) suggests that many routine X-rays, given to patients entering a hospital, may be unnecessary. A study of 294 chest X-rays at a Veterans' Administration Medical Center found that only one of the 294 revealed information that could have led to a different method of treatment or revealed a new medical problem. Suppose that you want to estimate the proportion of routine admission X-rays that will lead to a different treatment or reveal a new medical problem and you want your estimate correct to within .002. How many cases would have to be included in your sample?

HLTH **9.23** Exercise 9.13 described an experiment designed to compare the proportion of AIDS patients treated with Imuthiol who contract an opportunistic infection with the corresponding proportion for untreated patients (those receiving a placebo). If you want to estimate the difference in proportions correct to within .04, how many patients would have to be included in each sample? (You can see from the data given in Exercise 9.13 that $p_1 = .05$ and $p_2 = .11$. These proportions would be satisfactory for approximating p_1 and p_2 in the sample size formula.)

PSY **9.24** Suppose that you want to compare the probability of response to stimuli for two groups of subjects in a psychological experiment, one group subjected to stimulus 1 and the other group to stimulus 2. If you want your estimate correct to within .06, how many subjects would have to be included in each group?

9.6 A Contingency Table Analysis

Contingency Tables: Definitions

A table that gives the cell counts for the categorization of n elements of a sample for two qualitative variables is called a *contingency table*. Table 9.1, which is reproduced here as Table 9.6, is an example of a contingency table. The six cells of the table correspond to the six combinations of categories for the two qualitative variables, "community" and "reading." The count for a particular cell in the table gives the number of persons in the sample of 1,486 persons falling in that cell.

A contingency table that consists of r rows and c columns is called an $r \times c$ ("r by c") *contingency table*. For example, Table 9.6 is an example of a 3×2

TABLE 9.6 A Contingency Table for Characteristics of Potential Readers in a Newspaper Market Area

		Reading		
		Readers	Nonreaders	Totals
	Urban	529	121	650
Community	Rural	373	137	510
	Farm	237	89	326
Totals		1,139	347	1,486

contingency table because it contains $r = 3$ rows and $c = 2$ columns. The $r = 3$ rows correspond to the three categories for the qualitative variable "community" and the $c = 2$ columns correspond to the two categories of the qualitative variable "reading." Similarly, Table 9.5 in Section 9.4, which gives data on the use of lethal force by police in Philadelphia for three time periods, is a 4×3 contingency table because it contains $r = 4$ rows and $c = 3$ columns. The $r = 4$ rows correspond to the four ways that a case could be classified according to the qualitative variable "case type" and the $c = 3$ columns correspond to the three ways that a case could be classified according to the qualitative variable "time period."

The Purpose of a Contingency Table

The purpose of a contingency table is to investigate the relationship between two qualitative variables. Specifically, we want to see whether the proportions falling in the categories for one variable depend on (i.e., are contingent on) the category of the second variable. For example, in the contingency table, Table 9.6, we would want to know whether the "reading" qualitative variable is dependent on the "community" qualitative variable—that is, whether the proportion of readers varies depending on whether you are sampling people who live in urban areas, people who live in rural areas, or people who live on farms. The answer might provide information useful to the newspaper's advertisers or it might be useful in expanding the readership of the newspaper.

To test whether two qualitative variables are dependent, we hypothesize (the null hypothesis) that the variables are independent (analogous to an hypothesis of no correlation for two quantitative variables). Then we attempt to show with the data that the null hypothesis is false and that, in fact, the two variables are dependent. The test statistic that we will use compares the observed cell count, denoted by the symbol **Obs**, for each cell with the expected cell count **Exp**, the count that we would expect if the qualitative variables were independent. **The greater the differences between the observed and the expected cell counts, the greater is the evidence that the null hypothesis is false and that the variables are dependent.**

Notation

Observed cell count: Obs

Expected cell count: Exp

Step-by-Step: How a Chi-Square Test Statistic Is Calculated

Almost all contingency table analyses are done on a computer. However, before we show you a few printouts, we will walk through the computations and show how we calculate a test statistic to test for the dependence between two qualitative variables.

First, we need to find the expected cell counts, the number of observations that would fall in each cell if, in fact, the null hypothesis were true. The null hypothesis of independence between rows and columns of a contingency table implies that the probability that an observation will fall in a particular row–column cell of the table is equal to the product of the respective row and column probabilities. (See the definition of independent outcomes in Section 5.3.) For example, suppose that in Lynn's survey, Table 9.6, the probability that a person selected lives in a rural area (row 2) is p_2 and the probability that the person is a reader (column 1) is p_1. Then, if rows and columns are independent, the probability that a person selected will fall in both row 2 and column 1 (a reader living in a rural area) is equal to $p_2 \times p_1$. The expected number for the cell in row 2 and column 1 is the cell probability times the sample size n [e.g., if the cell probability is .1 and the sample size is 400, we would expect $.1(400) = 40$ observations to fall in that cell].

We cannot calculate the exact values of the expected cell counts for a contingency table (because the row and column probabilities are unknown) but we can estimate their values using the formula given in the box.

Expected Cell Count for Row *i* and Column *j*

$$\text{Exp}(i, j) = \frac{R_i C_j}{n}$$

where

R_i = **Total count for row** i

C_j = **Total count for column** j

n = **Total number of observations in the sample**

For example, in Table 9.6, the expected cell count for the cell in row 2 and column 1 is

$$\text{Exp}(2, 1) = \frac{R_2 C_1}{n} = \frac{(510)(1{,}139)}{1{,}486} = 390.91$$

In the same table, the corresponding observed cell count for row 2 and column 1 is 373, and the difference between the observed and the expected cell counts for row 2 and column 1 is

$$\text{Obs}(2, 1) - \text{Exp}(2, 1) = 373 - 390.91 = -17.91$$

After calculating the difference between the observed and expected cell counts for each cell of the contingency table, we calculate $(\text{Obs} - \text{Exp})^2/\text{Exp}$ for each cell and then sum the results. For example, $(\text{Obs} - \text{Exp})^2/\text{Exp}$ for cell (2, 1) is equal to $(-17.91)^2/390.91 = .82$. **The sum of $(\text{Obs} - \text{Exp})^2/\text{Exp}$ for all cells in a contingency table is called a *chi-square* (χ^2) *statistic*.** It is the test statistic used to test the null hypothesis that the two qualitative variables are independent.

The Chi-Square Test Statistic for a Contingency Table Analysis

$$\chi^2 = \sum \frac{(\text{Obs} - \text{Exp})^2}{\text{Exp}}$$

with $(r - 1)(c - 1)$ degrees of freedom.

r = Number of rows in the contingency table

c = Number of columns

The Sampling Distribution of the Chi-Square Statistic

A typical chi-square sampling distribution is shown in Figure 9.5 (page 360). As you can see, χ^2 can never be less than 0 and the distribution is skewed to the right. **The exact shape of a chi-square distribution depends on the number of *degrees of freedom* associated with the contingency table. This number will always equal**

$$\text{df} = (r - 1)(c - 1)$$

where r is the number of rows in the contingency table and c is the number of columns.

The Chi-Square Test for Contingency (Dependence)

The larger the deviations between the observed and expected cell counts, the larger χ^2 will be and the greater will be the weight of evidence to indicate that the

FIGURE 9.5 A chi-square distribution showing the location of the rejection region for a contingency table test

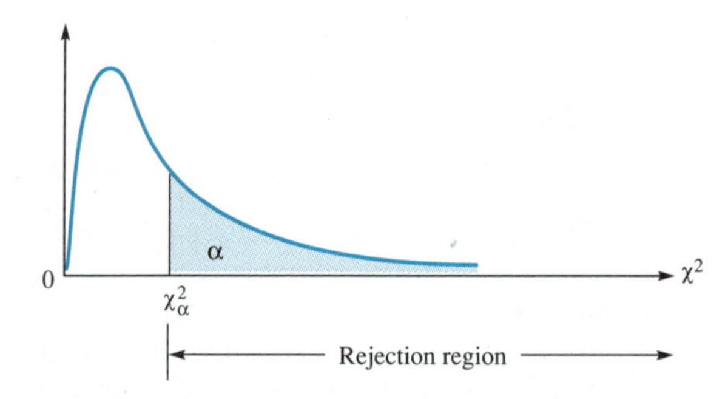

null hypothesis is false and the two qualitative variables are dependent. **Therefore, the rejection region for a contingency table test is always located in the upper tail of the chi-square distribution** (see Figure 9.5). The test is summarized in the box.

A Chi-Square Test to Detect Dependence Between Two Qualitative Variables

Null hypothesis: **H_0: The two qualitative variables are independent.**

Alternative hypothesis: **H_a : The two qualitative variables are dependent.**

$$\textit{Test statistic: } \chi^2 = \sum \frac{(\text{Obs} - \text{Exp})^2}{\text{Exp}}$$

Rejection region: **Choose the value of α that is acceptable to you. Then reject H_0 and accept H_a when the p-value for the test is less than or equal to α. The probability α is located in the upper tail of the chi-square distribution (see Figure 9.5).**

Assumptions:

1. **The sampling satisfies (*approximately*) the assumptions of a multinomial experiment given in Section 9.2. The method is also appropriate if the sampling is conducted in such a way that either the row totals or the column totals are fixed.**
2. **The sample size, n, is large enough so that all expected cell counts equal 5 or more.**

Minitab and SAS Printouts for the Telephone Survey Data

Figure 9.6 gives the Minitab and SAS printouts of the contingency table analysis for the telephone survey data of Table 9.6. The contingency table on both printouts is shaded in box 1. Minitab shows two numbers in each cell of the table. The top number is the **observed cell frequency** and the bottom number is the **expected cell frequency**. The SAS printout shows four numbers in each cell of the table. A key to their interpretation is shown in the top left corner of the table. The top number gives the observed cell frequency. The other numbers, which give the percentages of the overall, the row, and the column totals, are irrelevant to our test.

FIGURE 9.6 Minitab and SAS printouts of the contingency table analysis for the telephone survey data, Table 9.6

```
MTB > read cl c2
DATA> 529 121
DATA> 373 137
DATA> 237 89
DATA> end
      3 ROWS READ
MTB > chisquare cl-c2

Expected counts are printed below observed counts

             C1         C2     Total
    1       529        121       650
         498.22     151.78

    2       373        137       510
         390.91     119.09                1

    3       237         89       326
         249.87      76.13

Total      1139        347      1486

ChiSq =   1.902 +    6.243 +
          0.820 +    2.693 +              2
          0.663 +    2.177 = 14.499
df = 2 3

MTB > cdf 14.499 store in k1;
SUBC> chisquare with 2 df.
MTB > let k1=1-k1
MTB > print k1
K1         0.000710845 4
```

a. Minitab

FIGURE 9.6 *Continued*

TABLE OF READER BY COMM

READER Frequency Percent Row Pct Col Pct	COMM farm	rural	urban	Total	
nonreader	89 5.99 25.65 27.30	137 9.22 39.48 26.86	121 8.14 34.87 18.62	347 23.35	1
reader	237 15.95 20.81 72.70	373 25.10 32.75 73.14	529 35.60 46.44 81.38	1139 76.65	
Total	326 21.94	510 34.32	650 43.74	1486 100.00	

STATISTICS FOR TABLE OF READER BY COMM

Statistic	DF	Value	Prob	
Chi-Square	2	14.499	0.001	2
Likelihood Ratio Chi-Square	2	14.718	0.001	
Mantel-Haenszel Chi-Square	1	11.754	0.001	
Phi Coefficient		0.099		
Contingency Coefficient		0.098		
Cramer's V		0.099		

Sample Size = 1486

b. SAS

The computed value of chi-square, $\chi^2 = 14.499$, is shown in box 2 and the number of degrees of freedom for chi-square, $\text{df} = 2$, is shown in box 3 of the Minitab printout. **The Minitab printout does not give the *p*-value for the test directly but it can be programmed to compute it.** The Minitab *p*-value for this test is shown in box 4. We will show you how to find the approximate *p*-value for the test in Example 9.12.

The SAS printout gives the value of the chi-square statistic, its degrees of freedom, and the *p*-value for the test. These are shown in box 2 in the SAS printout. The other information shown on the SAS printout is irrelevant to our test.

Example 9.8

Do the data in Table 9.6 provide sufficient evidence to indicate that reading (i.e., the proportion of people who read the newspaper) is dependent on the type of community in which a person lives? Test using $\alpha = .05$.

Solution

We want to test the null hypothesis that the qualitative variables, reading and community, are independent against the alternative hypothesis that they are dependent. The p-value for the test, shown on the SAS printout, is .001. That means that the probability of observing a value of chi-square as large as or larger than $\chi^2 = 14.499$, assuming H_0 is true, is only .001. Since this p-value is less than $\alpha = .05$, we reject H_0 and conclude that the qualitative variables, reading and community, are dependent.

Contingency Table Analyses When Either Row or Column Totals Are Fixed

Up to this point, we have assumed that the data collection for a contingency table satisfies the assumptions of a multinomial experiment (see Section 9.2). **A contingency table analysis is also appropriate when either the row totals or the column totals are fixed.** For example, if the number of farms in the market area of the telephone survey, Table 9.6, is very small relative to the numbers of urban and rural homes, it is possible that no farms would appear in a random sample of all people in the market area. To make certain that we obtain information on each of the three types of communities, we could draw independent random samples of, say, 200 from each type of community. (The numbers of observations in the samples need not be equal.) The resulting table might appear as shown in Table 9.7. Since there are two cells in each row of the table, each sample of 200 people is a binomial experiment. Despite the sampling procedure, a contingency table analysis can be used to see whether reading is dependent on community.

TABLE 9.7 A Contingency Table with Fixed Row Totals

		Reading		
		Readers	Nonreaders	Totals
	Urban	168	32	200
Community	Rural	155	45	200
	Farm	122	78	200
Totals		445	155	600

Example 9.9

Nodules found on the deep-sea floor are a rich source of manganese (Menard, H. W., "Time, Chance and the Origin of Manganese Nodules," *American Scientist*, September–October 1976). Included in Menard's article are data (see Table 9.8a) that shed light on the relationship between the magnetic age of the earth's crust and the proportion of nodules that contain manganese (Mn). Column 1 of Table 9.8a lists seven magnetic ages of the earth's crust, column 2 gives the sample size (the number of nodules collected at each of the age locations), and column 3 gives the percentage of nodules at each age location that contain manganese.

TABLE 9.8 Percentage of Nodules Containing Manganese for Different Ages of the Deep-Sea Earth's Crust

a. Original Data			b. Contingency Table		
Magnetic Age	Sample Size	Percentage Containing Mn	Numbers Containing Mn	Numbers Containing No Mn	Total
Miocene-recent	389	5.9	23	366	389
Oligocene	140	17.9	25	115	140
Eocene	214	16.4	35	179	214
Paleocene	84	21.4	18	66	84
Late Cretaceous	247	21.1	52	195	247
Early–Middle Cretaceous	1,120	14.2	159	961	1,120
Jurassic	99	11.0	11	88	99
		Total	323	1,970	2,293

a. What practical value might be derived from an analysis of the data of Table 9.8a?

b. Does the sampling procedure used to produce the data of Table 9.8a permit a contingency table analysis of the data?

c. Do the data provide sufficient evidence to indicate that the percentage of nodules containing manganese varies from one magnetic age to another or is the variation in the percentages of Table 9.8a just due to sampling variation?

Solution

a. The United States needs unlimited access to a rich source of manganese. Is deep-sea mining of nodules the answer? How much manganese is contained in the nodules? Does the amount vary from one area of the deep-sea bottom to another? Is it feasible to mine the nodules? The data in Table 9.8 have been collected to provide information on the third question—namely, to determine whether the percentage of nodules that contain manganese varies depending on the magnetic age of the ocean bottom. To answer the question, we would want to conduct a contingency table analysis to see whether the qualitative variables, magnetic age and presence of manganese, are dependent.

b. In this particular problem, we would like to compare nodules in seven different universes, each universe representing all of the nodules at a location that correspond to a particular magnetic age. In order to compare the percentages of nodules in these universes that contain manganese, we select independent random samples of nodules from each universe. Each sample satisfies the characteristics of a multinomial (binomial, in this case) experiment.

Table 9.8b presents the data of Table 9.8a in the form of a 7×2 contingency table. The number of nodules, Table 9.8b, for a given magnetic age that contain manganese is obtained by multiplying the sample size for that age by the corresponding sample proportion. For example, the number of Miocene-recent nodules that contained manganese is $(389)(.059) = 23$. Subtracting this number from the sample size gives the number of nodules that did not contain manganese: $389 - 23 = 366$ nodules contained no manganese.

The seven rows of the contingency table, Table 9.8b, correspond to the seven categories of the qualitative variable, magnetic age. The two columns of the table correspond to the qualitative variable, presence of manganese. Each of the 2,293 nodules contained in the contingency table falls into one and only one cell corresponding to a magnetic age category and to a category that indicates whether it does or does not contain manganese. Since independent random samples were selected from the seven magnetic age locations, Table 9.8b is an example of a contingency table with fixed row totals.

c. Examine the percentages of Table 9.8a and note that they vary from a low of 5.9% for the nodules from deep-sea bottom of the Miocene-recent magnetic age to a high of 21.4% for those from the Paleocene magnetic age. Are the differences that we see in the seven percentages due solely to sampling variation (e.g., the percentages for two different samples from the *same* location will vary) or are the differences also due to the fact that the percentage of nodules containing manganese varies from one magnetic age to another? To answer this question, we test the null hypothesis that the proportions of nodules containing manganese are independent of magnetic age (i.e., the population proportions are identical) against the alternative that the proportions are dependent on magnetic age (i.e., that the population proportions vary from one magnetic age to another). The test results, given in the SAS contingency table analysis in Figure 9.7 (page 366), show that the computed value of the test statistic is $\chi^2 = 38.412$ and that the p-value for the test is less than .000. This tells us that we would reject H_0 for very small values of α, as small as .001. Therefore, there is very strong evidence to indicate that the proportion of nodules containing manganese is dependent on the magnetic age of the ocean bottom where they are found.

FIGURE 9.7 A SAS printout of the contingency table for the nodules data analysis

```
              TABLE OF AGE BY NODULES

AGE         NODULES

Frequency|
Percent  |
Row Pct  |
Col Pct  |no       |yes      |  Total
---------+---------+---------+
e_m_cret |    961  |    159  |   1120
         |  41.91  |   6.93  |  48.84
         |  85.80  |  14.20  |
         |  48.78  |  49.23  |
---------+---------+---------+
eo       |    179  |     35  |    214
         |   7.81  |   1.53  |   9.33
         |  83.64  |  16.36  |
         |   9.09  |  10.84  |
---------+---------+---------+
juras    |     88  |     11  |     99
         |   3.84  |   0.48  |   4.32
         |  88.89  |  11.11  |
         |   4.47  |   3.41  |
---------+---------+---------+
l_cret   |    195  |     52  |    247
         |   8.50  |   2.27  |  10.77
         |  78.95  |  21.05  |
         |   9.90  |  16.10  |
---------+---------+---------+
mio_rec  |    366  |     23  |    389
         |  15.96  |   1.00  |  16.96
         |  94.09  |   5.91  |
         |  18.58  |   7.12  |
---------+---------+---------+
oligo    |    115  |     25  |    140
         |   5.02  |   1.09  |   6.11
         |  82.14  |  17.86  |
         |   5.84  |   7.74  |
---------+---------+---------+
paleo    |     66  |     18  |     84
         |   2.88  |   0.78  |   3.66
         |  78.57  |  21.43  |
         |   3.35  |   5.57  |
---------+---------+---------+
Total        1970       323     2293
            85.91     14.09   100.00

Statistic                        DF      Value       Prob
---------------------------------------------------------
Chi-Square                        6     38.412      0.000
Likelihood Ratio Chi-Square       6     41.905      0.000
Mantel-Haenszel Chi-Square        1      0.162      0.688
Phi Coefficient                          0.129
Contingency Coefficient                  0.128
Cramer's V                               0.129

Sample Size = 2293
```

Are the Chi-Square and the z Tests Related? Yes, Sometimes . . .

Note that we used a contingency table analysis in Example 9.9 to test whether there was evidence to indicate differences among seven binomial proportions, the proportions of nodules at the seven locations that contained manganese. Can we use a contingency table analysis to test the equality of *two* binomial proportions (and, in the process, eliminate the necessity for Section 9.4)? Example 9.10 will show you that the answer is "Yes, but . . . ."

Example 9.10

In Example 9.4 we conducted a z test to see whether the proportion of cases of the unjustified use of force changed from the 1970–1972 time period to the 1974–1978 period. Use a chi-square contingency table analysis to test for a difference between these two proportions and explain the similarities and differences between the z and the chi-square tests.

Solution

Table 9.9 shows the lethal force data arranged in a 2×2 contingency table. Note that the "justified," the "unable to determine," and the "accidental" categories for each of the two time period samples of Table 9.5 have been collapsed into the single category, "other."

TABLE 9.9 A 2 x 2 Contingency Table for Example 9.10

	Time Period 1970–1972	1974–1978	Total
Not Justified	11	59	70
Other	100	238	338
Total	111	297	408

A chi-square test of the null hypothesis that the rows and columns in the 2 × 2 contingency table are independent is similar to a test of the equality of two binomial proportions; there is only one major difference. The p-value for the chi-square test is always calculated for a two-tailed test—that is, a test with the two-sided alternative, $H_a : p_1 \neq p_2$. This is confirmed by an examination of the SAS contingency table analysis, Figure 9.8 (page 368), for the data of Table 9.9. The printed value of the chi-square test statistic, based on

FIGURE 9.8 A SAS printout for the 2 x 2 contingency table, Table 9.9

```
                    TABLE OF STATE BY YEAR

             STATE      YEAR

             Frequency|
             Percent  |
             Row Pct  |
             Col Pct  |1970-72 |1974-78 |  Total
             ---------+--------+--------+
             nj       |     11 |     59 |     70
                      |   2.70 |  14.46 |  17.16
                      |  15.71 |  84.29 |
                      |   9.91 |  19.87 |
             ---------+--------+--------+
             other    |    100 |    238 |    338
                      |  24.51 |  58.33 |  82.84
                      |  29.59 |  70.41 |
                      |  90.09 |  80.13 |
             ---------+--------+--------+
             Total         111      297      408
                         27.21    72.79   100.00

            STATISTICS FOR TABLE OF STATE BY YEAR

     Statistic                        DF      Value        Prob
     -------------------------------------------------------------
     Chi-Square                        1      5.634       0.018
     Likelihood Ratio Chi-Square       1      6.171       0.013
     Continuity Adj. Chi-Square        1      4.956       0.026
     Mantel-Haenszel Chi-Square        1      5.621       0.018
     Fisher's Exact Test (Left)                        1.06E-02
                         (Right)                          0.996
                         (2-Tail)                      1.81E-02
     Phi Coefficient                         -0.118
     Contingency Coefficient                  0.117
     Cramer's V                              -0.118

     Sample Size = 408
```

$(r - 1)(c - 1) = (2 - 1)(2 - 1) = 1$ degree of freedom, is 5.634 and the p-value for the test is shown to be .018.

This agrees with our conclusion based on a two-tailed z test in Example 9.4, that the p-value for the test is "slightly less than .02." The equivalence of the two tests is also indicated by the fact that a chi-square statistic with 1 degree of freedom is equal to z^2. The calculated value of the test statistic, Example 9.4, is $z = -2.38$. You can see that $z^2 = (-2.38)^2 = 5.66$ is equal (except for rounding errors) to the value of the chi-square test statistic, $\chi^2 = 5.634$. **This tells us that we can use a chi-square contingency table analysis to test the equality of two binomial proportions as long as we remember that the p-value is for a two-tailed test.** It is possible to use it for a one-tailed test if we take into account the direction of the difference and halve the printed p-value. However, **to avoid complications, we recommend the z test of Section 9.4 for a one-tailed test.**

Finding the Approximate p-Value for a Chi-Square Test

Some statistical packages (e.g., SAS) will print the p-value for a chi-square contingency table test; others give only the computed value of the chi-square statistic. If we are given only the computed value of the test statistic and want to find the p-value for the test, we need a table that gives the values of χ_a^2 that locate an area a in the upper tail of the chi-square distribution (see Figure 9.9). These values of χ_a^2 for different values of a are given in Table 5 of Appendix 2.

FIGURE 9.9 Location of χ_a^2, Table 5 of Appendix 2

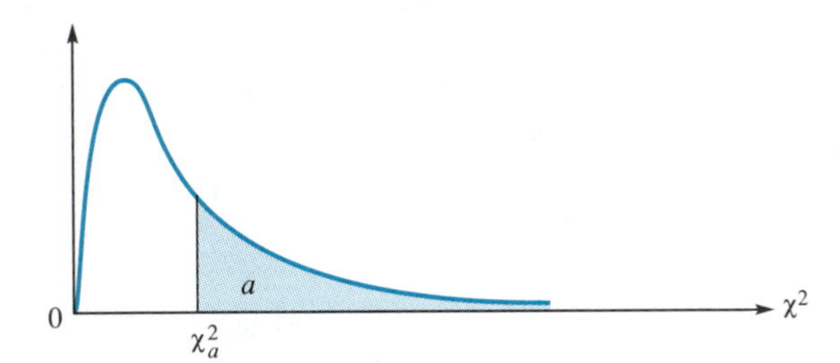

A portion of Table 5 of Appendix 2 is shown in Table 9.10 (page 370). The degrees of freedom (df) proceed down the left column and the value of a is shown across the top of the table. Suppose that your test statistic is based on 9 degrees of freedom. Then the value of chi-square that locates $a = .05$ in the upper tail of the chi-square distribution is located in the df = 9 row and the $a = .05$ column. This value, $\chi_{.05}^2 = 16.9190$, is shaded in Table 9.10.

Example 9.11

Find the value of χ^2 that locates $a = .10$ in the upper tail of the chi-square distribution with 6 degrees of freedom.

Solution

Looking for the value in Table 9.10 in the row corresponding to df = 6 and the column corresponding to $a = .10$, we find $\chi_{.10}^2 = 10.6446$. Therefore, the probability of observing a value of χ^2 as large as or larger than 10.6446 is $a = .10$.

Example 9.12

Suppose that the printout for the contingency table analysis of the telephone survey data, Table 9.6, gave the computed value of the chi-square test statistic but did not give the p-value for the test. The computed value of the test statistic (see the printout, Figure 9.6) is $\chi^2 = 14.499$. Use Table 9.10 (or Table 5 of Appendix 2) to find the approximate p-value for the test.

TABLE 9.10 A Portion of the Chi-Square Table, Table 5 of Appendix 2

	χ^2_a				
Degrees of Freedom	$a = .10$	$a = .05$	$a = .025$	$a = .01$	$a = .005$
1	2.70554	3.84146	5.02389	6.63490	7.87944
2	4.60517	5.99147	7.37776	9.21034	10.5966
3	6.25139	7.81473	9.34840	11.3449	12.8381
4	7.77944	9.48773	11.1433	13.2767	14.8602
5	9.23635	11.0705	12.8325	15.0863	16.7496
6	10.6446	12.5916	14.4494	16.8119	18.5476
7	12.0170	14.0671	16.0128	18.4753	20.2777
8	13.3616	15.5073	17.5346	20.0902	21.9550
9	14.6837	16.9190	19.0228	21.6660	23.5893
10	15.9871	18.3070	20.4831	23.2093	25.1882
⋮	⋮	⋮	⋮	⋮	⋮
60	74.3970	79.0819	83.2976	88.3794	91.9517
70	85.5271	90.5312	95.0231	100.425	104.215
80	96.5782	101.879	106.629	112.329	116.321
90	107.565	113.145	118.136	124.116	128.299
100	118.498	124.342	129.561	135.807	140.169

Solution

The chi-square test statistic for the telephone survey is based on 2 degrees of freedom. Looking in the df = 2 row of Table 9.10, we see that $\chi^2_{.05} = 5.99147$ locates $a = .05$ in the upper tail of the chi-square distribution, $\chi^2_{.01} = 9.21034$ corresponds to .01, and $\chi^2_{.005} = 10.5966$ corresponds to .005. Since the observed value, $\chi^2 = 14.449$ exceeds $\chi^2_{.005} = 10.5966$, the p-value for the test is less than .005. This agrees with the p-value, .001, given on the SAS printout in Figure 9.6b.

EXERCISES

9.25 Use Table 5 of Appendix 2 to find the value of χ^2 that locates an area a in the upper tail of the chi-square distribution for the following areas and degrees of freedom (df):

a. $a = .05$, df $= 6$
b. $a = .01$, df $= 3$
c. $a = .10$, df $= 1$

9.26 Follow the instructions of Exercise 9.25 for the following areas and df:

a. $a = .05$, df $= 1$
b. $a = .01$, df $= 4$
c. $a = .10$, df $= 2$

9.27 Follow the instructions of Exercise 9.25 for the following areas and df:

a. $a = .05$, df $= 3$
b. $a = .01$, df $= 1$
c. $a = .10$, df $= 4$

9.28 Suppose that the printout for a 2×3 contingency table analysis gave the computed value of the chi-square test statistic but did not give the p-value for the test. The computed value of the test statistic is $\chi^2 = 4.96$.

a. How many degrees of freedom are associated with the chi-square statistic?
b. Use Table 9.10 (or Table 5 of Appendix 2) to find the approximate p-value for the test.

9.29 Suppose that the printout for a 2×2 contingency table analysis gave the computed value of the chi-square test statistic but did not give the p-value for the test. The computed value of the test statistic is $\chi^2 = 2.67$.

a. How many degrees of freedom are associated with the chi-square statistic?
b. Use Table 9.10 (or Table 5 of Appendix 2) to find the approximate p-value for the test.

9.30 Suppose that the printout for a 2×5 contingency table analysis gave the computed value of the chi-square test statistic but did not give the p-value for the test. The computed value of the test statistic is $\chi^2 = 11.65$.

a. How many degrees of freedom are associated with the chi-square statistic?
b. Use Table 9.10 (or Table 5 of Appendix 2) to find the approximate p-value for the test.

9.31 Has the increase in the number of children involved in organized sports produced a decrease in the number of children involved in self-directed physical games? To obtain information on this question , D. A. Kleiber and G. C. Roberts interviewed a sample of 78 male and 77 female fourth- and fifth-grade students from the public schools in the Champaign-Urbana area of Illinois ("The Relationship Between Game and Sport Involvement in Later Childhood, A Preliminary Investigation," *Research Quarterly for Exercise and Sport,* Vol. 54, No. 2, 1983). Each child was categorized according to whether they participate in one or more organized sports (yes or no) and according to how often they participate in self-directed physical games (three levels of activity). The numbers of female students falling in the six categories are shown in the 3×2 contingency table, Table 9.11. A SAS contingency table analysis for the data is shown in Figure 9.10 (page 372).

TABLE 9.11 A Contingency Table: Participation in Organized Sports Versus Participation in Self-Directed Sports

Frequency of Participation in Self-Directed Sports	Participation in Organized Sports Yes	No	Total
Less than once a week	4	16	20
More than once a week but less than daily	15	20	35
Every day	7	15	22
Totals	26	51	77

FIGURE 9.10 A SAS contingency table analysis for the data of Table 9.11

TABLE OF FREQ BY PART

FREQ / PART Frequency Percent Row Pct Col Pct	nonparti cipate	particip ate	Total
<1	16 20.78 80.00 31.37	4 5.19 20.00 15.38	20 25.97
>1	20 25.97 57.14 39.22	15 19.48 42.86 57.69	35 45.45
daily	15 19.48 68.18 29.41	7 9.09 31.82 26.92	22 28.57
Total	51 66.23	26 33.77	77 100.00

STATISTICS FOR TABLE OF FREQ BY PART

Statistic	DF	Value	Prob
Chi-Square	2	3.025	0.220
Likelihood Ratio Chi-Square	2	3.138	0.208
Mantel-Haenszel Chi-Square	1	0.569	0.451
Phi Coefficient		0.198	
Contingency Coefficient		0.194	
Cramer's V		0.198	

Sample Size = 77

a. Find the computed value of chi-square.

b. Table 5 in Appendix 2 gives upper-tail values of chi-square. Use it to find the approximate p-value for the test.

c. The p-value for the test is shown on the SAS printout in Figure 9.10. Find it and compare it with your approximation in part **b**.

d. In their paper, Kleiber and Roberts show $\chi^2 = 3.02$, df $= 2$, and a p-value equal to .22. Do you agree?

GEN **9.32** One hundred ten people faced with the hazard of flood damage were asked (1) whether they expected the damage to their property to be major or minor and (2) whether their preparation for a flood would result in low, moderate, or high costs. The objective of the study was to determine whether a person's expected cost of preparation for flood damage depends on the person's expectation of the level of damage. The data are shown in the contingency table, Table 9.12, and a Minitab contingency table analysis of the data is shown in Figure 9.11.

a. Find the value of the test statistic on the printout. How many degrees of freedom does it have?

b. Find the approximate p-value for the test.

c. State your conclusions. Assume that you have chosen $\alpha = .01$.

TABLE 9.12 A Contingency Table: Expected Cost of Preparation Versus Expected Flood Damage

Cost	Expected Damage Major	Expected Damage Minor	Total
Low	43	16	59
Moderate	10	28	38
High	4	9	13
Total	57	53	110

FIGURE 9.11 A Minitab contingency table analysis: Expected cost of preparation versus expected flood damage

```
MTB > read c1 c2
      3 ROWS READ
MTB > end
MTB > chisquare c1-c2

Expected counts are printed below observed counts

              C1        C2     Total
    1         43        16        59
           30.57     28.43

    2         10        28        38
           19.69     18.31

    3          4         9        13
            6.74      6.26

Total         57        53       110

ChiSq =    5.051 +   5.433 +
           4.769 +   5.129 +
           1.112 +   1.195 = 22.690
df = 2

MTB > cdf 22.69 store in k1;
SUBC> chisquare with 2 df.
MTB > let k1=1-k1
MTB > print k1
K1         0.000012279
```

TABLE 9.13 A Contingency Table: Period of the Year Versus Prison Type, Exercise 9.33

Prison	June–Sept	Period of the Year Oct, Nov, Apr, May	Dec–Mar	Total
DCI	32	20	37	89
UCI	241	216	191	648
Total	273	236	228	737

SOURCE: "Violence in Prison," *Environment and Behavior,* Vol. 16, No. 3 (1984). © 1984 Sage Publications, Inc. with permission.

FIGURE 9.12 A SAS contingency table analysis: Period of the year versus prison type

```
              TABLE OF PERIOD BY PRISON

         PERIOD      PRISON

         Frequency|
         Percent  |
         Row Pct  |
         Col Pct  |dci     |uci     |  Total
         ---------+--------+--------+
         dm       |     37 |    191 |    228
                  |   5.02 |  25.92 |  30.94
                  |  16.23 |  83.77 |
                  |  41.57 |  29.48 |
         ---------+--------+--------+
         j_s      |     32 |    241 |    273
                  |   4.34 |  32.70 |  37.04
                  |  11.72 |  88.28 |
                  |  35.96 |  37.19 |
         ---------+--------+--------+
         onam     |     20 |    216 |    236
                  |   2.71 |  29.31 |  32.02
                  |   8.47 |  91.53 |
                  |  22.47 |  33.33 |
         ---------+--------+--------+
         Total          89      648      737
                     12.08    87.92   100.00

       STATISTICS FOR TABLE OF PERIOD BY PRISON

Statistic                      DF      Value        Prob
------------------------------------------------------
Chi-Square                      2      6.617       0.037
Likelihood Ratio Chi-Square     2      6.598       0.037
Mantel-Haenszel Chi-Square      1      6.544       0.011
Phi Coefficient                        0.095
Contingency Coefficient                0.094
Cramer's V                             0.095

Sample Size = 737
```

SOC **9.33** Is the frequency of prison assaults related to environmental conditions? To study this problem, Randy Atlas collected data on prison assaults during three periods of the year at two Florida prisons, the air-conditioned minimum-security Dade Correctional Institution (DCI) and the non-air-conditioned maximum-security Union Security Institution (UCI). The 737 cases, classified according to the institution in which they occurred and in one of three periods of the year, are shown in Table 9.13. The months of June through September are hot and humid; the months of October, November, April, and May are temperate; and the period from December through March contains a mixture of temperate and cold days.

a. If the frequency of prison assaults is dependent on the weather, how should it affect the distribution of cases in the cells of Table 9.13?

b. If you want to test to determine whether the data support the hypothesis that the frequency of prison assaults is dependent on the weather, state the null and alternative hypotheses for the test.

c. A SAS printout of the contingency table analysis for the data of Table 9.13 is shown in Figure 9.12. Identify the important elements in the printout. Explain the relevance of these elements to your test.

HLTH **9.34** A study by J. E. Brush and associates suggests that the initial electrocardiogram for a suspected heart attack victim may allow physicians to forecast the likelihood of in-hospital life-threatening complications ("Use of the Initial Electrocardiogram to Predict In-Hospital Complications of Acute Myocardial Infarction," *New England Journal of Medicine*, May 2, 1985). Table 9.14 categorizes 469 patients who entered the hospital with suspected myocardial infarction according to whether their electrocardiogram was positive or negative and whether or not they subsequently suffered in-hospital life-threatening complications.

a. Let p_1 represent the probability of an in-hospital life-threatening complication for a suspected heart attack victim whose initial electrocardiogram is negative and let p_2 represent the corresponding probability for a victim whose electrocardiogram is positive. Suppose that we want to test H_0: $p_1 = p_2$ against the alternative hypothesis H_a: $p_1 \neq p_2$. Explain why this test is equivalent to a contingency table test.

b. Figure 9.13 (page 376) gives the SAS printout for a contingency table analysis for the data of Table 9.14. Find the value of the chi-square statistic.

c. Find the approximate p-value for the test and explain how you arrived at your answer.

d. Find the exact p-value on the printout and compare it with your approximation. What is its interpretation?

e. If you have chosen $\alpha = .05$, what are your conclusions?

9.35 Test the null hypothesis of Exercise 9.34a using the z test of Section 9.4.

a. State the null and alternative hypotheses.

b. Calculate the value of the test statistic. Show that z^2 equals the value of the chi-square test statistic of Exercise 9.34.

TABLE 9.14 Contingency Table: Electrocardiogram Results Versus the Occurrence or Nonoccurrence of Life-Threatening Results

Electrocardiogram Result	In-Hospital Occurrence of Life-Threatening Complications: No	Yes	Total
Negative	166	1	167
Positive	260	42	302
Total	426	43	469

FIGURE 9.13 A SAS contingency table analysis: Electrocardiogram results versus the occurrence or nonoccurrence of life-threatening results, Exercise 9.34

```
              TABLE OF LIFETHR BY RESULTS

        LIFETHR      RESULTS

        Frequency|
        Percent  |
        Row Pct  |
        Col Pct  |neg     |pos     |  Total
        ---------+--------+--------+
        no       |    166 |    260 |    426
                 |  35.39 |  55.44 |  90.83
                 |  38.97 |  61.03 |
                 |  99.40 |  86.09 |
        ---------+--------+--------+
        yes      |      1 |     42 |     43
                 |   0.21 |   8.96 |   9.17
                 |   2.33 |  97.67 |
                 |   0.60 |  13.91 |
        ---------+--------+--------+
        Total         167      302      469
                    35.61    64.39   100.00


       STATISTICS FOR TABLE OF LIFETHR BY RESULTS

Statistic                        DF      Value          Prob
-------------------------------------------------------------
Chi-Square                        1     22.870         0.000
Likelihood Ratio Chi-Square       1     31.611         0.000
Continuity Adj. Chi-Square        1     21.300         0.000
Mantel-Haenszel Chi-Square        1     22.822         0.000
Fisher's Exact Test (Left)                             1.000
                    (Right)                         5.43E-08
                    (2-Tail)                        1.15E-07
Phi Coefficient                          0.221
Contingency Coefficient                  0.216
Cramer's V                               0.221

Sample Size = 469
```

c. Find the approximate p-value for the test and compare it with the p-value in Exercise 9.34.

d. If you have chosen $\alpha = .05$, what are your conclusions?

9.36 If sampling results in a contingency table with either fixed row or column totals, how will it differ from the sampling implied in Exercise 9.34? Why would you want to select your data in such a way that row (or column) totals were fixed?

9.37 When is a chi-square contingency table test equivalent to the z test of Section 9.4?

GEN **9.38** Left-handed people may be more prone to accidents than right-handed people. Stanley Coren reports on a sample of 1,896 students from the University of British Columbia ("Left-Handedness and Accident-Related Injury Risk," *American Journal of Public Health*, Vol. 79, No. 8, August 1989). Each student was classified as either left-handed or right-handed and each recorded the number of accident-related injuries requiring medical attention that they had sustained in the previous 2 years. Table 9.15 (extracted from

TABLE 9.15 Students Classified According to Handedness and Number of Accidents

	Left-Handed	Right-Handed
No Injuries	87	1,097
At Least One Injury	93	619
Total	180	1,716

Coren's Table 1) shows a 2 × 2 contingency table for all 1,896 students.

a. The students in Table 9.15 were classified according to two variables, handedness and number of accident-related injuries. Are these variables qualitative or quantitative? Can quantitative variables be classified?

b. A Minitab contingency table analysis for the data of Table 9.15 is shown in Figure 9.14. Do the data provide sufficient evidence to indicate a dependence between handedness and number of accident-related injuries? State the null and alternative hypotheses you would use to answer this question. Give the test statistic and rejection region for the test assuming that you have chosen $\alpha = .01$.

c. Give the approximate (or exact) *p*-value for the test.

d. State your conclusions.

9.39 Refer to Exercise 9.38 and find a 95% confidence interval for the proportion of the UBC student body who are left-handed. What assumption(s) must you make in order that the confidence interval be valid?

FIGURE 9.14 A Minitab contingency table analysis of Coren's handedness data

```
Expected counts are printed below observed counts

            C1        C2     Total
    1       87      1097      1184
        112.41   1071.59

    2       93       619       712
         67.59    644.41

Total      180      1716      1896

ChiSq =  5.742 +  0.602 +
         9.548 +  1.002 = 16.894
df = 1

MTB > cdf 16.894 store in k1;
SUBC> chisquare with 1 df.
MTB > let k1=1-k1
MTB > print k1
K1       0.000039577
```

9.7 Key Words and Concepts

- This chapter tells you how to make inferences from qualitative data. The most common types of qualitative data sets are those generated in public opinion or consumer preference polls, sets of "yes" and "no" answers, or answers to questions that fall into a predetermined set of categories.
- The statistical methods used in this chapter assume that the data collection satisfies the requirements of a multinomial experiment.
- A ***multinomial experiment*** consists of n identical and independent trials, each of which can result, with a given probability, in one of a set of k outcomes. A summary of the sample observations gives the number of outcomes falling in each of the k categories.
- A ***binomial experiment*** (Chapter 5) is a multinomial experiment with $k = 2$ categories.
- Public opinion polls satisfy (for all practical purposes) the requirements of a multinomial experiment when the number N of elements in the universe is large and when the sample size n is small relative to N.
- We sample a population of qualitative data in order to make inferences about the proportions, $p_1, p_2, \ldots, p_k$, of elements in the population that fall into the k categories.
- We used the z statistic in Section 9.3 to find a confidence interval for the proportion p of elements from a *single* population that fall into a particular category and to test hypotheses (i.e., make decisions) about its value.
- In Section 9.4, we used the z statistic to estimate and test hypotheses about the difference between the proportions of elements in two *different* populations, a proportion p_1 from population 1 and a proportion p_2 from population 2, based on independent random samples from the two populations.
- We learned in Section 9.6 how to determine whether two qualitative variables are related—in particular, whether the proportions of responses falling in the categories for one variable depend on the categories of the second variable. A test of the null hypothesis that "the two qualitative variables are independent" against the alternative hypothesis that "the two variables are dependent" is performed using a contingency table analysis with a chi-square test.
- We learned Section 9.5 how to decide on the number of people that we should include in an opinion poll or a consumer preference poll. As in estimating population means, the answer depends on how large an error of estimation we are willing to tolerate. We can then find the sample size required to estimate a proportion p correct to within a bound B.

SUPPLEMENTARY EXERCISES

9.40 A *New York Times*–WCBS News Poll of New Yorkers' beliefs about race relations involved a telephone sample of 1,047 adults (*New York Times*, June 27, 1990). If the sample was a random sample, what are the limits on the error of estimating a population proportion? Explain what your answer means.

SOC **9.41** Does the conviction of people for a criminal offense after treatment for drug abuse depend on the level of a person's education? Table 9.16, part of a much larger study by Stephen Wilson and Bertram Mandelbrote, categorizes 60 graduates of a drug rehabilitation program according to years of education and according to whether they were convicted after treatment ("Drug Rehabilitation and Criminality," *British Journal of Criminality*, Vol. 18, No. 4, 1978). Do the data provide sufficient evidence to indicate that the probability of conviction after treatment is dependent on the amount of education a person has?

a. Test using the z test of Section 9.4. Give each part of the test, including the approximate p-value, and state your conclusions. Use $\alpha = .10$.

b. Perform the test using a chi-square test statistic. A Minitab contingency table analysis for the data is shown in Figure 9.15 (page 380). Give each part of the test and state your conclusions.

c. Explain in detail how the tests of parts **a** and **b** are related. Use the results of the tests to support your argument.

9.42 Refer to Exercise 9.41. Find a 95% confidence interval for the difference in the probability of conviction for those with 15 years or less of education versus those with 16 years or more.

9.43 Note that "years of education" in Exercise 9.41 is a quantitative variable that has been treated as a qualitative variable for purposes of analysis. Would it be possible to study the relationship between status of conviction (y) and years of education (x) by performing a simple linear regression analysis on the data and testing to see if x contributes information for the prediction of y? Explain.

SOC **9.44** Refer to Exercise 9.41. Suppose we believed that the more education a rehabilitated drug addict had, the less likely it would be that he or she would be convicted after rehabilitation. Then, if p_1 is the probability of conviction for persons with 15 years of education or less and p_2 is the corresponding probability for those with 16 years or more, we would want to determine whether p_1 is greater than p_2; i.e., we would want to conduct a one-sided test. Do the data provide sufficient evidence to indicate that $p_1 > p_2$ if you have chosen $\alpha = .10$? What are the implications of choosing $\alpha = .10$? Under what circumstances would you choose $\alpha = .05$ or $\alpha = .01$?

PSY **9.45** Are hard-driving, intense people—people said to possess Type A behavior—more inclined to heart problems than those with low-key, easy-going Type B behavior? Robert B. Case et al. administered a Jenkins' Activity Survey (JAS) to

TABLE 9.16 Contingency Table: Years of Education Versus Status of Conviction

Education	Convicted	Not Convicted	Total
16 Years or More	6	18	24
15 Years or Less	16	20	36
Total	22	38	60

FIGURE 9.15 A Minitab printout of the contingency table analysis for years of education versus status of conviction, Exercise 9.41

```
Expected counts are printed below observed counts

            C1         C2     Total
    1        6         18        24
          8.80      15.20

    2       16         20        36
         13.20      22.80

Total       22         38        60

ChiSq =  0.891 +  0.516 +
         0.594 +  0.344 = 2.344
df = 1

MTB > cdf 2.344 store in k1;
SUBC> chisquare with 1 df.
MTB > let k1=1-k1
MTB > print k1
K1       0.125766
```

516 heart attack patients ("Type A Behavior and Survival After Acute Myocardial Infarction," *New England Journal of Medicine*, Vol. 312, No. 12, March 21, 1985). A score of −5 or below on the test is supposed to indicate Type B behavior, −5 to 5 is supposed to indicate neutral behavior, and 5 or above is supposed to indicate Type A behavior. Each person's case was followed for 3 years. They were then categorized according to whether they were alive or dead and according to their JAS score category. The data are shown in the contingency table, Table 9.17.

a. Identify the two variables in the contingency table and state whether they are quantitative or qualitative.

b. Do the data provide sufficient evidence to indicate that the survival of a heart attack patient is dependent on their JAS score? To answer

TABLE 9.17 A Contingency Table: Survival Versus JAS Score

	JAS Score			
Survival	Less than −5	−5 to 5	Larger than 5	Total
Died	21	17	11	49
Survived	159	154	154	467
Total	180	171	165	516

FIGURE 9.16 A SAS printout of the contingency table analysis for survival versus JAS score

```
                TABLE OF DIED BY SCORE

DIED        SCORE

Frequency|
Percent  |
Row Pct  |
Col Pct  |-5to5    |<-5     |>5       |  Total
---------+---------+---------+---------+
no       |    154  |    159  |    154  |    467
         |  29.84  |  30.81  |  29.84  |  90.50
         |  32.98  |  34.05  |  32.98  |
         |  90.06  |  88.33  |  93.33  |
---------+---------+---------+---------+
yes      |     17  |     21  |     11  |     49
         |   3.29  |   4.07  |   2.13  |   9.50
         |  34.69  |  42.86  |  22.45  |
         |   9.94  |  11.67  |   6.67  |
---------+---------+---------+---------+
Total         171       180       165      516
            33.14     34.88     31.98   100.00


        STATISTICS FOR TABLE OF DIED BY SCORE

Statistic                      DF       Value          Prob
-----------------------------------------------------------
Chi-Square                      2       2.563         0.278
Likelihood Ratio Chi-Square     2       2.664         0.264
Mantel-Haenszel Chi-Square      1       1.019         0.313
Phi Coefficient                         0.070
Contingency Coefficient                 0.070
Cramer's V                              0.070

Sample Size = 516
```

this question, see the SAS printout of the contingency table analysis for the data shown in Figure 9.16. Give the value of the chi-square statistic and the number of degrees of freedom associated with it. Give the p-value for the test. If you have chosen $\alpha = .05$, what are your conclusions?

c. Use Table 5 of Appendix 2 to find the approximate p-value for the test. Compare your answer with the exact value shown on the printout.

d. Can you perform the test in part **b** using a z statistic and the method of Section 9.4? Explain.

EXERCISES FOR YOUR COMPUTER

MED **9.46** An article in *The New York Times* (November 13, 1991), based on a study of nearly 5,000 men and women, states that, "Men are twice as likely as women to receive newer, life-saving treatments for heart attacks." The article goes on to state that, in the study, 26% of the 3,232 men with heart attacks received injections of "clotbuster" drugs such as streptokinase and TPA that can stop heart attacks in progress. Of the 1,659 women, 14% received treatment with the drugs.

a. Use the percentages to complete the contingency table shown here.

	Received Drug Yes	No	Total
Men			
Women			
Total			

b. Use your computer to perform a contingency table analysis on the data.

c. Test to see whether the data support the suggestion that receipt of newer life-saving treatments for heart attacks is dependent on the sex of the recipient. Write a brief report describing all aspects of the test and state your conclusions.

SOC **9.47** An article titled "Crime Up in New York in Elementary Schools" (*The New York Times,* April 24, 1990) gives the number of reported crimes against members of the United Federation of Teachers during the first term of the academic years 1988–1989 and 1989–1990. The numbers, distributed among ten categories, are shown in Table 9.18. Suppose that these reported crimes represent random samples from the total of all crimes, reported and unreported, committed in the elementary schools of New York during the years 1988–1989 and 1989–1990. Do the data provide sufficient information to indicate a change in the proportion of assaults from the 1988–1989 year to 1989–1990? Write a statement describing all as-

TABLE 9.18 Distribution of Reported Crimes in New York Elementary Schools

	First Term of the School Year		
	1988–1989	1989–1990	Change
Assault	344	464	+34.9%
Harassment	397	441	+16.4
Larceny	222	275	+23.9
Disorderly conduct (like fighting)	16	49	+206.3
Robbery	16	30	+87.5
Sex offense (includes public lewdness)	6	3	−50.0
Menacing (threatening action)	31	25	−19.4
Criminal mischief (destroying property)	97	104	+7.2
Reckless endangerment	116	160	+37.9
Other	12	15	+25.0%
Total	1,257	1,566	+24.6%

SOURCE: United Federation of Teachers. Copyright © 1990 by The New York Times Company. Reprinted by permission.

pects of your test and state your conclusions. Use your computer to perform the analysis.

9.48 Refer to Exercise 9.47. Do the data provide sufficient information to indicate a change in the distribution of crimes in the ten categories from 1988–1989 to 1989–1990? Write a statement describing all aspects of your test and state your conclusions. Use your computer to perform the analysis.

GEN **9.49** Return to Coren's data on left-handedness and accident-related injuries that we discussed in Exercise 9.38. Table 9.19 gives the accident injury data for 1,086 females in the original sample of 1,896 students. Use your computer to perform a contingency table analysis on the data.

a. Do the data provide sufficient evidence to indicate a dependence between handedness and number of accident-related injuries? State the null and alternative hypotheses you would use to answer this question. Give the test statistic and rejection region for the test assuming that you have chosen $\alpha = .05$.

b. Give the approximate (or exact) p-value for the test.

c. State your conclusions.

GEN **9.50** Table 9.20 gives the accident injury data for 810 males in Coren's original sample of 1,896 students. Use your computer to perform a contingency table analysis on the data.

a. Do the data provide sufficient evidence to indicate a dependence between handedness and number of accident-related injuries? State the null and alternative hypotheses you would use to answer this question. Give the test statistic and rejection region for the test assuming that you have chosen $\alpha = .05$.

b. Give the approximate (or exact) p-value for the test.

c. State your conclusions.

9.51 Estimate the proportion of women in the UBC student body, assuming that the 1,896 students in Coren's sample were randomly selected from among all students at UBC. How accurate is your estimate? Explain.

TABLE 9.19 Female Students Classified According to Handedness and Number of Accidents

	Left-Handed	Right-Handed
No Injuries	55	687
At Least One Injury	41	303
Total	96	990

TABLE 9.20 Male Students Classified According to Handedness and Number of Accidents*

	Left-Handed	Right-Handed
No Injuries	31	410
At Least One Injury	53	316
Total	84	726

*The frequency counts for males and females shown in Coren's Table 1 do not sum to his frequencies for "All Cases." Despite this discrepancy, we have used Coren's frequencies for Tables 9.15, 9.18, and 9.19.

References

1. Fleiss, J. L., *Statistical Methods for Rates and Proportions.* New York: Wiley, 1973.
2. Freedman, D., Pisani, R., Purves, R., and Adhikari, A., *Statistics*, 2nd edition. New York: W. W. Norton & Co., 1991.
3. McClave, J. and Dietrich, F., *Statistics,* 5th edition. San Francisco: Dellen Publishing Co., 1991.
4. Mendenhall, W. and Beaver, R., *Introduction to Probability and Statistics,* 8th edition. Boston: PWS-Kent, 1991.
5. Moore, D. and McCabe, G., *Introduction to the Practice of Statistics,* 2nd edition. New York: W. H. Freeman & Co., 1993.
6. Ott, L., *An Introduction to Statistical Methods and Data Analysis*, 4th edition. Boston: Duxbury Press, 1991.
7. Ott, L., Larson, R. F., and Mendenhall, W., *Statistics, A Tool for the Social Sciences*, 5th edition. Boston: Duxbury Press, 1992.
8. Sincich, T., *Statistics by Example*, 4th edition. San Francisco: Dellen Publishing Co., 1990.

ten

An Analysis of Variance for Designed Experiments

In a Nutshell

Does your T-shirt shrink when you wash and dry it? Researchers often vary one or more variables—say, the type of material, water temperature, etc.—to observe changes in mean shrinkage. This chapter tells us how to use a designed experiment and an analysis of variance to compare the means for two or more sets of experimental conditions.

10.1 The Problem
10.2 How an Analysis of Variance Works
10.3 An ANOVA Table for Comparing Two or More Population Means
10.4 An ANOVA F Test for Comparing Two or More Population Means
10.5 Tests and Confidence Intervals for Individual Means
10.6 Multiple Comparisons: Tukey's HSD Procedure
10.7 Experimental Design: Factorial Experiments
10.8 The Analysis of Variance for a Two-Factor Factorial Experiment
10.9 Other Analyses of Variance
10.10 Key Words and Concepts

10.1 The Problem

In Chapter 8 we learned how a quantitative independent variable x can be used to predict the value of a quantitative dependent variable y, and in Chapter 9 we learned how to test for the dependence of two qualitative variables. This chapter is similar to Chapter 8. **We want to know whether one or more independent variables affect a response variable y.** It differs from Chapter 8 because the **independent "predictor" variables** may now be either quantitative or qualitative. (If they were all quantitative, we could dispense with this chapter and analyze the data using a regression analysis.) It also differs from Chapter 8 in assuming that the response measurements were obtained for specific settings of the independent variable(s). Selecting the settings of the independent variables is another aspect of experimental design. It enables us to tell whether changes in the independent variable(s) *cause* changes in the mean response and it permits us to analyze the data using a method called an **analysis of variance** or **ANOVA** (or, sometimes, AOV).

For example, suppose that a pharmaceutical company wants to compare the potencies of a new drug prepared according to four different procedures. The potency y of drug specimens prepared using any one particular procedure will vary from one specimen to another. If an experimenter were to prepare and test many, many specimens, she would generate a conceptual population of potency measurements, one measurement for each experimental specimen that the experimenter prepares.

Suppose that independent random samples of $n_1 = n_2 = n_3 = n_4 = 5$ specimens of the drug are prepared for each procedure and the data appear as shown in Table 10.1. The experimenter would want to know whether the qualitative independent variable, procedure, affects the potency variable y. Or, putting it an-

TABLE 10.1 Drug Potency Measurements for Four Preparation Procedures

	Procedure			
	1	2	3	4
	1.32	2.15	2.64	2.10
	2.25	1.86	1.70	1.68
	1.74	2.68	2.05	1.42
	1.05	2.31	2.51	2.49
	1.55	1.73	2.77	2.35
$\bar{y}$	1.582	2.146	2.334	2.008

other way, do the data provide sufficient evidence to indicate differences in mean potency for the four different procedure populations? Look at Table 10.1. What do you think?

Why do we need a new method to test for differences among the four procedure population means? Why not use the t test of Chapter 7? We could test to see whether there is sufficient evidence to indicate a difference between each pair of means, procedure 1 versus procedure 2, 1 versus 3, 1 versus 4, 2 versus 3, 2 versus 4, and 3 versus 4. If any one of these tests leads to the rejection of the hypothesis of equal means, we might conclude that at least two of the four population means differ.

The problem with this method is that our final decision is based on the results of six t tests, any one of which can lead us to the conclusion that a difference exists between a pair of means when, in fact, all the means are equal! If, for each of the six tests, the probability is $\alpha = .10$ of being wrong (concluding that a difference exists between a specific pair of means when, in fact, they are all equal), then the probability that *at least one* of the six tests leads to this conclusion is much larger than .10. In other words, if the probability of losing on a single poker hand is .10, then the probability of losing *at least once* in six poker hands is more than .10. To avoid this problem, we test the null hypothesis that there are no differences among the four population means using an analysis of variance, a method for which the probability α can be chosen in advance.

Now that we know why an analysis of variance can be useful, let us see how it works.

10.2 How an Analysis of Variance Works

We will return to the drug potency data, but first let us examine the logic behind an analysis of variance. See Tables 10.2a and 10.2b (page 388). Each shows independent random samples from three populations, 1, 2, and 3. The sample means are shown below the sample data. Figure 10.1a shows the dot diagram for the three samples in Table 10.2a and Figure 10.1b shows the dot diagram for Table 10.2b. Now compare the sample means for the two tables and note that corresponding sample means are identical; that is, $\bar{y}_1 = 9.2$, $\bar{y}_2 = 9.5$, and $\bar{y}_3 = 9.3$ for samples 1, 2, and 3 in both data sets. Look carefully at both data sets and at their dot diagrams. Use your intuition. Which of the two data sets, Table 10.2a or Table 10.2b, provides more evidence to indicate a difference between at least two of the population means?

We think that you will choose the data set in Table 10.2b because the **variation among the sample means**, from 9.2 to 9.5, appears quite large compared with the very small amount of **variation of the measurements within each of the three samples**. In contrast, the variation among the sample means in Table 10.2a appears to be relatively small when compared with the large amount of variation

TABLE 10.2 An Intuitive Explanation: How an Analysis of Variance Works

a.

Sample 1	Sample 2	Sample 3
8.1	8.0	14.8
4.2	15.1	5.3
14.7	4.7	11.1
9.9	10.4	7.9
12.1	9.0	9.3
6.2	9.8	7.4
$\bar{y}_1 = 9.2$	$\bar{y}_2 = 9.5$	$\bar{y}_3 = 9.3$

b.

Sample 1	Sample 2	Sample 3
9.2	9.5	9.4
9.1	9.5	9.3
9.2	9.5	9.3
9.2	9.6	9.3
9.3	9.5	9.2
9.2	9.4	9.3
$\bar{y}_1 = 9.2$	$\bar{y}_2 = 9.5$	$\bar{y}_3 = 9.3$

FIGURE 10.1 Dot diagrams for the data, Tables 10.2a and 10.2b

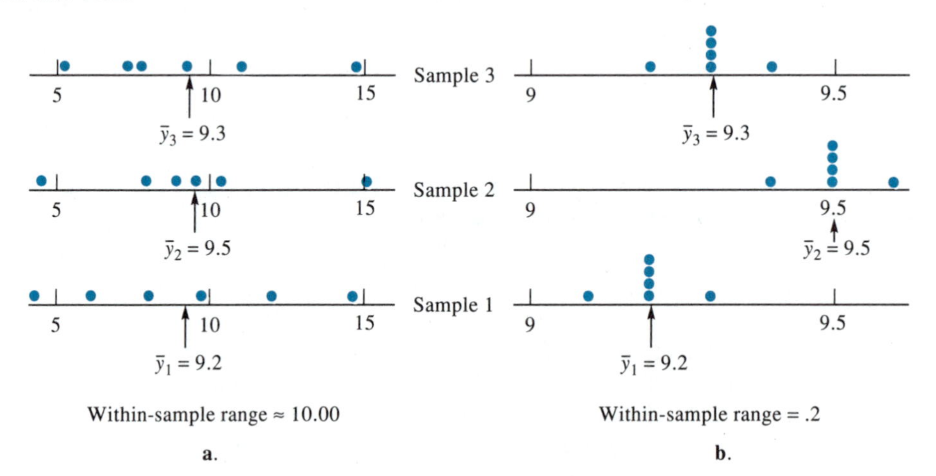

within the samples of the data set. (Look at Table 10.2 and at Figure 10.1. Do you see it?)

The two data sets of Table 10.2 illustrate the reasoning employed in an analysis of variance. **The variation among the sample means is compared with the random variation of the measurements within the samples. The larger the variation among sample means compared with the variation of the measurements within samples, the greater is the evidence to indicate a difference among population means.**

How can we measure the within-sample and between-sample variation? Surprisingly, the total sum of squares of the deviations of the y-values (all four samples) can be partitioned into two sums of squares of deviations. One, the **Sum of Squares for Error**, represented by the symbol **SSE**, measures the within-sample variation of the y-values. The other, the **Sum of Squares for Treatments**, represented by the symbol **SST**, measures the variation among the sample means. See Figure 10.2.

FIGURE 10.2 A diagram showing the partitioning of the Total SS for an experiment to compare procedure means

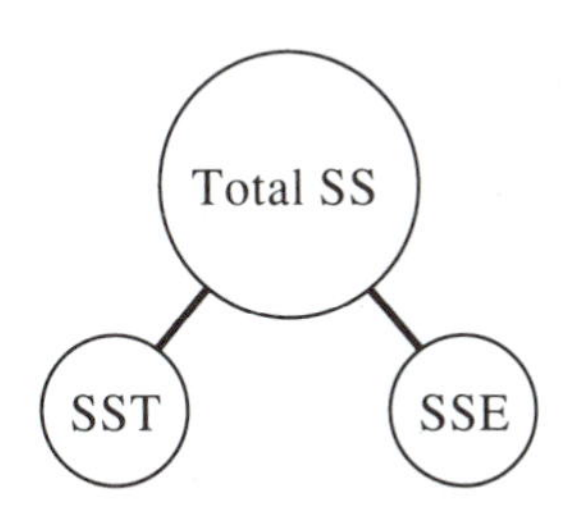

We will not concern ourselves with the formulas for these two sums of squares. With a reputable software package, they will be computed for us. Our problem will be to interpret the output.

Additivity of Sums of Squares of Deviations

Total SS = SSE + SST

The quantities Total SS, SSE, and SST, as well as others needed to conduct an analysis of variance, are presented in a table. Section 10.3 will give the format of an analysis of variance (ANOVA) table for comparing two or more population means and will give the computer printouts of the ANOVA table for the drug potency data of Table 10.1. Section 10.4 will explain how to draw inferences from the quantities presented in the table.

10.3 An ANOVA Table for Comparing Two or More Population Means

Table 10.3 (page 390) gives the format of an analysis of variance table for comparing two or more population means or, alternatively, for investigating the effect of a single

TABLE 10.3 An ANOVA Table for an Independent Random Samples Design

Source	df	SS	MS	F
Treaments	$k-1$	SST	MST	MST/MSE
Error	$n-k$	SSE	MSE	
Total	$n-1$	Total SS		

FIGURE 10.3 Minitab and SAS printouts of the ANOVA table for the drug potency data, Table 10.1

```
MTB > read c1 c2 c3 c4
      5 ROWS READ
MTB > end
MTB > aovoneway c1 c2 c3 c4

ANALYSIS OF VARIANCE
SOURCE      DF        SS        MS         F        p
FACTOR       3     1.532     0.511      2.72    0.079
ERROR       16     2.999     0.187
TOTAL       19     4.531
                                    INDIVIDUAL 95 PCT CI'S FOR MEAN
                                    BASED ON POOLED STDEV
LEVEL        N      MEAN     STDEV  -------+---------+---------+---------
C1           5    1.5820    0.4538  (---------*--------)
C2           5    2.1460    0.3765              (--------*-------)
C3           5    2.3340    0.4465                  (---------*--------)
C4           5    2.0080    0.4503           (--------*--------)
                                    -------+---------+---------+---------
POOLED STDEV =    0.4330                  1.50      2.00      2.50
```

a. Minitab

```
                    Analysis of Variance Procedure

Dependent Variable: RESPONSE
                                     Sum of            Mean
Source                     DF       Squares          Square    F Value      Pr > F

Model                       3    1.53217500      0.51072500       2.72      0.0787

Error                      16    2.99920000      0.18745000

Corrected Total            19    4.53137500

                 R-Square              C.V.        Root MSE           RESPONSE Mean

                 0.338126          21.45997        0.432955              2.01750000


Source                     DF       Anova SS     Mean Square    F Value      Pr > F

GROUP                       3    1.53217500      0.51072500       2.72      0.0787
```

b. SAS

qualitative (or quantitative) variable on a response variable y. Figure 10.3 gives the Minitab and the SAS printouts of the ANOVA table for the drug potency data of Table 10.1.

1. The first column of Table 10.3 identifies the three sources of data variation. The first source, representing the variation among sample means, is identified as "Treatments" in Table 10.3, as "Factor" on the Minitab printout, and as "Model" on the SAS printout. The variation within samples is identified as "Error" in Table 10.3 and on both printouts. The total variation of the y-values about their mean is identified as "Total SS" in Table 10.3 and on the Minitab printout, and as "Corrected Total Sum of Squares" on the SAS printout.

2. The second column of Table 10.3 and the printouts in Figure 10.3 gives the number of degrees of freedom associated with each of the sums of squares that appear in column 3. The number of degrees of freedom associated with SST, the sum of squares for treatments, will always equal $(k - 1)$, one less than the number k of treatments (or, equivalently, one less than the number of sample means). Since we are comparing the means for $k = 4$ treatments (populations) for the drug potency data, the number of degrees of freedom for treatments is $(k - 1) = (4 - 1) = 3$. This number is shown in the treatments ("Factor" or "Model") row and the DF column on the printouts.

The degrees of freedom for error is equal to $(n - k)$, where n is the total number of measurements in the k samples: $n = n_1 + n_2 + \cdots + n_k$. For the drug potency data, $n = n_1 + n_2 + n_3 + n_4 = 5 + 5 + 5 + 5 = 20$ and the degrees of freedom for error is $(n - k) = (20 - 4) = 16$. This number is shown on both printouts in the "Error" row and the "DF" column.

The degrees of freedom for the Total SS will always equal $(n - 1)$. For the drug potency data of Table 10.1, $n = 20$ and $(n - 1) = 19$. This agrees with number of degrees of freedom for Total SS shown in the DF column on both printouts. **The sum of the degrees of freedom for treatments and error will always equal the number of degrees of freedom for Total SS.** For the drug potency data, $3 + 16 = 19$.

3. The third column of the ANOVA tables shown in Table 10.3 and the printouts in Figure 10.3 gives the sums of squares of deviations for the three sources of variation. The sums of squares for the drug potency data are

$$\text{SST} = 1.532 \qquad \text{SSE} = 2.999 \qquad \text{Total SS} = 4.531$$

Note that SST + SSE = Total SS:

$$1.532 + 2.999 = 4.531$$

4. Column 4 of the ANOVA tables contains the "Mean Square" for each source of variation. **The mean square for a source is obtained by dividing the source sum of squares by its degrees of freedom.** For the drug potency data,

$$\text{Mean Square for Treatments} = \text{MST} = \frac{\text{SST}}{3} = \frac{1.532}{3} = .511$$

$$\text{Mean Square for Error} = \text{MSE} = \frac{\text{SSE}}{16} = \frac{2.999}{16} = .187$$

The MST and MSE are used to calculate an F statistic that will be used to test for differences among the population means. We will explain the F test in Section 10.4.

Formulas for MST and MSE for an Independent Random Samples Design

$$\textbf{Mean Square for Treatments: MST} = \frac{\text{SST}}{k-1}$$

where $(k - 1)$ is the number of degrees of freedom for treatments

$$\textbf{Mean Square for Error: MSE} = \frac{\text{SSE}}{n-k}$$

where $(n - k)$ is the number of degrees of freedom for error

$$F = \frac{\text{MST}}{\text{MSE}}$$

EXERCISES

10.1 What is the purpose of an analysis of variance?

10.2 If we want to determine whether differences exist among five population means, why do we not reach our conclusion by running t tests to compare the 10 different pairs of means?

10.3 How does an analysis of variance work?

10.4 What do we mean by Total SS? What do the symbols SST and SSE represent?

10.5 How are Total SS, SST, and SSE related?

10.6 If an analysis of variance is conducted to compare k population means based on a total of $n_1 + n_2 + \cdots + n_k = n$ measurements, how many degrees of freedom are associated with treatments? How many will be associated with error?

10.7 Refer to Exercise 10.6. If you want to compare five population means based on independent random samples of 10 observations from each population, how many degrees of freedom will be associated with treatments? How many will be associated with error?

10.8 How are the degrees of freedom for error, treatments, and Total SS related?

10.9 Suppose that an analysis of variance is conducted to compare k population means based on a total of $n_1 + n_2 + \cdots + n_k = n$ measurements. Give the analysis of variance table for these data using the appropriate symbols to represent the entries.

10.10 If you have performed an analysis of variance to compare three population means and if

$n_1 = 4$, $n_2 = 6$, $n_3 = 6$, SSE $= 2.1$, and Total SS $= 15.6$, give the complete ANOVA table for the data showing all source, df, SS, and MS entries.

10.11 If you have performed an analysis of variance to compare six population means and if $n_1 = n_2 = \cdots = n_6 = 4$, SST $= 2.1$, and SSE $= .9$, give the complete ANOVA table for the data showing all source, df, SS, and MS entries.

10.12 If you have performed an analysis of variance to compare five population means and if $n_1 = n_2 = \cdots = n_5 = 3$, SST $= 2.1$, and MSE $= .1$, give the complete ANOVA table for the data showing all source, df, SS, and MS entries.

10.13 If you have performed an analysis of variance to compare four population means and if $n_1 = n_2 = n_3 = n_4 = 4$, MST $= 10.5$, and Total SS $= 38.2$, give the complete ANOVA table for the data showing all source, df, SS, and MS entries.

10.4 An ANOVA *F* Test for Comparing Two or More Population Means

Assumptions for an ANOVA

All tests and confidence intervals for an analysis of variance to compare k population means are based on the same assumptions used in Chapter 7 for comparing two population means:

1. **The k samples have been collected using an independent random samples design.** For example, the sample drug specimens for the four drug preparation procedures were prepared independently of each other. No outcome of one method of preparation could affect the outcome of another. The populations from which the samples were selected are conceptual. They consist of the millions of potency measurements that the researcher *could* theoretically make for each procedure but which he or she cannot do in a practical setting. Despite the fact that the populations are conceptual, the researcher expects the drug potency measurements in Table 10.1 to approximate random samples from the four populations.

2. **The k populations are normally distributed with a common variance σ^2.** This assumption is portrayed graphically in Figure 10.4. The population

FIGURE 10.4 A graphical description of the assumptions of normality and equal variances

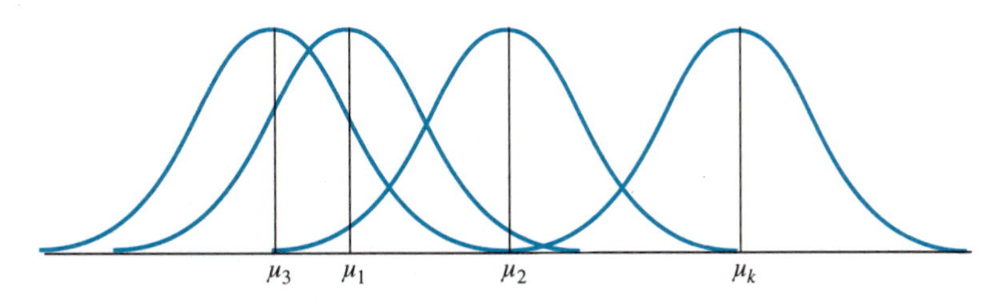

distributions are represented by the bell-shaped normal curves. They have different means but all have the same spread. There is no way to know whether the drug potency data come from normally distributed populations, but it is likely that the distributions are at least mound-shaped. The four samples in Table 10.1 do not appear to contradict the assumption that the populations have variances that are (approximately) equal. We can see that the ranges for the four samples are of the same magnitude.

An Analysis of Variance *F* Test

(*Proof of the following facts is omitted.*) If there are no differences among the k population means, the mean square for treatments, MST, is an unbiased estimate of the common population variance σ^2. However, if the population means differ, MST will tend to be larger than σ^2. The greater the differences, the larger MST will tend to be. Therefore, values of MST that are "too large" provide evidence of differences among the population means.

How large is "too large"? In contrast to MST, the mean square for error MSE, which is also represented by the symbol s^2, will *always* be an unbiased estimator of σ^2. Therefore, we can use MSE as a yardstick to tell whether MST is too large. If the population means are equal, the ratio

$$F = \frac{\text{MST}}{\text{MSE}}$$

will be near 1. If the population means differ, MST will tend to be larger than MSE and the F statistic will tend to be larger than 1. **Therefore, we will reject the null hypothesis that the population means are equal (and conclude that at least two differ) for large values of *F*.** The sampling distribution for the F statistic and the location of the rejection region are shown in Figure 10.5.

The value of F to test for differences in a set of population means is usually shown in column 5 of an ANOVA table. For example, the value of the F statistic

FIGURE 10.5 The *F* distribution and the location of the rejection region for an *F* test

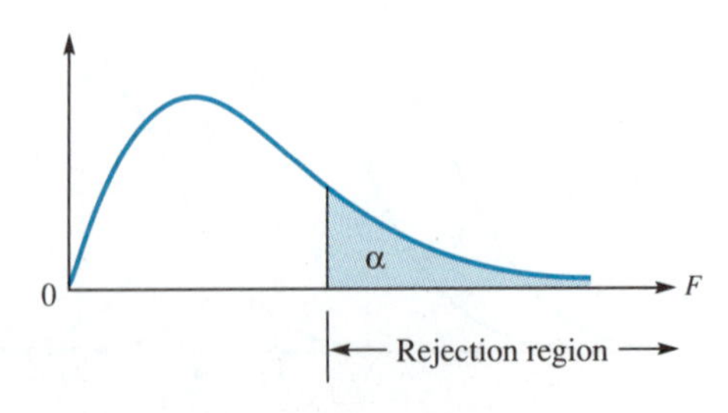

for testing for differences in drug potencies is shown on the Minitab and SAS printouts, Figure 10.3, as 2.72.

The p-value for the test is often shown in column 6 of an ANOVA table. It is shown on the Minitab printout as .079 and on the SAS printout as .0787. This means that the probability of observing an F value as large as or larger than 2.72, assuming no differences among the four population means, is only .079. We would conclude that the population means differ if we were to choose $\alpha = .10$ but would not conclude that they differ for $\alpha = .05$. What should we conclude? It appears that there is some evidence to indicate differences in mean potency among the four procedures. If differences of the observed magnitude are economically worth achieving, the company may want to invest in further experimentation.

The Tabulated Values of *F*

If your statistical computer package does not print the p-value for the F test, you can use a table to find the value of F_a that locates an area a in the upper tail of the F distribution (see Figure 10.6).

FIGURE 10.6 Location of the tabulated value of *F*

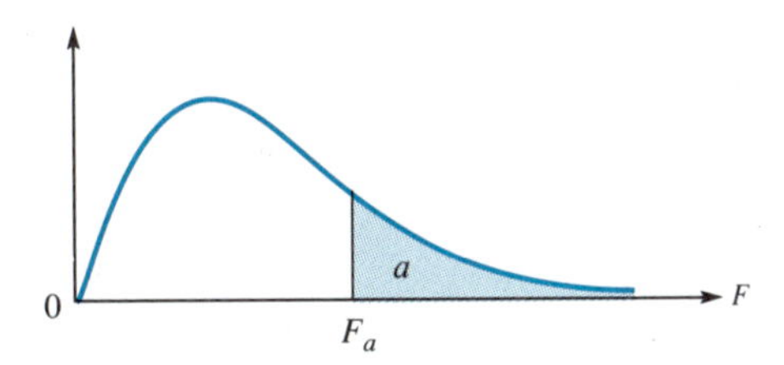

The tabulated values of F_a are given in Table 6 of Appendix 2. An abbreviated version of this table is shown in Table 10.4 on page 396. The shape of the F distribution (and, therefore, the value of F_a) depends on the numbers of degrees of freedom associated with the numerator mean square, MST, and the denominator mean square, MSE. Therefore, an F table must give values of F_a for different combinations of the numerator and denominator degrees of freedom.

Now look at Table 10.4 and you will see that the numerator degrees of freedom, df = 1, 2, . . . , 40, 60, . . . , run across the top of the table and the denominator degrees of freedom, df = 1, 2, . . . , 16, . . . , ∞, run down the left side of the table. For each combination of numerator and denominator degrees of freedom, the table gives the value of F_a for a = .10, .050, .025, .010, and .005. For example, the F statistic in the Minitab and SAS printouts has 3 numerator degrees of freedom and 16 denominator degrees of freedom. The value of F that locates a = .05 in the upper tail of this F distribution is $F_{.05} = 3.24$.

TABLE 10.4 An Abbreviated Version of the *F* Table, Table 6 of Appendix 2

Denominator d.f.	a	1	2	3	...	40	60	120	∞	a	d.f.
1	.100	39.86	49.50	53.59	...	62.53	62.79	63.06	63.33	.100	1
	.050	161.4	199.5	215.7	...	251.1	252.2	253.3	254.3	.050	
	.025	647.8	799.5	864.2	...	1006	1010	1014	1018	.025	
	.010	4052	4999.5	5403	...	6287	6313	6339	6366	.010	
	.005	16211	20000	21615	...	25148	25253	25359	25465	.005	
2	.100	8.53	9.00	9.16	...	9.47	9.47	9.48	9.49	.100	2
	.050	18.51	19.00	19.16	...	19.47	19.48	19.49	19.50	.050	
	.025	38.51	39.00	39.17	...	39.47	39.48	39.49	39.50	.025	
	.010	98.50	99.00	99.17	...	99.47	99.48	99.49	99.50	.010	
	.005	198.5	199.0	199.2	...	199.5	199.5	199.5	199.5	.005	
⋮	⋮	⋮	⋮	⋮	⋮	⋮	⋮	⋮	⋮	⋮	
16	.100	3.05	2.67	2.46	...	1.81	1.78	1.75	1.72	.100	16
	.050	4.49	3.63	3.24	...	2.15	2.11	2.06	2.01	.050	
	.025	6.12	4.69	4.08	...	2.51	2.45	2.38	2.32	.025	
	.010	8.53	6.23	5.29	...	3.02	2.93	2.84	2.75	.010	
	.005	10.58	7.51	6.30	...	3.44	3.33	3.22	3.11	.005	
⋮	⋮	⋮	⋮	⋮	⋮	⋮	⋮	⋮	⋮	⋮	
∞	.100	2.71	2.30	2.08	...	1.30	1.24	1.17	1.00	.100	∞
	.050	3.84	3.00	2.60	...	1.39	1.32	1.22	1.00	.050	
	.025	5.02	3.69	3.12	...	1.48	1.39	1.27	1.00	.025	
	.010	6.63	4.61	3.78	...	1.59	1.47	1.32	1.00	.010	
	.005	7.88	5.30	4.28	...	1.67	1.53	1.36	1.00	.005	

(Numerator d.f. spans columns 1 through ∞.)

Example 10.1

Suppose that the ANOVA printouts for the drug potency data did not contain the p-value for the F test. Use Table 6 of Appendix 2 to find the approximate p-value for the test. Compare your approximate p-value with the exact value given on the printouts.

Solution

The observed value of the F test statistic shown on the printouts is $F = 2.72$. Looking in Table 6 of Appendix 2, for the values of F_a for 3 numerator and 16 denominator degrees of freedom, we read $F_{.10} = 2.46$, $F_{.05} = 3.24$, $F_{.025} = 4.08$, etc. Since the observed value of the test statistic falls between $F_{.10} = 2.46$ and $F_{.05} = 3.24$, the area to the right of $F = 2.72$ must lie between .05 and .10. Therefore, the p-value for the test falls between .05 and .10. This agrees with the exact p-value, which was shown on the printouts as $p = .079$.

The analysis of variance F test for comparing two or more population means, based on an independent random samples design, is summarized in the box.

> **Analysis of Variance F Test for an Independent Random Samples Design**
>
> ***Null hypothesis:*** H_0: $\mu_1 = \mu_2 = \mu_3 = \cdots = \mu_k$
>
> ***Alternative hypothesis:*** H_a: **At least two of the population means differ.**
>
> ***Test statistic:*** $F = \dfrac{\text{MST}}{\text{MSE}}$
>
> **where F is based on $(k - 1)$ numerator and $(n - k)$ denominator degrees of freedom.**
>
> ***Rejection region:*** **Choose the value of α that is acceptable to you. Then reject H_0 and accept H_a if the p-value for the test is less than or equal to α.**

EXERCISES

10.14 If you conducted an analysis of variance to compare five population means based on independent random samples of 3 observations from each population, and if you found MST = 12.9 and MSE = 3.3, give the value of F for comparing the population means. Give the numerator and denominator degrees of freedom for F.

10.15 Suppose that you conducted an analysis of variance to compare three population means based on independent random samples of 6 observations from each population. If you found Total SS = 127.3 and SSE = 65.2, give the value of F for comparing the population means, and give the numerator and denominator degrees of freedom for F.

10.16 If you conducted an analysis of variance to compare four population means based on independent random samples of 2 observations from each population, and if you found SSE = 242.8 and Total SS = 393.3, give the value of F for comparing the population means. Give the numerator and denominator degrees of freedom for F.

10.17 From Table 6 of Appendix 2, give the values of F_a for the following values of a and numerator and denominator degrees of freedom:

a. $a = .05$, numerator df = 3, denominator df = 9
b. $a = .01$, numerator df = 2, denominator df = 12
c. $a = .10$, numerator df = 4, denominator df = 10

10.18 From Table 6 of Appendix 2, give the values of F_a for the following values of a and numerator and denominator degrees of freedom:

a. $a = .05$, numerator df = 3, denominator df = 8
b. $a = .01$, numerator df = 1, denominator df = 10
c. $a = .10$, numerator df = 4, denominator df = 11

10.19 If a computed value of F, based on 2 numerator and 7 denominator degrees of freedom, was equal to 4.10, give the approximate p-value for the test. If you have chosen $\alpha = .05$, what would be your test conclusion?

10.20 If a computed value of F, based on 4 numerator and 15 denominator degrees of freedom, was equal to 3.81, give the approximate p-value for the test. If you have chosen $\alpha = .05$, what would be your test conclusion?

10.21 If a computed value of F, based on 3 numerator and 16 denominator degrees of freedom, was equal to 2.95, give the approximate p-value for the test. If you have chosen $\alpha = .10$, what would be your test conclusion?

10.5 Tests and Confidence Intervals for Individual Means

We can estimate or test hypotheses about any individual treatment mean or the difference between a pair of means using the methods of Chapters 6 and 7. The only difference is that we now use the value of s^2, and its associated degrees of freedom, that we obtained in the analysis of variance. That is, we use

$$s^2 = \text{MSE} \quad \text{and} \quad s = \sqrt{\text{MSE}}$$

Example 10.2

Suppose that we had, prior to experimentation, a desire to estimate the mean drug potency for preparation procedure 2. Find a 95% confidence interval for μ_2.

Solution

The sample mean for preparation procedure 2 is $\bar{y}_2 = 2.146$, the number of measurements used in calculating the sample mean was 5, and, from the ANOVA printouts, Figure 10.3, MSE $= s^2 = .187$ based on 16 degrees of freedom. Then $s = \sqrt{s^2} = \sqrt{.187} = .432$ and $t_{.025}$, based on 16 degrees of freedom is (from Table 4 of Appendix 2) $t_{.025} = 2.120$.

Substituting into the formula for a small-sample confidence interval for μ_2, Section 6.3, gives the limits for the 95% confidence interval for μ_2:

$$\text{LCL} = \bar{y}_2 - t_{.025}\frac{s}{\sqrt{n}} = 2.146 - 2.120\frac{.432}{\sqrt{5}} = 1.736$$

$$\text{UCL} = \bar{y}_2 + t_{.025}\frac{s}{\sqrt{n}} = 2.146 + 2.120\frac{.432}{\sqrt{5}} = 2.556$$

Therefore, we estimate that the interval from 1.736 to 2.556 encloses the mean potency for preparation procedure 2 with confidence coefficient equal to .95.

Example 10.3

Suppose that we had, prior to experimentation, a desire to estimate the difference in mean drug potency between preparation procedures 2 and 1. Find a 95% confidence interval for $(\mu_2 - \mu_1)$.

Solution

The sample means for preparation procedures 2 and 1 are $\bar{y}_2 = 2.146$ and $\bar{y}_1 = 1.582$, the number of measurements used in calculating each of the sample means was 5, and, from Example 10.2, $s = .432$ and $t_{.025}$, based on 16 degrees of freedom is (from Table 4 of Appendix 2) $t_{.025} = 2.120$.

Substituting into the formula for a small-sample confidence interval for $(\mu_2 - \mu_1)$, Section 7.3, gives the limits for the 95% confidence interval for $(\mu_2 - \mu_1)$:

$$\text{LCL} = (\bar{y}_2 - \bar{y}_1) - t_{.025}s\sqrt{\frac{1}{n_1} + \frac{1}{n_2}}$$

$$= (2.146 - 1.582) - 2.120(.432)\sqrt{\frac{1}{5} + \frac{1}{5}} = -.015$$

$$\text{UCL} = (\bar{y}_2 - \bar{y}_1) + t_{.025}s\sqrt{\frac{1}{n_1} + \frac{1}{n_2}}$$

$$= (2.146 - 1.582) + 2.120(.432)\sqrt{\frac{1}{5} + \frac{1}{5}} = 1.143$$

Therefore, we estimate that the interval from $-.015$ to 1.143 encloses the difference in mean potency between preparation procedures 2 and 1 with confidence coefficient equal to .95. Note that 0 is included in the interval. Therefore it is possible that there is no difference in the mean drug potency betweem preparation procedures 2 and 1.

EXERCISES

MFG

10.22 An experiment was conducted to compare the effectiveness of three training programs, A, B, and C, in training assemblers of a piece of electronic equipment. Independent random samples of five employees were assigned to each of the three methods of assembly. After completing the courses, each assembler was asked to assemble four pieces of the equipment and the average length of time required to complete the assembly was recorded. Due to resignations from the company, only four employees completed program A and only three completed program B. The data are shown in Table 10.5 and a SAS printout of the ANOVA table for the data is shown in Figure 10.7 (page 400).

a. Give the p-value for the test and interpret it. Do the data provide sufficient evidence to indicate a difference in mean assembly times for people trained by the three programs?

b. Find the value of F for the ANOVA test and use it and Table 6 of Appendix 2 to find the approximate p-value for the test. Compare your answer with the exact p-value in part **a**.

c. Find a 90% confidence interval for the mean assembly time for persons trained in program A.

d. Do you think the assumption that the data have been selected from normally distributed populations is (approximately) satisfied? Why? (*Hint:* See the Central Limit Theorem, Section 5.8.)

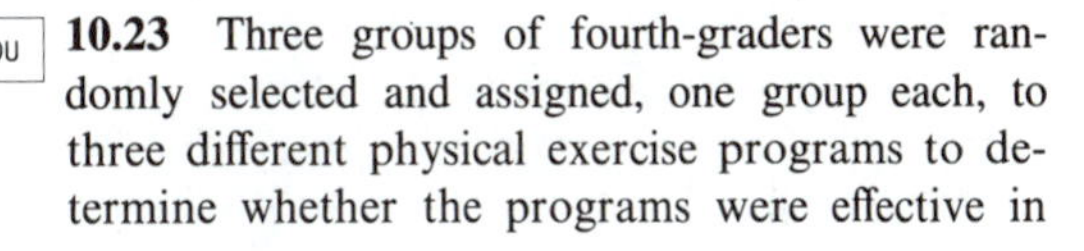

EDU

10.23 Three groups of fourth-graders were randomly selected and assigned, one group each, to three different physical exercise programs to determine whether the programs were effective in

TABLE 10.5 Data for Three Different Training Programs for Assemblers

Program	Average Assembly Time (minutes)
A	59 64 57 62
B	52 58 54
C	58 65 71 63 64

FIGURE 10.7 A SAS printout of the ANOVA for a comparison of mean assembly times

```
                    Analysis of Variance Procedure

Dependent Variable: Y
                                  Sum of          Mean
Source                  DF       Squares        Square    F Value      Pr > F

Model                    2   170.4500000    85.2250000       5.70      0.0251

Error                    9   134.4666667    14.9407407

Corrected Total         11   304.9166667

               R-Square             C.V.      Root MSE                 Y Mean

               0.559005         6.380180      3.865325             60.5833333

Source                  DF       Anova SS   Mean Square   F Value      Pr > F

TRTMENTS                 2   170.4500000    85.2250000       5.70      0.0251
```

increasing the children's abilities to throw a ball. Twenty-eight students were involved in the experiment, ten in a control group (no exercise) and nine each assigned to two different exercise regimes, each of which lasted 4 weeks. The velocity at which a child was able to throw a test ball was measured before and after the 4-week period and the gain (or loss) in velocity y (in feet per second) was recorded. A table of mean gains for the three groups and a partially completed ANOVA table are shown in Tables 10.6a and 10.6b, respectively.

a. Fill in the missing numbers in the ANOVA table, Table 10.6b.

b. Do the data provide sufficient evidence to indicate a difference in population means for the three groups? Find the approximate p-value for the test and explain the implications of the test results.

c. Find a 95% confidence interval for the difference in mean gain between children in the control group versus those on exercise regime B.

d. Find a 95% confidence interval for the mean gain for children on exercise regime B.

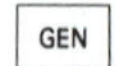

10.24 A study was conducted to determine the mercury concentration in the wing muscles of ten species of Australian waterfowl (Bacher, G. J. and Norman, F. I. "Mercury Concentrations in Ten Species of Australian Waterfowl [Family *Antidael*]," *Australian Wildlife Research*, Vol. 11, 1984). The sample size, sample mean, and standard deviation for three of these species are shown in

TABLE 10.6 Table of Means and the ANOVA Table for the Comparison of Exercise Programs, Exercise 10.23

a. Sample means

Control	Exercise Regime *A*	Exercise Regime *B*
−1.34	.32	3.69

b. ANOVA table

Source	df	SS	MS
Exercise regimes	—	64.31	—
Error	—	—	—
Total	—	402.33	

TABLE 10.7 An Analysis of Mercury Content: Sample Sizes, Means and Standard Deviations for Three Species of Waterfowl, Exercise 10.24

Species	Sample Size	Mean	Standard Deviation
Australian shelduck	6	.09	.06
Australian shoveler	3	.10	.05
Blue-billed duck	18	.12	.05

Table 10.7. An analysis of variance for the data would show SST = .004467 and SSE = .0655.

a. Construct an ANOVA table for the data.

b. Do the data present sufficient evidence to indicate differences in the mean concentration of mercury among the three species of waterfowl? Give the value of F for the test. Give the approximate p-value for the test. Test using $\alpha = .05$.

BIO **10.25** How does age affect the germination of seeds? Experimental data that shed light on this question were provided by E. F. Karlin and L. C. Bliss ("Germination Ecology of *Ledumgroenlandicum* and *Ledum palustre ssp. decumbens*," *Arctic and Alpine Research*, Vol. 15, No. 3, 1983). Each experimental unit, a set of 50 seeds of a specific age, was placed in a controlled moist environment and the percentage of germinating seeds was recorded. Experimental units were prepared for seeds aged 1, 8, 13, and 22 months. The number of experimental units, the sample mean, and the sample standard deviation are shown in Table 10.8 for each of the

TABLE 10.8 Sample Sizes, Means, and Standard Deviations of the Mean Percentage of Seeds Germinating

Seed Age	Sample Size	$\bar{y}$ (percent)	s (percent)
1	6	58	4
8	7	47	4
13	3	16	3
22	3	16	2

four seed ages. An analysis of variance for the data would show SST = 5,734.105263 and SSE = 202.0

a. Construct an ANOVA table for the data.

b. Do the data present sufficient evidence to indicate differences in the mean percentage of seeds germinating among the four age groups? Give the value of F for the test. Give the approximate p-value for the test. Test using $\alpha = .05$.

10.6 Multiple Comparisons: Tukey's HSD Procedure

We explained in Section 10.1 that if we compared all pairs of drug potency means using t tests, the probability of seeing at least one difference when none exist is relatively large. Likewise, you cannot look at the sample means, Table 10.1, and decide to use a t test to compare the largest and the smallest means. That is cheating! The difference between the largest and the smallest means in a set is *always* as large as or larger than the difference between two means chosen at random!

That creates a problem because many experiments are exploratory and we do not know in advance which means we want to compare. We usually want to look at the data and often want to compare the populations with the largest sample means. J. Tukey proposed a method, called the HSD (honestly significant difference) test, to cope with this problem. Based on the sampling distribution of the difference between the largest and smallest means in a set, it allows us to compare any and all pairs of means and to do so with a known probability α of being wrong for *at least one* of the comparisons. **The method is based on the assumption that samples of equal size have been selected using an independent random samples design and that the sampled populations are normally distributed with a common variance equal to σ^2.**

The HSD Procedure

The yardstick that determines whether two treatment means differ is based on the probability distribution of the difference between the largest and the smallest in a set of sample means. If we know that, with probability $(1 - \alpha)$, the difference between the largest and the smallest sample means cannot be greater than a number, call it W, then W is the yardstick. If any pair of sample means differ by more than W, we conclude that the corresponding population means differ. We can compare any and all pairs of sample means using this procedure.

The yardstick, W, is equal to

$$W = \frac{qs}{\sqrt{r}}$$

where

$$r = \text{Number of observations in a treatment sample (i.e., we assume that the sample sizes are equal)}$$

$$s = \sqrt{\text{MSE}}$$

and q, called the **Studentized range**, is a tabulated quantity that depends on the value of a that you have chosen for the test, the number k of sample means involved in the comparison, and the number of degrees of freedom (df) associated with $s^2 = \text{MSE}$.

SAS prints out the value of q appropriate for the number, k, of means that we wish to compare and for the number, df, of degrees of freedom associated with SSE and s^2. It also prints the value of W and identifies pairs of sample means that differ by more than W. The corresponding population means are judged to be different. Let us look at the SAS printout for the drug potency means of Table 10.1.

Applying the HSD Procedure to the Drug Potency Data

Figure 10.8 shows the SAS printout of the HSD procedure for the drug potency data of Table 10.1. The "Critical Value of Studentized Range," shown as 3.520, is the tabulated value of q for 16 degrees of freedom, $k = 4$, and $a = .10$. The quantity that we have called W, labeled as the "Minimum Significant Difference" on the printout, is .6816. The means are arranged in order, from the largest to the smallest. A

FIGURE 10.8 A SAS printout of the HSD procedure for the data of Table 10.1

```
                    Analysis of Variance Procedure

       Tukey's Studentized Range (HSD) Test for variable: RESPONSE

       NOTE: This test controls the type I experimentwise error rate, but
             generally has a higher type II error rate than REGWQ.

                    Alpha= 0.1  df= 16  MSE= 0.18745
                 Critical Value of Studentized Range= 3.520
                   Minimum Significant Difference= 0.6816

       Means with the same letter are not significantly different.

              Tukey Grouping              Mean       N  GROUP

                               A          2.334      5  3
                               A
                       B       A          2.146      5  2
                       B       A
                       B       A          2.008      5  4
                       B
                       B                  1.582      5  1
```

column of A (or B) letters connecting a set of means indicates that there is no evidence to indicate a difference between the largest mean in the set and the others. For example, the line of A's indicates that there is no evidence to indicate that mean 3 differs from means 2 and 4, but it does differ from mean 1.

A more common graphical presentation of the results of an HSD comparison of means is shown in Figure 10.9. The four means, shown by dots, are plotted along a horizontal line. Means not judged to be different are connected by a solid line in Figure 10.9. Those judged to be different are not connected. You can see in Figure 10.9 that there is no evidence of differences among means 1, 4, and 2 nor among 4, 2, and 3. There is evidence of a difference between the means for populations 1 and 3. You can see that Figure 10.9 provides essentially the same information as the SAS graphical presentation in Figure 10.8.

FIGURE 10.9 Graph showing how the HSD procedure is applied to the drug potency data

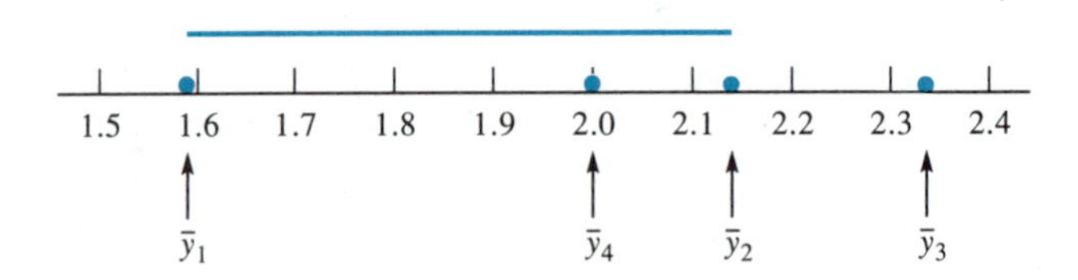

Calculating the Length of the Yardstick *W*

Some printouts do not give the HSD comparison of a set of means and you may want to calculate W and perform the comparisons yourself. The tabulated values of q are shown in Table 7 of Appendix 2, for $a = .10$, .05, and .01. A portion of Table 7a is reproduced in Table 10.9, which gives the tabulated values of q for $a = .10$. For example, you can see that for $a = .10$, $k = 4$, and 16 degrees of freedom, the tabulated value of q (shaded in the table) is 3.52.

To find the value of the yardstick W to compare the $k = 4$ drug potency means of Table 10.1, we need to know, in addition to q, the number r of measurements in each sample and the estimate s obtained from an analysis of variance of the data. Referring to Table 10.1, you can see that the number of measurements in each sample is $r = 5$. The Minitab ANOVA printout in Figure 10.3a gives MSE = .187. Therefore,

$$s = \sqrt{\text{MSE}} = \sqrt{.187} = .432$$

Substituting $q = 3.52$, $s = .432$, and $r = 5$ into the formula for W, we obtain

$$W = \frac{qs}{\sqrt{r}} = \frac{(3.52)(.432)}{\sqrt{5}} = .68$$

TABLE 10.9 A Portion of Table 7a of Appendix 2: Values of q for Use in Tukey's HSD Procedure, $a = .10$

df \ k	2	3	4	5	6	7	8	9	10
1	8.93	13.44	16.36	18.49	20.15	21.51	22.64	23.62	24.48
2	4.13	5.73	6.77	7.54	8.14	8.63	9.05	9.41	9.72
3	3.33	4.47	5.20	5.74	6.16	6.51	6.81	7.06	7.29
4	3.01	3.98	4.59	5.03	5.39	5.68	5.93	6.14	6.33
5	2.85	3.72	4.26	4.66	4.98	5.24	5.46	5.65	5.82
6	2.75	3.56	4.07	4.44	4.73	4.97	5.17	5.34	5.50
7	2.68	3.45	3.93	4.28	4.55	4.78	4.97	5.14	5.28
8	2.63	3.37	3.83	4.17	4.43	4.65	4.83	4.99	5.13
9	2.59	3.32	3.76	4.08	4.34	4.54	4.72	4.87	5.01
10	2.56	3.27	3.70	4.02	4.26	4.47	4.64	4.78	4.91
11	2.54	3.23	3.66	3.96	4.20	4.40	4.57	4.71	4.84
12	2.52	3.20	3.62	3.92	4.16	4.35	4.51	4.65	4.78
13	2.50	3.18	3.59	3.88	4.12	4.30	4.46	4.60	4.72
14	2.49	3.16	3.56	3.85	4.08	4.27	4.42	4.56	4.68
15	2.48	3.14	3.54	3.83	4.05	4.23	4.39	4.52	4.64
16	2.47	3.12	3.52	3.80	4.03	4.21	4.36	4.49	4.61
17	2.46	3.11	3.50	3.78	4.00	4.18	4.33	4.46	4.58
18	2.45	3.10	3.49	3.77	3.98	4.16	4.31	4.44	4.55
19	2.45	3.09	3.47	3.75	3.97	4.14	4.29	4.42	4.53
20	2.44	3.08	3.46	3.74	3.95	4.12	4.27	4.40	4.51
24	2.42	3.05	3.42	3.69	3.90	4.07	4.21	4.34	4.44
30	2.40	3.02	3.39	3.65	3.85	4.02	4.16	4.28	4.38
40	2.38	2.99	3.35	3.60	3.80	3.96	4.10	4.21	4.32
60	2.36	2.96	3.31	3.56	3.75	3.91	4.04	4.16	4.25
120	2.34	2.93	3.28	3.52	3.71	3.86	3.99	4.10	4.19
∞	2.33	2.90	3.24	3.48	3.66	3.81	3.93	4.04	4.13

You can see that this is equal to the value of W obtained on the SAS printout. If your ANOVA printout did not give the HSD comparison of the drug potency means, you could use this value of W and make the comparison yourself.

EXERCISES

10.26 How does Tukey's HSD procedure differ from a Student's t test?

10.27 Suppose that you compared 5 population means and that $\bar{y}_1 = 422$, $\bar{y}_2 = 495$, $\bar{y}_3 = 387$,

$\bar{y}_4 = 405$, $\bar{y}_5 = 369$, and $W = 48$. What does Tukey's HSD procedure tell you about the population means if you have chosen $\alpha = .05$? What type of risk does α pertain to?

10.28 Suppose that you compared 3 population means and that $\bar{y}_1 = 2.44$, $\bar{y}_2 = 2.31$, $\bar{y}_3 = 2.57$, and $W = .15$. What does Tukey's HSD procedure tell you about the population means if you have chosen $\alpha = .10$?

10.29 Suppose that you compared 5 population means based on independent random samples of 4 observations selected from each population and that $s^2 = 28.1$. Find the value of W that you would use for Tukey's HSD procedure. Use $\alpha = .05$.

10.30 Suppose that you compared 4 population means based on independent random samples of 3 observations selected from each population and that $s^2 = 1.44$. Find the value of W that you would use for Tukey's HSD procedure. Use $\alpha = .05$.

10.31 Refer to Exercise 10.25. Explain why Tukey's HSD procedure would or would not be appropriate for comparing the mean percentage of seeds that germinate for the four age groups.

10.7 Experimental Design: Factorial Experiments

A Single-Factor Experiment

Many experiments are conducted to investigate the effect of one or more variables, quantitative or qualitative, on a quantitative variable y. For example, in the drug potency experiment discussed in the preceding sections, the experimenter purposely varied the qualitative variable, procedure, to see whether it caused changes in the quantitative variable, drug potency y. Presumably, if the drug preparation procedure caused drug potency to vary, the experimenter might want to choose the procedure that produced the drug at the maximum potency.

In an experiment of the type described, **the variable that we purposely vary is often called a *factor*** and **the variable affected is called a *response variable*. The setting of a factor is called a factor *level*.** For example, the factor, Procedure, was set at four levels in the drug potency experiment; that is, we observed the response variable y, drug potency, for four different procedures.

As another example, a producer of a liquid floor wax might want to rate the appearance of different surfaces treated with different numbers of applications of wax. The experimenter plans to apply 1, 2, or 3 coats of wax to each of three surfaces, oak, ceramic tile, and quarry tile. The objective of the experiment is to see how the two factors, type of tile (a qualitative variable) and number of applications (a quantitative variable), affect the quantitative ratings (the response variable) produced by a panel of judges.

Multifactor Experiments: The One-at-a-Time Approach

Suppose that a drug is brewed in a vat and that an experimenter wants to determine whether changes in two variables, vat temperature and vat pressure, affect drug potency y. In our new language, the response variable is drug potency. The two factors are temperature and pressure. We want to vary the levels (settings) of the two factors and observe how these changes in levels affect the response variable y. For what combinations of factor levels (i.e., what levels of temperature and pressure) should the response variable, drug potency, be observed?

One way to answer the question is to use the one-at-a-time approach. Using this approach to design, we hold one variable constant and vary the levels of the other. For example, suppose that we want to observe the drug potency at three temperature levels, 80, 90, and 100 degrees Fahrenheit (°F), and the vat pressure at 12, 14, and 16 pounds per square inch (psi). We could hold the pressure at 12 psi and record the drug potency y for each of the three levels of temperature (see Figure 10.10). Suppose that a graph of the potency measurements appears as shown in Figure 10.11; that is, the potency appears to increase as temperature increases.

FIGURE 10.10 Factor level combinations used to investigate the effect of temperature on drug potency: A one-at-a-time approach

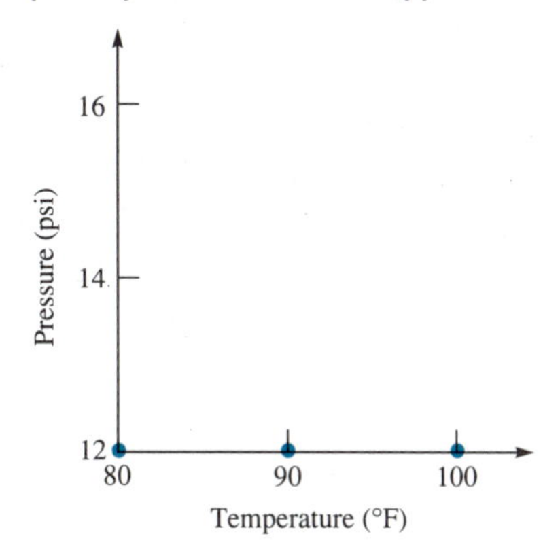

FIGURE 10.11 A graph showing drug potency increasing as vat temperature increases: Pressure = 12 psi

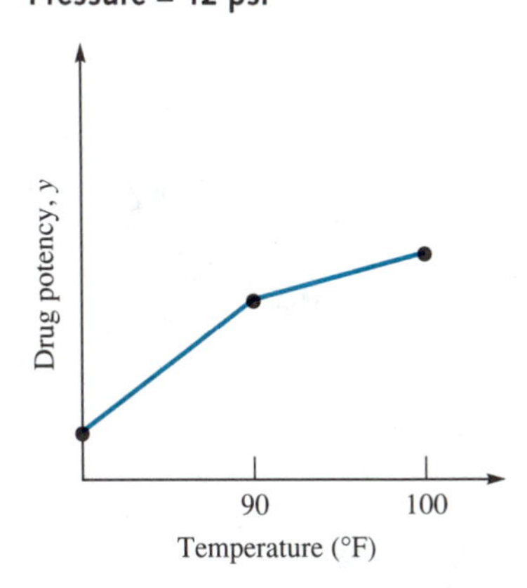

The next step in the one-at-a-time approach is to reverse the process. We hold the temperature constant at 80°F and observe the vat pressure at levels 14 and 16 psi. Adding these new factor level combinations to the ones in Figure 10.10 allows us to observe y for the five factor level combinations shown in Figure 10.12 (page 408). Suppose that for these new observations, drug potency increases as vat pressure increases (see Figure 10.13).

What do the two graphs in Figures 10.11 and 10.13 tell us about the effect of vat temperature and pressure on drug potency? It appears that drug potency can be increased by increasing vat temperature or vat pressure. But in drawing this conclusion, we are making a **big** assumption. We are assuming that the drug potency will also increase as temperature increases for any other level of pressure, say for

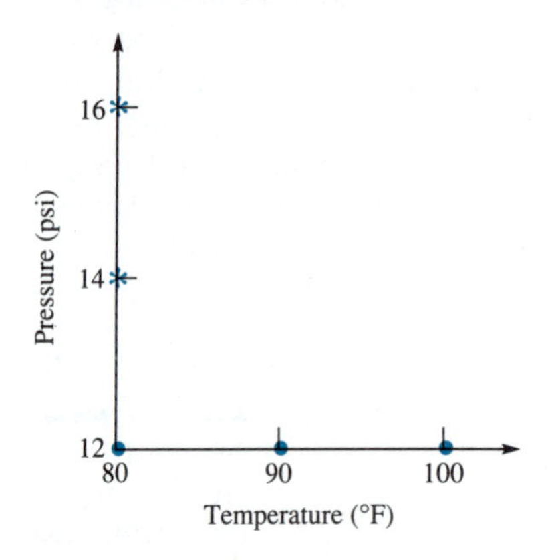

FIGURE 10.12 The treatments added (shown by asterisks) to investigate the effect of pressure on drug potency using the one-at-a-time approach

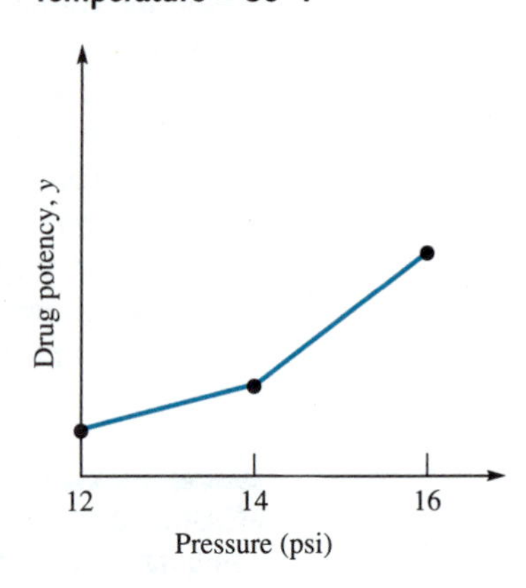

FIGURE 10.13 A graph showing drug potency increasing as vat pressure increases: Temperature = 80 °F

14 and for 16 psi. In other words, we are assuming that the effect of temperature on drug potency is **independent** of the pressure level.

Figure 10.14 shows graphs of drug potency y versus temperature for each of three pressure levels: $P = 12$, $P = 14$, and $P = 16$ pounds per square inch. These graphs depict the situation where the two factors, temperature and pressure, are

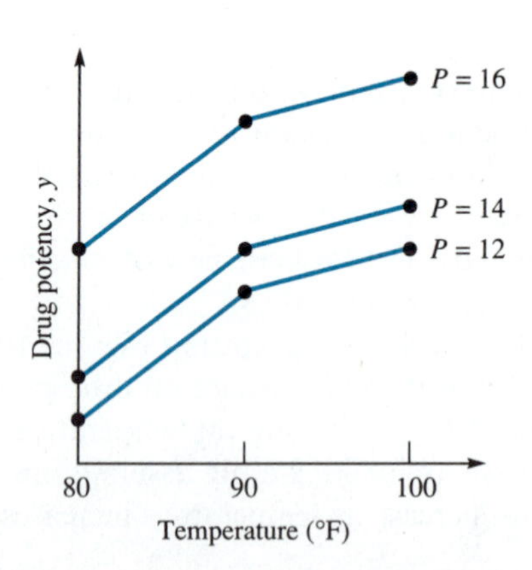

FIGURE 10.14 A graph showing the effect of temperature on drug potency when the effect of temperature is independent of pressure

independent. You can see that the *change* in drug potency, as temperature increases from 80°F to 90°F, is exactly the same, regardless of whether the pressure is 12, 14, or 16 psi. Similarly, the change in drug potency as temperature increases from 90°F to 100°F is the same, regardless of the pressure level. This tells us that the effect of changes in temperature on drug potency y is **independent** of the level of pressure. Summarizing, **two factors are said to be *independent* if changes in the mean value of y for changes in the levels of one factor are the same for all levels of the second factor.**

Factor Interactions

Suppose that when we observed drug potency for all nine combinations of temperature–pressure levels (see Figure 10.15), the plot of the potency measurements did not look like Figure 10.14, but instead looked like Figure 10.16. What does Figure 10.16 tell us about the effect of vat temperature and pressure on drug potency?

FIGURE 10.15 The nine factor level combinations of a 3 x 3 factorial experiment

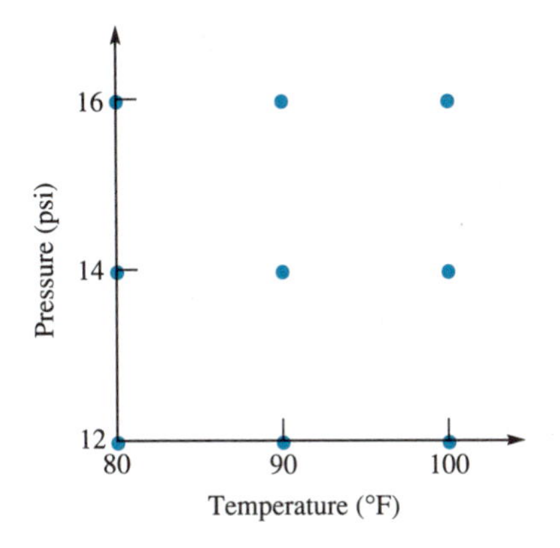

FIGURE 10.16 A graph showing the effect of temperature on drug potency when temperature and pressure interact

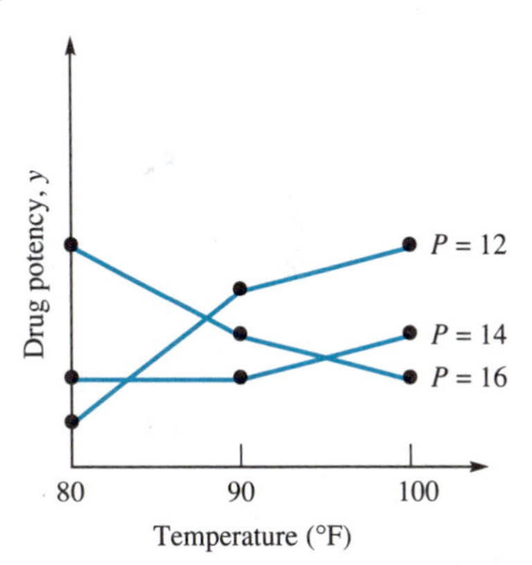

Examining Figure 10.16, you can see that drug potency y increases as temperature increases when the pressure is set at 12 psi, it changes very little as temperature increases when the pressure equals 14 psi, and it falls as temperature increases when the pressure equals 16 psi. The practical conclusions that we derive from Figure 10.16 are completely different from the conclusions based on Figure 10.14. Figure 10.16 tells us that the effect of temperature on drug potency *depends* on the pressure level. **When the effect of one factor on a response variable y depends on the level of a second factor, we say that the two factors *interact*.**

The first thing that we want to learn from a multifactor experiment is whether the factors interact. If the factors interact, we will want to find the particular factor level combination that yields the optimal response (the largest, the smallest, or whatever the "optimum" is for your practical problem). If the factors do not interact, we will want to determine whether and how each factor affects the response variable y independently of the others. **One-at-a-time experimentation will never provide information on factor interaction. Factorial experiments do.**

Factorial Experiments

In order to detect factor interaction, we decide which levels we want to investigate for each factor. Then we select the treatments—i.e., the factor level combinations—that we are going to run in the experiment. **A *complete factorial experiment* is one in which we observe y for every different factor level combination.** A factorial experiment designed to investigate the effects of the two factors, vat temperature and pressure, each at three levels, is called a 3×3 (three-by-three) factorial experiment. It would require that we observe drug potency y for each of the $(3)(3) = 9$ factor level combinations. A 2×3 factorial experiment would involve two factors, one factor at two levels and the other at three levels. Such an experiment would require that we observe y for each of the $(2)(3) = 6$ factor level combinations. A $2 \times 2 \times 2$ factorial experiment would involve three factors, each factor at two levels, and would require that we observe y for each of the $(2)(2)(2) = 8$ factor level combinations.

A *single replication* of a complete factorial experiment is one in which we take a single observation on y for each treatment (i.e., for each factor level combination). Two replications would require that we select two observations per treatment, and k replications would require that we select k observations per treatment.

A factorial experiment is a particular way of selecting the treatments for a multifactor experiment. After we have selected the treatments, we have to decide on how the samples, one for each treatment, will be selected. The most common method is to select independent random samples of equal sample size, one sample for each treatment, but the samples could also be selected in matched sets or by using other more complex sampling procedures.

10.8 The Analysis of Variance for a Two-Factor Factorial Experiment

In this section we will learn how to conduct an analysis of variance for a complete factorial experiment that has been replicated r times and for which the data have been collected according to an independent random samples design.

For example, Table 10.10 gives the data for $r = 2$ replications of a 3×3 factorial experiment. The objective of the experiment was to investigate the effect of two factors, temperature and pressure, on drug potency. Both temperature and pressure are at three levels: temperature at 80, 90, and 100°F, and pressure at 12, 14, and 16 psi.

TABLE 10.10 Data for Two Replications of a Factorial Experiment

		Temperature (°F)		
		80	90	100
Pressure (psi)	16	2.65 2.25	1.39 1.43	1.24 1.02
	14	1.73 1.44	1.55 1.30	1.68 1.47
	12	.86 1.10	1.91 2.07	3.05 2.69

TABLE 10.11 Treatment Means for the Nine Factor Level Combinations

		Temperature, T		
		80	90	100
Pressure, P	16	2.45	1.41	1.13
	14	1.59	1.43	1.58
	12	.98	1.99	2.87

Table 10.11 shows the mean drug potency for each pair of observations for the nine treatments (factor level combinations). Figure 10.17 (page 412) shows how the mean potency varies as the temperature varies for the three temperature settings, 80°, 90°, and 100°, when pressure equals 12, 14, and 16 psi. Examine the graph. What do you think? Does it appear that the factors affect drug potency in an independent manner, that the factors interact, or that the variation in sample means is due just to random variation?

The graph in Figure 10.17 strongly suggests that the factors, temperature and pressure, interact. Drug potency y appears to increase as temperature increases for a pressure $P = 12$ psi, stays relatively constant as temperature increases for $P = 14$ psi, and decreases as temperature increases for $P = 16$ psi. Therefore, it appears that the effect of temperature on the response y depends on the level of pressure. Or, is what we see just random variation in the sample means? To answer this question, we will run an analysis of variance on the data and test for factor interaction.

FIGURE 10.17 A graph showing how the sample mean drug potency varies as temperature varies for P = 12, 14, and 16 psi

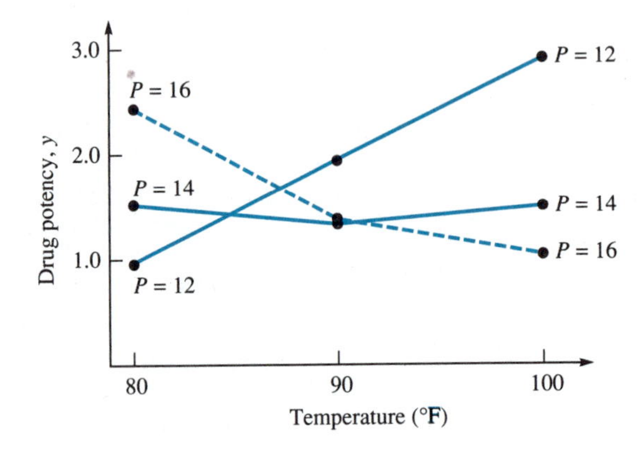

The ANOVA Table for *r* Replications of a Two-Factor Factorial Experiment

The ANOVA table for r replications of a two-factor factorial experiment, with factor C at c levels and factor D at d levels, is shown in Table 10.12. The Minitab printout of the analysis of variance for the drug potency data of Table 10.10 is shown in Figure 10.18.

1. The first column of the ANOVA Table, Table 10.12, titled "Source," tells us that the "Total" variation of the y-values is partitioned into four sources: variation due to factor C, variation due to factor D, variation attributable to the interaction ($C \times D$) of factors C and D, and the remaining variation (Error), which we

TABLE 10.12 ANOVA Table for *r* Replications of a Two-Factor Factorial Experiment: Factor *C* at *c* Levels and Factor *D* at *d* levels

Source	df	SS	MS	F	p-value
Factor C	$c - 1$	SS(C)	MS(C)	MS(C)/MSE	—
Factor D	$d - 1$	SS(D)	MS(D)	MS(D)/MSE	—
$C \times D$	$(c - 1)(d - 1)$	SS($C \times D$)	MS($C \times D$)	MS($C \times D$)/MSE	—
Error	$cd(r - 1)$	SSE	MSE		
Total	$n - 1$	Total SS			

FIGURE 10.18 A Minitab ANOVA printout for the 3 x 3 factorial experiment, Table 10.10

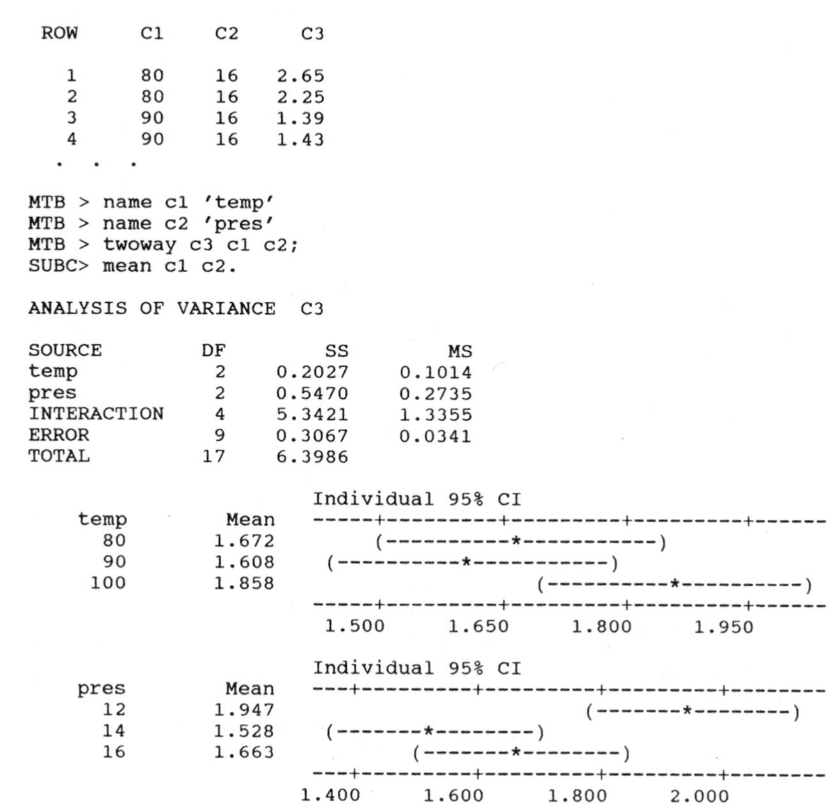

```
ROW     C1     C2      C3

  1     80     16    2.65
  2     80     16    2.25
  3     90     16    1.39
  4     90     16    1.43
 .   .   .

MTB > name c1 'temp'
MTB > name c2 'pres'
MTB > twoway c3 c1 c2;
SUBC> mean c1 c2.

ANALYSIS OF VARIANCE  C3

SOURCE          DF         SS         MS
temp             2     0.2027     0.1014
pres             2     0.5470     0.2735
INTERACTION      4     5.3421     1.3355
ERROR            9     0.3067     0.0341
TOTAL           17     6.3986

                           Individual 95% CI
    temp        Mean    -----+---------+---------+---------+------
      80       1.672         (-----------*-----------)
      90       1.608    (-----------*-----------)
     100       1.858                        (-----------*----------)
                        -----+---------+---------+---------+------
                          1.500     1.650     1.800     1.950

                           Individual 95% CI
    pres        Mean    ---+---------+---------+---------+--------
      12       1.947                              (-------*--------)
      14       1.528      (-------*--------)
      16       1.663             (-------*--------)
                        ---+---------+---------+---------+--------
                         1.400     1.600     1.800     2.000
```

attribute to random variation. For example, the Minitab printout in Figure 10.18, shows that the total variation partitions into four sources: variation attributable to temperature (temp), pressure (pres), the temperature–pressure INTERACTION, and ERROR.

2. The second column of the ANOVA table gives the number of degrees of freedom associated with each source of variation. The number of degrees of freedom for a factor will equal the number of levels for the factor minus 1. Since both of the factors involved in the drug potency experiment are at three levels, the Minitab printout in Figure 10.18 shows 2 degrees of freedom for both temperature and pressure: $(c-1) = 3-1 = 2$ and $(d-1) = 3-1 = 2$.

The number of degrees of freedom for the $C \times D$ interaction will equal $(c-1)(d-1)$. The Minitab printout shows $(c-1)(d-1)=(3-1)(3-1)=2\times 2=4$ degrees of freedom for the temperature–pressure interaction.

The number of degrees of freedom for Error will equal $cd(r-1)$, where c and d are the numbers of levels of factors C and D, respectively, and r is the number of replications of the complete factorial experiment. Therefore, the Minitab printout shows $cd(r-1) = 3(3)(2-1) = 9$ degrees of freedom for error. Note that if you perform only a single replication of a two-factor factorial experiment, the degrees of freedom for error will equal 0, MSE will equal 0, and you will be unable to test for interaction.

The number of degrees of freedom for Total SS, always equal to $(n-1)$, is shown as $(n-1) = (18-1) = 17$ on the Minitab printout. **The number of degrees of freedom for the four sources will always sum to the number $(n-1)$ of degrees of freedom for Total SS;** for the drug potency factorial experiment, $2 + 2 + 4 + 9 = 17$.

3. Column 3 of the ANOVA table gives the sum of squares of deviations for each source of variation. Looking at the Minitab printout, you can see that

$$\begin{aligned} \text{SS(Temperature)} &= .2027 \\ \text{SS(Pressure)} &= .5470 \\ \text{SS(Interaction)} &= 5.3421 \\ \text{SS(Error)} = \text{SSE} &= .3067 \\ \text{Total SS} &= 6.3986 \end{aligned}$$

4. The mean square for each source of variation is shown in column 4 of the ANOVA table. The mean square for a particular source is equal to the sum of squares of deviations for that source divided by its degrees of freedom. For example, the mean square for temperature shown on the Minitab printout is computed as

$$\text{MS(Temperature)} = \frac{\text{SS(Temperature)}}{\text{df}} = \frac{.2027}{2} = .1014$$

The other mean squares shown on the Minitab printout are

$$\begin{aligned} \text{MS(Pressure)} &= .2735 \\ \text{MS(Interaction)} &= 1.3355 \\ \text{MS(Error)} = \text{MSE} &= .0341 \end{aligned}$$

A Test for Factor Interaction

The test for factor interaction is almost identical to the F test of Section 10.4 except that we compare MS(Interaction) to MSE using the F statistic

$$F = \frac{\text{MS(Interaction)}}{\text{MSE}}$$

The numerator and denominator degrees of freedom for any F statistic are always those associated with the numerator and denominator mean squares. For a test for interaction, F will have $(c-1)(d-1)$ numerator and $cd(r-1)$ denominator degrees of freedom. We reject the null hypothesis that the factors do not interact (and conclude that they do interact) for large values of F. The larger the value of F, the greater is the evidence to indicate interaction.

The computed values of the F statistic are often shown in column 5 and the Interaction row of an ANOVA table but they are not shown on the Minitab printout in Figure 10.18. The F statistic needed to test for interaction of temperature and pressure for the drug potency data, Table 10.10, based on 4 numerator and 9 denominator degrees of freedom, is

$$F = \frac{\text{MS(Interaction)}}{\text{MSE}} = \frac{1.3355}{.0341} = 39.2$$

Examining Table 6 of Appendix 2, we find that for 4 numerator and 9 denominator degrees of freedom, $F_{.005} = 7.96$, i.e., the probability that F will exceed 7.96, if the null hypothesis is true, is only .005. Since the observed value, $F = 39.2$, is larger than 7.96, the p-value of the test is less than .005. This tells us that there is strong evidence to indicate a pressure–temperature interaction. The interaction that we thought we saw in Figure 10.17 appears to be real.

Tests for Factor Main Effects

If the factor interaction between two factors, call them C and D (as in Table 10.12), is negligible or nonexistent, we would want to know whether the two factors affect the response y independently of one another. The differences in mean response for changes in the levels of a factor, say factor C, are called the **main effects** for factor C. The degrees of freedom, sum of squares of deviations, and mean square for this source of variation, main effects for factor C, are shown in the C row of Table 10.12. The corresponding main effects information for factor D is shown in the D row of Table 10.12. The test for the main effects for a factor, say factor C, uses the F statistic

$$F = \frac{\text{MS}(C)}{\text{MSE}}$$

where F has $(c-1)$ numerator and $cd(r-1)$ denominator degrees of freedom.

As with the test for interaction, the larger the value of F, the greater is the weight of evidence indicating that changes in the levels of factor C affect the mean value of the response variable y. The test for the main effects for factor D is identical to the test for factor C except that it uses the test statistic

$$F = \frac{\text{MS}(D)}{\text{MSE}}$$

The computed values of these F statistics will usually appear in column 5 of an ANOVA table in their respective Source rows. For the ANOVA table shown in Figure 10.18, the F value to test for the main effects of temperature and pressure are, respectively,

$$F = \frac{\text{MS(Temperature)}}{\text{MSE}} = \frac{.1014}{.0341} = 2.97$$

and

$$F = \frac{\text{MS(Pressure)}}{\text{MSE}} = \frac{.2735}{.0341} = 8.02$$

Both of these F statistics are based on 2 numerator and 9 denominator degrees of freedom. Checking Table 6 of Appendix 2, we find that $F_{.10} = 3.01$, $F_{.05} = 4.26$, $F_{.025} = 5.71$, and $F_{.01} = 8.02$. Since the observed value, $F = 2.97$, for temperature main effects is slightly less than $F_{.10}$, the p-value for the test is close to but slightly larger than .10. A similar test for pressure main effects shows that the observed value of F falls exactly on $F_{.01} = 8.02$. Therefore, the p-value for that test is exactly .01. If we choose $\alpha = .10$, there is no evidence (p-value $> .10$) to indicate that temperature affects mean drug potency. There is evidence (p-value $= .01$) that pressure does. These tests for main effects are not very useful for the drug potency data. Why?

Tests for factor main effects are meaningful if factor interaction is negligible or nonexistent. Since we have already shown that there is evidence of a temperature–pressure interaction for the drug potency data, we already know that both temperature and pressure affect mean potency. Therefore, if we want to choose the temperature–pressure levels that will produce the most desirable mean drug potency, we need to look at the individual means for the various factor level combinations. This is a case where the HSD procedure for comparing and ranking the means is useful.

Comparing Treatment Means

We can estimate or test hypotheses about any individual treatment mean or the difference between a pair of means using the methods of Chapters 6 and 7. The only difference is that we would use the value of s^2, and its associated degrees of freedom, obtained in the analysis of variance. We can also use Tukey's HSD procedure to analyze the data.

For example, now that we know that the drug potency data show an interaction between temperature and pressure, Tukey's HSD method would be a good way to rank the nine treatment means, especially if we want to locate the population with the largest (or the smallest or whatever) treatment mean.

Figure 10.19 shows the SAS printout of the HSD procedure for the factorial experiment in Table 10.10. The treatments are numbered from 1 to 9. The groupings of treatments for which no evidence of differences exist are indicated at the left side of the printout. If we were looking for the treatment giving the highest drug potency, we would select treatment (T = 100°F, P = 12 psi). Although the analysis does not show this treatment mean as different from the mean for treatment (T = 80°F, P = 16 psi), there is evidence to indicate that it differs from the means of the other treatments.

FIGURE 10.19 A SAS HSD analysis of the nine treatment means shown in Table 10.11

```
                 General Linear Models Procedure

        Tukey's Studentized Range (HSD) Test for variable: Y

NOTE: This test controls the type I experimentwise error rate, but
      generally has a higher type II error rate than REGWQ.

             Alpha= 0.05  df= 9  MSE= 0.034083
         Critical Value of Studentized Range= 5.595
           Minimum Significant Difference= 0.7303

Means with the same letter are not significantly different.

        Tukey Grouping              Mean       N  GRP

                        A           2.870      2  7
                        A
                B       A           2.450      2  3
                B
                B       C           1.990      2  4
                        C
                D       C           1.585      2  2
                D       C
                D       C           1.575      2  8
                D       C
                D       C           1.425      2  5
                D       C
                D       C           1.410      2  6
                D
                D                   1.130      2  9
                D
                D                   0.980      2  1
```

OBS	TEMP	PRES	Y	GRP
1	80	16	2.65	3
2	80	16	2.25	3
3	90	16	1.39	6
4	90	16	1.43	6
5	100	16	1.24	9
6	100	16	1.02	9
7	80	14	1.73	2
8	80	14	1.44	2
9	90	14	1.55	5
10	90	14	1.30	5
11	100	14	1.68	8
12	100	14	1.47	8
13	80	12	0.86	1
14	80	12	1.10	1
15	90	12	1.91	4
16	90	12	2.07	4
17	100	12	3.05	7
18	100	12	2.69	7

EXERCISES

10.32 What is a factor?

10.33 What is meant by the *level* of a factor?

10.34 What is a two-factor factorial experiment?

10.35 What is a 3×5 factorial experiment?

10.36 Describe what we mean by three replications of a 2×3 factorial experiment.

10.37 What do we mean by factor interaction?

10.38 Explain why the treatments defined by a factorial experiment are preferred to a one-at-a-time approach when investigating the effects of two or more factors on a response variable y.

10.39 Suppose that you have performed three replications of a 3×5 factorial experiment and that Total SS $= 68.54$, SS$(A) = 4.38$, SS$(B) = 6.04$, and SS$(A \times B) = 25.12$.

a. Give the ANOVA table for the data.
b. Find the values of F to test for main effects for A, main effects for B, and the $A \times B$ interaction. Find the approximate p-value for each test. Choose a value of α that you consider appropriate and complete the test. State your conclusions.
c. Based on your test results, how do the factors affect the mean response?

10.40 Suppose that you have performed two replications of a 3×3 factorial experiment and that Total SS $= 10.76$, SS$(A) = 4.18$, SS$(B) = 1.04$, and SSE $= 3.59$.

a. Give the ANOVA table for the data.
b. Find the values of F to test for main effects for A, main effects for B, and the $A \times B$ interaction. Find the approximate p-value for each test. Choose a value of α that you consider appropriate and complete the test. State your conclusions.
c. Based on your test results, how do the factors affect the mean response?

BUS **10.41** A builder of speculative houses uses one of three designs and assigns the construction of each house to the supervision of one of four foremen. Noticing variation in profit per house, the builder decided to investigate the effect of two factors, house design and foreman, on profit per house. The builder used each foreman as supervisor for each house design and used three houses for each foreman–design combination. The data (in thousands of dollars profit per house) are shown in Table 10.13 and a SAS printout of the ANOVA table for the data is shown in Figure 10.20.

TABLE 10.13 Profit per House as a Function of Foreman and Design

Design	Foreman A_1	A_2	A_3	A_4
B_1	12.8	9.2	11.6	8.7
	9.4	7.8	12.9	7.4
	10.3	10.9	9.6	8.5
B_2	9.2	11.4	8.7	10.3
	7.4	9.6	7.5	10.9
	8.6	8.3	9.0	11.7
B_3	13.7	10.7	10.1	7.3
	12.0	10.2	8.7	8.6
	14.6	11.1	9.1	6.9

a. The twelve treatment means are shown on Figure 10.21. Construct a plot similar to the plot shown in Figure 10.17. Do the factors appear to interact?
b. Do the data provide sufficient evidence to indicate that A and B interact? See Figure 10.20. Give the value of the F statistic. Give the p-value for the test. If you have chosen $\alpha = .05$, what is your conclusion?
c. What are the practical implications of the test results in part **b**?
d. Do the data show evidence of main effects due to either A or B? What is the relevance of these results?
e. Figure 10.21 gives a SAS printout of Tukey's HSD analysis of the profit means for the twelve foreman–design combinations. If the builder

FIGURE 10.20 A SAS printout of the ANOVA table for the profit data of Table 10.13

```
                        Analysis of Variance Procedure

Dependent Variable: PROFIT
                                       Sum of             Mean
Source                    DF          Squares           Square    F Value      Pr > F

Model                     11      91.86972222       8.35179293       6.11      0.0001

Error                     24      32.82000000       1.36750000

Corrected Total           35     124.68972222

                    R-Square             C.V.         Root MSE              PROFIT Mean

                    0.736787         11.86875         1.169402               9.85277778

Source                    DF         Anova SS      Mean Square    F Value      Pr > F

DESIGN                     2       4.60055556       2.30027778       1.68      0.2072
FOREMAN                    3      17.72750000       5.90916667       4.32      0.0143
DESIGN*FOREMAN             6      69.54166667      11.59027778       8.48      0.0001
```

FIGURE 10.21 A SAS printout of the HSD procedure for the means for the twelve design–foreman combinations

```
                  Analysis of Variance Procedure

 Tukey's Studentized Range (HSD) Test for variable: PROFIT

NOTE: This test controls the type I experimentwise error rate, but
      generally has a higher type II error rate than REGWQ.

                Alpha= 0.05  df= 24  MSE= 1.3675
           Critical Value of Studentized Range= 5.099
             Minimum Significant Difference= 3.4427

Means with the same letter are not significantly different.

       Tukey Grouping              Mean       N   GROUP

                    A            13.433       3   9
                    A
            B       A            11.367       3   3
            B       A
            B       A    C       10.967       3   8
            B       A    C
            B       A    C       10.833       3   1
            B       A    C
            B       A    C       10.667       3   10
            B            C
            B            C        9.767       3   6
            B            C
            B            C        9.300       3   11
            B            C
            B            C        9.300       3   2
            B            C
            B            C        8.400       3   5
            B            C
            B            C        8.400       3   7
            B            C
            B            C        8.200       3   4
                         C
                         C        7.600       3   12
```

wants to make money on a house, which design–foreman combination should be chosen to yield the maximum profit?

BUS **10.42** A chain of jewelry stores conducted an experiment to investigate the relationship between price and location on the demand for its diamonds. Six small-town stores were selected for the study as well as six stores located in large suburban city malls. Two stores in each of these locations were assigned to each of three item percentage markups. The percentage gain (or loss) in sales for each store was recorded at the end of 1 month. The data are shown in Table 10.14. The SAS analysis of variance printout for the data is shown in Figure 10.22.

a. Do the data present sufficient evidence to indicate an interaction between markup and location? Give the value of F and the p-value for the test. If you have chosen $\alpha = .05$, what do you conclude?
b. What are the practical implications if the factors, markup and location, interact?
c. Find a 95% confidence interval for the difference in mean change in sales for stores in small towns versus those in suburban malls if the stores are using price markup A_3.

TABLE 10.14 Percentage Gain (or Loss) for Jewelry Stores

Location	Markup A_1	A_2	A_3
Small Towns, B_1	10 4	−3 7	−10 −24
Suburban Malls, B_2	14 18	8 3	−4 3

FIGURE 10.22 A SAS printout of the ANOVA table for the jewelry store data

```
                  Analysis of Variance Procedure

Dependent Variable: PCTGAIN
                                   Sum of             Mean
Source                  DF        Squares           Square    F Value      Pr > F

Model                    5    1200.666667       240.133333       6.83      0.0183

Error                    6     211.000000        35.166667

Corrected Total         11    1411.666667

              R-Square              C.V.          Root MSE           PCTGAIN Mean

              0.850531          273.6992          5.930149             2.16666667

Source                  DF       Anova SS      Mean Square    F Value      Pr > F

LOCATION                 1    280.3333333      280.3333333       7.97      0.0302
MARKUP                   2    835.1666667      417.5833333      11.87      0.0082
MARKUP*LOCATION          2     85.1666667       42.5833333       1.21      0.3616
```

GEN **10.43** Helena F. Barsam and Zita M. Simutis, at the Army Research Institute for the Behavioral and Social Sciences, conducted a study to determine the effect of two factors on terrain visualization for soldiers. During the training programs, participants viewed contour maps of various terrains and then were permitted to view a computer reconstruction of the terrain as it would appear from a specified angle. The two factors investigated in the experiment were the participants' spatial abilities (abilities to visualize in three dimensions) and the viewing procedure, active or passive. Active participation permitted participants to view the computer-generated reconstructions of the terrain from any angle. Passive participation gave the participants a set of preselected reconstructions of the terrain. Participants were tested for spatial ability and, from the test scores, 20 were categorized as possessing high spatial ability, 20 medium, and 20 low. Then 10 participants within each of these groups were assigned to each of the two training modes, active and passive. Each group was then tested to evaluate its terrain visualization ability. Table 10.15 shows the six treatment means computed by Barsam and Simutis, and Figure 10.23 shows the printout of their ANOVA table. Note that the Error source in their ANOVA table is called "Within cells." This is because the Error source in a replicated factorial experiment is "within-sample" variation—i.e., variation within the r observations assigned to each treatment.

TABLE 10.15 The Mean Visualization Score for Six Spatial Ability–Training Condition Combinations

	Training Condition	
Spatial Ability	Active	Passive
High	17.895	9.508
Medium	5.031	5.648
Low	1.728	1.610

a. Explain how the authors arrived at the degrees of freedom shown in their ANOVA table.

b. Show that the computed values of F are correct.

c. Plot the mean terrain visualization scores for the data of Table 10.15. What do you think that they tell us about the two training procedures?

d. Find the p-value for the test for factor interaction. For $\alpha = .05$, what do you conclude? Do the test results support your answer to part **c**?

e. Find the p-value for the test for training condition main effects. For $\alpha = .05$, what do you conclude? Do the test results support your answer to part **c**?

f. Calculate W and use the HSD procedure to compare the six treatment means.

g. Based on the results of the analysis of variance and the HSD procedure, which training condition would you recommend that the U.S. Army use?

FIGURE 10.23 ANOVA table for the terrain visualization data

Source	df	MS	Error df	F	p
Main effects:					
Training condition	1	103.7009	54	3.66	.061
Ability	2	760.5889	54	26.87	.0005
Interaction:					
Training condition × ability	2	124.9905	54	4.42	.017
Within cells	54	28.3015			

10.9 Other Analyses of Variance

There are many other experimental designs that can be used to perform multifactor experiments and there are many different designs for assigning the treatments to the experimental units. Cookbook formulas to calculate the sums of squares of deviations and perform an analysis of variance are available for many of these designs. They are also options in many statistical computer software packages.

Statistical design and the analysis of experiments is a broad subject. For additional information, see the references at the end of this chapter.

10.10 Key Words and Concepts

- The objective of this chapter is to investigate the effects of one or more ***factors*** on a quantitative ***response variable*** y.
- The factors may be either quantitative or qualitative, or both.
- We assume that the data have been collected according to an experimental design. The level x of a factor is purposely varied. If a change in x results in a change in the mean value of y, we can deduce that the change in x *caused* y to change. This differs from many regression analyses where the independent variables, $x_1, x_2, \ldots, x_k$, are simply observed, rather than controlled, and where correlation does not imply a causal relationship.
- The design of an experiment involves three steps: (1) choosing the factors to be included in the experiment, (2) deciding on the factor levels and factor level combinations for which y is to be observed, and (3) deciding how to assign the factor level combinations (i.e., the treatments) to the elements in a sample.
- A factorial experiment is a plan for selecting the factor level combinations to be included in an experiment (i.e., step 3 in the design process). A ***complete factorial experiment*** is one in which we observe y for every different factor level combination.
- A ***single replication*** of a complete factorial experiment is one in which we take a single observation on y for each treatment (i.e., for each factor level combination).
- We assume in this chapter that all data have been collected using an ***independent random samples design***.
- If a design is appropriate, the data can be analyzed using an ***analysis of variance***.
- An analysis of variance attempts to explain why the data vary. Some variation is explainable. It arises because we have purposely varied the levels of one or more factors. The remainder—the unexplainable portion due to many unknown and uncontrolled variables—is called random error.
- An analysis of variance partitions Total SS into sums of squares of deviations, one

for each explainable source (***factor main effects and interactions***) and one for random error. A mean square for each source is then calculated by dividing the source sum of squares by its degrees of freedom. We then compare the mean square for each explainable source against the mean square for error using an F test. Large values of F indicate that a change in the source variable produces a change in the mean value of y.

- We presented analyses of variance for two types of experiments. The first involved the comparison of a number, say k, of population means using an independent random samples design. The second was for a replicated factorial experiment. The samples for this experiment (one sample for each factor level combination) were collected using an independent samples design.
- ***Tukey's HSD procedure*** enables us to test for differences between any and all pairs of means for samples of equal size collected according to an independent random samples design. The probability of incorrectly concluding that *at least one pair* of means differ, when in fact they are equal, is α. Note that the value of α for a Student's t test applies to only one comparison. In contrast, the value of α for the HSD procedure covers all of the test comparisons that we choose to make.

SUPPLEMENTARY EXERCISES

GEN **10.44** In Exercise 7.42, we presented data on the mean number of rat markings when exposed to three different odor sources (Birke, L. I. A. and Saddler, D., "Scent Marking Behavior to Conspecific Odors by the Rat, *Rattus norvegicus*," *Animal Behavior*, Vol. 32, 1984). Ten male rats were exposed to each of the odor environments: no odor, a female odor, and a male odor. The mean number of rat markings and their standard errors for the three samples are shown in Table 10.16.

TABLE 10.16 Mean Number (and Standard Error of the Mean) of Rat Markings for Three Odor Conditions

Odor Source		
None (Control)	Female	Male
9.0 (1.1)	19.4 (2.85)	24.2 (3.40)

a. An analysis of variance for Birke and Saddler's data would show SST = 1,207.4667 and SSE = 1,880.3250. Construct an ANOVA table for the data.

b. Does the mean number of markings vary depending on the odor source? Find the approximate p-value for the test. Test using $\alpha = .05$.

c. Calculate W and employ the HSD procedure to compare the two treatment means with the mean of the control. What do you conclude concerning the frequency of marking and the presence of an odor source?

d. The data suggest that one of the assumptions required for validity of the analysis of variance F test may not be satisfied. Which assumption is it?

MFG **10.45** What factors affect the production (number of items per shift) in a manufacturing plant? The choice of the foreman to supervise production? The shift (day, evening, or late) during which time production is occurring? To answer this question, a plant superintendent decided to conduct a 2×3 factorial experiment. He recorded the number y of items per shift for each of two foremen assigned to supervise each of the three (day, evening, or late) shifts. Three production measurements were recorded for each of the six foreman–shift combinations. The data are shown in Table 10.17.

a. Figure 10.24 shows a SAS ANOVA printout for the data. Find the F-value to test for each source of variation.

b. Test for each factor and interaction source. Find the approximate p-value for each test. Then choose an appropriate value for α and state your conclusions.

c. The six treatment means are shown in Table 10.18. Use the HSD procedure to compare the means.

d. Based on your answers to parts **b** and **c**, how should the superintendent assign foremen to shifts?

TABLE 10.17 Production per Shift for a 2 x 3 Factorial Experiment

Foreman (Factor A)	Shift (Factor B) B_1 8 A.M. to 4 P.M.	B_2 4 P.M. to 12 A.M.	B_3 12 A.M. to 8 A.M.
A_1	570	480	470
	610	475	430
	625	540	450
A_2	480	625	630
	515	600	680
	465	580	660

FIGURE 10.24 A SAS ANOVA printout for the productivity data, Table 10.17

```
                     Analysis of Variance Procedure

Dependent Variable: OUTPUT
                                     Sum of              Mean
Source                   DF         Squares            Square    F Value       Pr > F

Model                     5     100179.1667        20035.8333      27.85       0.0001

Error                    12       8633.3333          719.4444

Corrected Total          17     108812.5000

                   R-Square            C.V.          Root MSE            OUTPUT Mean

                   0.920659        4.884212          26.82246             549.166667


Source                   DF         Anova SS      Mean Square    F Value       Pr > F

FOREMAN                   1      19012.50000      19012.50000      26.43       0.0002
SHIFT                     2        258.33333        129.16667       0.18       0.8379
FOREMAN*SHIFT             2      80908.33333      40454.16667      56.23       0.0001
```

TABLE 10.18 Sample Means for the 2 x 3 Factorial Experiment

Foreman (Factor A)	Shift (Factor B) B_1 8 A.M. to 4 P.M.	B_2 4 P.M. to 12 A.M.	B_3 12 A.M. to 8 A.M.
A_1	601.67	498.33	450.00
A_2	486.67	601.67	656.67

EXERCISES FOR YOUR COMPUTER

GEN **10.46** Water samples were taken at four different locations in a river to determine whether the quantity of dissolved oxygen, a measure of water pollution, varied from one location to another. Locations 1 and 2 were selected above an industrial plant, one near the shore and the other in midstream; location 3 was adjacent to the industrial water discharge for the plant, and location 4 was slightly downriver in midstream. Five water specimens were randomly selected at each location, but one specimen from location 4 was lost in the laboratory. The data are shown in Table 10.19 (the greater the pollution, the lower the dissolved oxygen readings will be).

a. Describe the populations of dissolved oxygen content measurements that we wish to compare. What do the population means measure? What is the difference between a sample mean and a population mean?

b. Calculate the sample means. Plot them on graph paper. Do you think that the data suggest differences among the population means?

c. Use your computer to perform an analysis of variance on the data.

d. Do the data provide sufficient evidence to indicate differences among the population means? Find the value of F for the test. Find the p-value (exact or approximate) for the test. If you have chosen $\alpha = .05$, what do you conclude?

e. Use the HSD procedure to test for differences between pairs of means. What do your F test and the HSD analysis tell you about dissolved oxygen content in the river?

BUS **10.47** A large food products company conducted an experiment to investigate the effects of two factors, package wrapper material and color of wrapper, on sales of one of the company's products. Two types of wrapping material were employed, a waxed paper and a plastic, in three colors. Eighteen supermarkets were chosen for the experiment and

TABLE 10.19 Mean Dissolved Oxygen Content at Four Locations in a River

Location	Mean Dissolved Oxygen Content				
1	5.9	6.1	6.3	6.1	6.0
2	6.3	6.6	6.4	6.4	6.5
3	4.8	4.3	5.0	4.7	5.1
4	6.0	6.2	6.1	5.8	

three were assigned to each of the six factor level combinations. After the product had been in the supermarkets for 1 week, the company recorded the percentage gain (or loss) in weekly sales over each supermarket's average weekly sales of the product over the past year. The data are shown in Table 10.20.

a. Use your computer to produce an ANOVA table for the data.
b. Perform an ANOVA test for each factor and interaction source. Find the approximate p-value for each test. Then choose an appropriate value for α and state your conclusions.
c. Compare the six treatment means using the HSD procedure.
d. Based on your answers to parts **b** and **c**, what package color and wrapping material would you recommend to the manufacturer?

TABLE 10.20 Percentage Change in Weekly Sales

	Package Color		
Wrapping Material	A_1	A_2	A_3
B_1	6	−3	7
	−2	7	3
	4	−2	10
B_2	5	3	12
	2	6	7
	5	4	10

References

Box, G., Hunter, W., and Hunter, S., *Statistics for Experimenters.* New York: Wiley, 1978.

Cochran, W. G. and Cox, G. M. *Experimental Designs*, 2nd edition. New York: Wiley, 1957.

Cox, D. R., *Planning of Experiments.* New York: Wiley, 1958.

Davies, O. L., *The Design and Analysis of Industrial Experiments*, 2nd edition. New York: Hafner, 1956.

Duncan, D. B., "Multiple Range and Multiple *F* Tests," *Biometrics*, Vol. 11, 1955.

Keuls, M., "The Use of the Studentized Range in Connection with an Analysis of Variance," *Euphytica*, Vol. 1., 1952.

Mendenhall, W. and Sincich, T., *Statistics for the Engineering and Computer Sciences*, 3rd edition. San Francisco: Dellen Publishing Co., 1992.

Neter, J., Wasserman, W., and Kutner, M. H., *Applied Linear Statistical Models*, 2nd edition. Homewood, Ill.: Richard D. Irwin, 1985.

Ott, L., *Introduction to Statistical Methods and Data Analysis.* Boston: Duxbury Press, 1978.

Scheaffer, R., Mendenhall, W., and Ott, L., *Elementary Survey Sampling*, 4th edition. Boston: PWS-Kent, 1990.

1 2 4 5 6 7 8 9 10 11

eleven

Nonparametric Statistical Methods

In a Nutshell

What if the assumptions of Chapters 6–10 are not satisfied? The population distributions are not only nonnormal, but they are also highly skewed. Or worse, perhaps we cannot read the exact values of the measurements; we can only rank them. Nonparametric statistical methods may help.

11.1 The Problem

11.2 The Mann–Whitney *U* Test for Independent Random Samples

11.3 The Wilcoxon Signed Ranks Test for a Matched-Pairs Design

11.4 The Kruskal–Wallis *H* Test

11.5 The Spearman Rank Correlation Coefficient

11.6 Key Words and Concepts

11.1 The Problem

Another Look at a Student's *t* Test

In Example 7.5, we used a Student's *t* test to test for a difference in mean permeability between two types of asphalt concrete (see Table 11.1). The test concluded (p-value $< .10$) that there is evidence of a difference in permeability between 3% and 7% asphalt concrete. This conclusion was based on the unsubstantiated assumptions that the sampled populations were normally distributed with a common variance.

TABLE 11.1 Permeability Data for Asphalt Concrete

Asphalt Content	
3%	7%
1,189	853
840	900
1,020	733
980	785

Are the results of a *t* test valid? In most cases they are because the *t* test is a robust test—one that is insensitive to moderate departures from assumptions. If, however, we have reason to suspect that the population distributions are highly skewed or that they possess unequal variances, we can resort to a test that requires few, if any, assumptions about the sampled population distributions. It is called a **nonparametric statistical test.**

How a Nonparametric Test Works

A nonparametric test for the difference in permeability between 3% and 7% concrete does not use the exact values of the observations. Instead, we use the ranks of the measurements. Table 11.2 shows the $n_1 + n_2 = 4 + 4 = 8$ permeability measurements ranked from the smallest (rank 1) to the largest (rank 8).

The logic behind a nonparametric test is similar to the logic underlying a Student's *t* test, except that we compare the rank sums (rather than the averages of the observation ranks). If the two population distributions of permeability measurements are identical (see Figure 11.1a), the eight ranked observations should be randomly distributed between the two samples. Each sample would likely contain some observations with small ranks and some with large ranks, and the rank sums for the two samples would tend to be of the same size. In contrast, if all of the observations with large ranks fall in one sample and all observations with small ranks fall in the other, there

TABLE 11.2 Ranked Permeability Data for Asphalt Concrete

Asphalt Concrete			
3%		7%	
Inches/hour	(rank)	Inches/hour	(rank)
1,189	(8)	853	(4)
840	(3)	900	(5)
1,020	(7)	733	(1)
980	(6)	785	(2)

FIGURE 11.1 Identical and shifted population distributions

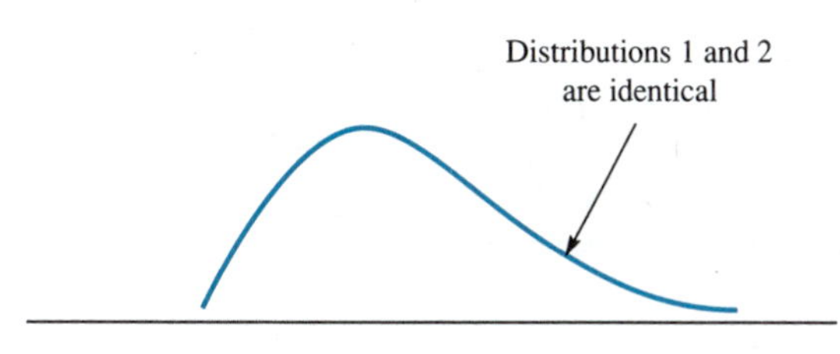

a. The two population distributions are identical.

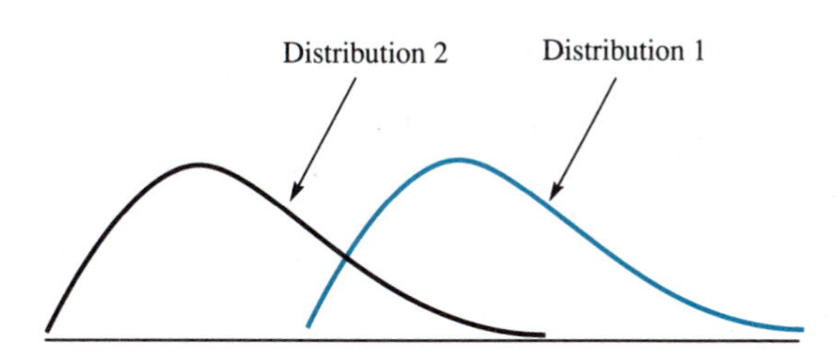

b. The two population distributions are different.

would be strong evidence to indicate that one distribution was shifted to the right or left of the other (see Figure 11.1b).

The Difference Between Parametric and Nonparametric Tests

Now we have two tests for comparing the locations of two population relative frequency distributions based on independent random samples: the Student's t test of

Section 7.3 and the nonparametric test that we have just described. How do they differ?

First, they differ in the assumptions that must be satisfied for the test results to be valid. **Parametric tests** require that we make assumptions about the sampled population relative frequency distributions. Everything about the population distribution(s) is assumed to be known except the values of one or more population parameters. For example, all of the Student t tests that we have discussed are based on the assumption that the sampled populations are normally distributed. The object of the tests is to make inferences about the unknown parameters of the distributions. In contrast, **nonparametric tests** require few, if any, assumptions about the sampled population relative frequency distributions. Second, nonparametric tests differ from parametric tests because they usually use the ranks rather than the exact values of the sample measurements to construct a test statistic.

Nonparametric tests have been devised to answer the same practical questions as were answered by the parametric tests of Chapters 6–10. We will present a few of these tests in the following sections. For more information on nonparametric methods, please see the references.

11.2 The Mann–Whitney *U* Test for Independent Random Samples

The test to detect a shift in the locations of two populations, based on rank sums, was developed by Frank Wilcoxon, a statistician at the Lederle Laboratories (a division of the American Cyanamid Company). A second test was developed by H. B. Mann and D. R. Whitney. The two tests, published independently as the Wilcoxon rank sum test and the Mann–Whitney U test, are algebraically equivalent.

Both tests require that we rank the complete set of $(n_1 + n_2)$ observations, n_1 from sample 1 and n_2 from sample 2, from the smallest (rank 1) to the largest (rank $n_1 + n_2$). We then calculate the rank sums, T_1 and T_2, for the two samples. The Wilcoxon test uses either T_1 or T_2 as the test statistic. The Mann–Whitney U test uses either U_1 or U_2, where

$$U_1 = n_1 n_2 + \frac{n_1(n_1 + 1)}{2} - T_1$$

and

$$U_2 = n_1 n_2 + \frac{n_2(n_2 + 1)}{2} - T_2$$

You can see from these formulas that U_1 will be small when T_1 is large and U_1 will be large when T_1 is small. U_2 bears the same relationships to T_2. Therefore, large values of T_1 (or small values of U_1) will indicate that the relative frequency distribution for population 1 is shifted to the right of the distribution for population 2 (see

FIGURE 11.2 Locations of the population frequency distributions for large values of T and small values of U

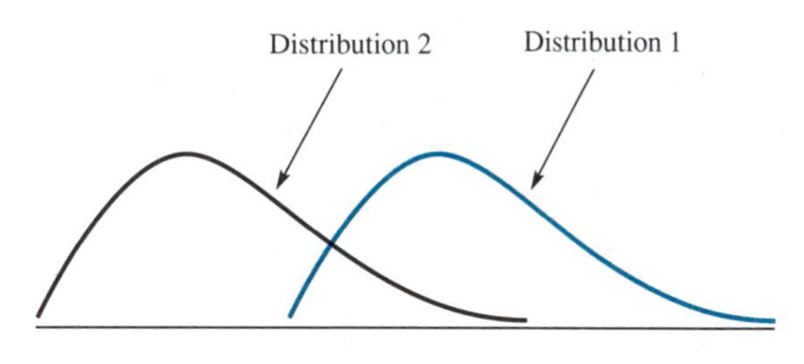

a. T_1 is large; U_1 is small.

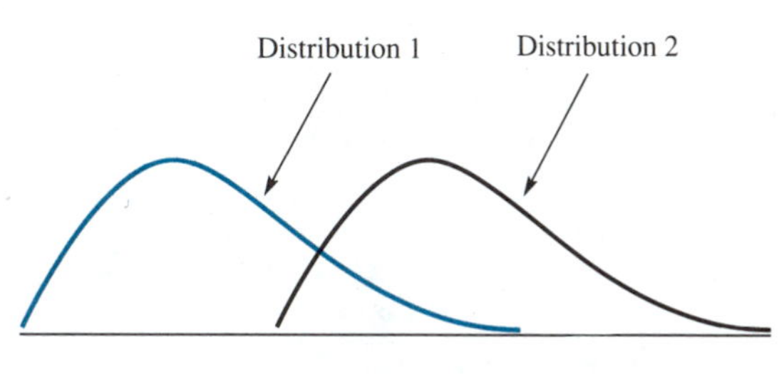

b. T_2 is large; U_2 is small.

Figure 11.2a). Large values of T_2 (or small values of U_2) will indicate that the relative frequency distribution for population 2 is shifted to the right of the distribution for population 1 (see Figure 11.2b).

One of these tests, the Mann–Whitney U test, is summarized in the box and is illustrated in the following example.

The Mann–Whitney U Test for an Independent Random Samples Design

Null hypothesis: **H_0: The relative frequency distributions for populations 1 and 2 are identical.**

Alternative hypothesis:

1. **H_a: Population distribution 1 is shifted to the right of population distribution 2 (a one-tailed test)**

   or

   H_a: Population distribution 2 is shifted to the right of population distribution 1 (a one-tailed test).

(continued)

2. H_a: Population 1 is shifted either to the left or to the right of population distribution 2 (a two-tailed test).

Test statistic: **Rank the $(n_1 + n_2)$ observations in the two samples from the smallest (rank 1) to the largest (rank $n_1 + n_2$) and calculate the rank sums, T_1 and T_2, the sums of the ranks for samples 1 and 2, respectively. Tied observations are assigned ranks equal to the average of the ranks that would have been assigned to the observations if they had not been tied. The test statistic U will be**

For a one-tailed test:

1. ***To detect population*** **1** ***shifted to the right of population*** **2: U will equal**

$$U_1 = n_1 n_2 + \frac{n_1(n_1 + 1)}{2} - T_1$$

2. ***To detect population*** **2** ***shifted to the right of population*** **1: U will equal**

$$U_2 = n_1 n_2 + \frac{n_2(n_2 + 1)}{2} - T_2$$

For a two-tailed test: U is the smaller of U_1 and U_2.

Rejection region:

Choose the value of α that is acceptable to you. Then reject H_0 and accept H_a if the p-value for the test is less than α.

The p-value for the one-tailed test is the probability that U is less than or equal to the observed value of the test statistic. This probability can be determined directly from Table 8 of Appendix 2.*

The p-value for the two-tailed test is double the probability that U is less than or equal to the observed value of the test statistic. This probability can be determined from Table 8 of Appendix 2.*

Example 11.1

Refer to the permeability data of Table 11.1. Use the Mann–Whitney U test to test the null hypothesis that the two population relative frequency distributions are identical, i.e., that there is no difference in the permeability of 3% and 7% asphalt concrete.

a. State the alternative hypothesis for the test.

b. Find the rank sums for samples 1 and 2.

c. Find U_1 and U_2.

d. Give the value of the test statistic.

*The procedure for using Table 8, Appendix 2, is explained in Example 11.1.

e. Find the p-value for the test.

f. What do we conclude if we choose $\alpha = .10$?

Solution

a. The null hypothesis is that the two population relative frequency distributions are identical. The alternative hypothesis is that they are different: either that population distribution 1 is shifted to the right of population distribution 2, or vice versa.

b. The ranks of the eight permeability measurements are shown in parentheses in Table 11.2. Summing the ranks for the two samples, we find the rank sums are

$$T_1 = 8 + 3 + 7 + 6 = 24$$

and

$$T_2 = 4 + 5 + 1 + 2 = 12$$

c. The values of U_1 and U_2 are

$$U_1 = n_1 n_2 + \frac{n_1(n_1 + 1)}{2} - T_1 = 4(4) + \frac{4(4 + 1)}{2} - 24 = 2$$

and

$$U_2 = n_1 n_2 + \frac{n_2(n_2 + 1)}{2} - T_2 = 4(4) + \frac{4(4 + 1)}{2} - 12 = 14$$

d. Since this is a two-tailed test (i.e., we want to determine whether distribution 1 is shifted either to the right or to the left of distribution 2), the test statistic (see the box) is U, the smaller of U_1 and U_2, i.e., $U = 2$.

e. The p-value for the test is twice the probability that U is less than or equal to the observed value of the test statistic, i.e., the probability that $U \leqslant 2$. Table 8 of Appendix 2 gives the probability that U is less than or equal to a value, say U_0, for different combinations of n_1 and n_2. For example, the table for $n_2 = 4$ is reproduced in Table 11.3 on page 434. To find the probability that $U \leqslant 2$, go to the $n_1 = 4$ column of Table 11.3 and move down to the row corresponding to $U_0 = 2$. The probability that $U \leqslant 2$, shaded in Table 11.3, is .0571. Therefore, the p-value for the test, the probability that *either* U_1 or U_2 is less than or equal to 2, is $2(.0571) = .1142$.

f. We have chosen $\alpha = .10$. Therefore, we will reject the null hypothesis if the p-value for the test is less than .10. The p-value for the test, .1142, is not less than .10 (but it is close to it!). Therefore, for $\alpha = .10$, we conclude that there is insufficient evidence to indicate a difference in permeability for 3% and 7% asphalt concrete. Because the sample sizes were small and the p-value was so close to .10, it would probably be desirable to repeat the experiment using larger sample sizes. It is likely that larger samples would reveal a difference in location for the two population distributions if such a difference exists.

TABLE 11.3 A Portion of Table 8 of Appendix 2: Tabulated Values of the Probability That $U \leq U_0$ for $n_2 = 4$

	n_1			
U_0	1	2	3	4
0	.2000	.0667	.0286	.0143
1	.4000	.1333	.0571	.0286
2	.6000	.2667	.1143	.0571
3		.4000	.2000	.1000
4		.6000	.3143	.1714
5			.4286	.2429
6			.5714	.3429
7				.4429
8				.5571

Comparing Results for the Mann–Whitney *U* and the Student's *t* Tests

Now that we have tested for a difference in permeability between 3% and 7% asphalt concrete using both the parametric Student's t test (Example 7.5) and the nonparametric Mann–Whitney U test (Example 11.1), how do the tests compare?

First, note that the two test conclusions appear to disagree. The Student's t test is statistically significant (i.e., it rejects H_0) for $\alpha = .10$. The Mann–Whitney U test does not. The reason for the disagreement identifies one of the major differences between a parametric test and its nonparametric counterpart. If the assumptions of the t test are satisfied, the parametric test (which is based on the actual numerical values of the sample measurements) is more likely to detect a difference in location for the population distributions if, indeed, a difference exists. For the asphalt permeability data, the t test detects a difference; the nonparametric Mann–Whitney U test does not.

Looking at the p-values for the two tests, you can see that the test results are not so different. The p-value for the t test (Example 7.5) was shown to fall between .05 and .10. The p-value for the Mann–Whitney U test, .1142, is only slightly larger, indicating that the Mann–Whitney U test detects slightly less evidence of a difference in permeability than does the t test.

The Student's t test will always be more likely to detect a difference in location between two relative frequency distributions than the Mann–Whitney U test if the assumptions of normality and a common variance are satisfied. In contrast, the Mann–Whitney U test may be more likely to detect a shift in location if the assumptions for the t test are not satisfied.

A Computer Printout for a Mann–Whitney U Test

Example 11.2

Refer to the permeability data of Table 11.1 and Example 11.1. Use the Minitab output for the Mann–Whitney U test to test the null hypothesis that the two population relative frequency distributions of permeability readings are identical, i.e., that there is no difference in the permeability of 3% and 7% asphalt concrete. Interpret the printout.

Solution

Most computer statistical software packages will perform a Mann–Whitney U test or a Wilcoxon rank sum test. To use a package, we enter the data into the computer; the computer ranks the observations and then performs the test. The Minitab printout shown in Figure 11.3 gives a nonparametric confidence interval as well as the Mann–Whitney U test for the difference in location for the population permeability distributions. The test results are shaded in Figure 11.3. The W shown on the printout is T_1, the rank sum for the sample of 3% asphalt measurements. The printout shows that the p-value for the Mann–Whitney test is .1124 and notes that we cannot reject H_0 for $\alpha = .10$.* Thus, for $\alpha = .10$, there is insufficient evidence to indicate a difference in permeability between 3% and 7% asphalt concrete. Note that the t test in Example 7.5 detected a difference in locations for the two distributions. The U test, which is based on less information (ranks instead of the original measurements) than the t test, does not.

FIGURE 11.3 A Minitab printout for a Mann–Whitney U test

```
Mann-Whitney Confidence Interval and Test

three       N =    4      Median =       1000.0
seven       N =    4      Median =        819.0
Point estimate for ETA1-ETA2 is           181.0
97.0 pct c.i. for ETA1-ETA2 is (-60.1,455.9)
W = 24.0
Test of ETA1 = ETA2  vs.  ETA1 n.e. ETA2 is significant at 0.1124

Cannot reject at alpha = 0.10
```

The Mann–Whitney U Test for Large Samples

Table 8 of Appendix 2 gives lower-tail areas for the U distribution for values of n_1 and n_2 less than or equal to 10. When n_1 and n_2 are both equal to 10 or larger, the

*The slight difference between the p-value calculated in Example 11.1 and the value shown on the printout in Figure 11.3 is due to rounding errors.

sampling distribution for U can be approximated by a normal distribution with the following mean and standard deviation:

$$\text{Mean value of } U: \quad \text{Mean}(U) = \frac{n_1 n_2}{2}$$

$$\text{Standard error of } U: \quad \text{SE}(U) = \sqrt{\frac{n_1 n_2 (n_1 + n_2 + 1)}{12}}$$

The following example will show how to find the p-value for a Mann–Whitney U test when n_1 and n_2 are equal to or larger than 10.

Example 11.3

Suppose that $n_1 = n_2 = 10$ and $T_1 = 125$, and we want to conduct a one-tailed test to see whether distribution 1 is shifted to the right of distribution 2.

a. Find the value of the U statistic.

b. Find the p-value for the test using the normal approximation to the U distribution.

c. Find the p-value for the test using Table 8 of Appendix 2. Compare this exact value with the normal approximation obtained in part **b.**

Solution

a. Since we want to detect a shift in distribution 1 to the right of distribution 2, we will reject the null hypothesis that the two distributions are identical for large values of T_1 (i.e., small values of U_1). The test statistic is

$$U_1 = n_1 n_2 + \frac{n_1(n_1 + 1)}{2} - T_1$$

Substituting $n_1 = n_2 = 10$ and $T_1 = 125$ into the formula for U_1, we obtain

$$U_1 = 10(10) + \frac{10(10 + 1)}{2} - 125$$
$$= 30$$

b. The p-value for this one-tailed test is the probability that U is less than or equal to $U_1 = 30$. The normal approximation to the p-value is the area in the lower tail of a normal distribution (see Figure 11.4a) with mean and standard deviation given by

$$\text{Mean}(U): \quad \frac{n_1 n_2}{2} = \frac{10(10)}{2} = 50$$

and

$$\text{SE}(U) = \sqrt{\frac{n_1 n_2 (n_1 + n_2 + 1)}{12}}$$
$$= \sqrt{\frac{10(10)(10 + 10 + 1)}{12}} = \sqrt{175} = 13.23$$

FIGURE 11.4 A normal approximation to the p-value for the test in Example 11.3

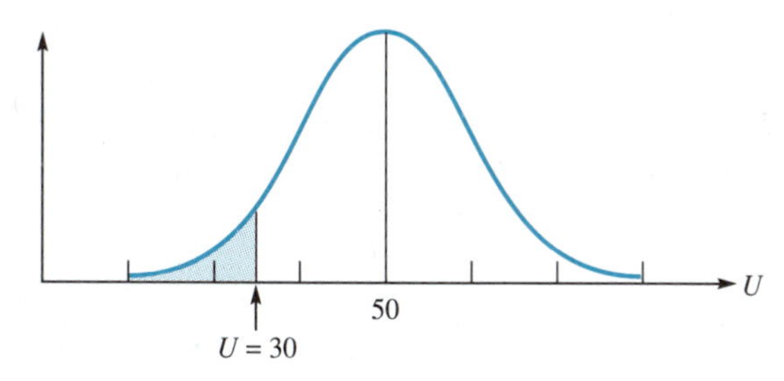

a. The U distribution

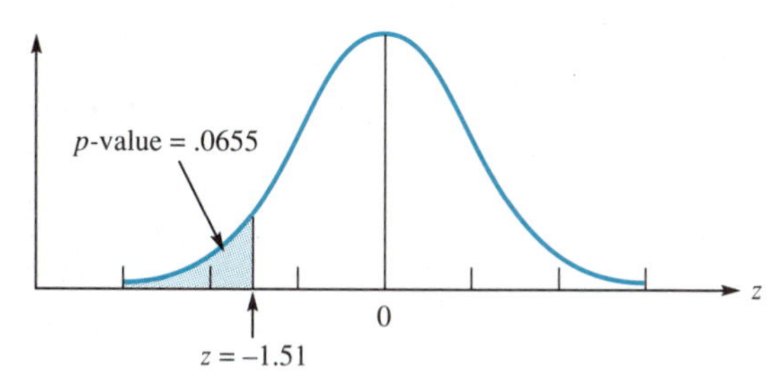

b. The z distribution

To find this lower-tail area, we find the z value corresponding to $U = 30$:

$$z = \frac{U - \text{Mean}(U)}{\text{SE}(U)} = \frac{30 - 50}{13.23} = -1.51$$

The area below $z = -1.51$ is equal to the area above $z = 1.51$ (see Figure 11.4b). The area A between $z = 0$ and $z = 1.51$, given in Table 3 of Appendix 2, is .4345. Since the total area to the right of $z = 0$ is .5, the area in the upper tail above $z = 1.51$ is .0655. **Therefore, the normal approximation to the p-value for the test is .0655.**

c. The exact p-value for the test is the probability that U is less than or equal to 30. This value, given in Table 8 of Appendix 2, is .0716. You can see that the normal approximation, .0655 (obtained in part **b**) is close to the exact p-value. The normal approximation becomes better for larger values of n_1 and n_2.

EXERCISES

11.1 What is the objective of the Student's t test of Section 7.3 and the Mann–Whitney U test of this section?

11.2 Why is the Student's t test called a parametric test, whereas the Mann–Whitney U test is called a nonparametric test?

11.3 In what major respect do the test statistics for the two tests differ?

11.4 Suppose that you wanted to use a Mann–Whitney U test to determine whether the relative frequency distribution for population 1 is shifted to the right of population 2.

a. Would you employ a one- or a two-tailed test? Explain.

b. Give the test statistic for the test. What values (small, large, or both) of the test statistic tend to disagree with the null hypothesis?

c. Suppose that $n_1 = 5$, $n_2 = 7$, and that the observed value of the test statistic was 6. Give the p-value for the test. What does this p-value mean?

d. If you had decided to use $\alpha = .10$, what would be your test conclusion?

11.5 Suppose that you wanted to use a Mann–Whitney U test to determine whether the relative frequency distribution for population 2 is shifted to the right of population 1.

a. Would you employ a one- or a two-tailed test? Explain.

b. Give the test statistic for the test. What values (small, large, or both) of the test statistic tend to disagree with the null hypothesis?

c. Suppose that $n_1 = 7$, $n_2 = 8$, and that the observed value of the test statistic was 15. Give the p-value for the test. What does this p-value mean?

d. If you had decided to use $\alpha = .10$, what would be your test conclusion?

11.6 Suppose that you wanted to use a Mann–Whitney U test to determine whether the relative frequency distribution for population 1 is shifted either to the right or to the left of population 2.

a. Would you employ a one- or a two-tailed test? Explain.

b. Give the test statistic for the test. What values (small, large, or both) of the test statistic tend to disagree with the null hypothesis?

c. Suppose that $n_1 = 6$, $n_2 = 6$, and that the observed value of the test statistic was 7. Give the p-value for the test. What does this p-value mean?

d. If you had decided to use $\alpha = .05$, what would be your test conclusion?

11.7 Table 8 of Appendix 2 gives the probability that $U \leq U_0$ for values of n_1 less than or equal to n_2. For example, the table corresponding to $n_2 = 8$ gives the probability that $U \leq U_0$ for $n_1 = 1, 2, 3, \ldots, 8$. How would you use the table if $n_1 = 8$ and $n_2 = 7$?

BIO **11.8** In Exercises 7.13 and 7.15, we presented data on the wing stroke frequencies of two species of Euglossine bees (Casey, T. M., May, M. L., and Morgan, K. R., "Flight Energetics of Euglossine Bees in Relation to Morphology and Wing Stroke Velocity," *Journal of Experimental Biology,* Vol. 116, 1985). The data, along with their ranks, are reproduced in

TABLE 11.4 Wing Stroke Frequencies (in hertz)

Sample 1 (*Euglossa mandibularis Friese*)	Sample 2 (*Euglossa imperialis Cockerell*)
235 (10)	180 (3.5)
225 (9)	169 (1)
190 (8)	180 (3.5)
188 (7)	185 (6)
	178 (2)
	182 (5)

Table 11.4. Do the data present sufficient evidence to indicate a difference in location between the two population distributions of wing stroke frequency?

a. State the null and alternative hypotheses.

b. Figure 11.5 gives the Minitab printout of the Mann–Whitney U test. Interpret the output.

FIGURE 11.5 A Minitab printout of the Mann–Whitney *U* test for the wing stroke frequency data of Exercise 11.8

```
Mann-Whitney Confidence Interval and Test

friese       N =   4     Median =        207.50
cockerell    N =   6     Median =        180.00
Point estimate for ETA1-ETA2 is           30.50
95.7 pct c.i. for ETA1-ETA2 is (5.99,56.01)
W = 34.0
Test of ETA1 = ETA2  vs.  ETA1 n.e. ETA2 is significant at 0.0142
The test is significant at 0.0139 (adjusted for ties)
```

11.9 Refer to Exercise 11.8. Suppose that you did not have a statistical software package that included the Mann–Whitney U test. Perform the test.

a. Find the value of the test statistic.

b. Find the p-value for the test.

c. If you have chosen $\alpha = .10$, what do you conclude?

d. Compare your test results with the printout shown in Exercise 11.8.

e. Why might you prefer to use the U test rather than the t test of Exercise 7.15 to test for a difference in locations for the two population distributions of wing stroke frequencies?

f. How does the result for the U test compare with the result for the t test of Exercise 7.15? Are the p-values for the tests comparable?

GEN **11.10** Air pollution, emanating from urban areas, may be spreading and creating even heavier pollution in rural areas. Data collected on air pollution at eight Ohio locations are shown in Table 11.5, which gives

TABLE 11.5 Air Pollution Data Collected at $n_1 = 3$ Rural and $n_2 = 5$ Urban Locations

City	Designation	Maximum concentration (ppm)	Days exceeding standard (%)	Number of violations (total)
Canton	Urban	.14	44	148
Cincinnati	Urban	.18	44	54
Cleveland	Urban	.14	26	51
Columbus	Urban	.15	27	113
Dayton	Urban	.13	35	114
McConnelsville	Rural	.16	56	239
Wilmington	Rural	.18	58	259
Wooster	Rural	.17	55	262

Source: Environment Midwest, December 1976.

the maximum amount of ozone recorded at each of eight Ohio air pollution monitoring stations as well as the number of violations of the maximum allowable limit at each location recorded over the period from June 14 to August 31, 1974. Do the data provide sufficient evidence to indicate that levels of the "number of violations" for rural locations tend to be higher than for urban locations? Test using a Mann–Whitney U test.

a. State the null and alternative hypotheses.

b. Figure 11.6 gives the Minitab printout of the Mann–Whitney U test. Interpret the output.

FIGURE 11.6 A Minitab printout of the Mann–Whitney *U* test for the air pollution data of Exercise 11.10

```
MTB > set c5
DATA> 239 259 262
DATA> end
MTB > name c5 'rural'
MTB > set c6
DATA> 148 54 51 113 114
DATA> end
MTB > name c6 'urban'
MTB > mann whitney c5 c6;
SUBC> alternative=+1.

Mann-Whitney Confidence Interval and Test

rural       N =   3     Median =          259.00
urban       N =   5     Median =          113.00
Point estimate for ETA1-ETA2 is           148.00
96.3 pct c.i. for ETA1-ETA2 is (90.98,210.98)
W = 21.0
Test of ETA1 = ETA2  vs.  ETA1 g.t. ETA2 is significant at 0.0184
```

11.11 Refer to Exercise 11.10. Suppose that you did not have a statistical software package that included the Mann–Whitney U test. Perform the test.

a. Find the value of the test statistic.

b. Find the p-value for the test.

c. If you have chosen $\alpha = .10$, what do you conclude?

d. Compare your test results with the printout given in Exercise 11.10.

11.12 How can the normal probability distribution be of assistance in conducting a Mann–Whitney U test?

MED **11.13** Cancer treatment with chemicals—chemotherapy—utilizes chemicals that kill both cancerous and normal cells. In some instances, the toxic effect of chemotherapy on noncancerous cells can be reduced by the simultaneous injection of a second drug, an antitoxin. A test was conducted to compare the mean survival time of rats treated only with chemotherapy and those treated with the chemotherapy plus an antitoxin. Twenty-four rats were randomly assigned, twelve rats to each of the two treatments, and the length of survival time was recorded for each rat. The data are shown in Table 11.6. Do the data provide sufficient evidence to indicate that rats receiving the antitoxin tend to survive longer after chemotherapy than those not receiving the antitoxin? Use the Mann–Whitney U test with $\alpha = .05$.

a. State the null and alternative hypotheses.

b. Identify the test statistic and find its value.

c. Find the p-value for the test and interpret it.

d. If you have chosen $\alpha = .05$ for the test, what do you conclude?

TABLE 11.6 Survival Time (in hours) for Rats on Two Treatments

Chemotherapy only A		Chemotherapy plus drug B	
84	76	140	480
128	104	184	244
168	72	368	440
92	180	96	380
184	144	480	480
92	120	188	196

11.14 Samples of $n_1 = n_2 = 10$ measurements were randomly and independently selected from two populations to determine whether the two distributions differ in location. The sample rank sums were computed to be $T_1 = 84$ and $T_2 = 126$. Suppose we want to know whether distribution 2 is shifted to the right of distribution 1.

a. Find the value of the test statistic.
b. Find the p-value for the test using Table 8 of Appendix 2.
c. Find the approximate p-value for the test using the normal approximation to the distribution.
d. Based on the p-value for the test, what would you conclude?

11.15 Samples of $n_1 = n_2 = 10$ measurements were randomly and independently selected from two populations to determine whether the two distributions differ in location. The sample rank sums were computed to be $T_1 = 82$ and $T_2 = 128$. Suppose we want to know whether distribution 2 is shifted either to the right or to the left of distribution 1.

a. Find the value of the test statistic.
b. Find the p-value for the test using Table 8 of Appendix 2.
c. Find the approximate p-value for the test using the normal approximation to the distribution.
d. Based on the p-value for the test, what would you conclude?

11.3 The Wilcoxon Signed Ranks Test for a Matched-Pairs Design

Recall that we can often obtain more information on the difference in two population means by using a matched-pairs design (Section 7.2). The parametric t test for comparing two population means is based on an analysis of the differences of the matched pairs.

Frank Wilcoxon proposed a nonparametric test to detect a difference in location between two populations based on the signs and the ranks of the differences. The logic behind his test can be seen by examining the matched-pairs data of Table 7.5. For that experiment, twenty 6-year-old children were matched on IQ to form $n = 10$ matched pairs. One member of each pair was randomly selected and assigned to a new teaching method, Method 1. The other member of the pair was assigned to the teaching method currently in use, Method 2. The student achievement test scores for the two methods of teaching are reproduced in Table 11.7. The fifth column of Table 11.7 gives the ranks (in parentheses) of the absolute (i.e., unsigned) values of the ten differences.

TABLE 11.7 Achievement Test Scores for a Matched-Pairs Design

Pair	Method 1	Method 2	Difference d	Rank of d
1	78	69	9	(5)
2	63	56	7	(4)
3	95	77	18	(9)
4	75	62	13	(6.5)
5	65	60	5	(3)
6	79	59	20	(10)
7	82	85	−3	(2)
8	85	72	13	(6.5)
9	67	51	16	(8)
10	72	71	1	(1)
Sample means	76.10	66.20	9.90	

The Logic Behind the Wilcoxon Signed Ranks Test

Wilcoxon reasoned that if the two population distributions were identical (i.e., the null hypothesis is true), the numbers of positive and negative differences should be approximately equal. Further, the large and small differences should be randomly distributed between negative and positive. In contrast, the greater the difference in location for the two distributions, the larger will be the number of differences with the same sign and the larger will be the difference in the rank sum of the negative and the rank sum of the positive differences.

An examination of the ranks of the positive and the negative differences in Table 11.7 illustrates Wilcoxon's reasoning. All but one of the differences is positive (i.e., the score for Method 1 was higher than for Method 2) and the rank sum $T^+ = 53$ for the positive differences is very large relative to the rank sum $T^- = 2$ for the single negative difference. It does seem improbable (doesn't it?) that the difference in positive and negative rank sums would be so large if, in fact, there is no difference in the distributions of scores obtained for the two teaching methods.

Wilcoxon calculated the probabilities associated with various positive and negative rank sums. His test is summarized in the box and its use is illustrated in the following example.

The Wilcoxon Signed Ranks Test for a Matched-Pairs Design

Null hypothesis: **H_0: The relative frequency distributions for populations 1 and 2 are identical.**

Alternative hypothesis:

1. H_a: Population distribution 1 is shifted to the right of population 2 (a one-tailed test)

or

H_a: Population distribution 2 is shifted to the right of population 1 (a one-tailed test)

2. H_a: Population 1 is shifted either to the left or to the right of population 2 (a two-tailed test).

Test statistic: **Rank the absolute values of the n differences from the smallest (rank 1) to the largest (rank n). Differences equal to 0 are deleted and the number of differences is reduced accordingly. Tied differences are assigned ranks equal to the average of the ranks that would have been assigned to the differences if they had not been tied. Then calculate the rank sums, T^+ and T^-, the sums of the ranks for the positive and for the negative differences, respectively. The test statistic is**

For a one-tailed test:

1. To detect population 1 shifted to the right of population 2: Use T^-.

2. To detect population 2 shifted to the right of population 1: Use T^+.

For a two-tailed test: Use the smaller of T^+ and T^-.

Rejection region: **Choose the value of α that is acceptable to you. Then reject H_0 and accept H_a if the test statistic is less than or equal to the tabulated value for T shown in Table 9 of Appendix 2.***

Example 11.4

Do the data given in Table 11.7 provide sufficient evidence to indicate that teaching Method 1 (the new method) tends to produce higher test scores than the test scores for teaching Method 2? Test using the Wilcoxon signed ranks test.

a. State the alternative hypothesis.

b. Find the value of the test statistic.

c. Use Table 9 of Appendix 2 to find the values of the test statistic that fall in the rejection region for $\alpha = .05$. State your conclusions.

d. Give the approximate p-value for the test.

*The procedure for using Table 9 of Appendix 2 is explained in Example 11.4.

Solution

a. We want to determine whether teaching Method 1 is "better"—i.e., tends to produce higher test scores than teaching Method 1. Therefore, the alternative hypothesis is that the population distribution of test scores for Method 1 is shifted to the right of the distribution of scores for Method 2.

b. The alternative hypothesis stated in part **a** implies a one-tailed test. The test statistic is T^-, the rank sum of the negative differences. We found (see Table 11.7) that $T^- = 2$.

c. The rejection region for the Wilcoxon signed ranks test always includes small values—say, values less than a number, call it T_0. The values of T_0 are given in Table 9 of Appendix 2 that locate different values of the tail area a.

A portion of Table 9 of Appendix 2 is shown in Table 11.8. Since our test is based on $n = 10$ differences, the values of T_0 are given in the column of Table 11.8 corresponding to $n = 10$. To decide which T_0 value in the column to choose, see the values of a at the left side of the table. The first column, identified as **one-sided,** gives the values of a for a one-tailed test. The next column to the right, identified as **two-sided,** gives values of a for a two-tailed test. Since we want to conduct a one-tailed test with $\alpha = .05$, T_0 is given in the first row of Table 11.8 and the $n = 10$ column: $T_0 = 11$. Therefore, we will reject H_0 and conclude that distribution 1 is shifted to the right of distribution 2 when the test statistic, T^-, is less than or equal to 11. Since the observed value of T^- is 2, we reject H_0 and conclude that teaching Method 1 tends to produce higher test scores than teaching Method 2.

d. We cannot get the exact p-value for a test using Wilcoxon's tables but we can approximate it. We know (from Table 9 of Appendix 2) that we will reject H_0 for

TABLE 11.8 A Portion of Table 9 of Appendix 2

One-sided	Two-sided	$n = 5$	$n = 6$	$n = 7$	$n = 8$	$n = 9$	$n = 10$
$a = .05$	$a = .10$	1	2	4	6	8	11
$a = .025$	$a = .05$		1	2	4	6	8
$a = .01$	$a = .02$			0	2	3	5
$a = .005$	$a = .01$				0	2	3

One-sided	Two-sided	$n = 11$	$n = 12$	$n = 13$	$n = 14$	$n = 15$	$n = 16$
$a = .05$	$a = .10$	14	17	21	26	30	36
$a = .025$	$a = .05$	11	14	17	21	25	30
$a = .01$	$a = .02$	7	10	13	16	20	24
$a = .005$	$a = .01$	5	7	10	13	16	19

$\alpha = .05$ if $T^- = 11$ or less, and we will reject H_0 for $\alpha = .025$ if $T^- = 8$ or less. Therefore, if $T^- = 11$, 10, or 9, we know that the p-value for the test is between .025 and .05. By similar reasoning, if $T^- = 8$, 7, or 6, the p-value is between .01 and .025. Table 9 also tells us that we should reject H_0 for $\alpha = .005$ if T^- is equal to or less than 3. Since the observed value of T^- for our test is 2, the p-value for the test is less than .005. This is certainly strong evidence to indicate that Method 1 tends to produce higher test scores than Method 2.

A Computer Printout for a Wilcoxon Signed Ranks Test

Example 11.5

The Minitab printout for the Wilcoxon signed ranks test for the data of Example 11.4 is shown in Figure 11.7. Interpret the printout.

FIGURE 11.7 A Minitab printout of the Wilcoxon signed ranks test for the teaching methods data of Example 11.4

```
MTB > set c7
DATA> 78 63 95 75 65 79 82 85 67 72
DATA> end
MTB > set c8
DATA> 69 56 77 62 60 59 85 72 51 71
DATA> end
MTB > subtact c8 c7 c9
MTB > name c9 'diff'
MTB > wtest c9;
SUBC> alternative k=+1.

TEST OF MEDIAN = 0.000000 VERSUS MEDIAN G.T. 0.000000

                N FOR   WILCOXON              ESTIMATED
            N   TEST   STATISTIC   P-VALUE      MEDIAN
diff       10     10        53.0     0.005       10.00
```

Solution

The relevant portion of the Wilcoxon signed ranks test is shaded in Figure 11.7. The statistic shown on the printout will always be T^+, the rank sum of the positive differences. The printout indicates the type of test, one-sided or two-sided, and gives the p-value for this one-sided test as .005. This value, computed by Minitab, is very close to the p-value found in part **d** of Example 11.4. As noted in Example 11.4, this p-value provides strong evidence to indicate that new teaching Method 1 tends to produce higher achievement test scores than teaching Method 2.

EXERCISES

11.16 Suppose that you wanted to use a Wilcoxon signed ranks test to determine whether the relative frequency distribution for population 1 is shifted to the right of population 2.

a. Would you use a one- or a two-tailed test? Explain.
b. Give the test statistic for the test. What values (small, large, or both) of the test statistic tend to disagree with the null hypothesis?
c. Suppose that $n = 7$. What values of the test statistic would lead to rejection of the null hypothesis if $\alpha = .05$?
d. What can you say about the p-value for the test if the test statistic equals 5?
e. What do you conclude?

11.17 Suppose that you wanted to determine whether the relative frequency distribution for population 2 is shifted to the right of population 1.

a. Would you employ a one- or a two-tailed test? Explain.
b. Give the test statistic for the test. What values (small, large, or both) of the test statistic tend to disagree with the null hypothesis?
c. Suppose that $n = 12$. What values of the test statistic would lead to rejection of the null hypothesis if $\alpha = .05$?
d. What can you say about the p-value for the test if the test statistic equals 14?
e. What do you conclude?

11.18 Suppose that you wanted to determine whether the relative frequency distribution for population 1 is shifted either to the right or to the left of population 2.

a. Would you employ a one- or a two-tailed test? Explain.
b. Give the test statistic for the test. What values (small, large, or both) of the test statistic tend to disagree with the null hypothesis?
c. Suppose that $n = 20$. What values of the test statistic would lead to rejection of the null hypothesis if $\alpha = .05$?
d. What can you say about the p-value for the test if the test statistic equals 54?
e. What do you conclude?

MFG **11.19** Table 11.9 reproduces the data from Exercise 7.29 on the mean power level readings (in watts) on a

TABLE 11.9 Power Output Measured by Two Pieces of Test Equipment

Tube Number	Tester 1	Tester 2	Difference
1	2,563	2,556	7
2	2,665	2,479	186
3	2,460	2,426	34
4	2,650	2,619	31
5	2,610	2,617	−7
6	2,657	2,491	166
7	2,529	2,590	−61
8	2,427	2,466	−39
9	2,448	2,516	−68
10	2,480	2,428	52

Source: Unpublished report by Burnett Tyson, Williamsport, Pennsylvania.

type of military electronic tube. The output of each tube was measured by each of two identical pieces of test equipment. The purpose of the experiment was to determine whether the pieces of test equipment were reading essentially the same (except for experimental error) or whether one piece of test equipment was reading higher than the other. The ranks of the signed differences are also shown in Table 11.9. Do the data present sufficient evidence to indicate a difference in location between the two population distributions of power output? Test using the Wilcoxon signed ranks test.

a. State the null and alternative hypotheses.

b. Figure 11.8 gives the Minitab printout for the Wilcoxon signed ranks test. Interpret the output.

c. Compare with the results of the parametric t test of Exercise 7.29.

11.20 Refer to Exercise 11.19. Suppose that you did not have a statistical software package that included the Wilcoxon signed ranks test. Perform the test.

a. Find the value of the test statistic.

b. Find the approximate p-value for the test.

c. If you have chosen $\alpha = .10$, what do you conclude?

d. Compare your test results with the printout shown in Figure 11.8.

GEN **11.21** The earth's temperature, which affects seed germination, crop survival, etc., can be measured using either ground monitors or satellite-based infrared sensors. Ground-based sensoring is tedious and requires many replications to obtain an accurate estimate of the earth's temperature. Satellite sensoring may produce a bias in the measurements; that is, they may tend to read higher or lower than the unbiased ground-based readings. To determine whether a bias exists, the earth's temperature was measured at five different locations using both ground-based and satellite-based sensors. The ground- and air-based temperature readings (in degrees Celsius) for the five locations are shown in Table 11.10 on page 448.

a. Explain why we believe that the data were collected using a matched-pairs design; in other words, explain how the pairing occurred.

b. Suppose that you want to test whether the data provide sufficient evidence to indicate that the satellite readings tend to read higher (or lower) than the ground-based readings. State the null and the alternative hypotheses for the test.

c. Figure 11.9 gives the Minitab printout of the Wilcoxon signed ranks test. Interpret the output.

FIGURE 11.8 A Minitab printout of the Wilcoxon signed ranks test for the test equipment data of Table 11.9

```
MTB > set c7
DATA> 2563 2665 2460 2650 2610 2657 2529 2427 2448 2480
DATA> end
MTB > set c8
DATA> 2556 2479 2426 2619 2617 2491 2590 2466 2516 2428
DATA> end
MTB > name c7 't1'
MTB > name c8 't2'
MTB > subtract c8 c7 c9
MTB > name c9 't1t2'
MTB > wtest c9

TEST OF MEDIAN = 0.000000 VERSUS MEDIAN N.E. 0.000000

                 N FOR   WILCOXON            ESTIMATED
            N    TEST   STATISTIC   P-VALUE     MEDIAN
t1t2       10      10        33.5     0.575      20.50
```

TABLE 11.10 Earth's Temperature Readings at Five Locations (degrees Celsius)

Location	Ground	Air
1	46.9	47.3
2	45.4	48.1
3	36.3	37.9
4	31.0	32.7
5	24.7	26.2

FIGURE 11.9 A Minitab printout of the Wilcoxon signed ranks test for the earth's temperature data of Exercise 11.21

```
MTB > set c10
DATA> 46.9 45.4 36.3 31 24.7
DATA> end
MTB > set c11
DATA> 47.3 48.1 37.9 32.7 26.2
DATA> end
MTB > name c10 'ground'
MTB > name c11 'air'
MTB > subtract c11 c10 c12

MTB > name c12 'grair'
MTB > wtest c12

TEST OF MEDIAN = 0.000000 VERSUS MEDIAN N.E. 0.000000

                   N FOR   WILCOXON              ESTIMATED
              N    TEST   STATISTIC   P-VALUE      MEDIAN
grair         5       5         0.0     0.059      -1.600
```

11.22 Refer to Exercise 11.21. Suppose that you did not have a statistical software package that included the Wilcoxon signed ranks test. Perform the test.

a. Find the value of the test statistic.

b. Find the *p*-value for the test.

c. If you have chosen $\alpha = .10$, what do you conclude?

d. Compare your test results with the printout shown in Exercise 11.21.

11.4 The Kruskal–Wallis *H* Test

In Chapter 10 we used an analysis of variance F statistic to test for differences among two or more population means. A Kruskal–Wallis H test is the nonparametric counterpart of the analysis of variance F test for data collected according to an independent random samples design.

Table 11.11 reproduces the drug potency data of Table 10.1. It also shows the ranks of the combined set of $n = 20$ measurements along with the rank sums for the four drug preparation procedures. There were no ties among the 20 measurements. If there had been ties, tied observations would have received a rank equal to the average of the ranks they would have occupied had they not been tied.

TABLE 11.11 Drug Potency Measurements for Four Preparation Procedures

	Procedure			
	1	2	3	4
	1.32 (2)	2.15 (12)	2.64 (18)	2.10 (11)
	2.25 (13)	1.86 (9)	1.70 (6)	1.68 (5)
	1.74 (8)	2.68 (19)	2.05 (10)	1.42 (3)
	1.05 (1)	2.31 (14)	2.51 (17)	2.49 (16)
	1.55 (4)	1.73 (7)	2.77 (20)	2.35 (15)
Rank sums	$T_1 = 28$	$T_2 = 61$	$T_3 = 71$	$T_4 = 50$

The Kruskal–Wallis H test tests the null hypothesis that all four populations are identical against the alternative hypothesis that the population relative frequency distributions differ in location. The details of the test are summarized in the box and explained in the following examples.

The Kruskal–Wallis *H* Test for Comparing More Than Two Populations: Independent Random Samples

Null hypothesis: **H_0: The k population relative frequency distributions are identical.**

Alternative hypothesis: **H_a: At least two of the population distributions differ in location.**

(continued)

Test statistic:

$$H = \frac{12}{n(n+1)}\left(\frac{T_1^2}{n_1} + \frac{T_2^2}{n_2} + \cdots + \frac{T_k^2}{n_k}\right) - 3(n+1)$$

where

k = Number of samples;

$T_1, T_2, \ldots, T_k$ are the rank sums for the k samples;

$n_1, n_2, \ldots, n_k$ are the sample sizes;

n = Total number of observations = $n_1 + n_2 + \cdots + n_k$.

Rejection region: Choose the value of α that is acceptable to you. Then reject H_0 and accept H_a if the p-value for the test is less than α.

When all of the sample sizes are equal to or greater than 5, the sampling distribution for the H statistic is approximately a chi-square distribution with $(k - 1)$ degrees of freedom (see Chapter 9). Like the contingency table test of Chapter 9, we reject H_0 for large values of the H statistic.

The p-value for the test is the probability that H is greater than or equal to the value of H calculated from the data. The procedure for finding the p-value is illustrated in Example 11.8.

Assumptions:

1. All sample sizes are greater than or equal to 5.
2. Ties assume the average of the ranks that they would have occupied if they had not been tied.

A Computer Printout for a Kruskal–Wallis *H* Test

Example 11.6

Interpret Figure 11.10, a Minitab printout of the Kruskal–Wallis H test for the drug potency data of Table 11.1.

FIGURE 11.10 A Minitab printout of the Kruskal–Wallis *H* test for the drug potency data of Table 11.11

```
LEVEL      NOBS     MEDIAN   AVE. RANK    Z VALUE
    1         5      1.550         5.6      -2.14
    2         5      2.150        12.2       0.74
    3         5      2.510        14.2       1.61
    4         5      2.100        10.0      -0.22
OVERALL      20                   10.5

H = 5.83  d.f. = 3   p = 0.121
```

Solution

The relevant information (shaded) gives the computed value of H as 5.83, the degrees of freedom for the approximating chi-square distribution as df = 3, and the p-value of the test as .121. Therefore, the Wilcoxon signed ranks test does not find sufficient evidence to indicate differences in drug potency for the four different methods of drug preparation.

Example 11.7

Compare the results of the Kruskal–Wallis H Test, Example 11.6, with the results for the comparable parametric F test in Section 10.4.

Solution

The F test to detect differences in mean drug potency was discussed in Section 10.4, and the Minitab and SAS printouts for the test are shown in Figure 10.3. The p-value for the test, given on the SAS printout, is .0787. Thus the parametric F test provides evidence of a difference between at least two of the population means for $\alpha = .10$. The p-value for the comparable Kruskal–Wallis H test, shown on the printout in Figure 11.10, is .121. It finds less evidence to indicate differences in levels of potency for the four methods of drug preparation.

Example 11.8

Suppose that your computer package did not contain the Kruskal–Wallis H test.

a. Use the rank sums in Table 11.11 to calculate H for the drug potency data. Compare this value of H with the value shown in Figure 11.10.

b. Find the approximate p-value for the test and compare it with the p-value on the printout in Figure 11.10.

Solution

a. For the data of Table 11.11, $n_1 = n_2 = n_3 = n_4 = 5$, $n = 20$, $T_1 = 28$, $T_2 = 61$, $T_3 = 71$, and $T_4 = 50$. Substituting these numbers into the formula for H gives

$$\begin{aligned} H &= \frac{12}{n(n+1)}\left(\frac{T_1^2}{n_1} + \frac{T_2^2}{n_2} + \cdots + \frac{T_4^2}{n_4}\right) - 3(n+1) \\ &= \frac{12}{20(21)}\left[\frac{(28)^2}{5} + \frac{(61)^2}{5} + \frac{(71)^2}{5} + \frac{(50)^2}{5}\right] - 3(20+1) \\ &= 5.83 \end{aligned}$$

This calculated value of H agrees with the value, $H = 5.83$, shown on the Minitab printout.

b. The p-value for the test is the probability that a chi-square variable with $(k - 1) = (4 - 1) = 3$ degrees of freedom is larger than the observed value of H, i.e., the probability that $\chi^2 > 5.83$. The upper-tail values of the chi-square distribution for $a = .10, .05, .025, .01$, and $.005$ are given in Table 5 of Appendix 2. Consulting Table 5, we see that for 3 degrees of freedom, $\chi^2_{.10} = 6.25139$ and $\chi^2_{.05} = 7.81473$. Since the observed value of H is a little less than $\chi^2_{.10} = 6.25139$, the p-value of the test is slightly larger than .10. This agrees with the exact p-value for the test, .121, given on the Minitab printout in Figure 11.10.

EXERCISES

BIO **11.23** Table 11.12 presents data on the growth of vegetation at four swampy undeveloped sites. The purpose of the experiment was to determine whether the different soil types produced different rates of plant growth. Six plants of the same species were randomly selected at each of the four sites to be used in the comparison. Each measurement in Table 11.12 represents the mean leaf length per plant, in centimeters, for a random sample of ten leaves per plant. A Minitab printout of the Kruskal–Wallis H test for the data is shown in Figure 11.11. Do the data present sufficient evidence to indicate differences in the leaf lengths of plants at the four locations?

a. State the null and alternative hypotheses you would use for your H test.

b. Find the value of the H statistic on the printout in Figure 11.11. Find the p-value for the test.

c. If you have chosen $\alpha = .10$, what would you conclude? What are the practical implications of the test conclusions?

11.24 Rank the 24 observations of Table 11.12.

a. Calculate the rank sums and the value of the H test statistic. Compare with the value shown on the printout in Figure 11.11.

b. Use the value of the H test statistic from part **a** to find the approximate p-value for the test. Compare this value with the exact p-value shown on the printout in Figure 11.11.

11.25 Refer to Exercise 11.23 and the data of Table 11.12. Figure 11.12 gives the Minitab analysis of variance for the data. Interpret the ANOVA output. Briefly compare the two outputs shown in Figures 11.11 and 11.12. Explain how they do or do not agree.

BIO **11.26** A sampling of the acidity of rain for ten randomly selected rainfalls was recorded at three different locations in the United States: the Northeast, the Middle Atlantic region, and the Southeast. The pH readings for the 30 rainfalls, which can range from 0 (acid) to 14 (alkaline), are shown in Table 11.13. Do

TABLE 11.12 Mean Leaf Lengths at Four Locations

Location	Mean leaf length (cm)					
1	5.7	6.3	6.1	6.0	5.8	6.2
2	6.2	5.3	5.7	6.0	5.2	5.5
3	5.4	5.0	6.0	5.6	4.9	5.2
4	3.7	3.2	3.9	4.0	3.5	3.6

FIGURE 11.11 A Minitab printout for the Kruskal–Wallis *H* test for the leaf length data

```
LEVEL     NOBS      MEDIAN  AVE. RANK   Z VALUE
    1        6       6.050       19.8      2.93
    2        6       5.600       15.1      1.03
    3        6       5.300       11.6     -0.37
    4        6       3.650        3.5     -3.60
OVERALL     24                   12.5

H = 17.08  d.f. = 3  p = 0.001
H = 17.13  d.f. = 3  p = 0.001 (adj. for ties)
```

FIGURE 11.12 A Minitab ANOVA for the mean leaf length data of Exercise 11.23

```
MTB > anova c10=c9

Factor      Type Levels Values
loc        fixed      4      1      2      3      4

Analysis of Variance for length

Source      DF          SS          MS        F      P
loc          3     19.7400      6.5800    57.38  0.000
Error       20      2.2933      0.1147
Total       23     22.0333
```

TABLE 11.13 Acidity of Rainfall at Four Different Locations

Northeast	Middle Atlantic	Southeast
4.45	4.60	4.55
4.02	4.27	4.31
4.13	4.31	4.84
3.51	3.88	4.67
4.42	4.49	4.28
3.89	4.22	4.95
4.18	4.54	4.72
3.95	4.76	4.63
4.07	4.36	4.36
4.29	4.21	4.47

the data present sufficient evidence to indicate differences in the levels of acidity for rainfalls for the three regions?

a. State the null and alternative hypotheses you would use for a Kruskal–Wallis H test.

b. Figure 11.13 shows the SAS printout for a Kruskal–Wallis H test for the data. Find the value of the H statistic, identified as CHISQ on the printout in Figure 11.13. Find the p-value for the test.

c. If you have chosen $\alpha = .10$, what would you conclude? What are the practical implications of the test conclusions?

11.27 Rank the 30 observations of Table 11.13.

a. Calculate the rank sums and the value of the H test statistic. Compare with the value shown on the printout in Figure 11.13.

b. Use the value of the H test statistic from part **a** to find the approximate p-value for the test. Compare this value with the exact p-value shown on the printout in Figure 11.13.

11.28 Refer to Exercise 11.26 and the data of Table 11.13. Figure 11.14 gives the SAS analysis of variance for the data. Interpret the ANOVA output. Briefly compare the outputs in Figures 11.13 and 11.14. Explain how they do or do not agree.

11.29 What requirements must the data satisfy to ensure the validity of the Kruskal–Wallis H test?

11.30 Suppose that you are comparing $k = 3$ population relative frequency distributions based on independent random samples of $n_1 = n_2 = n_3 = 8$ observations. If $H = 4.65$, what is the approximate p-value for the test?

11.31 Suppose that you are comparing $k = 4$ population relative frequency distributions based on independent random samples of $n_1 = n_2 = n_3 = 8$ and $n_4 = 10$ observations. If $H = 10.3$, what is the approximate p-value for the test?

11.32 Suppose that you are comparing $k = 5$ population relative frequency distributions based on independent random samples of $n_1 = n_2 = \cdots = n_5 = 8$ observations. If $H = 7.3$, what is the approximate p-value for the test?

FIGURE 11.13 A SAS printout of the Kruskal–Wallis *H* test for the rainfall acidity data, Table 11.13

```
                     Sum of      Expected       Std Dev          Mean
LOC            N     Scores      Under HO       Under HO         Score

ne            10       84.0         155.0     22.7252455      8.4000000
ma            10      158.0         155.0     22.7252455     15.8000000
se            10      223.0         155.0     22.7252455     22.3000000
                 Average Scores were used for Ties

        Kruskal-Wallis Test (Chi-Square Approximation)
        CHISQ=   12.488        DF=   2       Prob > CHISQ=       0.0019
```

FIGURE 11.14 A SAS ANOVA for the rainfall acidity data in Table 11.13

```
Dependent Variable: ACIDITY
                                    Sum of            Mean
Source                   DF        Squares          Square    F Value      Pr > F

Model                     2     1.19164667      0.59582333       9.47      0.0008

Error                    27     1.69829000      0.06289963

Corrected Total          29     2.88993667

                   R-Square           C.V.        Root MSE          ACIDITY Mean

                   0.412344       5.772991        0.250798            4.34433333


Source                   DF       Anova SS     Mean Square    F Value      Pr > F

LOC                       2     1.19164667      0.59582333       9.47      0.0008
```

11.5 The Spearman Rank Correlation Coefficient

We can use ranks not only to detect the difference in location for two or more population relative frequency distributions, but also to measure the strength of the relationship between two variables x and y.

Spearman's Rank Correlation Coefficient: What Is It?

Spearman's rank correlation coefficient r_s is calculated in the same way as the sample coefficient of correlation r of Chapter 8,

$$r_s = \frac{SS_{xy}}{\sqrt{SS_x SS_y}}$$

except that the measurements in the formula for r are replaced by their ranks. Therefore, r_s has the same numerical properties as r (i.e., it always assumes a value between -1 and $+1$) but the interpretation of r_s differs from the interpretation of r.

Interpreting the Value of r_s

If the simple linear coefficient of correlation r is equal to 1 in a regression analysis, it means that all of the data points fall on a straight line that rises as we move from left to right (see Figure 11.15a). In contrast, if $r_s = 1$, it means that **the ranks of x and y are perfectly associated** (1, 1), (2, 2), (3, 3), etc., but **it does not mean that**

FIGURE 11.15 Graphs of relationships between y and x for which r_s equals 1

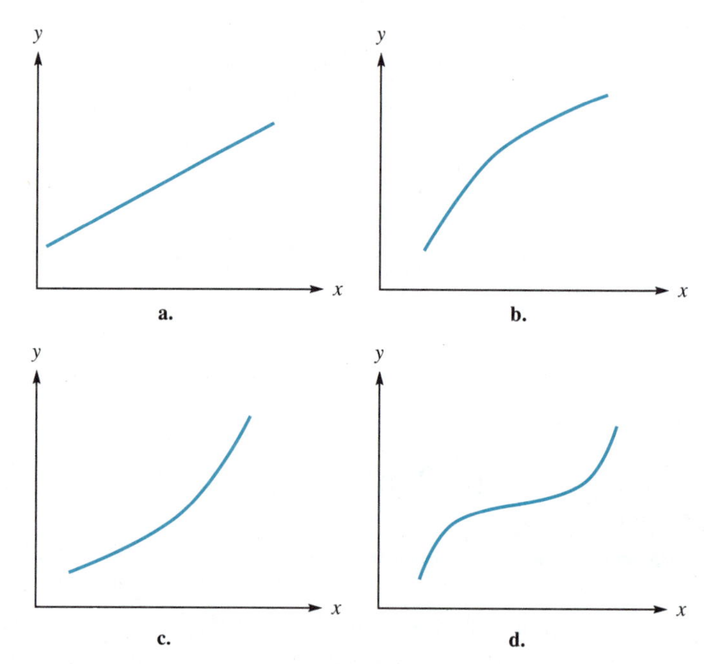

the original (x, y) data points fall on a straight line. In fact, r_s will equal 1 when the quantitative data points (x, y) fall on any curve that rises as we move to the right along the curve.

For example, Figure 11.15 shows curves that describe four relationships between the quantitative variables x and y. In each case, y increases as x increases. Pick any four data points on each curve. You can see for each curve that the ranks of the data points are paired: (1, 1), (2, 2), (3, 3), and (4, 4). The value of the sample correlation coefficient r will equal 1 only if the data points fall on the straight line, as in Figure 11.15a. In contrast, the value of r_s will equal 1 for *all four* relationships shown in Figure 11.15. It equals 1 even when the relationship between y and x does not graph as a straight line, as shown in Figures 11.15b–d. Therefore, **$r_s = 1$ tells us only that y increases as x increases.**

By a similar argument, it follows that **$r_s = -1$ means that y decreases as x increases.** A value of r_s near 0 tells us that y is neither an increasing nor a decreasing function of x.

A test to detect population values of r_s that differ from 0 is shown in the accompanying box. Its use is illustrated in Examples 11.9–11.11.

Spearman's Rank Correlation Test

Null hypothesis: **There is no association between the rank pairs for the data points in the sampled population.**

Alternative hypothesis:

1. **H_a: There is an association between the rank pairs (a two-tailed test)**
2. **H_a: The correlation between the rank pairs is positive (or negative) (a one-tailed test)**

Test statistic: **Rank the n values of x in the data set from the smallest (rank 1) to the largest (rank n) and repeat this process for the n values of y. Tied observations in each set are assigned ranks equal to the average of the ranks that would have been assigned to the observations if they had not been tied. The test statistic is**

$$r_s = \frac{SS_{xy}}{\sqrt{SS_x\, SS_y}}$$

where the x and y values used to calculate SS_x, SS_y, and SS_{xy} are the ranks of the original (x, y) measurements.

Rejection region: **Choose the value of α that is acceptable to you. Then reject H_0 and accept H_a if the p-value for the test is less than α.**

The p-value for an upper one-tailed test is the probability that r_s is greater than or equal to the observed value of the test statistic. Table 10 in Appendix 2 gives the values of r_s that locate areas a = .05, .025, .01, and .005 in the upper tail of the sampling distribution of r_s.* The negative of a tabulated value of r_s gives the value of r_s that locates an area a in the lower tail of the distribution.

The p-value for the two-tailed test is double the probability that r_s is less than or equal to the observed value of the test statistic (if the observed value is negative), or double the probability that r_s is greater than or equal to the observed value of the test statistic (if the observed value is positive). This probability can be determined from Table 10 of Appendix 2.*

Computer Printouts for a Spearman's Rank Correlation Analysis

SAS calculates the rank correlation coefficient r_s and tests the null hypothesis that the population rank correlation coefficient equals 0. Example 11.9 shows the SAS output for the quantitative aptitude and achievement test scores of Table 8.1.

*The procedure for using Table 10 of Appendix 2 is explained in Example 11.10.

Example 11.9

Table 11.14 reproduces the quantitative aptitude and achievement test scores of Table 8.1 along with the ranks of the x and the y measurements.

TABLE 11.14 Quantitative Aptitude and Achievement Test Scores

Student	1	2	3	4	5	6	7	8	9	10	11	12
x	88	57	76	97	71	90	66	58	92	85	51	85
rank (x)	(9)	(2)	(6)	(12)	(5)	(10)	(4)	(3)	(11)	(7.5)	(1)	(7.5)
y	620	495	549	635	480	568	570	437	655	547	395	662
rank(y)	(9)	(4)	(6)	(10)	(3)	(7)	(8)	(2)	(11)	(5)	(1)	(12)

a. We would expect a student's achievement test score to be positively correlated with his or her quantitative aptitude test score, if they are related at all. State the null and alternative hypotheses that we want to test.

b. Identify the test statistic and the rejection region for the test.

c. Figure 11.16 gives the SAS printout for a Spearman's rank correlation analysis. Describe the output and state your test conclusions.

FIGURE 11.16 A SAS printout of the test for rank correlation, Example 11.9

```
                                  Simple Statistics

Variable                 N              Mean           Std Dev            Median

X                       12          76.33333          15.47040          80.50000
Y                       12         551.08333          85.88201         558.50000

                    Simple Statistics

Variable           Minimum           Maximum

X                 51.00000          97.00000
Y                395.00000         662.00000

Spearman Correlation Coefficients / Prob > |R| under Ho: Rho=0 / N = 12

                                   X                   Y

                  X          1.00000             0.77408
                              0.0                0.0031

                  Y          0.77408             1.00000
                              0.0031              0.0
```

Solution

a. We would expect the correlation between a student's quantitative aptitude and quantitative achievement to be positive, if it differs from 0. Therefore, we would test the null hypothesis that the population rank correlation coefficient equals 0 against the alternative that it is greater than 0. This would be an upper one-tailed test.

b. The test statistic is r_s. We would reject H_0 for large positive values of r_s.

c. The relevant portion of the SAS printout in Figure 11.16 is shaded. SAS uses the symbol R (instead of r_s) to denote the sample rank correlation coefficient. It shows $r_s = .77408$ and states the p-value as "Prob $> |R|$." **Thus, SAS prints out the *p*-value for a two-tailed test**—that is, for a test designed to detect either large negative or large positive values of the population rank correlation coefficient. This value, .0031, is shown directly below $r_s = .77408$ on the printout. Since we want the p-value for an upper one-tailed test, our p-value will be equal to $.0031/2 = .0016$. This small p-value provides strong evidence to indicate a positive rank correlation between a student's quantitative aptitude test score and his or her mathematics achievement test score. It tells us that a student's aptitude test score provides information for predicting his or her mathematics achievement test score. As the aptitude score increases, so does the mathematics achievement test score.

Example 11.10

Suppose that your statistical software package gave you the value of r_s but did not provide the p-value. Use Table 10 of Appendix 2 to find the approximate p-value for the test in Example 11.9. Compare with the computer output shown in Figure 11.16.

Solution

We need to know how much r_s will deviate from 0 if, in fact, the null hypothesis is true—that is, if the population rank correlation coefficient is equal to 0. Table 10 of Appendix 2 gives the upper-tail values of the sampling distribution of r_s for $a = .05$, .025, .01, and .005 for different sample sizes and for a population rank correlation coefficient equal to 0. A portion of Table 10 of Appendix 2 is reproduced in Table 11.15 on page 460.

Because we are conducting an upper one-tailed test, the p-value for the test is the probability that r_s is greater than or equal to the observed value, $r_s = .774$ (see Figure 11.17). Our data set contained $n = 12$ data points. Table 11.15 gives the upper-tail values of r_s for $n = 12$ and $a = .05$, .025, .01, and .005 as .497, .591, .703, and .780, respectively. Since the observed value of r_s, .774, is very close to .780, the upper-tail value of r_s for $a = .005$, the p-value for the test is approximately equal to .005. This value is close to but not exactly equal to the p-value of .0031 shown on the SAS printout in Figure 11.16. The discrepancy is due to the fact that the tail probabilities of Table 10 in Appendix 2 are calculated on the assumption that the data include no ties. Our data set, Table 11.14, contained one tie.

TABLE 11.15 A Partial Reproduction of Table 10 of Appendix 2

n	$a = .05$	$a = .025$	$a = .01$	$a = .005$
5	.900	—	—	—
6	.829	.886	.943	—
7	.714	.786	.893	—
8	.643	.738	.833	.881
9	.600	.683	.783	.833
10	.564	.648	.745	.794
11	.523	.623	.736	.818
12	.497	.591	.703	.780

FIGURE 11.17 The *p*-value for a test of rank correlation between quantitative aptitude and mathematics achievement test scores

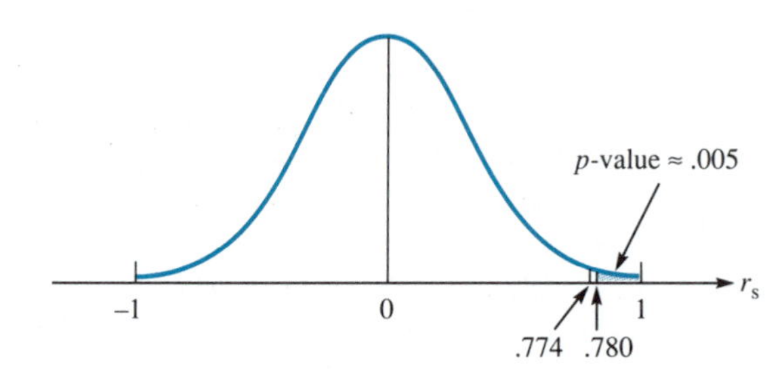

Consumer Preference Experiments

Some quantitative variables cannot be measured. Typical are the results of consumer preference experiments. We can rank four cola drinks according to taste preference but we cannot give an exact measurement of the "taste" for one particular cola. Similarly, we can rank five different sports cars in order of our preference but we cannot attach a measure of our preference for one particular car. Parametric methods of estimation and testing cannot be used if we cannot measure the quantitative variable. Then nonparametric procedures such as the Mann–Whitney U test and the Wilcoxon signed ranks test not only provide alternative tests—they provide the *only* tests.

Example 11.11

Two art critics each ranked ten paintings by contemporary (but anonymous) artists in accordance with their appeal to the respective critics. The ratings are shown in Table 11.16. Do the critics seem to agree on their ratings of contemporary art? That is, do the data present sufficient evidence to indicate a rank correlation between critics A and B? Test using $\alpha = .05$.

TABLE 11.16 Two Critics' Ratings of 10 Contemporary Paintings

Painting	Critic A	Critic B
1	6	5
2	4	6
3	9	10
4	1	2
5	2	3
6	7	8
7	3	1
8	8	7
9	5	4
10	10	9

FIGURE 11.18 A SAS printout of the test for rank correlation in Example 11.11

```
                              Simple Statistics

Variable            N            Mean         Std Dev          Median

CRIT_A             10         5.50000         3.02765         5.50000
CRIT_B             10         5.50000         3.02765         5.50000

                Simple Statistics

Variable       Minimum       Maximum

CRIT_A         1.00000      10.00000
CRIT_B         1.00000      10.00000

Spearman Correlation Coefficients / Prob > |R| under Ho: Rho=0 / N = 10

                            CRIT_A             CRIT_B

            CRIT_A         1.00000            0.90303
                            0.0                0.0003

            CRIT_B         0.90303            1.00000
                            0.0003             0.0
```

Solution

The ten paintings represent a sample from the large (conceptual) universe of contemporary paintings that the critics could conceivably rank. We want to determine whether the critics tend to agree—that is, whether the value of r_s is large and positive. Therefore, our null hypothesis is that the critics do not agree (i.e., the population value of r_s equals 0) and the one-sided alternative is that the critics tend to agree (i.e., the population value of r_s is greater than 0). Thus, we will conduct an upper one-tailed test and reject H_0 for large positive values of r_s.

Figure 11.18 shows a SAS printout of the rank correlation test for Spearman's r_s. The computed value of Spearman's rank correlation coefficient shown on the printout is $r_s = .90303$. The printout gives the p-value for a two-tailed test as .0003. Therefore, the p-value for our one-tailed test is half this value, or .00015. This extremely small p-value provides strong evidence of a rank correlation between the critics' ratings. The critics tend to agree in their ratings of contemporary paintings.

EXERCISES

11.33 Suppose that x and y are quantitative variables and that we have observed a sample of $n = 20$ data points. If the value of Spearman's rank correlation coefficient is positive, what does it tell you about the relationship between x and y?

11.34 Refer to Exercise 11.33. What does it tell you if r_s is negative?

11.35 Refer to Exercise 11.33. What does it tell you if r_s is near 0?

11.36 Refer to Exercise 11.33. What does it tell you if $r_s = +1$?

11.37 Refer to Exercise 11.33. What does it tell you if $r_s = -1$?

11.38 Suppose that the rank correlation between two variables, x and y, is equal to 1. Does this imply that the (x, y) data points fall on a straight line? Explain.

11.39 Find the value of r_s, based on $n = 12$ data points, that locates an area equal to .05 in the upper tail of its sampling distribution, if in fact the true population Spearman's rank correlation coefficient is equal to 0.

11.40 Find the value of r_s, based on $n = 12$ data points, that locates an area equal to .01 in the upper tail of its sampling distribution, if in fact the true population Spearman's rank correlation coefficient is equal to 0.

11.41 Find the value of r_s, based on $n = 15$ data points, that locates an area equal to .05 in the lower tail of its sampling distribution, if in fact the true population Spearman's rank correlation coefficient is equal to 0.

11.42 Find the values of r_s, based on $n = 8$ data points, that locate an area equal to .025 in both the lower and the upper tail of its sampling distribution, if in fact the true population Spearman's rank correlation coefficient is equal to 0.

BIO **11.43** In Exercise 8.34, we examined the relationship between the height x and the diameter y for ten fossil specimens. The fossils were specimens of a shellfish unearthed in a mapping expedition near the Antarctic Peninsula. The data of Table 8.9 are reproduced in Table 11.17. Column 1 of the table gives an identification symbol for the fossil specimen, column 2 gives the diameter y of the fossil (in millimeters, mm), and column 3 gives its height (in mm). A Spearman's rank correlation analysis is shown on the SAS printout in Figure 11.19.

TABLE 11.17 Diameter and Height for Ten Fossil Specimens of *Rotularia (Annelida) fallax*

Specimen	Diameter	Height	D/H
OSU 36651	185	78	2.37
OSU 36652	194	65	2.98
OSU 36653	173	77	2.25
OSU 36654	200	76	2.63
OSU 36655	179	72	2.49
OSU 36656	213	76	2.80
OSU 36657	134	75	1.79
OSU 36658	191	77	2.48
OSU 36659	177	69	2.57
OSU 36660	199	65	3.06

Source: Macellari, Carlos E., "Revision of Serpulids of the Genus *Rotularia (Annelida)* at Seymour Island (Antarctic Peninsula) and Their Value in Stratigraphy," *Journal of Paleontology,* Vol. 58, No. 4, July 1984.

FIGURE 11.19 A SAS printout of the test for rank correlation: Fossil data, Exercise 11.43

```
                              Simple Statistics

Variable             N          Mean         Std Dev          Median

DIA                 10     184.50000        21.51098       188.00000
HT                  10      73.00000         4.98888        75.50000

               Simple Statistics

Variable          Minimum        Maximum

DIA             134.00000      213.00000
HT               65.00000       78.00000

Spearman Correlation Coefficients / Prob > |R| under Ho: Rho=0 / N = 10

                                 DIA                HT

               DIA           1.00000          -0.10398
                             0.0               0.7750

               HT           -0.10398           1.00000
                             0.7750            0.0
```

a. Find the value of r_s and explain what it tells you about the relationship between x and y for the ten data points.
b. Describe the population from which the ten data points were selected. Do the data provide sufficient evidence to indicate that a rank correlation exists between x and y in the population of data points? Explain.

GEN **11.44** Hooke's Law states that when a pulling force is applied to a body that is long relative to its cross-sectional area, the change y (i.e., the stretch) in its length is proportional to the force x:

$$y = \beta_1 x$$

where β_1 is the constant of proportionality. Exercise 8.38 presented data from a college student's sophomore physics laboratory experiment. Six lengths of steel wire, each .34 millimeter (mm) in diameter and 2 meters (m) long, were used to obtain the six force–length change measurements shown in Table 11.18. A SAS Spearman's rank correlation analysis of the data is shown in Figure 11.20.

a. Describe the population of data points sampled by the student.
b. If Hooke's Law is correct, describe a graph of the population of data points.
c. Write a brief statement explaining whether, how, and to what extent the experiment and the rank correlation analysis support Hooke's Law.

11.45 We explained in this section that r_s is calculated in the same way that we calculated the simple linear coefficient of correlation except that we replace x and y in the formula for r by their corresponding ranks:

$$r_s = \frac{\text{SS}_{xy}}{\sqrt{\text{SS}_x \text{SS}_y}}$$

There is a shortcut method for calculating r_s that gives exactly the same result as the formula above if there are no ties in either the x observations or the y observations. It will also give an answer that is an adequate approximation to r_s if the number of ties is small relative to the number of data points.

Let d equal the difference in the rank of x and the rank of y for a data point. Calculate d for each data point. Then calculate Σd^2, the sum of the d^2 values for the n data points. Then the shortcut formula for r_s is given by

$$r_s = 1 - \frac{6 \Sigma d^2}{n(n^2 - 1)}$$

Use the shortcut method to calculate r_s for the quantitative aptitude–achievement test score data of Table 11.14. Compare your answer with the value of r_s shown on the Minitab printout in Figure 11.16.

TABLE 11.18 Laboratory Data to Verify Hooke's Law

Force x (kg)	Change in Length y (mm)
29.4	4.25
39.2	5.25
49.0	6.50
58.8	7.85
68.6	8.75
78.4	10.00

Source: C. M. Mendenhall, physics laboratory notes, University of Florida, 1972–1973.

FIGURE 11.20 A SAS printout of the test for rank correlation: Hooke's Law experiment, Exercise 11.44

```
                              Simple Statistics

Variable                  N            Mean         Std Dev          Median

FORCE                     6        53.90000        18.33412        53.90000
CH_LEN                    6         7.10000         2.17256         7.17500

               Simple Statistics

Variable           Minimum         Maximum

FORCE             29.40000        78.40000
CH_LEN             4.25000        10.00000

Spearman Correlation Coefficients / Prob > |R| under Ho: Rho=0 / N = 6

                                  FORCE              CH_LEN

               FORCE             1.00000            1.00000
                                 0.0                0.0

               CH_LEN            1.00000            1.00000
                                 0.0                0.0
```

11.6 Key Words and Concepts

- ***Nonparametric statistical tests*** generally use the ranks of quantitative measurements, rather than the measurements themselves, to test hypotheses about the sampled populations.
- Nonparametric tests are especially useful because they require few, if any, assumptions about the sampled population. They are essential if one or more of the quantitative variables can be ranked but cannot be measured.
- The ***Mann–Whitney U test*** is the nonparametric counterpart of a Student's t test for comparing two population means. The U test is used to detect a difference in location between two population relative frequency distributions based on independent random samples.
- The ***Wilcoxon signed ranks test*** is the nonparametric counterpart of a Student's paired-difference t test. The Wilcoxon signed ranks test is used to detect a difference in location between two population relative frequency distributions based on data collected according to a matched-pairs design.
- A ***Kruskal–Wallis H test*** is the nonparametric counterpart of an analysis of variance F test for comparing the means of more than two populations based on independent

random samples. It tests for differences in location among three or more population relative frequency distributions.

- A ***Spearman's rank correlation coefficient*** r_s is a measure of the strength of the relationship between the ranks of two quantitative variables. It is calculated using the formula for r, the sample coefficient of linear correlation, except that the ranks of the original measurements, rather than the x and y measurements themselves, are used in the calculation.
- A rank correlation coefficient r_s will assume a value between -1 and $+1$. If $r_s = 1$, the quantitative variable y will increase as the quantitative variable x increases. If $r_s = -1$, the quantitative variable y will decrease as the quantitative variable x increases. A value of r_s near 0 tells us that y is neither an increasing nor a decreasing function of x.

SUPPLEMENTARY EXERCISES

GEN **11.46** In an investigation of the visual scanning behavior of deaf children, measurements of eye-movement rate were taken on nine deaf and nine hearing children. The rate measurements, along with their ranks, are shown in Table 11.19. Do the data provide sufficient evidence to indicate a difference in location for the two population distributions of rate measurements?

TABLE 11.19 Eye-Movement Rate for Deaf and for Hearing Children

Deaf Children	Hearing Children
2.75	.89
2.14	1.43
3.23	1.06
2.07	1.01
2.49	.94
2.18	1.79
3.16	1.12
2.93	2.01
2.20	1.12

a. State the null and alternative hypotheses that you would use to answer the question.

b. Do the hypotheses in part **a** imply a one- or a two-tailed test?

c. The Minitab printout for a Mann–Whitney U test is shown in Figure 11.21. Write a brief statement that interprets the printout. State the practical conclusions to be derived from the test.

11.47 Refer to Exercise 11.46.

a. What should you use as the test statistic? Find the rank sums and the value of the test statistic for the data in Table 11.19.

b. Find the approximate p-value for the test and interpret it. What are the practical implications of this p-value?

GEN **11.48** Exercise 7.43 compares the density of cakes prepared from two different cake mixes, A and B. Six cake pans received batter A and six received batter B. Then, to eliminate the effect of variation in oven temperatures from one spot in the oven to another, the experimenter placed an A and a B side-by-side at six different locations. The six paired cake density readings of Table 7.12 are reproduced in Table 11.20 and a Minitab printout of a Wilcoxon signed ranks test is shown in Figure 11.22 on page 468. Do the data present sufficient evidence to indicate a difference in the density levels for the two cake mixes? Write a brief statement that interprets the printout. State the practical conclusions to be derived from the test.

11.49 Find the approximate p-value for the test in Exercise 11.48, and compare it with the p-value for

FIGURE 11.21 A Minitab printout of a Mann–Whitney test for the eye-movement rate data of Table 11.19, Exercise 11.46

```
MTB > set c10
DATA> 2.75 2.14 3.23 2.07 2.49 2.18 3.16 2.93 2.20
DATA> end
MTB > name c10 'deaf'
MTB > set c11
DATA> .89 1.43 1.06 1.01 .94 1.79 1.12 2.01 1.12
DATA> end
MTB > name c11 'hearing'
MTB > mann whitney c10 c11

Mann-Whitney Confidence Interval and Test

deaf         N =    9      Median =        2.4900
hearing      N =    9      Median =        1.1200
Point estimate for ETA1-ETA2 is            1.2400
95.8 pct c.i. for ETA1-ETA2 is (0.9499,1.8101)
W = 126.0
Test of ETA1 = ETA2  vs.  ETA1 n.e. ETA2 is significant at 0.0004
The test is significant at 0.0004 (adjusted for ties)
```

TABLE 11.20 Paired Cake Density Measurements for Two Cake Mixes, Exercise 11.48

Mix	Density (oz/in.^{3})					
A	.135	.102	.098	.141	.131	.144
B	.129	.120	.112	.152	.135	.163

the parametric test obtained in Exercise 7.43. Explain whether the p-values for the two tests do or do not agree.

GEN **11.50** Table 8.13 gives the number x of robberies and the number y of robbery arrests per month in New York City during the first half of 1988. The data are reproduced in Table 11.21 on page 468. Figure 11.23 (page 469) gives a SAS printout that shows the Spearman's rank correlation coefficient for the data and tests whether the rank correlation for x and y differs from 0. Describe the content of the printout and explain what it tells us about the relation between x and y.

FIGURE 11.22 A Minitab printout of a Wilcoxon signed ranks test for the cake density experiment of Exercise 11.48

```
MTB > set c4
DATA> .135 .102 .098 .141 .131 .144
DATA> end
MTB > set c5
DATA> .129 .120 .112 .152 .135 .163
DATA> end
MTB > subtract c5 c4 c6
MTB > name c4 'mixa'
MTB > name c5 'mixb'
MTB > name c6 'mixab'
MTB > wtest c6

TEST OF MEDIAN = 0.000000 VERSUS MEDIAN N.E. 0.000000

                  N FOR   WILCOXON            ESTIMATED
              N    TEST  STATISTIC  P-VALUE     MEDIAN
mixab         6       6        2.0    0.093   -0.01100
```

TABLE 11.21 Robberies and Robbery Arrests in New York City, Exercise 11.50

Month	Robberies	Robbery arrests
January	475	180
February	465	155
March	470	160
April	500	190
May	550	225
June	600	220
July	602	223

Source: Data estimated from graph in *The New York Times,* October 2, 1988.

FIGURE 11.23 A SAS Spearman's rank correlation analysis for the New York robbery data, Exercise 11.50

```
                              Simple Statistics

Variable                 N          Mean         Std Dev          Median

ROB                      7     523.14286        60.42193       500.00000
ARRESTS                  7     193.28571        29.89824       190.00000

            Simple Statistics

Variable        Minimum        Maximum

ROB           465.00000      602.00000
ARRESTS       155.00000      225.00000

Spearman Correlation Coefficients / Prob > |R| under Ho: Rho=0 / N = 7

                                 ROB            ARRESTS

             ROB             1.00000            0.89286
                              0.0                0.0068

             ARRESTS         0.89286            1.00000
                              0.0068             0.0
```

EXERCISES FOR YOUR COMPUTER

11.51 Refer to Exercise 11.46 and the eye-movement rate data of Table 11.19. Enter the data of Table 11.19 into your computer and use your statistical software package to perform a Mann–Whitney test for a difference in location for the respective population distributions. Compare your printout with the one shown in Figure 11.21.

11.52 Refer to Exercise 11.48 and the cake density data of Table 11.20. Enter the data of Table 11.20 into your computer and use your statistical software package to perform a Wilcoxon signed ranks test for a difference in location for the respective population distributions. Compare your printout with the one shown in Figure 11.22.

11.53 Refer to Exercise 11.50 and the New York City robbery arrest data of Table 11.21. Enter the data of Table 11.21 into your computer and use your statistical software package to perform a Spearman's rank correlation analysis for the data. Compare your printout with the one given in Figure 11.23.

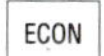

11.54 Exercise 8.7 and Example 8.9, Chapter 8, discuss data on the relationship between a nation's savings rate and its investment income tax rate. The data, originally shown in Table 8.2, are reproduced in Table 11.22 on page 470. Enter the data of Table 11.22 into your computer and use your statistical software package to perform a Spearman's rank correlation analysis for the data. What does your analysis tell you about the relationship between a nation's saving rate and its investment tax rate?

TABLE 11.22 Personal Savings Rate and Investment Income Tax Liability for Eight Countries

Country	y Personal savings rate (%)	Rank of y	x Investment income tax liability (%)	Rank of x
Italy	23.1	(8)	6.4	(1)
Japan	21.5	(7)	14.4	(4)
France	17.2	(6)	7.3	(2)
W. Germany	14.5	(5)	11.8	(3)
United Kingdom	12.2	(4)	32.5	(6)
Canada	10.3	(3)	30.0	(5)
Sweden	9.1	(2)	52.7	(8)
United States	6.3	(1)	33.5	(7)

Source: New York Stock Exchange with assistance of Price Waterhouse. From Blotnick, S., "Psychology and Investing," *Forbes,* May 25, 1981.

References

1. Beyer, W. H., *Handbook of Tables for Probability and Statistics.* Cleveland, Ohio: The Chemical Rubber Co., 1966.
2. Conover, W. J., *Practical Nonparametric Statistics,* 2nd edition. New York: Wiley, 1980.
3. Freedman, D., Pisani, R., Purves, R., and Adhikari, A., *Statistics,* 2nd edition. New York: W. W. Norton & Co., 1991.
4. Hollander, M. and Wolfe, D. A., *Nonparametric Statistical Methods.* New York: Wiley, 1973.
5. McClave, J. and Dietrich, F., *Statistics,* 5th edition. San Francisco: Dellen Publishing Co., 1991.
6. Mendenhall, W. and Beaver, R., *Introduction to Probability and Statistics,* 8th edition. Boston: PWS-Kent, 1991.
7. Moore, D. and McCabe, G., *Introduction to the Practice of Statistics,* 2nd edition. New York: W. H. Freeman & Co., 1993.
8. Sincich, T., *Statistics by Example,* 4th edition. San Francisco: Dellen Publishing Co., 1990.

appendix 1

Data Sets

Table 1 Iron Content (Percentage) of 390 1.5-Kilogram Specimens of Iron Ore from a 20,000-Ton Consignment of Canadian Ore

Table 2 DDT Analyses on Fish in the Tennessee River, 1980

Table 3 University of Florida Study of Body Cholesterol for 107 People

TABLE 1 Iron Content (Percentage) of 390 1.5-Kilogram Specimens of Iron Ore from a 20,000-Ton Consignment of Canadian Ore

Number	% Iron	Number	% Iron	Number	% Iron	Number	% Iron	Number	% Iron
1	66.08	79	64.66	157	65.81	235	65.57	313	65.33
2	65.92	80	65.21	158	66.30	236	65.79	314	65.00
3	65.83	81	64.50	159	65.68	237	65.51	315	65.03
4	65.81	82	64.18	160	65.40	238	65.97	316	64.95
5	65.77	83	64.25	161	65.83	239	66.05	317	65.16
6	65.83	84	64.24	162	65.57	240	66.27	318	65.39
7	65.82	85	64.41	163	65.39	241	65.23	319	65.55
8	66.06	86	64.01	164	65.57	242	66.00	320	65.91
9	65.85	87	64.39	165	65.57	243	65.14	321	65.50
10	65.75	88	64.89	166	65.89	244	65.85	322	65.72
11	66.02	89	64.89	167	65.79	245	65.66	323	65.76
12	65.81	90	65.09	168	65.60	246	66.50	324	65.50
13	65.85	91	64.02	169	65.62	247	65.69	325	65.38
14	65.66	92	64.82	170	65.96	248	65.30	326	65.95
15	65.97	93	64.35	171	66.46	249	65.78	327	65.48
16	65.80	94	64.86	172	66.34	250	65.81	328	65.40
17	65.97	95	64.51	173	65.98	251	65.42	329	65.38
18	66.08	96	64.95	174	65.69	252	65.13	330	65.55
19	65.79	97	65.48	175	65.83	253	65.50	331	65.87
20	66.13	98	65.24	176	66.28	254	65.98	332	65.92
21	66.23	99	64.93	177	66.05	255	65.38	333	66.58
22	65.81	100	65.19	178	66.18	256	65.83	334	66.04
23	65.99	101	65.30	179	65.83	257	65.64	335	66.00
24	66.18	102	64.95	180	66.09	258	65.07	336	65.74
25	65.97	103	64.96	181	66.29	259	65.37	337	65.84
26	66.30	104	64.98	182	66.04	260	65.32	338	63.48
27	66.23	105	65.04	183	66.36	261	65.27	339	63.29
28	65.99	106	65.73	184	66.16	262	65.22	340	63.62
29	65.99	107	66.18	185	66.37	263	65.13	341	63.89
30	65.78	108	65.61	186	66.25	264	64.98	342	64.14
31	65.97	109	66.53	187	66.86	265	65.49	343	63.53
32	66.32	110	65.48	188	66.16	266	65.34	344	63.57
33	65.73	111	65.75	189	66.09	267	64.70	345	63.52
34	66.06	112	65.95	190	66.36	268	65.52	346	63.85
35	65.67	113	65.98	191	66.29	269	65.33	347	64.44
36	65.98	114	65.96	192	66.57	270	65.44	348	63.75
37	66.32	115	65.86	193	66.50	271	65.29	349	63.95
38	65.56	116	66.25	194	66.31	272	65.72	350	63.60
39	65.85	117	65.80	195	65.81	273	65.51	351	62.77

TABLE 1 (Continued)

Number	% Iron	Number	% Iron	Number	% Iron	Number	% Iron	Number	% Iron
40	66.11	118	65.77	196	65.98	274	65.07	352	66.43
41	66.02	119	65.75	197	65.97	275	65.28	353	66.74
42	65.88	120	65.98	198	66.16	276	65.44	354	66.40
43	66.18	121	66.49	199	66.23	277	65.40	355	66.28
44	66.11	122	66.41	200	66.18	278	65.63	356	65.92
45	66.03	123	66.38	201	65.88	279	65.72	357	64.44
46	66.11	124	66.11	202	66.07	280	65.47	358	66.28
47	66.28	125	65.88	203	65.93	281	65.09	359	65.34
48	66.09	126	66.39	204	65.50	282	65.04	360	66.55
49	66.08	127	66.75	205	66.35	283	65.26	361	66.72
50	66.24	128	66.56	206	66.24	284	65.65	362	66.50
51	66.01	129	66.83	207	66.33	285	65.47	363	66.62
52	66.33	130	66.34	208	66.27	286	65.24	364	66.02
53	66.59	131	66.78	209	66.41	287	66.23	365	65.53
54	65.82	132	66.67	210	66.67	288	65.74	366	66.19
55	66.21	133	66.86	211	66.62	289	65.66	367	65.78
56	66.32	134	66.82	212	65.99	290	65.73	368	65.92
57	66.10	135	66.82	213	66.23	291	65.41	369	66.10
58	65.77	136	66.86	214	66.77	292	65.20	370	65.80
59	65.86	137	66.68	215	66.29	293	65.24	371	64.66
60	65.80	138	66.80	216	66.20	294	66.26	372	65.80
61	66.07	139	66.86	217	66.54	295	65.50	373	65.55
62	65.94	140	66.81	218	66.45	296	65.73	374	65.54
63	66.16	141	66.69	219	66.57	297	65.28	375	65.90
64	65.86	142	66.68	220	66.69	298	64.98	376	66.30
65	65.79	143	66.71	221	66.19	299	65.19	377	66.70
66	65.77	144	66.34	222	66.01	300	65.68	378	66.75
67	65.69	145	66.18	223	66.02	301	65.68	379	66.80
68	65.87	146	65.78	224	65.88	302	65.81	380	66.70
69	65.88	147	65.63	225	66.30	303	65.85	381	66.29
70	65.31	148	66.81	226	66.56	304	65.73	382	66.26
71	65.30	149	65.80	227	66.06	305	65.69	383	66.34
72	65.61	150	65.80	228	66.27	306	65.75	384	66.64
73	65.00	151	66.54	229	65.94	307	65.46	385	66.54
74	64.76	152	66.51	230	65.14	308	65.73	386	66.60
75	64.45	153	63.95	231	65.41	309	65.21	387	66.58
76	65.02	154	65.86	232	65.18	310	65.48	388	66.49
77	65.25	155	65.95	233	65.50	311	64.85	389	66.70
78	64.75	156	65.56	234	65.70	312	65.10	390	66.38

Source: Takahashi, U. and Imaizami, M. "Sampling Experiment of Fine Iron Ore," *Reports of Statistical Application Research,* Union of Japanese Scientists and Engineers. Vol. 18, No. 1, 1971.

TABLE 2 DDT Analyses on Fish in the Tennessee River, 1980

Obs.	Location	Species	Length	Weight	DDT
1	FCM5	Channel Catfish	42.5	732	10.00
2	FCM5	Channel Catfish	44.0	795	16.00
3	FCM5	Channel Catfish	41.5	547	23.00
4	FCM5	Channel Catfish	39.0	465	21.00
5	FCM5	Channel Catfish	50.5	1252	50.00
6	FCM5	Channel Catfish	52.0	1255	150.00
7	LCM3	Channel Catfish	40.5	741	28.00
8	LCM3	Channel Catfish	48.0	1151	7.70
9	LCM3	Channel Catfish	48.0	1186	2.00
10	LCM3	Channel Catfish	43.5	754	19.00
11	LCM3	Channel Catfish	40.5	679	16.00
12	LCM3	Channel Catfish	47.5	985	5.40
13	SCM1	Channel Catfish	44.5	1133	2.60
14	SCM1	Channel Catfish	46.0	1139	3.10
15	SCM1	Channel Catfish	48.0	1186	3.50
16	SCM1	Channel Catfish	45.0	984	9.10
17	SCM1	Channel Catfish	43.0	965	7.80
18	SCM1	Channel Catfish	45.0	1084	4.10
19	TRM275	Channel Catfish	48.0	986	8.40
20	TRM275	Channel Catfish	45.0	1023	15.00
21	TRM275	Channel Catfish	49.0	1266	25.00
22	TRM275	Channel Catfish	50.0	1086	5.60
23	TRM275	Channel Catfish	46.0	1044	4.60
24	TRM275	Channel Catfish	52.0	1770	8.20
25	TRM280	Channel Catfish	48.0	1048	6.10
26	TRM280	Channel Catfish	51.0	1641	13.00
27	TRM280	Channel Catfish	48.5	1331	6.00
28	TRM280	Channel Catfish	51.0	1728	6.60
29	TRM280	Channel Catfish	44.0	917	5.50
30	TRM280	Channel Catfish	51.0	1398	11.00
31	TRM280	Small-Mouth Buffalo	49.0	1763	4.50
32	TRM280	Small-Mouth Buffalo	46.0	1459	4.20
33	TRM280	Small-Mouth Buffalo	52.0	2302	3.00
34	TRM280	Small-Mouth Buffalo	46.0	1614	2.30
35	TRM280	Small-Mouth Buffalo	46.0	1444	2.50
36	TRM280	Small-Mouth Buffalo	48.0	2006	6.80
37	TRM285	Channel Catfish	44.0	936	19.00
38	TRM285	Channel Catfish	42.0	1058	7.20
39	TRM285	Channel Catfish	42.5	800	6.00
40	TRM285	Channel Catfish	45.5	1087	10.00
41	TRM285	Channel Catfish	48.0	1329	12.00
42	TRM285	Channel Catfish	44.0	897	2.80
43	TRM285	Large-Mouth Bass	28.5	778	0.48
44	TRM285	Large-Mouth Bass	26.0	532	0.18
45	TRM285	Large-Mouth Bass	25.5	441	0.34
46	TRM285	Large-Mouth Bass	25.0	544	0.11
47	TRM285	Large-Mouth Bass	23.0	393	0.22
48	TRM285	Large-Mouth Bass	28.0	733	0.80

TABLE 2 (Continued)

Obs.	Location	Species	Length	Weight	DDT
49	TRM290	Channel Catfish	41.0	961	8.70
50	TRM290	Channel Catfish	44.0	886	22.00
51	TRM290	Channel Catfish	41.0	678	13.00
52	TRM290	Channel Catfish	42.0	1011	3.50
53	TRM290	Channel Catfish	42.5	947	9.30
54	TRM290	Channel Catfish	44.0	989	21.00
55	TRM290	Small-Mouth Buffalo	43.5	1291	3.40
56	TRM290	Small-Mouth Buffalo	46.5	1186	13.00
57	TRM290	Small-Mouth Buffalo	43.0	1293	5.60
58	TRM290	Small-Mouth Buffalo	47.0	1709	12.00
59	TRM290	Small-Mouth Buffalo	46.0	1425	21.00
60	TRM290	Small-Mouth Buffalo	41.0	1176	8.00
61	TRM295	Channel Catfish	36.0	980	12.00
62	TRM295	Channel Catfish	47.5	1176	6.00
63	TRM295	Channel Catfish	41.5	989	4.70
64	TRM295	Channel Catfish	49.5	1084	31.00
65	TRM295	Channel Catfish	46.0	1115	5.20
66	TRM295	Channel Catfish	46.5	724	27.00
67	TRM300	Channel Catfish	36.0	847	18.00
68	TRM300	Channel Catfish	37.0	876	7.50
69	TRM300	Channel Catfish	35.0	844	3.00
70	TRM300	Channel Catfish	36.0	908	13.00
71	TRM300	Channel Catfish	48.0	1358	7.30
72	TRM300	Channel Catfish	49.0	1019	15.00
73	TRM300	Small-Mouth Buffalo	35.5	1300	1.30
74	TRM300	Small-Mouth Buffalo	46.0	1365	4.80
75	TRM300	Small-Mouth Buffalo	45.0	1437	5.10
76	TRM300	Small-Mouth Buffalo	44.5	1460	5.10
77	TRM300	Small-Mouth Buffalo	49.0	1671	4.00
78	TRM300	Small-Mouth Buffalo	47.5	1717	10.00
79	TRM305	Channel Catfish	35.0	613	12.00
80	TRM305	Channel Catfish	51.0	353	22.00
81	TRM305	Channel Catfish	42.5	909	10.00
82	TRM305	Channel Catfish	38.0	886	11.00
83	TRM305	Channel Catfish	41.0	890	17.00
84	TRM305	Channel Catfish	47.0	1031	9.70
85	TRM310	Channel Catfish	45.0	1083	12.00
86	TRM310	Channel Catfish	45.5	864	4.70
87	TRM310	Channel Catfish	45.0	886	6.00
88	TRM310	Channel Catfish	45.0	965	3.80
89	TRM310	Channel Catfish	39.0	537	17.00
90	TRM310	Channel Catfish	40.5	630	12.00
91	TRM310	Small-Mouth Buffalo	46.0	1486	1.40
92	TRM310	Small-Mouth Buffalo	47.0	1743	6.10
93	TRM310	Small-Mouth Buffalo	48.5	2061	2.80
94	TRM310	Small-Mouth Buffalo	48.0	1707	4.80
95	TRM310	Small-Mouth Buffalo	38.0	862	5.70
96	TRM310	Small-Mouth Buffalo	38.5	911	3.30

TABLE 2 (Continued)

Obs.	Location	Species	Length	Weight	DDT
97	TRM315	Channel Catfish	29.5	476	3.30
98	TRM315	Channel Catfish	42.0	743	3.70
99	TRM315	Channel Catfish	47.5	1128	9.90
100	TRM315	Channel Catfish	43.5	848	6.80
101	TRM315	Channel Catfish	47.5	1091	13.00
102	TRM315	Channel Catfish	43.5	715	8.80
103	TRM320	Channel Catfish	47.5	983	57.00
104	TRM320	Channel Catfish	51.5	1251	96.00
105	TRM320	Channel Catfish	49.5	1255	360.00
106	TRM320	Channel Catfish	47.0	1152	130.00
107	TRM320	Channel Catfish	47.5	1085	13.00
108	TRM320	Channel Catfish	47.0	1118	61.00
109	TRM320	Small-Mouth Buffalo	36.0	1285	12.00
110	TRM320	Small-Mouth Buffalo	34.5	1178	33.00
111	TRM320	Small-Mouth Buffalo	44.5	1492	48.00
112	TRM320	Small-Mouth Buffalo	46.0	1524	10.00
113	TRM320	Small-Mouth Buffalo	46.0	1473	44.00
114	TRM320	Small-Mouth Buffalo	32.5	520	0.43
115	TRM325	Channel Catfish	46.0	863	1100.00
116	TRM325	Channel Catfish	40.0	549	9.40
117	TRM325	Channel Catfish	43.5	810	4.10
118	TRM325	Channel Catfish	46.5	908	2.80
119	TRM325	Channel Catfish	43.0	804	0.74
120	TRM325	Channel Catfish	47.5	1179	14.00
121	TRM330	Channel Catfish	32.0	556	22.00
122	TRM330	Channel Catfish	40.5	659	9.10
123	TRM330	Channel Catfish	51.5	1229	140.00
124	TRM330	Channel Catfish	48.0	1050	4.20
125	TRM330	Channel Catfish	47.0	952	12.00
126	TRM330	Channel Catfish	41.0	826	2.00
127	TRM330	Small-Mouth Buffalo	33.5	599	0.30
128	TRM330	Small-Mouth Buffalo	47.0	1704	1.20
129	TRM340	Channel Catfish	50.0	1207	7.10
130	TRM340	Channel Catfish	45.0	911	180.00
131	TRM340	Channel Catfish	49.0	1498	1.50
132	TRM340	Channel Catfish	49.5	1496	2.40
133	TRM340	Channel Catfish	50.0	1142	4.30
134	TRM340	Channel Catfish	45.0	879	3.90
135	TRM340	Small-Mouth Buffalo	32.5	525	0.99
136	TRM340	Small-Mouth Buffalo	38.0	806	0.45
137	TRM340	Small-Mouth Buffalo	38.5	694	2.50
138	TRM340	Small-Mouth Buffalo	36.0	643	0.25
139	TRM345	Large-Mouth Bass	26.5	514	0.58
140	TRM345	Large-Mouth Bass	23.5	358	2.00
141	TRM345	Large-Mouth Bass	30.0	856	2.20
142	TRM345	Large-Mouth Bass	29.0	793	7.40
143	TRM345	Large-Mouth Bass	17.5	173	0.35
144	TRM345	Large-Mouth Bass	36.0	1433	1.90

TABLE 3 University of Florida Study of Body Cholesterol for 107 People

ID	Age	Sex	TCHOL	TRI	HDL	LDL	VLDL	RATIO
1	42	M	178.0	87.0	42.0	118.60	17.40	2.82381
2	38	M	195.9	74.4	48.3	132.72	14.88	2.74783
3	50	M	236.0	410.0	23.0	131.00	82.00	5.69565
4	22	M	158.0	138.0	35.0	95.40	27.60	2.72571
5	44	F	277.0	117.0	82.0	171.60	23.40	2.09268
6	45	F	267.0	225.0	35.0	187.00	45.00	5.34286
7	62	M	237.0	189.0	39.0	160.20	37.80	4.10769
8	42	F	248.0	211.0	31.4	174.40	42.20	5.55414
9	38	M	230.0	158.0	47.0	151.40	31.60	3.22128
10	53	M	181.0	123.0	40.0	116.40	24.60	2.91000
11	50	F	172.0	80.0	77.0	79.00	16.00	1.02597
12	46	F	161.0	102.0	65.0	75.60	20.40	1.16308
13	42	F	223.0	124.0	54.0	144.20	24.80	2.67037
14	46	M	126.0	134.0	29.0	70.20	26.80	2.42069
15	44	M	202.0	234.0	32.0	123.20	46.80	3.85000
16	22	M	157.0	66.0	46.0	97.80	13.20	2.12609
17	48	F	284.0	233.0	78.0	159.40	46.60	2.04359
18	48	M	159.0	73.0	51.0	93.40	14.60	1.83137
19	50	F	223.0	88.0	68.0	137.40	17.60	2.02059
20	26	M	311.0	253.0	27.2	233.20	50.60	8.57353
21	24	F	154.0	74.0	57.0	82.20	14.80	1.44211
22	45	F	218.0	210.0	77.0	99.00	42.00	1.28571
23	48	M	237.0	162.0	44.0	160.60	32.40	3.65000
24	45	M	237.0	143.0	52.0	156.40	28.60	3.00769
25	54	F	228.0	119.0	95.0	109.20	23.80	1.14947
26	54	M	225.0	122.0	65.0	135.60	24.40	2.08615
27	50	M	289.0	74.0	38.0	236.20	14.80	6.21579
28	45	F	279.0	266.0	54.0	171.80	53.20	3.18148
29	40	M	223.0	277.0	34.0	133.60	55.40	3.92941
30	50	F	200.0	128.0	65.0	109.40	25.60	1.68308
31	52	M	249.0	108.0	49.7	177.70	21.60	3.57545
32	24	F	171.0	104.0	57.2	93.00	20.80	1.62587
33	52	F	131.0	154.0	43.0	57.20	30.80	1.33023
34	50	F	250.0	160.0	51.0	167.00	32.00	3.27451
35	23	F	264.0	110.0	53.2	188.80	22.00	3.54887
36	27	M	176.0	179.0	27.0	113.20	35.80	4.19259

(*continued*)

TABLE 3 (Continued)

ID	Age	Sex	TCHOL	TRI	HDL	LDL	VLDL	RATIO
37	38	F	216.0	201.0	52.0	123.80	40.20	2.38077
38	38	M	205.0	88.0	47.3	140.10	17.60	2.96195
39	47	F	153.0	113.0	51.0	79.40	22.60	1.55686
40	50	M	399.0	876.0	62.0	161.80	175.20	2.60968
41	50	F	158.0	80.0	97.0	45.00	16.00	0.46392
42	35	M	193.0	147.0	38.0	125.60	29.40	3.30526
43	41	F	259.0	82.5	79.6	162.90	16.50	2.04648
44	68	M	274.0	114.0	77.0	174.20	22.80	2.26234
45	38	F	170.0	94.0	53.0	98.20	18.80	1.85283
46	45	F	196.0	70.0	78.0	104.00	14.00	1.33333
47	48	M	205.0	191.0	48.0	118.80	38.20	2.47500
48	55	F	272.0	153.0	35.0	206.40	30.60	5.89714
49	44	M	120.0	90.0	44.0	58.00	18.00	1.31818
50	45	F	250.0	180.0	44.0	170.00	36.00	3.86364
51	40	F	337.0	149.0	43.6	263.60	29.80	6.04587
52	45	F	194.0	139.0	62.3	103.90	27.80	1.66774
53	31	F	264.0	148.0	57.0	177.40	29.60	3.11228
54	62	F	224.0	65.0	70.0	141.00	13.00	2.01429
55	48	M	234.0	81.0	47.0	170.80	16.20	3.63404
56	45	M	242.0	189.0	37.0	167.20	37.80	4.51892
57	28	F	190.0	77.0	67.0	107.60	15.40	1.60597
58	35	M	226.0	97.0	56.0	150.60	19.40	2.68929
59	38	M	241.0	79.0	61.0	164.20	15.80	2.69180
60	47	M	269.0	277.0	54.0	159.60	55.40	2.95556
61	50	M	196.0	93.0	45.0	132.40	18.60	2.94222
62	50	M	226.0	96.0	33.0	173.80	19.20	5.26667
63	32	M	185.0	117.0	46.9	114.70	23.40	2.44563
64	48	M	210.0	97.0	67.0	123.60	19.40	1.84478
65	45	M	298.0	276.0	37.0	205.80	55.20	5.56216
66	40	F	278.0	326.0	43.2	169.60	65.20	3.92593
67	40	F	205.0	170.0	41.0	130.00	34.00	3.17073
68	45	M	187.0	149.0	43.0	114.20	29.80	2.65581
69	45	F	269.0	142.0	35.0	205.60	28.40	5.87429
70	22	M	205.0	203.0	33.0	131.40	40.60	3.98182
71	50	M	221.0	144.0	49.0	143.20	28.80	2.92245
72	50	M	203.0	64.0	58.0	132.20	12.80	2.27931

TABLE 3 (Continued)

ID	Age	Sex	TCHOL	TRI	HDL	LDL	VLDL	RATIO
73	45	F	156.0	124.0	31.0	100.20	24.80	3.23226
74	40	M	198.0	264.0	31.0	114.20	52.80	3.68387
75	38	F	312.0	112.0	51.5	238.10	22.40	4.62330
76	24	F	126.0	80.0	43.0	67.00	16.00	1.55814
77	45	F	193.0	141.0	46.0	118.80	28.20	2.58261
78	55	M	169.0	164.0	37.6	98.60	32.80	2.62234
79	30	M	285.0	336.0	34.0	183.80	67.20	5.40588
80	40	F	181.0	262.0	40.0	88.60	52.40	2.21500
81	30	F	123.0	106.0	59.0	42.80	21.20	0.72542
82	30	M	234.0	124.0	46.0	163.20	24.80	3.54783
83	19	F	130.0	45.0	44.0	77.00	9.00	1.75000
84	38	M	219.0	186.0	32.0	149.80	37.20	4.68125
85	45	F	248.0	338.0	80.0	100.40	67.60	1.25500
86	45	M	235.0	184.0	73.0	125.20	36.80	1.71507
87	44	F	235.0	76.0	65.0	154.80	15.20	2.38154
88	35	M	212.0	98.0	54.0	138.40	19.60	2.56296
89	45	M	257.0	171.0	48.0	174.80	34.20	3.64167
90	50	M	215.0	161.0	42.0	140.80	32.20	3.35238
91	36	M	217.0	308.0	33.0	122.40	61.60	3.70909
92	50	M	202.0	457.0	16.8	93.80	91.40	5.58333
93	17	F	295.0	162.0	55.0	207.60	32.40	3.77455
94	17	M	187.0	114.0	43.0	121.20	22.80	2.81860
95	22	M	168.0	94.0	57.0	92.20	18.80	1.61754
96	48	F	235.0	185.0	60.0	138.00	37.00	2.30000
97	50	M	229.0	152.0	48.0	150.60	30.40	3.13750
98	60	F	152.0	90.0	64.0	70.00	18.00	1.09375
99	48	M	209.0	114.0	39.0	147.20	22.80	3.77436
100	34	M	155.0	173.0	26.0	94.40	34.60	3.63077
101	25	M	178.0	139.0	38.0	112.20	27.80	2.95263
102	19	M	168.0	96.0	39.0	109.80	19.20	2.81538
103	65	F	414.0	93.4	52.8	342.52	18.68	6.48712
104	22	M	250.0	134.0	48.0	175.20	26.80	3.65000
105	22	F	191.0	84.0	90.0	84.20	16.80	0.93556
106	34	F	138.0	75.0	45.0	78.00	15.00	1.73333
107	29	F	257.0	143.0	73.0	155.40	28.60	2.12877

Source: J. J. Cerda, M.D., University of Florida. Acknowledgment is given to grants 4910290542712 (Florida Citrus Commission) and RR-00082 (General Clinical Research Branch, Division of Research Resources, National Institutes of Health).

appendix 2

Statistical Tables

Table 1 **Random Numbers**

Table 2 **Binomial Probabilities (n = 5, 10, 15, 20, 25)**

Table 3 **Areas Under the Normal Curve**

Table 4 **Upper-Tail Values for the Student's *t* Distribution**

Table 5 **Upper-Tail Values of the Chi-Square Distribution**

Table 6 **Upper-Tail Values for an *F* Distribution**

Table 7 **Upper-Tail Values for the Studentized Range**

Table 8 **Lower-Tail Values for the Mann–Whitney *U* Statistic**

Table 9 **Lower-Tail Values for the Wilcoxon Signed Ranks Test**

Table 10 **Upper-Tail Values of Spearman's Rank Correlation Coefficient**

TABLE 1 Random Numbers

Row \ Column	1	2	3	4	5	6	7	8	9	10	11	12	13	14
1	10480	15011	01536	02011	81647	91646	69179	14194	62590	36207	20969	99570	91291	90700
2	22368	46573	25595	85393	30995	89198	27982	53402	93965	34095	52666	19174	39615	99505
3	24130	48360	22527	97265	76393	64809	15179	24830	49340	32081	30680	19655	63348	58629
4	42167	93093	06243	61680	07856	16376	39440	53537	71341	57004	00849	74917	97758	16379
5	37570	39975	81837	16656	06121	91782	60468	81305	49684	60672	14110	06927	01263	54613
6	77921	06907	11008	42751	27756	53498	18602	70659	90655	15053	21916	81825	44394	42880
7	99562	72905	56420	69994	98872	31016	71194	18738	44013	48840	63213	21069	10634	12952
8	96301	91977	05463	07972	18876	20922	94595	56869	69014	60045	18425	84903	42508	32307
9	89579	14342	63661	10281	17453	18103	57740	84378	25331	12566	58678	44947	05585	56941
10	85475	36857	53342	53988	53060	59533	38867	62300	08158	17983	16439	11458	18593	64952
11	28918	69578	88231	33276	70997	79936	56865	05859	90106	31595	01547	85590	91610	78188
12	63553	40961	48235	03427	49626	69445	18663	72695	52180	20847	12234	90511	33703	90322
13	09429	93969	52636	92737	88974	33488	36320	17617	30015	08272	84115	27156	30613	74952
14	10365	61129	87529	85689	48237	52267	67689	93394	01511	26358	85104	20285	29975	89868
15	07119	97336	71048	08178	77233	13916	47564	81056	97735	85977	29372	74461	28551	90707
16	51085	12765	51821	51259	77452	16308	60756	92144	49442	53900	70960	63990	75601	40719
17	02368	21382	52404	60268	89368	19885	55322	44819	01188	65255	64835	44919	05944	55157
18	01011	54092	33362	94904	31273	04146	18594	29852	71585	85030	51132	01915	92747	64951
19	52162	53916	46369	58586	23216	14513	83149	98736	23495	64350	94738	17752	35156	35749
20	07056	97628	33787	09998	42698	06691	76988	13602	51851	46104	88916	19509	25625	58104
21	48663	91245	85828	14346	09172	30168	90229	04734	59193	22178	30421	61666	99904	32812
22	54164	58492	22421	74103	47070	25306	76468	26384	58151	06646	21524	15227	96909	44592
23	32639	32363	05597	24200	13363	38005	94342	28728	35806	06912	17012	64161	18296	22851
24	29334	27001	87637	87308	58731	00256	45834	15398	46557	41135	10367	07684	36188	18510
25	02488	33062	28834	07351	19731	92420	60952	61280	50001	67658	32586	86679	50720	94953
26	81525	72295	04839	96423	24878	82651	66566	14778	76797	14780	13300	87074	79666	95725
27	29676	20591	68086	26432	46901	20849	89768	81536	86645	12659	92259	57102	80428	25280
28	00742	57392	39064	66432	84673	40027	32832	61362	98947	96067	64760	64584	96096	98253
29	05366	04213	25669	26422	44407	44048	37937	63904	45766	66134	75470	66520	34693	90449
30	91921	26418	64117	94305	26766	25940	39972	22209	71500	64568	91402	42416	07844	69618
31	00582	04711	87917	77341	42206	35126	74087	99547	81817	42607	43808	76655	62028	76630
32	00725	69884	62797	56170	86324	88072	76222	36086	84637	93161	76038	65855	77919	88006
33	69011	65795	95876	55293	18988	27354	26575	08625	40801	59920	29841	80150	12777	48501
34	25976	57948	29888	88604	67917	48708	18912	82271	65424	69774	33611	54262	85963	03547
35	09763	83473	73577	12908	30883	18317	28290	35797	05998	41688	34952	37888	38917	88050
36	91576	42595	27958	30134	04024	86385	29880	99730	55536	84855	29080	09250	79656	73211
37	17955	56349	90999	49127	20044	59931	06115	20542	18059	02008	73708	83517	36103	42791
38	46503	18584	18845	49618	02304	51038	20655	58727	28168	15475	56942	53389	20562	87338
39	92157	89634	94824	78171	84610	82834	09922	25417	44137	48413	25555	21246	35509	20468
40	14577	62765	35605	81263	39667	47358	56873	56307	61607	49518	89656	20103	77490	18062
41	98427	07523	33362	64270	01638	92477	66969	98420	04880	45585	46565	04102	46880	45709
42	34914	63976	88720	82765	34476	17032	87589	40836	32427	70002	70663	88863	77775	69348
43	70060	28277	39475	46473	23219	53416	94970	25832	69975	94884	19661	72828	00102	66794
44	53976	54914	06990	67245	68350	82948	11398	42878	80287	88267	47363	46634	06541	97809
45	76072	29515	40980	07391	58745	25774	22987	80059	39911	96189	41151	14222	60697	59583
46	90725	52210	83974	29992	65831	38857	50490	83765	55657	14361	31720	57375	56228	41546
47	64364	67412	33339	31926	14883	24413	59744	92351	97473	89286	35931	04110	23726	51900
48	08962	00358	31662	25388	61642	34072	81249	35648	56891	69352	48373	45578	78547	81788
49	95012	68379	93526	70765	10592	04542	76463	54328	02349	17247	28865	14777	62730	92277
50	15664	10493	20492	38391	91132	21999	59516	81652	27195	48223	46751	22923	32261	85653

TABLE 1 (Continued)

Row \ Column	1	2	3	4	5	6	7	8	9	10	11	12	13	14
51	16408	81899	04153	53381	79401	21438	83035	92350	36693	31238	59649	91754	72772	02338
52	18629	81953	05520	91962	04739	13092	97662	24822	94730	06496	35090	04822	86774	98289
53	73115	35101	47498	87637	99016	71060	88824	71013	18735	20286	23153	72924	35165	43040
54	57491	16703	23167	49323	45021	33132	12544	41035	80780	45393	44812	12515	98931	91202
55	30405	83946	23792	14422	15059	45799	22716	19792	09983	74353	68668	30429	70735	25499
56	16631	35006	85900	98275	32388	52390	16815	69298	82732	38480	73817	32523	41961	44437
57	96773	20206	42559	78985	05300	22164	24369	54224	35083	19687	11052	91491	60383	19746
58	38935	64202	14349	82674	66523	44133	00697	35552	35970	19124	63318	29686	03387	59846
59	31624	76384	17403	53363	44167	64486	64758	75366	76554	31601	12614	33072	60332	92325
60	78919	19474	23632	27889	47914	02584	37680	20801	72152	39339	34806	08930	85001	87820
61	03931	33309	57047	74211	63445	17361	62825	39908	05607	91284	68833	25570	38818	46920
62	74426	33278	43972	10119	89917	15665	52872	73823	73144	88662	88970	74492	51805	99378
63	09066	00903	20795	95452	92648	45454	09552	88815	16553	51125	79375	97596	16296	66092
64	42238	12426	87025	14267	20979	04508	64535	31355	86064	29472	47689	05974	52468	16834
65	16153	08002	26504	41744	81959	65642	74240	56302	00033	67107	77510	70625	28725	34191
66	21457	40742	29820	96783	29400	21840	15035	34537	33310	06116	95240	15957	16572	06004
67	21581	57802	02050	89728	17937	37621	47075	42080	97403	48626	68995	43805	33386	21597
68	55612	78095	83197	33732	05810	24813	86902	60397	16489	03264	88525	42786	05269	92532
69	44657	66999	99324	51281	84463	60563	79312	93454	68876	25471	93911	25650	12682	73572
70	91340	84979	46949	81973	37949	61023	43997	15263	80644	43942	89203	71795	99533	50501
71	91227	21199	31935	27022	84067	05462	35216	14486	29891	68607	41867	14951	91696	85065
72	50001	38140	66321	19924	72163	09538	12151	06878	91903	18749	34405	56087	82790	70925
73	65390	05224	72958	28609	81406	39147	25549	48542	42627	45233	57202	94617	23772	07896
74	27504	96131	83944	41575	10573	08619	64482	73923	36152	05184	94142	25299	84387	34925
75	37169	94851	39117	89632	00959	16487	65536	49071	39782	17095	02330	74301	00275	48280
76	11508	70225	51111	38351	19444	66499	71945	05422	13442	78675	84081	66938	93654	59894
77	37449	30362	06694	54690	04052	53115	62757	95348	78662	11163	81651	50245	34971	52924
78	46515	70331	85922	38329	57015	15765	97161	17869	45349	61796	66345	81073	49106	79860
79	30986	81223	42416	58353	21532	30502	32305	86482	05174	07901	54339	58861	74818	46942
80	63798	64995	46583	09785	44160	78128	83991	42865	92520	83531	80377	35909	81250	54238
81	82486	84846	99254	67632	43218	50076	21361	64816	51202	88124	41870	52689	51275	83556
82	21885	32906	92431	09060	64297	51674	64126	62570	26123	05155	59194	52799	28225	85762
83	60336	98782	07408	53458	13564	59089	26445	29789	85205	41001	12535	12133	14645	23541
84	43937	46891	24010	25560	86355	33941	25786	54990	71899	15475	95434	98227	21824	19585
85	97656	63175	89303	16275	07100	92063	21942	18611	47348	20203	18534	03862	78095	50136
86	03299	01221	05418	38982	55758	92237	26759	86367	21216	98442	08303	56613	91511	75928
87	79626	06486	03574	17668	07785	76020	79924	25651	83325	88428	85076	72811	22717	50585
88	85636	68335	47539	03129	65651	11977	02510	26113	99447	68645	34327	15152	55230	93448
89	18039	14367	61337	06177	12143	46609	32989	74014	64708	00533	35398	58408	13261	47908
90	08362	15656	60627	36478	65648	16764	53412	09013	07832	41574	17639	82163	60859	75567
91	79556	29068	04142	16268	15387	12856	66227	38358	22478	73373	88732	09443	82558	05250
92	92608	82674	27072	32534	17075	27698	98204	63863	11951	34648	88022	56148	34925	57031
93	23982	25835	40055	67006	12293	02753	14827	23235	35071	99704	37543	11601	35503	85171
94	09915	96306	05908	97901	28395	14186	00821	80703	70426	75647	76310	88717	37890	40129
95	59037	33300	26695	62247	69927	76123	50842	43834	86654	70959	79725	93872	28117	19233
96	42488	78077	69882	61657	34136	79180	97526	43092	04098	73571	80799	76536	71255	64239
97	46764	86273	63003	93017	31204	36692	40202	35275	57306	55543	53203	18098	47625	88684
98	03237	45430	55417	63282	90816	17349	88298	90183	36600	78406	06216	95787	42579	90730
99	86591	81482	52667	61582	14972	90053	89534	76036	49199	43716	97548	04379	46370	28672
100	38534	01715	94964	87288	65680	43772	39560	12918	86537	62738	19636	51132	25739	56947

Source: Abridged from W. H. Beyer (ed.) *CRC Standard Mathematical Tables,* 24th edition. (Cleveland: *The Chemical Rubber Company*), 1976.

TABLE 2 Binomial Probabilities.
Tabulated values are $p(x)$.

a. n=5

	p												
x	.01	.05	.1	.2	.3	.4	.5	.6	.7	.8	.9	.95	.99
0	.9510	.7738	.5905	.3277	.1681	.0778	.0313	.0102	.0024	.0003	.0000	.0000	.0000
1	.0480	.2036	.3280	.4096	.3601	.2592	.1563	.0768	.0283	.0064	.0005	.0000	.0000
2	.0010	.0214	.0729	.2048	.3087	.3456	.3125	.2304	.1323	.0512	.0081	.0011	.0000
3	.0000	.0011	.0081	.0512	.1323	.2304	.3125	.3456	.3087	.2048	.0729	.0214	.0010
4	.0000	.0000	.0004	.0064	.0283	.0768	.1563	.2592	.3601	.4096	.3280	.2036	.0480
5	.0000	.0000	.0000	.0003	.0024	.0102	.0313	.0778	.1681	.3277	.5905	.7738	.9510

b. n=10

	p												
x	.01	.05	.1	.2	.3	.4	.5	.6	.7	.8	.9	.95	.99
0	.9044	.5987	.3487	.1074	.0282	.0060	.0010	.0001	.0000	.0000	.0000	.0000	.0000
1	.0914	.3151	.3874	.2684	.1211	.0403	.0098	.0016	.0001	.0000	.0000	.0000	.0000
2	.0042	.0746	.1937	.3020	.2335	.1209	.0439	.0106	.0014	.0001	.0000	.0000	.0000
3	.0001	.0105	.0574	.2013	.2668	.2150	.1172	.0425	.0090	.0008	.0000	.0000	.0000
4	.0000	.0010	.0112	.0881	.2001	.2508	.2051	.1115	.0368	.0055	.0001	.0000	.0000
5	.0000	.0001	.0015	.0264	.1029	.2007	.2461	.2007	.1029	.0264	.0015	.0001	.0000
6	.0000	.0000	.0001	.0055	.0368	.1115	.2051	.2508	.2001	.0881	.0112	.0010	.0000
7	.0000	.0000	.0000	.0008	.0090	.0425	.1172	.2150	.2668	.2013	.0574	.0105	.0001
8	.0000	.0000	.0000	.0001	.0014	.0106	.0439	.1209	.2335	.3020	.1937	.0746	.0042
9	.0000	.0000	.0000	.0000	.0001	.0016	.0098	.0403	.1211	.2684	.3874	.3151	.0914
10	.0000	.0000	.0000	.0000	.0000	.0001	.0010	.0060	.0282	.1074	.3487	.5987	.9044

c. n=15

	p												
x	.01	.05	.1	.2	.3	.4	.5	.6	.7	.8	.9	.95	.99
0	.8601	.4633	.2059	.0352	.0047	.0005	.0000	.0000	.0000	.0000	.0000	.0000	.0000
1	.1303	.3658	.3432	.1319	.0305	.0047	.0005	.0000	.0000	.0000	.0000	.0000	.0000
2	.0092	.1348	.2669	.2309	.0916	.0219	.0032	.0003	.0000	.0000	.0000	.0000	.0000
3	.0004	.0307	.1285	.2501	.1700	.0634	.0139	.0016	.0001	.0000	.0000	.0000	.0000
4	.0000	.0049	.0428	.1876	.2186	.1268	.0417	.0074	.0006	.0000	.0000	.0000	.0000
5	.0000	.0006	.0105	.1032	.2061	.1859	.0916	.0245	.0030	.0001	.0000	.0000	.0000
6	.0000	.0000	.0019	.0430	.1472	.2066	.1527	.0612	.0116	.0007	.0000	.0000	.0000
7	.0000	.0000	.0003	.0138	.0811	.1771	.1964	.1181	.0348	.0035	.0000	.0000	.0000
8	.0000	.0000	.0000	.0035	.0348	.1181	.1964	.1771	.0811	.0138	.0003	.0000	.0000
9	.0000	.0000	.0000	.0007	.0116	.0612	.1527	.2066	.1472	.0430	.0019	.0000	.0000
10	.0000	.0000	.0000	.0001	.0030	.0245	.0916	.1859	.2061	.1032	.0105	.0006	.0000
11	.0000	.0000	.0000	.0000	.0006	.0074	.0417	.1268	.2186	.1876	.0428	.0049	.0000
12	.0000	.0000	.0000	.0000	.0001	.0016	.0139	.0634	.1700	.2501	.1285	.0307	.0004
13	.0000	.0000	.0000	.0000	.0000	.0003	.0032	.0219	.0916	.2309	.2669	.1348	.0092
14	.0000	.0000	.0000	.0000	.0000	.0000	.0005	.0047	.0305	.1319	.3432	.3658	.1303
15	.0000	.0000	.0000	.0000	.0000	.0000	.0000	.0005	.0047	.0352	.2059	.4633	.8601

TABLE 2 (Continued)
d. $n=20$

						p							
x	.01	.05	.1	.2	.3	.4	.5	.6	.7	.8	.9	.95	.99
0	.8179	.3585	.1216	.0115	.0008	.0000	.0000	.0000	.0000	.0000	.0000	.0000	.0000
1	.1652	.3774	.2702	.0576	.0068	.0005	.0000	.0000	.0000	.0000	.0000	.0000	.0000
2	.0159	.1887	.2852	.1369	.0278	.0031	.0002	.0000	.0000	.0000	.0000	.0000	.0000
3	.0010	.0596	.1901	.2054	.0716	.0123	.0011	.0000	.0000	.0000	.0000	.0000	.0000
4	.0000	.0133	.0898	.2182	.1304	.0350	.0046	.0003	.0000	.0000	.0000	.0000	.0000
5	.0000	.0022	.0319	.1746	.1789	.0746	.0148	.0013	.0000	.0000	.0000	.0000	.0000
6	.0000	.0003	.0089	.1091	.1916	.1244	.0370	.0049	.0002	.0000	.0000	.0000	.0000
7	.0000	.0000	.0020	.0545	.1643	.1659	.0739	.0146	.0010	.0000	.0000	.0000	.0000
8	.0000	.0000	.0004	.0222	.1144	.1797	.1201	.0355	.0039	.0001	.0000	.0000	.0000
9	.0000	.0000	.0001	.0074	.0654	.1597	.1602	.0710	.0120	.0005	.0000	.0000	.0000
10	.0000	.0000	.0000	.0020	.0308	.1171	.1762	.1171	.0308	.0020	.0000	.0000	.0000
11	.0000	.0000	.0000	.0005	.0120	.0710	.1602	.1597	.0654	.0074	.0001	.0000	.0000
12	.0000	.0000	.0000	.0001	.0039	.0355	.1201	.1797	.1144	.0222	.0004	.0000	.0000
13	.0000	.0000	.0000	.0000	.0010	.0146	.0739	.1659	.1643	.0545	.0020	.0000	.0000
14	.0000	.0000	.0000	.0000	.0002	.0049	.0370	.1244	.1916	.1091	.0089	.0003	.0000
15	.0000	.0000	.0000	.0000	.0000	.0013	.0148	.0746	.1789	.1746	.0319	.0022	.0000
16	.0000	.0000	.0000	.0000	.0000	.0003	.0046	.0350	.1304	.2182	.0898	.0133	.0000
17	.0000	.0000	.0000	.0000	.0000	.0000	.0011	.0123	.0716	.2054	.1901	.0596	.0010
18	.0000	.0000	.0000	.0000	.0000	.0000	.0002	.0031	.0278	.1369	.2852	.1887	.0159
19	.0000	.0000	.0000	.0000	.0000	.0000	.0000	.0005	.0068	.0576	.2702	.3774	.1652
20	.0000	.0000	.0000	.0000	.0000	.0000	.0000	.0000	.0008	.0115	.1216	.3585	.8179

e. $n=25$

						p							
x	.01	.05	.1	.2	.3	.4	.5	.6	.7	.8	.9	.95	.99
0	.7778	.2774	.0718	.0038	.0001	.0000	.0000	.0000	.0000	.0000	.0000	.0000	.0000
1	.1964	.3650	.1994	.0236	.0014	.0000	.0000	.0000	.0000	.0000	.0000	.0000	.0000
2	.0238	.2305	.2659	.0708	.0074	.0004	.0000	.0000	.0000	.0000	.0000	.0000	.0000
3	.0018	.0930	.2265	.1358	.0243	.0019	.0001	.0000	.0000	.0000	.0000	.0000	.0000
4	.0001	.0269	.1384	.1867	.0572	.0071	.0004	.0000	.0000	.0000	.0000	.0000	.0000
5	.0000	.0060	.0646	.1960	.1030	.0199	.0016	.0000	.0000	.0000	.0000	.0000	.0000
6	.0000	.0010	.0239	.1633	.1472	.0442	.0053	.0002	.0000	.0000	.0000	.0000	.0000
7	.0000	.0001	.0072	.1108	.1712	.0800	.0143	.0009	.0000	.0000	.0000	.0000	.0000
8	.0000	.0000	.0018	.0623	.1651	.1200	.0322	.0031	.0001	.0000	.0000	.0000	.0000
9	.0000	.0000	.0004	.0294	.1336	.1511	.0609	.0088	.0004	.0000	.0000	.0000	.0000
10	.0000	.0000	.0001	.0118	.0916	.1612	.0974	.0212	.0013	.0000	.0000	.0000	.0000
11	.0000	.0000	.0000	.0040	.0536	.1465	.1328	.0434	.0042	.0001	.0000	.0000	.0000
12	.0000	.0000	.0000	.0012	.0268	.1140	.1550	.0760	.0115	.0003	.0000	.0000	.0000
13	.0000	.0000	.0000	.0003	.0115	.0760	.1550	.1140	.0268	.0012	.0000	.0000	.0000
14	.0000	.0000	.0000	.0001	.0042	.0434	.1328	.1465	.0536	.0040	.0000	.0000	.0000
15	.0000	.0000	.0000	.0000	.0013	.0212	.0974	.1612	.0916	.0118	.0001	.0000	.0000
16	.0000	.0000	.0000	.0000	.0004	.0088	.0609	.1511	.1336	.0294	.0004	.0000	.0000
17	.0000	.0000	.0000	.0000	.0001	.0031	.0322	.1200	.1651	.0623	.0018	.0000	.0000
18	.0000	.0000	.0000	.0000	.0000	.0009	.0143	.0800	.1712	.1108	.0072	.0001	.0000
19	.0000	.0000	.0000	.0000	.0000	.0002	.0053	.0442	.1472	.1633	.0239	.0010	.0000
20	.0000	.0000	.0000	.0000	.0000	.0000	.0016	.0199	.1030	.1960	.0646	.0060	.0000
21	.0000	.0000	.0000	.0000	.0000	.0000	.0004	.0071	.0572	.1867	.1384	.0269	.0001
22	.0000	.0000	.0000	.0000	.0000	.0000	.0001	.0019	.0243	.1358	.2265	.0930	.0018
23	.0000	.0000	.0000	.0000	.0000	.0000	.0000	.0004	.0074	.0708	.2659	.2305	.0238
24	.0000	.0000	.0000	.0000	.0000	.0000	.0000	.0000	.0014	.0236	.1994	.3650	.1964
25	.0000	.0000	.0000	.0000	.0000	.0000	.0000	.0000	.0001	.0038	.0718	.2774	.7778

TABLE 3 Areas Under the Normal Curve

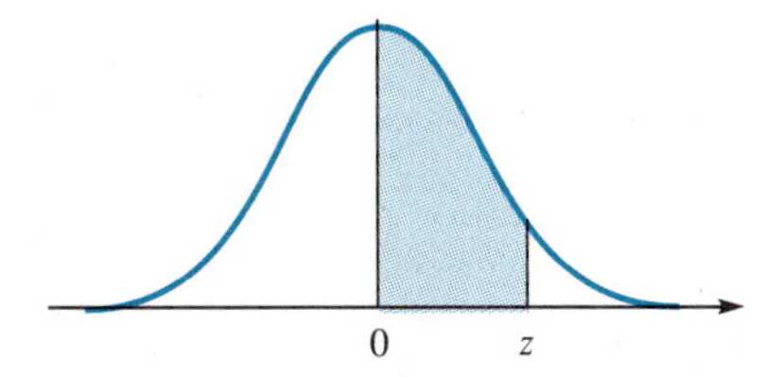

z	.00	.01	.02	.03	.04	.05	.06	.07	.08	.09
.0	.0000	.0040	.0080	.0120	.0160	.0199	.0239	.0279	.0319	.0359
.1	.0398	.0438	.0478	.0517	.0557	.0596	.0636	.0675	.0714	.0753
.2	.0793	.0832	.0871	.0910	.0948	.0987	.1026	.1064	.1103	.1141
.3	.1179	.1217	.1255	.1293	.1331	.1368	.1406	.1443	.1480	.1517
.4	.1554	.1591	.1628	.1664	.1700	.1736	.1772	.1808	.1844	.1879
.5	.1915	.1950	.1985	.2019	.2054	.2088	.2123	.2157	.2190	.2224
.6	.2257	.2291	.2324	.2357	.2389	.2422	.2454	.2486	.2517	.2549
.7	.2580	.2611	.2642	.2673	.2704	.2734	.2764	.2794	.2823	.2852
.8	.2881	.2910	.2939	.2967	.2995	.3023	.3051	.3078	.3106	.3133
.9	.3159	.3186	.3212	.3238	.3264	.3289	.3315	.3340	.3365	.3389
1.0	.3413	.3438	.3461	.3485	.3508	.3531	.3554	.3577	.3599	.3621
1.1	.3643	.3665	.3686	.3708	.3729	.3749	.3770	.3790	.3810	.3830
1.2	.3849	.3869	.3888	.3907	.3925	.3944	.3962	.3980	.3997	.4015
1.3	.4032	.4049	.4066	.4082	.4099	.4115	.4131	.4147	.4162	.4177
1.4	.4192	.4207	.4222	.4236	.4251	.4265	.4279	.4292	.4306	.4319
1.5	.4332	.4345	.4357	.4370	.4382	.4394	.4406	.4418	.4429	.4441
1.6	.4552	.4463	.4474	.4484	.4495	.4505	.4515	.4525	.4535	.4545
1.7	.4554	.4564	.4573	.4582	.4591	.4599	.4608	.4616	.4625	.4633
1.8	.4641	.4649	.4656	.4664	.4671	.4678	.4686	.4693	.4699	.4706
1.9	.4713	.4719	.4726	.4732	.4738	.4744	.4750	.4756	.4761	.4767
2.0	.4772	.4778	.4783	.4788	.4793	.4798	.4803	.4808	.4812	.4817
2.1	.4821	.4826	.4830	.4834	.4838	.4842	.4846	.4850	.4854	.4857
2.2	.4861	.4864	.4868	.4871	.4875	.4878	.4881	.4884	.4887	.4890
2.3	.4893	.4896	.4898	.4901	.4904	.4906	.4909	.4911	.4913	.4916
2.4	.4918	.4920	.4922	.4925	.4927	.4929	.4931	.4932	.4934	.4936
2.5	.4938	.4940	.4941	.4943	.4945	.4946	.4948	.4949	.4951	.4952
2.6	.4953	.4955	.4956	.4957	.4959	.4960	.4961	.4962	.4963	.4964
2.7	.4965	.4966	.4967	.4968	.4969	.4970	.4971	.4972	.4973	.4974
2.8	.4974	.4975	.4976	.4977	.4977	.4978	.4979	.4979	.4980	.4981
2.9	.4981	.4982	.4982	.4983	.4984	.4984	.4985	.4985	.4986	.4986
3.0	.4987	.4987	.4987	.4988	.4988	.4989	.4989	.4989	.4990	.4990

Source: Abridged from Table I of A. Hald, *Statistical Tables and Formulas* (New York: John Wiley & Sons, Inc.), 1952. Reproduced by permission of A. Hald and the publisher.

TABLE 4 Upper-Tail Values for the Student's t Distribution

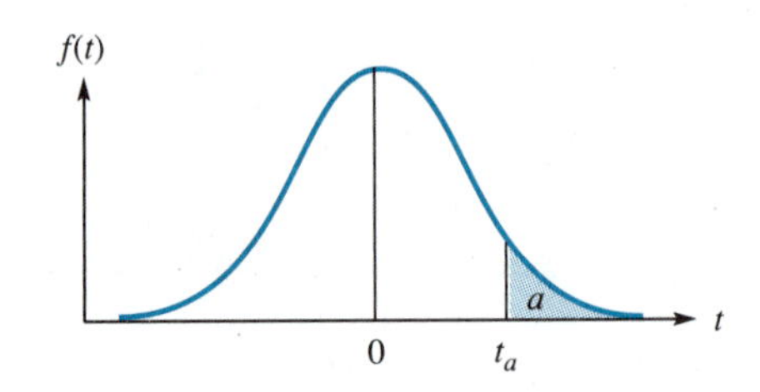

df	$t_{.100}$	$t_{.050}$	$t_{.025}$	$t_{.010}$	$t_{.005}$
1	3.078	6.314	12.706	31.821	63.657
2	1.886	2.920	4.303	6.965	9.925
3	1.638	2.353	3.182	4.541	5.841
4	1.533	2.132	2.776	3.747	4.604
5	1.476	2.015	2.571	3.365	4.032
6	1.440	1.943	2.447	3.143	3.707
7	1.415	1.895	2.365	2.998	3.499
8	1.397	1.860	2.306	2.896	3.355
9	1.383	1.833	2.262	2.821	3.250
10	1.372	1.812	2.228	2.764	3.169
11	1.363	1.796	2.201	2.718	3.106
12	1.356	1.782	2.179	2.681	3.055
13	1.350	1.771	2.160	2.650	3.012
14	1.345	1.761	2.145	2.624	2.977
15	1.341	1.753	2.131	2.602	2.947
16	1.337	1.746	2.120	2.583	2.921
17	1.333	1.740	2.110	2.567	2.898
18	1.330	1.734	2.101	2.552	2.878
19	1.328	1.729	2.093	2.539	2.861
20	1.325	1.725	2.086	2.528	2.845
21	1.323	1.721	2.080	2.518	2.831
22	1.321	1.717	2.074	2.508	2.819
23	1.319	1.714	2.069	2.500	2.807
24	1.318	1.711	2.064	2.492	2.797
25	1.316	1.708	2.060	2.485	2.787
26	1.315	1.706	2.056	2.479	2.779
27	1.314	1.703	2.052	2.473	2.771
28	1.313	1.701	2.048	2.467	2.763
29	1.311	1.699	2.045	2.462	2.756
30	1.310	1.697	2.042	2.457	2.750
40	1.303	1.684	2.021	2.423	2.704
60	1.296	1.671	2.000	2.390	2.660
120	1.289	1.658	1.980	2.358	2.617
∞	1.282	1.645	1.960	2.326	2.576

Source: This table is reproduced with the kind permission of the Trustees of Biometrika from E. S. Pearson and H. O. Hartley (eds.), *The Biometrika Tables for Statisticians,* Vol. 1, 3rd ed., *Biometrika,* 1966.

TABLE 5 Upper-Tail Values of the Chi-Square Distribution

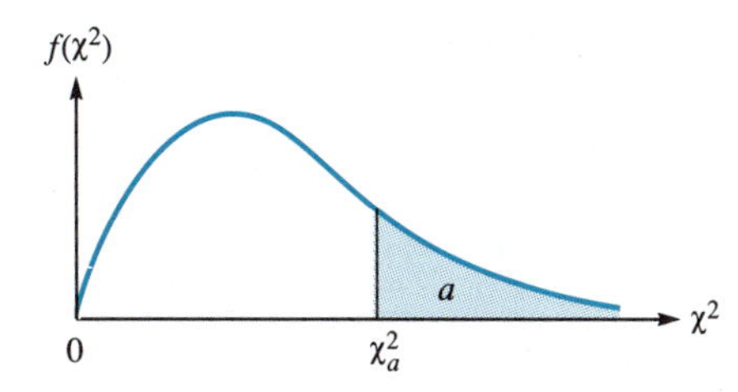

Degrees of Freedom (df)			χ^2_a		
	a=.10	a=.05	a=.025	a=.01	a=.005
1	2.70554	3.84146	5.02389	6.63490	7.87944
2	4.60517	5.99147	7.37776	9.21034	10.5966
3	6.25139	7.81473	9.34840	11.3449	12.8381
4	7.77944	9.48773	11.1433	13.2767	14.8602
5	9.23635	11.0705	12.8325	15.0863	16.7496
6	10.6446	12.5916	14.4494	16.8119	18.5476
7	12.0170	14.0671	16.0128	18.4753	20.2777
8	13.3616	15.5073	17.5346	20.0902	21.9550
9	14.6837	16.9190	19.0228	21.6660	23.5893
10	15.9871	18.3070	20.4831	23.2093	25.1882
11	17.2750	19.6751	21.9200	24.7250	26.7569
12	18.5494	21.0261	23.3367	26.2170	28.2995
13	19.8119	22.3621	24.7356	27.6883	29.8194
14	21.0642	23.6848	26.1190	29.1413	31.3193
15	22.3072	24.9958	27.4884	30.5779	32.8013
16	23.5418	26.2962	28.8454	31.9999	34.2672
17	24.7690	27.5871	30.1910	33.4087	35.7185
18	25.9894	28.8693	31.5264	34.8053	37.1564
19	27.2036	30.1435	32.8523	36.1908	38.5822
20	28.4120	31.4104	34.1696	37.5662	39.9968
21	29.6151	32.6705	35.4789	38.9321	41.4010
22	30.8133	33.9244	36.7807	40.2894	42.7956
23	32.0069	35.1725	38.0757	41.6384	44.1813
24	33.1963	36.4151	39.3641	42.9798	45.5585
25	34.3816	37.6525	40.6465	44.3141	46.9278
26	35.5631	38.8852	41.9232	45.6417	48.2899
27	36.7412	40.1133	43.1944	46.9630	49.6449
28	37.9159	41.3372	44.4607	48.2782	50.9933
29	39.0875	42.5569	45.7222	49.5879	52.3356
30	40.2560	43.7729	46.9792	50.8922	53.6720
40	51.8050	55.7585	59.3417	63.6907	66.7659
50	63.1671	67.5048	71.4202	76.1539	79.4900
60	74.3970	79.0819	83.2976	88.3794	91.9517
70	85.5271	90.5312	95.0231	100.425	104.215
80	96.5782	101.879	106.629	112.329	116.321
90	107.565	113.145	118.136	124.116	128.299
100	118.498	124.342	129.561	135.807	140.169

TABLE 6 Upper-Tail Values for an *F* Distribution

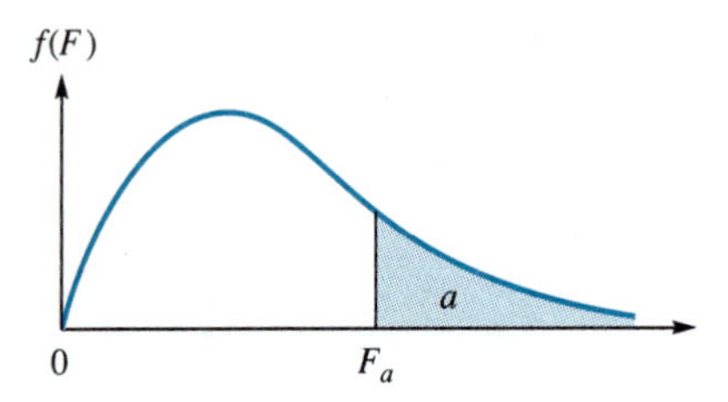

Denominator df	a	Numerator df 1	2	3	4	5	6	7	8	9
1	.100	39.86	49.50	53.59	55.83	57.24	58.20	58.91	59.44	59.86
	.050	161.4	199.5	215.7	224.6	230.2	234.0	236.8	238.9	240.5
	.025	647.8	799.5	864.2	899.6	921.8	937.1	948.2	956.7	963.3
	.010	4052	4999.5	5403	5625	5764	5859	5928	5982	6022
	.005	16211	20000	21615	22500	23056	23437	23715	23925	24091
2	.100	8.53	9.00	9.16	9.24	9.29	9.33	9.35	9.37	9.38
	.050	18.51	19.00	19.16	19.25	19.30	19.33	19.35	19.37	19.38
	.025	38.51	39.00	39.17	39.25	39.30	39.33	39.36	39.37	39.39
	.010	98.50	99.00	99.17	99.25	99.30	99.33	99.36	99.37	99.39
	.005	198.5	199.0	199.2	199.2	199.3	199.3	199.4	199.4	199.4
3	.100	5.54	5.46	5.39	5.34	5.31	5.28	5.27	5.25	5.24
	.050	10.13	9.55	9.28	9.12	9.01	8.94	8.89	8.85	8.81
	.025	17.44	16.04	15.44	15.10	14.88	14.73	14.62	14.54	14.47
	.010	34.12	30.82	29.46	28.71	28.24	27.91	27.67	27.49	27.35
	.005	55.55	49.80	47.47	46.19	45.39	44.84	44.43	44.13	43.88
4	.100	4.54	4.32	4.19	4.11	4.05	4.01	3.98	3.95	3.94
	.050	7.71	6.94	6.59	6.39	6.26	6.16	6.09	6.04	6.00
	.025	12.22	10.65	9.98	9.60	9.36	9.20	9.07	8.98	8.90
	.010	21.20	18.00	16.69	15.98	15.52	15.21	14.98	14.80	14.66
	.005	31.33	26.28	24.26	23.15	22.46	21.97	21.62	21.35	21.14
5	.100	4.06	3.78	3.62	3.52	3.45	3.40	3.37	3.34	3.32
	.050	6.61	5.79	5.41	5.19	5.05	4.95	4.88	4.82	4.77
	.025	10.01	8.43	7.76	7.39	7.15	6.98	6.85	6.76	6.68
	.010	16.26	13.27	12.06	11.39	10.97	10.67	10.46	10.29	10.16
	.005	22.78	18.31	16.53	15.56	14.94	14.51	14.20	13.96	13.77
6	.100	3.78	3.46	3.29	3.18	3.11	3.05	3.01	2.98	2.96
	.050	5.99	5.14	4.76	4.53	4.39	4.28	4.21	4.15	4.10
	.025	8.81	7.26	6.60	6.23	5.99	5.82	5.70	5.60	5.52
	.010	13.75	10.92	9.78	9.15	8.75	8.47	8.26	8.10	7.98
	.005	18.63	14.54	12.92	12.03	11.46	11.07	10.79	10.57	10.39
7	.100	3.59	3.26	3.07	2.96	2.88	2.83	2.78	2.75	2.72
	.050	5.59	4.74	4.35	4.12	3.97	3.87	3.79	3.73	3.68
	.025	8.07	6.54	5.89	5.52	5.29	5.12	4.99	4.90	4.82
	.010	12.25	9.55	8.45	7.85	7.46	7.19	6.99	6.84	6.72
	.005	16.24	12.40	10.88	10.05	9.52	9.16	8.89	8.68	8.51

Numerator df											
10	12	15	20	24	30	40	60	120	∞	*a*	df
60.19	60.71	61.22	61.74	62.00	62.26	62.53	62.79	63.06	63.33	.100	1
241.9	243.9	245.9	248.0	249.1	250.1	251.1	252.2	253.3	254.3	.050	
968.6	976.7	984.9	993.1	997.2	1001	1006	1010	1014	1018	.025	
6056	6106	6157	6209	6235	6261	6287	6313	6339	6366	.010	
24224	24426	24630	24836	24940	25044	25148	25253	25359	25465	.005	
9.39	9.41	9.42	9.44	9.45	9.46	9.47	9.47	9.48	9.49	.100	2
19.40	19.41	19.43	19.45	19.45	19.46	19.47	19.48	19.49	19.50	.050	
39.40	39.41	39.43	39.45	39.46	39.46	39.47	39.48	39.49	39.50	.025	
99.40	99.42	99.43	99.45	99.46	99.47	99.47	99.48	99.49	99.50	.010	
199.4	199.4	199.4	199.4	199.5	199.5	199.5	199.5	199.5	199.5	.005	
5.23	5.22	5.20	5.18	5.18	5.17	5.16	5.15	5.14	5.13	.100	3
8.79	8.74	8.70	8.66	8.64	8.62	8.59	8.57	8.55	8.53	.050	
14.42	14.34	14.25	14.17	14.12	14.08	14.04	13.99	13.95	13.90	.025	
27.23	27.05	26.87	26.69	26.60	26.50	26.41	26.32	26.22	26.13	.010	
43.69	43.39	43.08	42.78	42.62	42.47	42.31	42.15	41.99	41.83	.005	
3.92	3.90	3.87	3.84	3.83	3.82	3.80	3.79	3.78	3.76	.100	4
5.96	5.91	5.86	5.80	5.77	5.75	5.72	5.69	5.66	5.63	.050	
8.84	8.75	8.66	8.56	8.51	8.46	8.41	8.36	8.31	8.26	.025	
14.55	14.37	14.20	14.02	13.93	13.84	13.75	13.65	13.56	13.46	.010	
20.97	20.70	20.44	20.17	20.03	19.89	19.75	19.61	19.47	19.32	.005	
3.30	3.27	3.24	3.21	3.19	3.17	3.16	3.14	3.12	3.10	.100	5
4.74	4.68	4.62	4.56	4.53	4.50	4.46	4.43	4.40	4.36	.050	
6.62	6.52	6.43	6.33	6.28	6.23	6.18	6.12	6.07	6.02	.025	
10.05	9.89	9.72	9.55	9.47	9.38	9.29	9.20	9.11	9.02	.010	
13.62	13.38	13.15	12.90	12.78	12.66	12.53	12.40	12.27	12.14	.005	
2.94	2.90	2.87	2.84	2.82	2.80	2.78	2.76	2.74	2.72	.100	6
4.06	4.00	3.94	3.87	3.84	3.81	3.77	3.74	3.70	3.67	.050	
5.46	5.37	5.27	5.17	5.12	5.07	5.01	4.96	4.90	4.85	.025	
7.87	7.72	7.56	7.40	7.31	7.23	7.14	7.06	6.97	6.88	.010	
10.25	10.03	9.81	9.59	9.47	9.36	9.24	9.12	9.00	8.88	.005	
2.70	2.67	2.63	2.59	2.58	2.56	2.54	2.51	2.49	2.47	.100	7
3.64	3.57	3.51	3.44	3.41	3.38	3.34	3.30	3.27	3.23	.050	
4.76	4.67	4.57	4.47	4.42	4.36	4.31	4.25	4.20	4.14	.025	
6.62	6.47	6.31	6.16	6.07	5.99	5.91	5.82	5.74	5.65	.010	
8.38	8.18	7.97	7.75	7.65	7.53	7.42	7.31	7.19	7.08	.005	

Table 6 (Continued)

Denominator df	a	Numerator df 1	2	3	4	5	6	7	8	9
8	.100	3.46	3.11	2.92	2.81	2.73	2.67	2.62	2.59	2.56
	.050	5.32	4.46	4.07	3.84	3.69	3.58	3.50	3.44	3.39
	.025	7.57	6.06	5.42	5.05	4.82	4.65	4.53	4.43	4.36
	.010	11.26	8.65	7.59	7.01	6.63	6.37	6.18	6.03	5.91
	.005	14.69	11.04	9.60	8.81	8.30	7.95	7.69	7.50	7.34
9	.100	3.36	3.01	2.81	2.69	2.61	2.55	2.51	2.47	2.44
	.050	5.12	4.26	3.86	3.63	3.48	3.37	3.29	3.23	3.18
	.025	7.21	5.71	5.08	4.72	4.48	4.32	4.20	4.10	4.03
	.010	10.56	8.02	6.99	6.42	6.06	5.80	5.61	5.47	5.35
	.005	13.61	10.11	8.72	7.96	7.47	7.13	6.88	6.69	6.54
10	.100	3.29	2.92	2.73	2.61	2.52	2.46	2.41	2.38	2.35
	.050	4.96	4.10	3.71	3.48	3.33	3.22	3.14	3.07	3.02
	.025	6.94	5.46	4.83	4.47	4.24	4.07	3.95	3.85	3.78
	.010	10.04	7.56	6.55	5.99	5.64	5.39	5.20	5.06	4.94
	.005	12.83	9.43	8.08	7.34	6.87	6.54	6.30	6.12	5.97
11	.100	3.23	2.86	2.66	2.54	2.45	2.39	2.34	2.30	2.27
	.050	4.84	3.98	3.59	3.36	3.20	3.09	3.01	2.95	2.90
	.025	6.72	5.26	4.63	4.28	4.04	3.88	3.76	3.66	3.59
	.010	9.65	7.21	6.22	5.67	5.32	5.07	4.89	4.74	4.63
	.005	12.23	8.91	7.60	6.88	6.42	6.10	5.86	5.68	5.54
12	.100	3.18	2.81	2.61	2.48	2.39	2.33	2.28	2.24	2.21
	.050	4.75	3.89	3.49	3.26	3.11	3.00	2.91	2.85	2.80
	.025	6.55	5.10	4.47	4.12	3.89	3.73	3.61	3.51	3.44
	.010	9.33	6.93	5.95	5.41	5.06	4.82	4.64	4.50	4.39
	.005	11.75	8.51	7.23	6.52	6.07	5.76	5.52	5.35	5.20
13	.100	3.14	2.76	2.56	2.43	2.35	2.28	2.23	2.20	2.16
	.050	4.67	3.81	3.41	3.18	3.03	2.92	2.83	2.77	2.71
	.025	6.41	4.97	4.35	4.00	3.77	3.60	3.48	3.39	3.31
	.010	9.07	6.70	5.74	5.21	4.86	4.62	4.44	4.30	4.19
	.005	11.37	8.19	6.93	6.23	5.79	5.48	5.25	5.08	4.94
14	.100	3.10	2.73	2.52	2.39	2.31	2.24	2.19	2.15	2.12
	.050	4.60	3.74	3.34	3.11	2.96	2.85	2.76	2.70	2.65
	.025	6.30	4.86	4.24	3.89	3.66	3.50	3.38	3.29	3.21
	.010	8.86	6.51	5.56	5.04	4.69	4.46	4.28	4.14	4.03
	.005	11.06	7.92	6.68	6.00	5.56	5.26	5.03	4.86	4.72

Numerator df											
10	12	15	20	24	30	40	60	120	∞	a	df
2.54	2.50	2.46	2.42	2.40	2.38	2.36	2.34	2.32	2.29	.100	8
3.35	3.28	3.22	3.15	3.12	3.08	3.04	3.01	2.97	2.93	.050	
4.30	4.20	4.10	4.00	3.95	3.89	3.84	3.78	3.73	3.67	.025	
5.81	5.67	5.52	5.36	5.28	5.20	5.12	5.03	4.95	4.86	.010	
7.21	7.01	6.81	6.61	6.50	6.40	6.29	6.18	6.06	5.95	.005	
2.42	2.38	2.34	2.30	2.28	2.25	2.23	2.21	2.18	2.16	.100	9
3.14	3.07	3.01	2.94	2.90	2.86	2.83	2.79	2.75	2.71	.050	
3.96	3.87	3.77	3.67	3.61	3.56	3.51	3.45	3.39	3.33	.025	
5.26	5.11	4.96	4.81	4.73	4.65	4.57	4.48	4.40	4.31	.010	
6.42	6.23	6.03	5.83	5.73	5.62	5.52	5.41	5.30	5.19	.005	
2.32	2.28	2.24	2.20	2.18	2.16	2.13	2.11	2.08	2.06	.100	10
2.98	2.91	2.85	2.74	2.77	2.70	2.66	2.62	2.58	2.54	.050	
3.72	3.62	3.52	3.42	3.37	3.31	3.26	3.20	3.14	3.08	.025	
4.85	4.71	4.56	4.41	4.33	4.25	4.17	4.08	4.00	3.91	.010	
5.85	5.66	5.47	5.27	5.17	5.07	4.97	4.86	4.75	4.64	.005	
2.25	2.21	2.17	2.12	2.10	2.08	2.05	2.03	2.00	1.97	.100	11
2.85	2.79	2.72	2.65	2.61	2.57	2.53	2.49	2.45	2.40	.050	
3.53	3.43	3.33	3.23	3.17	3.12	3.06	3.00	2.94	2.88	.025	
4.54	4.40	4.25	4.10	4.02	3.94	3.86	3.78	3.69	3.60	.010	
5.42	5.24	5.05	4.86	4.76	4.65	4.55	4.44	4.34	4.23	.005	
2.19	2.15	2.10	2.06	2.04	2.01	1.99	1.96	1.93	1.90	.100	12
2.75	2.69	2.62	2.54	2.51	2.47	2.43	2.38	2.34	2.30	.050	
3.37	3.28	3.18	3.07	3.02	2.96	2.91	2.85	2.79	2.72	.025	
4.30	4.16	4.01	3.86	3.78	3.70	3.62	3.54	3.45	3.36	.010	
5.09	4.91	4.72	4.53	4.43	4.33	4.23	4.12	4.01	3.90	.005	
2.14	2.10	2.05	2.01	1.98	1.96	1.93	1.90	1.88	1.85	.100	13
2.67	2.60	2.53	2.46	2.42	2.38	2.34	2.30	2.25	2.21	.050	
3.25	3.15	3.05	2.95	2.89	2.84	2.78	2.72	2.66	2.60	.025	
4.10	3.96	3.82	3.66	3.59	3.51	3.43	3.34	3.25	3.17	.010	
4.82	4.64	4.46	4.27	4.17	4.07	3.97	3.87	3.76	3.65	.005	
2.10	2.05	2.01	1.96	1.94	1.91	1.89	1.86	1.83	1.80	.100	14
2.60	2.53	2.46	2.39	2.35	2.31	2.27	2.22	2.18	2.13	.050	
3.15	3.05	2.95	2.84	2.79	2.73	2.67	2.61	2.55	2.49	.025	
3.94	3.80	3.66	3.51	3.43	3.35	3.27	3.18	3.09	3.00	.010	
4.60	4.43	4.25	4.06	3.96	3.86	3.76	3.66	3.55	3.44	.005	

Table 6 (Continued)

Denominator df	a	Numerator df 1	2	3	4	5	6	7	8	9
15	.100	3.07	2.70	2.49	2.36	2.27	2.21	2.16	2.12	2.09
	.050	4.54	3.68	3.29	3.06	2.90	2.79	2.71	2.64	2.59
	.025	6.20	4.77	4.15	3.80	3.58	3.41	3.29	3.20	3.12
	.010	8.68	6.36	5.42	4.89	4.56	4.32	4.14	4.00	3.89
	.005	10.80	7.70	6.48	5.80	5.37	5.07	4.85	4.67	4.54
16	.100	3.05	2.67	2.46	2.33	2.24	2.18	2.13	2.09	2.06
	.050	4.49	3.63	3.24	3.01	2.85	2.74	2.66	2.59	2.54
	.025	6.12	4.69	4.08	3.73	3.50	3.34	3.22	3.12	3.05
	.010	8.53	6.23	5.29	4.77	4.44	4.20	4.03	3.89	3.78
	.005	10.58	7.51	6.30	5.64	5.21	4.91	4.69	4.52	4.38
17	.100	3.03	2.64	2.44	2.31	2.22	2.15	2.10	2.06	2.03
	.050	4.45	3.59	3.20	2.96	2.81	2.70	2.61	2.55	2.49
	.025	6.04	4.62	4.01	3.66	3.44	3.28	3.16	3.06	2.98
	.010	8.40	6.11	5.18	4.67	4.34	4.10	3.93	3.79	3.68
	.005	10.38	7.35	6.16	5.50	5.07	4.78	4.56	4.39	4.25
18	.100	3.01	2.62	2.42	2.29	2.20	2.13	2.08	2.04	2.00
	.050	4.41	3.55	3.16	2.93	2.77	2.66	2.58	2.51	2.46
	.025	5.98	4.56	3.95	3.61	3.38	3.22	3.10	3.01	2.93
	.010	8.29	6.01	5.09	4.58	4.25	4.01	3.84	3.71	3.60
	.005	10.22	7.21	6.03	5.37	4.96	4.66	4.44	4.28	4.14
19	.100	2.99	2.61	2.40	2.27	2.18	2.11	2.06	2.02	1.98
	.050	4.38	3.52	3.13	2.90	2.74	2.63	2.54	2.48	2.42
	.025	5.92	4.51	3.90	3.56	3.33	3.17	3.05	2.96	2.88
	.010	8.18	5.93	5.01	4.50	4.17	3.94	3.77	3.63	3.52
	.005	10.07	7.09	5.92	5.27	4.85	4.56	4.34	4.18	4.04
20	.100	2.97	2.59	2.38	2.25	2.16	2.09	2.04	2.00	1.96
	.050	4.35	3.49	3.10	2.87	2.71	2.60	2.51	2.45	2.39
	.025	5.87	4.46	3.86	3.51	3.29	3.13	3.01	2.91	2.84
	.010	8.10	5.85	4.94	4.43	4.10	3.87	3.70	3.56	3.46
	.005	9.94	6.99	5.82	5.17	4.76	4.47	4.26	4.09	3.96
21	.100	2.96	2.57	2.36	2.23	2.14	2.08	2.02	1.98	1.95
	.050	4.32	3.47	3.07	2.84	2.68	2.57	2.49	2.42	2.37
	.025	5.83	4.42	3.82	3.48	3.25	3.09	2.97	2.87	2.80
	.010	8.02	5.78	4.87	4.37	4.04	3.81	3.64	3.51	3.40
	.005	9.83	6.89	5.73	5.09	4.68	4.39	4.18	4.01	3.88

Numerator df											
10	12	15	20	24	30	40	60	120	∞	*a*	df
2.06	2.02	1.97	1.92	1.90	1.87	1.85	1.82	1.79	1.76	.100	15
2.54	2.48	2.40	2.33	2.29	2.25	2.20	2.16	2.11	2.07	.050	
3.06	2.96	2.86	2.76	2.70	2.64	2.59	2.52	2.46	2.40	.025	
3.80	3.67	3.52	3.37	3.29	3.21	3.13	3.05	2.96	2.87	.010	
4.42	4.25	4.07	3.88	3.79	3.69	3.58	3.48	3.37	3.26	.005	
2.03	1.99	1.94	1.89	1.87	1.84	1.81	1.78	1.75	1.72	.100	16
2.49	2.42	2.35	2.28	2.24	2.19	2.15	2.11	2.06	2.01	.050	
2.99	2.89	2.79	2.68	2.63	2.57	2.51	2.45	2.38	2.32	.025	
3.69	3.55	3.41	3.26	3.18	3.10	3.02	2.93	2.84	2.75	.010	
4.27	4.10	3.92	3.73	3.64	3.54	3.44	3.33	3.22	3.11	.005	
2.00	1.96	1.91	1.86	1.84	1.81	1.78	1.75	1.72	1.69	.100	17
2.45	2.38	2.31	2.23	2.19	2.15	2.10	2.06	2.01	1.96	.050	
2.92	2.82	2.72	2.62	2.56	2.50	2.44	2.38	2.32	2.25	.025	
3.59	3.46	3.31	3.16	3.08	3.00	2.92	2.83	2.75	2.65	.010	
4.14	3.97	3.79	3.61	3.51	3.41	3.31	3.21	3.10	2.98	.005	
1.98	1.93	1.89	1.84	1.81	1.78	1.75	1.72	1.69	1.66	.100	18
2.41	2.34	2.27	2.19	2.15	2.11	2.06	2.02	1.97	1.92	.050	
2.87	2.77	2.67	2.56	2.50	2.44	2.38	2.32	2.26	2.19	.025	
3.51	3.37	3.23	3.08	3.00	2.92	2.84	2.75	2.66	2.57	.010	
4.03	3.86	3.68	3.50	3.40	3.30	3.20	3.10	2.99	2.87	.005	
1.96	1.91	1.86	1.81	1.79	1.76	1.73	1.70	1.67	1.63	.100	19
2.38	2.31	2.23	2.16	2.11	2.07	2.03	1.98	1.93	1.88	.050	
2.82	2.72	2.62	2.51	2.45	2.39	2.33	2.27	2.20	2.13	.025	
3.43	3.30	3.15	3.00	2.92	2.84	2.76	2.67	2.58	2.49	.010	
3.93	3.76	3.59	3.40	3.31	3.21	3.11	3.00	2.89	2.78	.005	
1.94	1.89	1.84	1.79	1.77	1.74	1.71	1.68	1.64	1.61	.100	20
2.35	2.28	2.20	2.12	2.08	2.04	1.99	1.95	1.90	1.84	.050	
2.77	2.68	2.57	2.46	2.41	2.35	2.29	2.22	2.16	2.09	.025	
3.37	3.23	3.09	2.94	2.86	2.78	2.69	2.61	2.52	2.42	.010	
3.85	3.68	3.50	3.32	3.22	3.12	3.02	2.92	2.81	2.69	.005	
1.92	1.87	1.83	1.78	1.75	1.72	1.69	1.66	1.62	1.59	.100	21
2.32	2.25	2.18	2.10	2.05	2.01	1.96	1.92	1.87	1.81	.050	
2.73	2.64	2.53	2.42	2.37	2.31	2.25	2.18	2.11	2.04	.025	
3.31	3.17	3.03	2.88	2.80	2.72	2.64	2.55	2.46	2.36	.010	
3.77	3.60	3.43	3.24	3.15	3.05	2.95	2.84	2.73	2.61	.005	

Table 6 (Continued)

Denominator df	a	Numerator df 1	2	3	4	5	6	7	8	9
22	.100	2.95	2.56	2.35	2.22	2.13	2.06	2.01	1.97	1.93
	.050	4.30	3.44	3.05	2.82	2.66	2.55	2.46	2.40	2.34
	.025	5.79	4.38	3.78	3.44	3.22	3.05	2.93	2.84	2.76
	.010	7.95	5.72	4.82	4.31	3.99	3.76	3.59	3.45	3.35
	.005	9.73	6.81	5.65	5.02	4.61	4.32	4.11	3.94	3.81
23	.100	2.94	2.55	2.34	2.21	2.11	2.05	1.99	1.95	1.92
	.050	4.28	3.42	3.03	2.80	2.64	2.53	2.44	2.37	2.32
	.025	5.75	4.35	3.75	3.41	3.18	3.02	2.90	2.81	2.73
	.010	7.88	5.66	4.76	4.26	3.94	3.71	3.54	3.41	3.30
	.005	9.63	6.73	5.58	4.95	4.54	4.26	4.05	3.88	3.75
24	.100	2.93	2.54	2.33	2.19	2.10	2.04	1.98	1.94	1.91
	.050	4.26	3.40	3.01	2.78	2.62	2.51	2.42	2.36	2.30
	.025	5.72	4.32	3.72	3.38	3.15	2.99	2.87	2.78	2.70
	.010	7.82	5.61	4.72	4.22	3.90	3.67	3.50	3.36	3.26
	.005	9.55	6.66	5.52	4.89	4.49	4.20	3.99	3.83	3.69
25	.100	2.92	2.53	2.32	2.18	2.09	2.02	1.97	1.93	1.89
	.050	4.24	3.39	2.99	2.76	2.60	2.49	2.40	2.34	2.28
	.025	5.69	4.29	3.69	3.35	3.13	2.97	2.85	2.75	2.68
	.010	7.77	5.57	4.68	4.18	3.85	3.63	3.46	3.32	3.22
	.005	9.48	6.60	5.46	4.84	4.43	4.15	3.94	3.78	3.64
26	.100	2.91	2.52	2.31	2.17	2.08	2.01	1.96	1.92	1.88
	.050	4.23	3.37	2.98	2.74	2.59	2.47	2.39	2.32	2.27
	.025	5.66	4.27	3.67	3.33	3.10	2.94	2.82	2.73	2.65
	.010	7.72	5.53	4.64	4.14	3.82	3.59	3.42	3.29	3.18
	.005	9.41	6.54	5.41	4.79	4.38	4.10	3.89	3.73	3.60
27	.100	2.90	2.51	2.30	2.17	2.07	2.00	1.95	1.91	1.87
	.050	4.21	3.35	2.96	2.73	2.57	2.46	2.37	2.31	2.25
	.025	5.63	4.24	3.65	3.31	3.08	2.92	2.80	2.71	2.63
	.010	7.68	5.49	4.60	4.11	3.78	3.56	3.39	3.26	3.15
	.005	9.34	6.49	5.36	4.74	4.34	4.06	3.85	3.69	3.56
28	.100	2.89	2.50	2.29	2.16	2.06	2.00	1.94	1.90	1.87
	.050	4.20	3.34	2.95	2.71	2.56	2.45	2.36	2.29	2.24
	.025	5.61	4.22	3.63	3.29	3.06	2.90	2.78	2.69	2.61
	.010	7.64	5.45	4.57	4.07	3.75	3.53	3.36	3.23	3.12
	.005	9.28	6.44	5.32	4.70	4.30	4.02	3.81	3.65	3.52

Numerator df											
10	12	15	20	24	30	40	60	120	∞	a	df
1.90	1.86	1.81	1.76	1.73	1.70	1.67	1.64	1.60	1.57	.100	22
2.30	2.23	2.15	2.07	2.03	1.98	1.94	1.89	1.84	1.78	.050	
2.70	2.60	2.50	2.39	2.33	2.27	2.21	2.14	2.08	2.00	.025	
3.26	3.12	2.98	2.83	2.75	2.67	2.58	2.50	2.40	2.31	.010	
3.70	3.54	3.36	3.18	3.08	2.98	2.88	2.77	2.66	2.55	.005	
1.89	1.84	1.80	1.74	1.72	1.69	1.66	1.62	1.59	1.55	.100	23
2.27	2.20	2.13	2.05	2.01	1.96	1.91	1.86	1.81	1.76	.050	
2.67	2.57	2.47	2.36	2.30	2.24	2.18	2.11	2.04	1.97	.025	
3.21	3.07	2.93	2.78	2.70	2.62	2.54	2.45	2.35	2.26	.010	
3.64	3.47	3.30	3.12	3.02	2.92	2.82	2.71	2.60	2.48	.005	
1.88	1.83	1.78	1.73	1.70	1.67	1.64	1.61	1.57	1.53	.100	24
2.25	2.18	2.11	2.03	1.98	1.94	1.89	1.84	1.79	1.73	.050	
2.64	2.54	2.44	2.33	2.27	2.21	2.15	2.08	2.01	1.94	.025	
3.17	3.03	2.89	2.74	2.66	2.58	2.49	2.40	2.31	2.21	.010	
3.59	3.42	3.25	3.06	2.97	2.87	2.77	2.66	2.55	2.43	.005	
1.87	1.82	1.77	1.72	1.69	1.66	1.63	1.59	1.56	1.52	.100	25
2.24	2.16	2.09	2.01	1.96	1.92	1.87	1.82	1.77	1.71	.050	
2.61	2.51	2.41	2.30	2.24	2.18	2.12	2.05	1.98	1.91	.025	
3.13	2.99	2.85	2.70	2.62	2.54	2.45	2.36	2.27	2.17	.010	
3.54	3.37	3.20	3.01	2.92	2.82	2.72	2.61	2.50	2.38	.005	
1.86	1.81	1.76	1.71	1.68	1.65	1.61	1.58	1.54	1.50	.100	26
2.22	2.15	2.07	1.99	1.95	1.90	1.85	1.80	1.75	1.69	.050	
2.59	2.49	2.39	2.28	2.22	2.16	2.09	2.03	1.95	1.88	.025	
3.09	2.96	2.81	2.66	2.58	2.50	2.42	2.33	2.23	2.13	.010	
3.49	3.33	3.15	2.97	2.87	2.77	2.67	2.56	2.45	2.33	.005	
1.85	1.80	1.75	1.70	1.67	1.64	1.60	1.57	1.53	1.49	.100	27
2.20	2.13	2.06	1.97	1.93	1.88	1.84	1.79	1.73	1.67	.050	
2.57	2.47	2.36	2.25	2.19	2.13	2.07	2.00	1.93	1.85	.025	
3.06	2.93	2.78	2.63	2.55	2.47	2.38	2.29	2.20	2.10	.010	
3.45	3.28	3.11	2.93	2.83	2.73	2.63	2.52	2.41	2.29	.005	
1.84	1.79	1.74	1.69	1.66	1.63	1.59	1.56	1.52	1.48	.100	28
2.19	2.12	2.04	1.96	1.91	1.87	1.82	1.77	1.71	1.65	.050	
2.55	2.45	2.34	2.23	2.17	2.11	2.05	1.98	1.91	1.83	.025	
3.03	2.90	2.75	2.60	2.52	2.44	2.35	2.26	2.17	2.06	.010	
3.41	3.25	3.07	2.89	2.79	2.69	2.59	2.48	2.37	2.25	.005	

Table 6 (Continued)

Denominator df	a	Numerator df 1	2	3	4	5	6	7	8	9
29	.100	2.89	2.50	2.28	2.15	2.06	1.99	1.93	1.89	1.86
	.050	4.18	3.33	2.93	2.70	2.55	2.43	2.35	2.28	2.22
	.025	5.59	4.20	3.61	3.27	3.04	2.88	2.76	2.67	2.59
	.010	7.60	5.42	4.54	4.04	3.73	3.50	3.33	3.20	3.09
	.005	9.23	6.40	5.28	4.66	4.26	3.98	3.77	3.61	3.48
30	.100	2.88	2.49	2.28	2.14	2.05	1.98	1.93	1.88	1.85
	.050	4.17	3.32	2.92	2.69	2.53	2.42	2.33	2.27	2.21
	.025	5.57	4.18	3.59	3.25	3.03	2.87	2.75	2.65	2.57
	.010	7.56	5.39	4.51	4.02	3.70	3.47	3.30	3.17	3.07
	.005	9.18	6.35	5.24	4.62	4.23	3.95	3.74	3.58	3.45
40	.100	2.84	2.44	2.23	2.09	2.00	1.93	1.87	1.83	1.79
	.050	4.08	3.23	2.84	2.61	2.45	2.34	2.25	2.18	2.12
	.025	5.42	4.05	3.46	3.13	2.90	2.74	2.62	2.53	2.45
	.010	7.31	5.18	4.31	3.83	3.51	3.29	3.12	2.99	2.89
	.005	8.83	6.07	4.98	4.37	3.99	3.71	3.51	3.35	3.22
60	.100	2.79	2.39	2.18	2.04	1.95	1.87	1.82	1.77	1.74
	.050	4.00	3.15	2.76	2.53	2.37	2.25	2.17	2.10	2.04
	.025	5.29	3.93	3.34	3.01	2.79	2.63	2.51	2.41	2.33
	.010	7.08	4.98	4.13	3.65	3.34	3.12	2.95	2.82	2.72
	.005	8.49	5.79	4.73	4.14	3.76	3.49	3.29	3.13	3.01
120	.100	2.75	2.35	2.13	1.99	1.90	1.82	1.77	1.72	1.68
	.050	3.92	3.07	2.68	2.45	2.29	2.17	2.09	2.02	1.96
	.025	5.15	3.80	3.23	2.89	2.67	2.52	2.39	2.30	2.22
	.010	6.85	4.79	3.95	3.48	3.17	2.96	2.79	2.66	2.56
	.005	8.18	5.54	4.50	3.92	3.55	3.28	3.09	2.93	2.81
∞	.100	2.71	2.30	2.08	1.94	1.85	1.77	1.72	1.67	1.63
	.050	3.84	3.00	2.60	2.37	2.21	2.10	2.01	1.94	1.88
	.025	5.02	3.69	3.12	2.79	2.57	2.41	2.29	2.19	2.11
	.010	6.63	4.61	3.78	3.32	3.02	2.80	2.64	2.51	2.41
	.005	7.88	5.30	4.28	3.72	3.35	3.09	2.90	2.74	2.62

Source: From "Tables of Percentage Points of the Inverted Beta *(F)* Distribution." *Biometrika,* Vol. 33 (1943) by M. Merrington and C. M. Thompson and from Table 18 of *Biometrika Tables for Statisticians,* Vol. 1, Cambridge University Press, 1954, edited by E. S. Pearson and H. O. Hartley. Reproduced with permission of the authors, editors, and *Biometrika* trustees.

Numerator df											
10	12	15	20	24	30	40	60	120	∞	a	df
1.83	1.78	1.73	1.68	1.65	1.62	1.58	1.55	1.51	1.47	.100	29
2.18	2.10	2.03	1.94	1.90	1.85	1.81	1.75	1.70	1.64	.050	
2.53	2.43	2.32	2.21	2.15	2.09	2.03	1.96	1.89	1.81	.025	
3.00	2.87	2.73	2.57	2.49	2.41	2.33	2.23	2.14	2.03	.010	
3.38	3.21	3.04	2.86	2.76	2.66	2.56	2.45	2.33	2.21	.005	
1.82	1.77	1.72	1.67	1.64	1.61	1.57	1.54	1.50	1.46	.100	30
2.16	2.09	2.01	1.93	1.89	1.84	1.79	1.74	1.68	1.62	.050	
2.51	2.41	2.31	2.20	2.14	2.07	2.01	1.94	1.87	1.79	.025	
2.98	2.84	2.70	2.55	2.47	2.39	2.30	2.21	2.11	2.01	.010	
3.34	3.18	3.01	2.82	2.73	2.63	2.52	2.42	2.30	2.18	.005	
1.76	1.71	1.66	1.61	1.57	1.54	1.51	1.47	1.42	1.38	.100	40
2.08	2.00	1.92	1.84	1.79	1.74	1.69	1.64	1.58	1.51	.050	
2.39	2.29	2.18	2.07	2.01	1.94	1.88	1.80	1.72	1.64	.025	
2.80	2.66	2.52	2.37	2.29	2.20	2.11	2.02	1.92	1.80	.010	
3.12	2.95	2.78	2.60	2.50	2.40	2.30	2.18	2.06	1.93	.005	
1.71	1.66	1.60	1.54	1.51	1.48	1.44	1.40	1.35	1.29	.100	60
1.99	1.92	1.84	1.75	1.70	1.65	1.59	1.53	1.47	1.39	.050	
2.27	2.17	2.06	1.94	1.88	1.82	1.74	1.67	1.58	1.48	.025	
2.63	2.50	2.35	2.20	2.12	2.03	1.94	1.84	1.73	1.60	.010	
2.90	2.74	2.57	2.39	2.29	2.19	2.08	1.96	1.83	1.69	.005	
1.65	1.60	1.55	1.48	1.45	1.41	1.37	1.32	1.26	1.19	.100	120
1.91	1.83	1.75	1.66	1.61	1.55	1.50	1.43	1.35	1.25	.050	
2.16	2.05	1.94	1.82	1.76	1.69	1.61	1.53	1.43	1.31	.025	
2.47	2.34	2.19	2.03	1.95	1.86	1.76	1.66	1.53	1.38	.010	
2.71	2.54	2.37	2.19	2.09	1.98	1.87	1.75	1.61	1.43	.005	
1.60	1.55	1.49	1.42	1.38	1.34	1.30	1.24	1.17	1.00	.100	∞
1.83	1.75	1.67	1.57	1.52	1.46	1.39	1.32	1.22	1.00	.050	
2.05	1.94	1.83	1.71	1.64	1.57	1.48	1.39	1.27	1.00	.025	
2.32	2.18	2.04	1.88	1.79	1.70	1.59	1.47	1.32	1.00	.010	
2.52	2.36	2.19	2.00	1.90	1.79	1.67	1.53	1.36	1.00	.005	

TABLE 7 Upper-Tail Values for the Studentized Range
a. Upper 10% points

df \ k	2	3	4	5	6	7	8	9	10
1	8.93	13.44	16.36	18.49	20.15	21.51	22.64	23.62	24.48
2	4.13	5.73	6.77	7.54	8.14	8.63	9.05	9.41	9.72
3	3.33	4.47	5.20	5.74	6.16	6.51	6.81	7.06	7.29
4	3.01	3.98	4.59	5.03	5.39	5.68	5.93	6.14	6.33
5	2.85	3.72	4.26	4.66	4.98	5.24	5.46	5.65	5.82
6	2.75	3.56	4.07	4.44	4.73	4.97	5.17	5.34	5.50
7	2.68	3.45	3.93	4.28	4.55	4.78	4.97	5.14	5.28
8	2.63	3.37	3.83	4.17	4.43	4.65	4.83	4.99	5.13
9	2.59	3.32	3.76	4.08	4.34	4.54	4.72	4.87	5.01
10	2.56	3.27	3.70	4.02	4.26	4.47	4.64	4.78	4.91
11	2.54	3.23	3.66	3.96	4.20	4.40	4.57	4.71	4.84
12	2.52	3.20	3.62	3.92	4.16	4.35	4.51	4.65	4.78
13	2.50	3.18	3.59	3.88	4.12	4.30	4.46	4.60	4.72
14	2.49	3.16	3.56	3.85	4.08	4.27	4.42	4.56	4.68
15	2.48	3.14	3.54	3.83	4.05	4.23	4.39	4.52	4.64
16	2.47	3.12	3.52	3.80	4.03	4.21	4.36	4.49	4.61
17	2.46	3.11	3.50	3.78	4.00	4.18	4.33	4.46	4.58
18	2.45	3.10	3.49	3.77	3.98	4.16	4.31	4.44	4.55
19	2.45	3.09	3.47	3.75	3.97	4.14	4.29	4.42	4.53
20	2.44	3.08	3.46	3.74	3.95	4.12	4.27	4.40	4.51
24	2.42	3.05	3.42	3.69	3.90	4.07	4.21	4.34	4.44
30	2.40	3.02	3.39	3.65	3.85	4.02	4.16	4.28	4.38
40	2.38	2.99	3.35	3.60	3.80	3.96	4.10	4.21	4.32
60	2.36	2.96	3.31	3.56	3.75	3.91	4.04	4.16	4.25
120	2.34	2.93	3.28	3.52	3.71	3.86	3.99	4.10	4.19
∞	2.33	2.90	3.24	3.48	3.66	3.81	3.93	4.04	4.13

df \ k	11	12	13	14	15	16	17	18	19	20
1	25.24	25.92	26.54	27.10	27.62	28.10	28.54	28.96	29.35	29.71
2	10.01	10.26	10.49	10.70	10.89	11.07	11.24	11.39	11.54	11.68
3	7.49	7.67	7.83	7.98	8.12	8.25	8.37	8.48	8.58	8.68
4	6.49	6.65	6.78	6.91	7.02	7.13	7.23	7.33	7.41	7.50
5	5.97	6.10	6.22	6.34	6.44	6.54	6.63	6.71	6.79	6.86
6	5.64	5.76	5.87	5.98	6.07	6.16	6.25	6.32	6.40	6.47
7	5.41	5.53	5.64	5.74	5.83	5.91	5.99	6.06	6.13	6.19
8	5.25	5.36	5.46	5.56	5.64	5.72	5.80	5.87	5.93	6.00
9	5.13	5.23	5.33	5.42	5.51	5.58	5.66	5.72	5.79	5.85
10	5.03	5.13	5.23	5.32	5.40	5.47	5.54	5.61	5.67	5.73
11	4.95	5.05	5.15	5.23	5.31	5.38	5.45	5.51	5.57	5.63
12	4.89	4.99	5.08	5.16	5.24	5.31	5.37	5.44	5.49	5.55
13	4.83	4.93	5.02	5.10	5.18	5.25	5.31	5.37	5.43	5.48
14	4.79	4.88	4.97	5.05	5.12	5.19	5.26	5.32	5.37	5.43
15	4.75	4.84	4.93	5.01	5.08	5.15	5.21	5.27	5.32	5.38
16	4.71	4.81	4.89	4.97	5.04	5.11	5.17	5.23	5.28	5.33
17	4.68	4.77	4.86	4.93	5.01	5.07	5.13	5.19	5.24	5.30
18	4.65	4.75	4.83	4.90	4.98	5.04	5.10	5.16	5.21	5.26
19	4.63	4.72	4.80	4.88	4.95	5.01	5.07	5.13	5.18	5.23
20	4.61	4.70	4.78	4.85	4.92	4.99	5.05	5.10	5.16	5.20
24	4.54	4.63	4.71	4.78	4.85	4.91	4.97	5.02	5.07	5.12
30	4.47	4.56	4.64	4.71	4.77	4.83	4.89	4.94	4.99	5.03
40	4.41	4.49	4.56	4.63	4.69	4.75	4.81	4.86	4.90	4.95
60	4.34	4.42	4.49	4.56	4.62	4.67	4.73	4.78	4.82	4.86
120	4.28	4.35	4.42	4.48	4.54	4.60	4.65	4.69	4.74	4.78
∞	4.21	4.28	4.35	4.41	4.47	4.52	4.57	4.61	4.65	4.69

TABLE 7 (Continued)
b. Upper 5% points

df \ k	2	3	4	5	6	7	8	9	10
1	17.97	26.98	32.82	37.08	40.41	43.12	45.40	47.36	49.07
2	6.08	8.33	9.80	10.88	11.74	12.44	13.03	13.54	13.99
3	4.50	5.91	6.82	7.50	8.04	8.48	8.85	9.18	9.46
4	3.93	5.04	5.76	6.29	6.71	7.05	7.35	7.60	7.83
5	3.64	4.60	5.22	5.67	6.03	6.33	6.58	6.80	6.99
6	3.46	4.34	4.90	5.30	5.63	5.90	6.12	6.32	6.49
7	3.34	4.16	4.68	5.06	5.36	5.61	5.82	6.00	6.16
8	3.26	4.04	4.53	4.89	5.17	5.40	5.60	5.77	5.92
9	3.20	3.95	4.41	4.76	5.02	5.24	5.43	5.59	5.74
10	3.15	3.88	4.33	4.65	4.91	5.12	5.30	5.46	5.60
11	3.11	3.82	4.26	4.57	4.82	5.03	5.20	5.35	5.49
12	3.08	3.77	4.20	4.51	4.75	4.95	5.12	5.27	5.39
13	3.06	3.73	4.15	4.45	4.69	4.88	5.05	5.19	5.32
14	3.03	3.70	4.11	4.41	4.64	4.83	4.99	5.13	5.25
15	3.01	3.67	4.08	4.37	4.59	4.78	4.94	5.08	5.20
16	3.00	3.65	4.05	4.33	4.56	4.74	4.90	5.03	5.15
17	2.98	3.63	4.02	4.30	4.52	4.70	4.86	4.99	5.11
18	2.97	3.61	4.00	4.28	4.49	4.67	4.82	4.96	5.07
19	2.96	3.59	3.98	4.25	4.47	4.65	4.79	4.92	5.04
20	2.95	3.58	3.96	4.23	4.45	4.62	4.77	4.90	5.01
24	2.92	3.53	3.90	4.17	4.37	4.54	4.68	4.81	4.92
30	2.89	3.49	3.85	4.10	4.30	4.46	4.60	4.72	4.82
40	2.86	3.44	3.79	4.04	4.23	4.39	4.52	4.63	4.73
60	2.83	3.40	3.74	3.98	4.16	4.31	4.44	4.55	4.65
120	2.80	3.36	3.68	3.92	4.10	4.24	4.36	4.47	4.56
∞	2.77	3.31	3.63	3.86	4.03	4.17	4.29	4.39	4.47

df \ k	11	12	13	14	15	16	17	18	19	20
1	50.59	51.96	53.20	54.33	55.36	56.32	57.22	58.04	58.83	59.56
2	14.39	14.75	15.08	15.38	15.65	15.91	16.14	16.37	16.57	16.77
3	9.72	9.95	10.15	10.35	10.52	10.69	10.84	10.98	11.11	11.24
4	8.03	8.21	8.37	8.52	8.66	8.79	8.91	9.03	9.13	9.23
5	7.17	7.32	7.47	7.60	7.72	7.83	7.93	8.03	8.12	8.21
6	6.65	6.79	6.92	7.03	7.14	7.24	7.34	7.43	7.51	7.59
7	6.30	6.43	6.55	6.66	6.76	6.85	6.94	7.02	7.10	7.17
8	6.05	6.18	6.29	6.39	6.48	6.57	6.65	6.73	6.80	6.87
9	5.87	5.98	6.09	6.19	6.28	6.36	6.44	6.51	6.58	6.64
10	5.72	5.83	5.93	6.03	6.11	6.19	6.27	6.34	6.40	6.47
11	5.61	5.71	5.81	5.90	5.98	6.06	6.13	6.20	6.27	6.33
12	5.51	5.61	5.71	5.80	5.88	5.95	6.02	6.09	6.15	6.21
13	5.43	5.53	5.63	5.71	5.79	5.86	5.93	5.99	6.05	6.11
14	5.36	5.46	5.55	5.64	5.71	5.79	5.85	5.91	5.97	6.03
15	5.31	5.40	5.49	5.57	5.65	5.72	5.78	5.85	5.90	5.96
16	5.26	5.35	5.44	5.52	5.59	5.66	5.73	5.79	5.84	5.90
17	5.21	5.31	5.39	5.47	5.54	5.61	5.67	5.73	5.79	5.84
18	5.17	5.27	5.35	5.43	5.50	5.57	5.63	5.69	5.74	5.79
19	5.14	5.23	5.31	5.39	5.46	5.53	5.59	5.65	5.70	5.75
20	5.11	5.20	5.28	5.36	5.43	5.49	5.55	5.61	5.66	5.71
24	5.01	5.10	5.18	5.25	5.32	5.38	5.44	5.49	5.55	5.59
30	4.92	5.00	5.08	5.15	5.21	5.27	5.33	5.38	5.43	5.47
40	4.82	4.90	4.98	5.04	5.11	5.16	5.22	5.27	5.31	5.36
60	4.73	4.81	4.88	4.94	5.00	5.06	5.11	5.15	5.20	5.24
120	4.64	4.71	4.78	4.84	4.90	4.95	5.00	5.04	5.09	5.13
∞	4.55	4.62	4.68	4.74	4.80	4.85	4.89	4.93	4.97	5.01

TABLE 7 (Continued)
c. Upper 1% points

df \ k	2	3	4	5	6	7	8	9	10
1	90.03	135.0	164.3	185.6	202.2	215.8	227.2	237.0	245.6
2	14.04	19.02	22.29	24.72	26.63	28.20	29.53	30.68	31.69
3	8.26	10.62	12.17	13.33	14.24	15.00	15.64	16.20	16.69
4	6.51	8.12	9.17	9.96	10.58	11.10	11.55	11.93	12.27
5	5.70	6.98	7.80	8.42	8.91	9.32	9.67	9.97	10.24
6	5.24	6.33	7.03	7.56	7.97	8.32	8.61	8.87	9.10
7	4.95	5.92	6.54	7.01	7.37	7.68	7.94	8.17	8.37
8	4.75	5.64	6.20	6.62	6.96	7.24	7.47	7.68	7.86
9	4.60	5.43	5.96	6.35	6.66	6.91	7.13	7.33	7.49
10	4.48	5.27	5.77	6.14	6.43	6.67	6.87	7.05	7.21
11	4.39	5.15	5.62	5.97	6.25	6.48	6.67	6.84	6.99
12	4.32	5.05	5.50	5.84	6.10	6.32	6.51	6.67	6.81
13	4.26	4.96	5.40	5.73	5.98	6.19	6.37	6.53	6.67
14	4.21	4.89	5.32	5.63	5.88	6.08	6.26	6.41	6.54
15	4.17	4.84	5.25	5.56	5.80	5.99	6.16	6.31	6.44
16	4.13	4.79	5.19	5.49	5.72	5.92	6.08	6.22	6.35
17	4.10	4.74	5.14	5.43	5.66	5.85	6.01	6.15	6.27
18	4.07	4.70	5.09	5.38	5.60	5.79	5.94	6.08	6.20
19	4.05	4.67	5.05	5.33	5.55	5.73	5.89	6.02	6.14
20	4.02	4.64	5.02	5.29	5.51	5.69	5.84	5.97	6.09
24	3.96	4.55	4.91	5.17	5.37	5.54	5.69	5.81	5.92
30	3.89	4.45	4.80	5.05	5.24	5.40	5.54	5.65	5.76
40	3.82	4.37	4.70	4.93	5.11	5.26	5.39	5.50	5.60
60	3.76	4.28	4.59	4.82	4.99	5.13	5.25	5.36	5.45
120	3.70	4.20	4.50	4.71	4.87	5.01	5.12	5.21	5.30
∞	3.64	4.12	4.40	4.60	4.76	4.88	4.99	5.08	5.16

df \ k	11	12	13	14	15	16	17	18	19	20
1	253.2	260.0	266.2	271.8	277.0	281.8	286.3	290.4	294.3	298.0
2	32.59	33.40	34.13	34.81	35.43	36.00	36.53	37.03	37.50	37.95
3	17.13	17.53	17.89	18.22	18.52	18.81	19.07	19.32	19.55	19.77
4	12.57	12.84	13.09	13.32	13.53	13.73	13.91	14.08	14.24	14.40
5	10.48	10.70	10.89	11.08	11.24	11.40	11.55	11.68	11.81	11.93
6	9.30	9.48	9.65	9.81	9.95	10.08	10.21	10.32	10.43	10.54
7	8.55	8.71	8.86	9.00	9.12	9.24	9.35	9.46	9.55	9.65
8	8.03	8.18	8.31	8.44	8.55	8.66	8.76	8.85	8.94	9.03
9	7.65	7.78	7.91	8.03	8.13	8.23	8.33	8.41	8.49	8.57
10	7.36	7.49	7.60	7.71	7.81	7.91	7.99	8.08	8.15	8.23
11	7.13	7.25	7.36	7.46	7.56	7.65	7.73	7.81	7.88	7.95
12	6.94	7.06	7.17	7.26	7.36	7.44	7.52	7.59	7.66	7.73
13	6.79	6.90	7.01	7.10	7.19	7.27	7.35	7.42	7.48	7.55
14	6.66	6.77	6.87	6.96	7.05	7.13	7.20	7.27	7.33	7.39
15	6.55	6.66	6.76	6.84	6.93	7.00	7.07	7.14	7.20	7.26
16	6.46	6.56	6.66	6.74	6.82	6.90	6.97	7.03	7.09	7.15
17	6.38	6.48	6.57	6.66	6.73	6.81	6.87	6.94	7.00	7.05
18	6.31	6.41	6.50	6.58	6.65	6.73	6.79	6.85	6.91	6.97
19	6.25	6.34	6.43	6.51	6.58	6.65	6.72	6.78	6.84	6.89
20	6.19	6.28	6.37	6.45	6.52	6.59	6.65	6.71	6.77	6.82
24	6.02	6.11	6.19	6.26	6.33	6.39	6.45	6.51	6.56	6.61
30	5.85	5.93	6.01	6.08	6.14	6.20	6.26	6.31	6.36	6.41
40	5.69	5.76	5.83	5.90	5.96	6.02	6.07	6.12	6.16	6.21
60	5.53	5.60	5.67	5.73	5.78	5.84	5.89	5.93	5.97	6.01
120	5.37	5.44	5.50	5.56	5.61	5.66	5.71	5.75	5.79	5.83
∞	5.23	5.29	5.35	5.40	5.45	5.49	5.54	5.57	5.61	5.65

TABLE 8 Lower-Tail Values for the Mann–Whitney U Statistic

a. $n_2 = 3$

	n_1		
U_0	1	2	3
0	.25	.10	.05
1	.50	.20	.10
2		.40	.20
3		.60	.35
4			.50

b. $n_2 = 4$

	n_1			
U_0	1	2	3	4
0	.2000	.0667	.0286	.0143
1	.4000	.1333	.0571	.0286
2	.6000	.2667	.1143	.0571
3		.4000	.2000	.1000
4		.6000	.3143	.1714
5			.4286	.2429
6			.5714	.3429
7				.4429
8				.5571

c. $n_2 = 5$

	n_1				
U_0	1	2	3	4	5
0	.1667	.0476	.0179	.0079	.0040
1	.3333	.0952	.0357	.0159	.0079
2	.5000	.1905	.0714	.0317	.0159
3		.2857	.1250	.0556	.0278
4		.4286	.1964	.0952	.0476
5		.5714	.2857	.1429	.0754
6			.3929	.2063	.1111
7			.5000	.2778	.1548
8				.3651	.2103
9				.4524	.2738
10				.5476	.3452
11					.4206
12					.5000

TABLE 8 (Continued)

d. $n_2 = 6$

	n_1					
U_0	1	2	3	4	5	6
0	.1429	.0357	.0119	.0048	.0022	.0011
1	.2857	.0714	.0238	.0095	.0043	.0022
2	.4286	.1429	.0476	.0190	.0087	.0043
3	.5714	.2143	.0833	.0333	.0152	.0076
4		.3214	.1310	.0571	.0260	.0130
5		.4286	.1905	.0857	.0411	.0206
6		.5714	.2738	.1286	.0628	.0325
7			.3571	.1762	.0887	.0465
8			.4524	.2381	.1234	.0660
9			.5476	.3048	.1645	.0898
10				.3810	.2143	.1201
11				.4571	.2684	.1548
12				.5429	.3312	.1970
13					.3961	.2424
14					.4654	.2944
15					.5346	.3496
16						.4091
17						.4686
18						.5314

e. $n_2 = 7$

	n_1						
U_0	1	2	3	4	5	6	7
0	.1250	.0278	.0083	.0030	.0013	.0006	.0003
1	.2500	.0556	.0167	.0061	.0025	.0012	.0006
2	.3750	.1111	.0333	.0121	.0051	.0023	.0012
3	.5000	.1667	.0583	.0212	.0088	.0041	.0020
4		.2500	.0917	.0364	.0152	.0070	.0035
5		.3333	.1333	.0545	.0240	.0111	.0055
6		.4444	.1917	.0818	.0366	.0175	.0087
7		.5556	.2583	.1152	.0530	.0256	.0131
8			.3333	.1576	.0745	.0367	.0189
9			.4167	.2061	.1010	.0507	.0265
10			.5000	.2636	.1338	.0688	.0364
11				.3242	.1717	.0903	.0487
12				.3939	.2159	.1171	.0641
13				.4636	.2652	.1474	.0825
14				.5364	.3194	.1830	.1043
15					.3775	.2226	.1297
16					.4381	.2669	.1588
17					.5000	.3141	.1914
18						.3654	.2279
19						.4178	.2675
20						.4726	.3100
21						.5274	.3552
22							.4024
23							.4508
24							.5000

TABLE 8 (Continued)

f. $n_2 = 8$

	n_1							
U_0	1	2	3	4	5	6	7	8
0	.1111	.0222	.0061	.0020	.0008	.0003	.0002	.0001
1	.2222	.0444	.0121	.0040	.0016	.0007	.0003	.0002
2	.3333	.0889	.0242	.0081	.0031	.0013	.0006	.0003
3	.4444	.1333	.0424	.0141	.0054	.0023	.0011	.0005
4	.5556	.2000	.0667	.0242	.0093	.0040	.0019	.0009
5		.2667	.0970	.0364	.0148	.0063	.0030	.0015
6		.3556	.1394	.0545	.0225	.0100	.0047	.0023
7		.4444	.1879	.0768	.0326	.0147	.0070	.0035
8		.5556	.2485	.1071	.0466	.0213	.0103	.0052
9			.3152	.1414	.0637	.0296	.0145	.0074
10			.3879	.1838	.0855	.0406	.0200	.0103
11			.4606	.2303	.1111	.0539	.0270	.0141
12			.5394	.2848	.1422	.0709	.0361	.0190
13				.3414	.1772	.0906	.0469	.0249
14				.4040	.2176	.1142	.0603	.0325
15				.4667	.2618	.1412	.0760	.0415
16				.5333	.3108	.1725	.0946	.0524
17					.3621	.2068	.1159	.0652
18					.4165	.2454	.1405	.0803
19					.4716	.2864	.1678	.0974
20					.5284	.3310	.1984	.1172
21						.3773	.2317	.1393
22						.4259	.2679	.1641
23						.4749	.3063	.1911
24						.5251	.3472	.2209
25							.3894	.2527
26							.4333	.2869
27							.4775	.3227
28							.5225	.3605
29								.3992
30								.4392
31								.4796
32								.5204

TABLE 8 (Continued)

g. $n_2 = 9$

	n_1								
U_0	1	2	3	4	5	6	7	8	9
0	.1000	.0182	.0045	.0014	.0005	.0002	.0001	.0000	.0000
1	.2000	.0364	.0091	.0028	.0010	.0004	.0002	.0001	.0000
2	.3000	.0727	.0182	.0056	.0020	.0008	.0003	.0002	.0001
3	.4000	.1091	.0318	.0098	.0035	.0014	.0006	.0003	.0001
4	.5000	.1636	.0500	.0168	.0060	.0024	.0010	.0005	.0002
5		.2182	.0727	.0252	.0095	.0038	.0017	.0008	.0004
6		.2909	.1045	.0378	.0145	.0060	.0026	.0012	.0006
7		.3636	.1409	.0531	.0210	.0088	.0039	.0019	.0009
8		.4545	.1864	.0741	.0300	.0128	.0058	.0028	.0014
9		.5455	.2409	.0993	.0415	.0180	.0082	.0039	.0020
10			.3000	.1301	.0559	.0248	.0115	.0056	.0028
11			.3636	.1650	.0734	.0332	.0156	.0076	.0039
12			.4318	.2070	.0949	.0440	.0209	.0103	.0053
13			.5000	.2517	.1199	.0567	.0274	.0137	.0071
14				.3021	.1489	.0723	.0356	.0180	.0094
15				.3552	.1818	.0905	.0454	.0232	.0122
16				.4126	.2188	.1119	.0571	.0296	.0157
17				.4699	.2592	.1361	.0708	.0372	.0200
18				.5301	.3032	.1638	.0869	.0464	.0252
19					.3497	.1942	.1052	.0570	.0313
20					.3986	.2280	.1261	.0694	.0385
21					.4491	.2643	.1496	.0836	.0470
22					.5000	.3035	.1755	.0998	.0567
23						.3445	.2039	.1179	.0680
24						.3878	.2349	.1383	.0807
25						.4320	.2680	.1606	.0951
26						.4773	.3032	.1852	.1112
27						.5227	.3403	.2117	.1290
28							.3788	.2404	.1487
29							.4185	.2707	.1701
30							.4591	.3029	.1933
31							.5000	.3365	.2181
32								.3715	.2447
33								.4074	.2729
34								.4442	.3024
35								.4813	.3332
36								.5187	.3652
37									.3981
38									.4317
39									.4657
40									.5000

TABLE 8 (Continued)

h. $n_2 = 10$

	n_1									
U_0	1	2	3	4	5	6	7	8	9	10
0	.0909	.0152	.0035	.0010	.0003	.0001	.0001	.0000	.0000	.0000
1	.1818	.0303	.0070	.0020	.0007	.0002	.0001	.0000	.0000	.0000
2	.2727	.0606	.0140	.0040	.0013	.0005	.0002	.0001	.0000	.0000
3	.3636	.0909	.0245	.0070	.0023	.0009	.0004	.0002	.0001	.0000
4	.4545	.1364	.0385	.0120	.0040	.0015	.0006	.0003	.0001	.0001
5	.5455	.1818	.0559	.0180	.0063	.0024	.0010	.0004	.0002	.0001
6		.2424	.0804	.0270	.0097	.0037	.0015	.0007	.0003	.0002
7		.3030	.1084	.0380	.0140	.0055	.0023	.0010	.0005	.0002
8		.3788	.1434	.0529	.0200	.0080	.0034	.0015	.0007	.0004
9		.4545	.1853	.0709	.0276	.0112	.0048	.0022	.0011	.0005
10		.5455	.2343	.0939	.0376	.0156	.0068	.0031	.0015	.0008
11			.2867	.1199	.0496	.0210	.0093	.0043	.0021	.0010
12			.3462	.1518	.0646	.0280	.0125	.0058	.0028	.0014
13			.4056	.1868	.0823	.0363	.0165	.0078	.0038	.0019
14			.4685	.2268	.1032	.0467	.0215	.0103	.0051	.0026
15			.5315	.2697	.1272	.0589	.0277	.0133	.0066	.0034
16				.3177	.1548	.0736	.0351	.0171	.0086	.0045
17				.3666	.1855	.0903	.0439	.0217	.0110	.0057
18				.4196	.2198	.1099	.0544	.0273	.0140	.0073
19				.4725	.2567	.1317	.0665	.0338	.0175	.0093
20				.5275	.2970	.1566	.0806	.0416	.0217	.0116
21					.3393	.1838	.0966	.0506	.0267	.0144
22					.3839	.2139	.1148	.0610	.0326	.0177
23					.4296	.2461	.1349	.0729	.0394	.0216
24					.4765	.2811	.1574	.0864	.0474	.0262
25					.5235	.3177	.1819	.1015	.0564	.0315
26						.3564	.2087	.1185	.0667	.0376
27						.3962	.2374	.1371	.0782	.0446
28						.4374	.2681	.1577	.0912	.0526
29						.4789	.3004	.1800	.1055	.0615
30						.5211	.3345	.2041	.1214	.0716
31							.3698	.2299	.1388	.0827
32							.4063	.2574	.1577	.0952
33							.4434	.2863	.1781	.1088
34							.4811	.3167	.2001	.1237
35							.5189	.3482	.2235	.1399
36								.3809	.2483	.1575
37								.4143	.2745	.1763
38								.4484	.3019	.1965
39								.4827	.3304	.2179
40								.5173	.3598	.2406
41									.3901	.2644
42									.4211	.2894
43									.4524	.3153
44									.4841	.3421
45									.5159	.3697
46										.3980
47										.4267
48										.4559
49										.4853
50										.5147

Computed by M. Pagano, Department of Statistics, University of Florida.

TABLE 9 Lower-Tail Values for the Wilcoxon Signed Ranks Test

One-sided	Two-sided	$n = 5$	$n = 6$	$n = 7$	$n = 8$	$n = 9$	$n = 10$
$a = .05$	$a = .10$	1	2	4	6	8	11
$a = .025$	$a = .05$		1	2	4	6	8
$a = .01$	$a = .02$			0	2	3	5
$a = .005$	$a = .01$				0	2	3
One-sided	**Two-sided**	$n = 11$	$n = 12$	$n = 13$	$n = 14$	$n = 15$	$n = 16$
$a = .05$	$a = .10$	14	17	21	26	30	36
$a = .025$	$a = .05$	11	14	17	21	25	30
$a = .01$	$a = .02$	7	10	13	16	20	24
$a = .005$	$a = .01$	5	7	10	13	16	19
One-sided	**Two-sided**	$n = 17$	$n = 18$	$n = 19$	$n = 20$	$n = 21$	$n = 22$
$a = .05$	$a = .10$	41	47	54	60	68	75
$a = .025$	$a = .05$	35	40	46	52	59	66
$a = .01$	$a = .02$	28	33	38	43	49	56
$a = .005$	$a = .01$	23	28	32	37	43	49
One-sided	**Two-sided**	$n = 23$	$n = 24$	$n = 25$	$n = 26$	$n = 27$	$n = 28$
$a = .05$	$a = .10$	83	92	101	110	120	130
$a = .025$	$a = .05$	73	81	90	98	107	117
$a = .01$	$a = .02$	62	69	77	85	93	102
$a = .005$	$a = .01$	55	61	68	76	84	92
One-sided	**Two-sided**	$n = 29$	$n = 30$	$n = 31$	$n = 32$	$n = 33$	$n = 34$
$a = .05$	$a = .10$	141	152	163	175	188	201
$a = .025$	$a = .05$	127	137	148	159	171	183
$a = .01$	$a = .02$	111	120	130	141	151	162
$a = .005$	$a = .01$	100	109	118	128	138	149
One-sided	**Two-sided**	$n = 35$	$n = 36$	$n = 37$	$n = 38$	$n = 39$	
$a = .05$	$a = .10$	214	228	242	256	271	
$a = .025$	$a = .05$	195	208	222	235	250	
$a = .01$	$a = .02$	174	186	198	211	224	
$a = .005$	$a = .01$	160	171	183	195	208	
One-sided	**Two-sided**	$n = 40$	$n = 41$	$n = 42$	$n = 43$	$n = 44$	$n = 45$
$a = .05$	$a = .10$	287	303	319	336	353	371
$a = .025$	$a = .05$	264	279	295	311	327	344
$a = .01$	$a = .02$	238	252	267	281	297	313
$a = .005$	$a = .01$	221	234	248	262	277	292
One-sided	**Two-sided**	$n = 46$	$n = 47$	$n = 48$	$n = 49$	$n = 50$	
$a = .05$	$a = .10$	389	408	427	446	466	
$a = .025$	$a = .05$	361	379	397	415	434	
$a = .01$	$a = .02$	329	345	362	380	398	
$a = .005$	$a = .01$	307	323	339	356	373	

Source: From "Some Rapid Approximate Statistical Procedures" (1964), 28, F. Wilcoxon and R. A. Wilcox. Reproduced with the kind permission of R. A. Wilcox and the Lederle Laboratories.

Table 10 Upper-Tail Values of Spearman's Rank Correlation Coefficient

n	$a = .05$	$a = .025$	$a = .01$	$a = .005$
5	.900	—	—	—
6	.829	.886	.943	—
7	.714	.786	.893	—
8	.643	.738	.833	.881
9	.600	.683	.783	.833
10	.564	.648	.745	.794
11	.523	.623	.736	.818
12	.497	.591	.703	.780
13	.475	.566	.673	.745
14	.457	.545	.646	.716
15	.441	.525	.623	.689
16	.425	.507	.601	.666
17	.412	.490	.582	.645
18	.399	.476	.564	.625
19	.388	.462	.549	.608
20	.377	.450	.534	.591
21	.368	.438	.521	.576
22	.359	.428	.508	.562
23	.351	.418	.496	.549
24	.343	.409	.485	.537
25	.336	.400	.475	.526
26	.329	.392	.465	.515
27	.323	.385	.456	.505
28	.317	.377	.448	.496
29	.311	.370	.440	.487
30	.305	.364	.432	.478

Source: From "Distribution of Sums of Squares of Rank Differences for Small Samples," E. G. Olds, *Annals of Mathematical Statistics,* Volume 9 (1938). Reproduced with the kind permission of the Editor, *Annals of Mathematical Statistics.*

appendix 3

Answers to Selected Exercises

Chapter 2

2.1 See definitions. Example: Suppose that we are interested in the shelf life of a shipment of breakfast cereal. Shelf life is the length of time until the quality of the packaged cereal begins to deteriorate. It is a variable that varies from box to box. The universe is the collection of all boxes of cereal in the shipment. An element of the universe is a single box of cereal. The population is the collection of the shelf lives of all elements in the universe.

2.3 (a) Quantitative (b) Quantitative (c) Qualitative (d) Quantitative (e) Quantitative (f) Qualitative

2.5 (a) One of the three portions of blood (b) The universe, which exists in our minds, is the collection of all portions into which the drawn blood could conceivably have been partitioned. (c) The population is the collection of measurements made on the portions of blood in the conceptual universe. (d) No (e) A sample from the conceptual population of measurements

2.7 (a) A cab driver (b) The collection of 11,787 cab drivers (c) Cab fare; quantitative (d) The collection of fares that the drivers would have charged had they been hired (one fare for each of the 11,787 drivers) (e) The collection of 50 fares, one for each of the drivers that were hired (f) To investigate the overcharging practices of all New York cab drivers for the trip from Kennedy to the WTC.

2.9 (a) A student loan account (b) Qualitative (c) The collection of all student loans in the United States (d) The collection of "yes" and "no" responses (loan in error or lacking documentation), one response associated with each loan in the universe (e) The collection of "yes" and "no" responses for the 2,038 audited accounts (f) Approximately 18% of all student loans are in error or lack documentation.

2.11 (a) Each fossil specimen is an element. (b) The collection of all fossils of that species in or near the Antarctic Peninsula (c) Diameter, quantitative; height, quantitative (d) To describe the size of the shellfish, *Rotularia (Annelida) fallax*

Chapter 3

3.5 (c) Banks, 32.4; FHA, 14.9

3.7 The qualitative variable is "preference"; categories are A, B, and "no preference"; relative frequencies are: A, .56; B, .28; and "no preference," 16.

3.9 See Section 3.2.

3.11 (b) The interval from 10 to and including 19 (c) 0 to and including 9, 10 to and including 19, 20 to and including 29, and 30 to and including 39 (d) 0 to and including 9 (e) Most fires in the data set resulted in 9 or fewer deaths per fire but a small proportion of fires resulted in as many as 30 to 39 deaths. The distribution is skewed to the right.

3.13 (a) Both channel catfish and small-mouth buffalo tend to be larger than the large-mouth bass; the channel catfish tend to be larger than the small-mouth buffalo; lengths of small-mouth buffalo appear to be the most variable. (b) Catfish, approximately .995; bass, .25; buffalo 1.00

3.15 All three distributions of DDT are highly skewed to the right; most of the DDT percentages for catfish are less than 37.5 ppm; for bass, 4 ppm; and for small-mouth buffalo, 16 ppm. Readings for catfish and small-mouth buffalo were as high as 175 ppm and 44 ppm, respectively. (b) Catfish, approximately .74; bass, .08; small-mouth buffalo, .55

3.21 109, 212, 254, 359, . . . , 1,126

3.23 (b) 0

Chapter 4

4.1 See Section 4.2; locates a point near the "center" of the data set's distribution

4.3 Preferred as an estimate of the population mean

4.5 See Section 4.3; measures the spread of the distribution of the data set

4.7 See Section 4.3; measures the spread of the distribution of the data set

4.9 The Empirical Rule, along with the mean and standard deviation, is used to describe the distribution of a data set.

4.11 (a) 108.16 (b) The distribution is centered over 89.7 and all or almost all of the measurements fall in the interval from 58.5 to 120.9 (b) Approximately 95% fall within the interval from 68.9 to 110.5

4.13 Approximately 95%; no, 40 lies more than 3 standard deviations away from the mean

4.15 Shifted to the right

4.17 The distribution is centered over 10.5 and all or almost all of the measurements fall in the interval from 5.1 to 15.9. Approximately 95% will fall in the interval from 6.9 to 14.1.

4.19 The distribution is centered over 152.1 and all or almost all of the measurements fall in the interval from 90.0 to 214.2. Approximately 95% will fall in the interval from 110.7 to 193.5.

4.21 (a) $\bar{x} = 8.357143$, $s = 8.940893$ (b) $s^2 = 79.93956$ (c) 0, 36, 36 (d) −.58 to 17.30, −9.52 to 26.24, −18.46 to 35.18 (e) .93, 1.0, 1.0. The proportion falling in $\bar{x} \pm s$ is larger than specified because the data set is small and the value of s is inflated by the single large observation (35). The proportions in the intervals $\bar{x} \pm 2s$ and $\bar{x} \pm 3s$ agree reasonably well with the Empirical Rule.

4.23 (a) $\bar{x} = 5{,}334$, $s = 1{,}387.791$ (b) $s^2 = 1{,}925{,}963.9$ (c) 3,100, 8,600, 5,500 (d) 3,946.2 to 6,721.8, 2,558.4 to 8,109.6, 1,170.6 to 9,497.4 (f) The exact proportions are .72, .94, 1.0. They agree reasonably well with the Empirical Rule.

4.25 (a) $\bar{x} = 74.530$, $s = 2.4412$ (b) $s^2 = 5.9595$ (c) 69.7, 79.1, 9.4 (d) 72.089 to 76.971, 69.648 to 79.412, 67.206 to 81.854 (f) The exact proportions are .67, 1.00, 1.00. The proportions agree reasonably well with the Empirical Rule.

4.27 Near 12: range/3 = 12. We would expect the data to spread over approximately 3 to 4 (see Table 4.6) standard deviations.

4.29 Without knowing the sample size, we would guess that the data would spread over 4 standard deviations.

4.31 Near .7; range/6 = .68; $s = .694$

4.33 Near 13; range/4 = 13.25; $s = 9.541403$

4.35 An operation or activity that produces a continuous flow of elements over time

4.37 The distribution(s) of the quality variable(s) for the process remain unchanged over time.

4.39 To monitor a process to detect possible departures from "in control"

4.41 See Section 4.6.

4.43 The median is the second quartile.

4.45 Eighty-four percent scored below your score; 16% scored above.

4.47 (a) Slightly less than 75 (b) Would expect the distribution to be skewed to the right (see Section 4.6)

4.49 See Section 4.7.

4.51 (a) 4, 9.25 (b) 6 (c) 5.25 (d) Inner gates are located at −3.875 and 17.125. (e) 36 (f) Yes, 36; check to see if there is reason to believe the observation is faulty. (g) Skewed to the right

4.53 They both measure the variation in a data set.

4.55 See Section 4.9; the sample mean

4.57 Center of the distribution is located near 11.60. Approximately 68% (say 2/3) of the measurements lie within 11.45 to 11.75, 95% within 11.30 to 11.90, and all or almost all within 11.15 to 12.05.

4.59 65.410, 66.230; 25% of the measurements are less than 65.410, 75% are less than 66.230 (b) 65.830; half the measurements are smaller than 65.830, half are larger (c) .82 (d) Inner gates are located at 64.18 and 67.46 (e) 64.18, 64.01, 64.02, 63.48, 63.29, 63.62, 63.89, 64.14, 63.53, 63.57, 63.52, 63.85, 63.95, 63.60, 62.77 (f) 62.77

4.61 Range/6 = (414 − 120)/6 = 49

Chapter 5

5.1 See Section 2.4.

5.3–5.9 See Section 5.3.

5.11 Mutually exclusive outcomes are always dependent. For example, the probability that the day will end in a conviction is some number between 0 and 1. In contrast, the probability that the day will end in a conviction, given that the trial has been continued, is 0. Since the probability of the occurrence of one outcome is dependent on the occurrence or nonoccurrence of the other, the outcomes are dependent.

5.13 .7

5.15 1/36

5.17–5.19 See Sections 5.3 and 5.4.

5.21–5.23 See Section 5.4.

5.25 1/10

5.27 See Section 5.5.

5.29 (a) Continuous (b) Continuous (c) Discrete (d) Discrete (e) Discrete

5.31 It satisfies, to a reasonable degree of approximation, the five characteristics of a binomial experiment (similar to Example 5.6).

5.33 No. This experiment does not involve a series of identical trials.

5.35 800, 21.91

5.37 $\mu = 40$, $\sigma = 6$. All or almost all of the distribution will lie in the interval $\mu \pm 3\sigma$, or 22 to 58. No. See Section 5.6.

5.39 $\mu = 400$, $\sigma = 15.5$. The distribution will center over 400 and fall within the interval $\mu \pm 3\sigma$, or 353.5 to 446.5.

5.41 (a) .201 (b) .166 (c) .633

5.45 (a) 4 (b) 1.79 (c) .37 (d) .42 to 7.58 (e) $p(1) + \cdots + p(7) = .957$

5.47 (a) .993 (b) .901 (c) .007

5.49 See Section 5.7.

5.51 (a) .01 (b) .02 (c) .05 (d) .10 (e) .20

5.53 10th percentile

5.55 When the interval $np \pm 3\sqrt{np(1-p)}$, i.e., $\mu \pm 3\sigma$, falls in the interval from $x = 0$ to $x = n$

5.57 2.7 and 5.0 would be improbable values if $\mu = 4$ and $\sigma = .5$. 2.7 lies 2.6 standard deviations below and 5.0 lies 2.0 standard deviations above $\mu = 4$.

5.59 $\mu = 1$, $\sigma = .99$; no; the interval $np \pm 3\sqrt{np(1-p)}$, i.e., $\mu \pm 3\sigma$, does not fall in the interval from $x = 0$ to $x = 100$.

5.61–5.65 See Section 5.8.

5.67–5.79 See Section 5.9.

5.81 Consult the sampling distribution for the statistic and see how far from the parameter the statistic is likely to fall.

5.83 The trials in a binomial experiment must result in independent outcomes.

Chapter 6

6.1 If we want to know how far a random variable x deviates from its mean μ, we must divide the deviation $(x - \mu)$ by the standard deviation of x. Similarly, if we want to know how far the sample mean $\bar{x}$ deviates from the mean of its sampling distribution (which is also μ), we must divide the deviation $(\bar{x} - \mu)$ by the standard deviation of $\bar{x}$. That standard deviation is SE($\bar{x}$).

6.3 SE($\bar{x}$) = .49 (a) .05 (b) $z = 1.96$; the probability is (from Table 6.1) .025, (c) .05 (d) $z = 1.43$; the probability is (from Table 6.1) between .10 and .20.

6.5 Both are symmetric distributions with mean equal to 0. The t distribution has more spread. The spread of the t distribution depends on the number of degrees of freedom for t. When the number of degrees of freedom is large, say $n \geq 30$, the two distributions are, for all practical purposes, identical.

6.7 (a) 2.086 (b) 1.397 (c) 2.764 (d) 3.182

6.9 A random sample from a normally distributed population of data

6.11 Between .01 and .025

6.13 See Section 6.4.

6.15 See Section 6.4.

6.17 t_a

6.19 It increases.

6.21 217.7 to 224.3

6.25 (a) 13.0 to 53.6 (b) 4.98 to 11.34 (c) .321 to 2.439

6.29 .95

6.31–6.35 See Section 6.6.

6.37 See Section 6.6.

6.39 Because the p-value is less than $\alpha = .05$, reject H_0 and conclude that H_a is true, i.e., that $\mu < 103$. The probability that we will reject H_0 (and conclude that H_a is true) when, in fact, H_0 is true is only .05.

6.41 (a) 23 (b) Less than .005 (c) Because the p-value for the test is less than $\alpha = .10$, reject H_0 and conclude that H_a is true, i.e., that $\mu > 3.12$. The very small p-value provides strong evidence to indicate that $\mu > 3.12$.

6.43 (a) 6 (b) Slightly less than .05 (it is a two-tailed test) (c) Because the p-value is less than $\alpha = .10$, reject H_0 and conclude that H_a is true, i.e., that $\mu \neq -1.0$. The probability that we will reject H_0 (and conclude that H_a is true) when, in fact, H_0 is true is only .10. (d) Now the p-value is not less than .05; do not reject H_0; no relevance

6.45 Unreasonable. The smaller the value of α, the greater is the risk of failing to detect a bias in the scales (i.e., reject H_0) when, in fact, a bias does exist. Since we want to know whether the scale is biased and we have little to lose (in a practical sense) if we wrongfully conclude that it is, we would choose a value for α that provides a modest protection of concluding that the scales are biased when, in fact, they are not—say, $\alpha = .10$.

6.47 $\bar{x} = .03287$, $s = .01630$, $\widehat{\text{SE}}(\bar{x}) = .00340$; the confidence interval is from .02703 to .03871.
(a) The probability that a 90% confidence interval will enclose μ is .90. Therefore, we are 90% confident that the mean concentration of lead is between .02703 and .03871.

6.49 $\bar{x} = .1687$, $s = .0515$, $\widehat{\text{SE}}(\bar{x}) = .0107$; the confidence interval is from .1502 to .1872.
(a) The probability that a 90% confidence interval will enclose μ is .90. Therefore, we are 90% confident that the mean concentration of iron is between .1502 and .1872.

6.51 (a) H_0: $\mu = 1.0$ mg/l; H_a: $\mu < 1.0$; one-tailed (b) Student's t test; sample size is small, i.e., less than 30 (c) $t = -112.793$; p-value $\approx .0000$ (d) The probability that t will equal -112.793 or less, assuming H_0 is true, is approximately .0000. This small p-value indicates either that we have observed a very rare event, assuming that H_0 is true, or H_0 is false and the mean copper concentration is less than 1.0. (e) Reject H_0 and conclude that the mean amount of copper in the water is less than 1.0 mg/l. (f) The probability of falsely rejecting H_0, when in fact it is true, is .10. Note that we would also have rejected H_0 had we chosen a value for α as small as .0000.

6.53 (1), the cost per observation in the collision tests is much greater than the cost per observation in testing flashlight batteries.

6.55 No. You can reduce the standard error of $\bar{x}$, and, consequently, the width of the confidence interval, by increasing the sample size.

6.57 38

6.59 Range = 220; $s \approx$ range/6 = 36.7; $n = 36$.

6.61 To cut the width of the confidence interval in half, we would need 4 times as many observations as were in the original sample, i.e., (4)(390) = 1,560.

6.63 (a) Yes, the mean 18.118 falls in the middle of the measurements and the interval $\bar{x} \pm 2s$, 11.872 to 24.364, includes all of the measurements. (b) 16.02 to 20.22; the probability that a 95% confidence interval will enclose μ is .95. Therefore, we are 95% confident that the mean weight of the fish is between 16.02 and 20.22 grams.

6.65 (a) Yes, the mean .06491 falls in the middle of the measurements and the interval $\bar{x} \pm 2s$, .02179 to .10803, includes 10 of the 11 measurements, i.e., 91%; all of the measurements fall within $3s$ of $\bar{x}$. (b) .05043 to .07939; the probability that a 95% confidence interval will enclose μ is .95. Therefore, we are 95% confident that the mean oxygen uptake by water is between .05043 and .07939.

Chapter 7

7.1 See Section 7.2.

7.3 Suppose that you wanted to compare the strength of concrete before and after a waterproofing additive has been mixed with the concrete. Concrete varies in strength from one mix to another, so pair by molding test specimens obtained from the same mix of concrete. Mix a batch of concrete. Divide it and add the waterproofing additive to one portion. Leave the other portion to serve as a standard. Repeat n times to obtain n matched pairs.

7.5 (a) The amount of ore excavated per day will depend on the skill of the equipment operator, the location and nature of the mined terrain, etc. (b) Randomly select a number of locations; then randomly assign half of the locations to Method 1 and the remainder to Method 2. (c) Since the amount of ore removed per day will depend on the location of the ore (its consistency, rock formations, etc.) and the skill of the machine operator, we will match on operator and location. Have each operator work one day using Method 1 and a second day using Method 2 at each location. The order should be chosen at random. Repeat at other locations until you have collected data on n pairs.

7.7–7.9 See the appropriate box in Section 7.3.

7.11 See the appropriate box in Section 7.3.

7.13 (a) 12.094 to 48.906; the probability that a 90% confidence interval will enclose $(\mu_1 - \mu_2)$ is .90. Therefore, we are 90% confident that the difference in mean wing stroke velocities is between 12.094 and 48.906. (b) The larger the confidence coefficient for a confidence interval, the wider it will tend to be. We are willing to reduce the confidence coefficient to obtain a narrower interval. (c) $\bar{x}_1 = 209.50$, $s_1 = 24.035$, $\bar{x}_2 = 179$, $s_2 = 5.4406$; the pooled value for s^2 is 235.1305. (d) See the printout. Note that the standard deviations for the two samples are substantially different. This casts some doubt on the validity of the confidence interval.

7.15 (a) Student's t test; the sample sizes are small, i.e., less than 30. (b) H_0: $(\mu_1 - \mu_2) = 0$; H_a: $(\mu_1 - \mu_2) \neq 0$ (b) Test is two-tailed (c) $\alpha/2$ in each tail of the t distribution (d) (See Table 4, Appendix 2) Between .010 and .020 (e) Since the p-value is less than .10, we reject H_0 and conclude that $(\mu_1 - \mu_2)$ differs from 0, i.e., that there is a difference in mean wing frequencies between the two species of bees. (f) The probability of rejecting H_0 (as we have done) when H_0 is true, is only .10.

7.17 (a) H_0: $(\mu_1 - \mu_2) = 0$; H_a: $(\mu_1 - \mu_2) > 0$ (b) A z test; the sample sizes are large. (c) Upper one-tailed; we only want to detect values of $(\mu_1 - \mu_2)$ that are positive. (d) 3.37 (e) The p-value for the test (see Table 6.1 or Table 2, Appendix 2) is less than .001. (f) We are willing to risk, with probability .10, wrongly concluding that a difference exists when it does not. (g) Since the p-value is less than $\alpha = .10$, we reject H_0 and conclude that $(\mu_1 - \mu_2)$ is positive, i.e., that the mean serum cholesterol level dropped from 1960 to 1970. (h) The probability of wrongly concluding that a difference exists when it does not is .10.

7.19 The sample of differences satisfies all of the requirements (listed in Sections 6.2 and 6.4). If both populations 1 and 2 are normally distributed, the population of differences will be normally distributed. The sample differences represent an independent random sample selected from that population.

7.21 Use a sample of n documents of comparable size. Pair on documents by having each document typed using each of the two word-processing packages. Or, you could pair on typists and documents by employing a set of n typists and requiring each typist to type a document using each of the two word-processing packages.

7.23 Match on tubes, testing each tube by both machines.

7.25 (a) 15 (b) Between .05 and .10 (c) Since the p-value for the test exceeds .05, there is insufficient evidence to indicate a difference in the population means for $\alpha = .05$.

7.27 (a) H_0: $(\mu_1 - \mu_2) = 0$; H_a: $(\mu_1 - \mu_2) > 0$ (b) A t test; the number of differences is small, i.e., less than 30. (c) $t = 29.55$; less than .005 (d) Since the p-value for the test is less than .05, there is sufficient evidence to indicate a difference in the population means (e) .05 (f) 34.38 to 39.62

7.29 (a) Matched pairs; power measurements were matched on tubes to try to detect bias in the readings given by the two pieces of test equipment. (b) H_0: $(\mu_1 - \mu_2) = 0$; H_a: $(\mu_1 - \mu_2) \neq 0$ (c) A t test; the number of differences is small, i.e., less than 30. (d) $t = 1.0960409$; the p-value is .3015. (e) Since the p-value for the test is larger than .10, there is insufficient evidence to indicate a difference in the population means. (e) No relevance because we did not reject H_0.

7.31 We need to know the approximate values of the population standard deviations. Use estimates from prior sampling or an estimate based on knowledge of the approximate range of the measurements.

7.33 62

7.35 25

7.37 377

7.39 (a) Matched pairs; power measurements were matched on tubes to try to detect bias in the readings given by the two pieces of test equipment. (b) H_0: $(\mu_1 - \mu_2) = 0$; H_a: $(\mu_1 - \mu_2) = 0$ (c) A t test; the number of differences is small, i.e., less than 30. (d) The values of s_1 and s_2 are not required for a matched-pairs analysis. (e) $t = 2.49$ (f) The p-value is between .02 and .05. (e) Since the p-value for the test is less than .10, there is sufficient evidence to indicate a difference in the population means. (e) Using this test, we will incorrectly reject H_0, when it is true, only 10% of the time.

7.41 (a) H_0: $(\mu_1 - \mu_2) = 0$; H_a: $(\mu_1 - \mu_2) > 0$ (b) A z test; the sample sizes are large. (c) Upper one-tailed; we only want to detect values of $(\mu_1 - \mu_2)$ that are positive. (d) .03267 (e) 4.59 (f) The p-value for the test (see Table 6.1 or Table 2, Appendix 2) is less than .001; since the p-value is less than $\alpha = .05$, we reject H_0 and conclude that $(\mu_1 - \mu_2)$ is positive, i.e., that the mean serum cholesterol level dropped from 1980–82 to 1985–87.

Chapter 8

8.1 Data consisting of observations on each of two variables for each element in the sample. This will result in pairs of observations, one pair for each sample.

8.3 The association between a student's quantitative SAT score and the student's college GPA in calculus; the universe would be the collection of all college students who have taken the SAT and college calculus; each student is an element; the bivariate data would be the pairs of scores, SAT and calculus, one pair for each student.

8.5 The plot of data points for the data set

8.7 (a) Each country is an element. (b) The collection of all developed capitalistic countries in the world (c) The investment income tax x and the personal savings rate y (d) The investment income tax and the personal savings rate (e) The population is the collection of all bivariate observations, one for each element in the universe. (g) The personal savings for a country decreases as the investment income tax increases.

8.9 $y = \beta_0 + \beta_1 x$; β_0, the y-intercept, is the value of y when $x = 0$; β_1 is the change in y for a one-unit increase in x.

8.11 The line would intersect the y-axis at $y = \frac{1}{2}$ and slope downward to the right, going down 1 unit for every one-unit increase in x.

8.13 The line would intersect the y-axis at $y = \frac{3}{2}$ and slope downward to the right, going down 2 units for every one-unit increase in x.

8.19 SSE is the sum of squares of deviations for a least squares line.

8.21 See Section 8.5.

8.23 $r^2 = (.8)^2 = .64$; therefore, 64% of the variation of the y values is explained by a linear relationship between y and x. The remainder of the variation of the y values could be due to the fact that the relationship is not linear (e.g., it may be curvilinear) or because other variables, uncontrolled in the data collection, also affect y and are causing it to vary. The value $r^2 = .64$ indicates that a straight-line model would be useful for predicting y for a value of x but there is room for improvement (this topic is discussed in Section 8.9).

8.25 $r^2 = (-.2)^2 = .04$; therefore, only 4% of the variability of the y values is explained by a linear relationship between y and x. For all practical purposes, there is little or no relationship between y and x.

8.27 It gives the proportion of the Total SS = SS_y that is explained by variation in x.

8.29 (a) $x = 2.5$, $y = 2.0$ (d) $SS_x = 5.0$; $SS_y = 8.0$; $SS_{xy} = 6.0$ (e) $r = .95$ (f) Yes

8.31 33.961 minutes; the estimate of mean time and the prediction of the time for a single runner are identical but the error of estimation is less than the error of prediction.

8.33 We might expect a coach's win record to increase as time (experience) increases; we would expect the relationship to be weak because a win record would be affected by many other variables. For Coach Bryant, a test of H_0: $\beta_1 = 0$ gives $t = 5.02$ with p-value for this one-tailed test equal to $\frac{1}{2}(.037) = .019$; that result, along with $r^2 = .96$, supports the theory that time and experience were related to success in his case.

8.39 (a) $\hat{y} = -.629 + 1.075x$ (b) Yes. For testing H_0: $\beta_1 = 0$, $t = 67.83$ and the p-value for the test is .0001. This small p-value provides strong evidence to indicate that $\beta_1 \neq 0$, i.e., that DBN provides information for the prediction of DBS.

8.41 (a) If we believe that the number y of arrests will increase as the number x of robberies increases, we would expect β_1 in the straight-line model to be positive. (b) $y = \beta_0 + \beta_1 x$ (c) $r^2 = .844$; 84.4% of the variation in the y values is explained by the first-order model, part (b). (d) For testing H_0: $\beta_1 = 0$, $t = 5.20$ and the p-value (see Table 4 of the Appendix for t values with 5 df) for this one-tailed test is less than .005. This small p-value provides strong evidence to indicate that $\beta_1 > 0$, i.e., that the number y of arrests increases as the number x of robberies increases.

Chapter 9

9.1 See Section 2.1.

9.3 Problems concerned with qualitative sets of data.

9.5 A binomial experiment is the special case of a multinomial experiment with $k = 2$ categories.

9.7 (a) .233 to .247 (b) Yes

9.9 (a) $p = .467$; correct to within .032 with probability .95; the assumptions of a binomial experiment

9.11 (a) $\alpha = .10$; we have little to lose if we conclude that $p < .07$ when, in fact, $p = .07$. (b) H_0: $p = .07$; H_a: $p < .07$; $z = -21.03$; the p-value for this one-tailed test is less than .001. This p-value is less than $\alpha = .10$; therefore, we reject H_0 and conclude that the proportion of acceptable donors is less than .07. (c) .10; the small p-value provides very strong evidence to indicate that $p < .07$. (d) Yes; $\mu \pm 3\sigma$, i.e., $np \pm 3\sqrt{np(1-p)}$, falls in the interval from 0 to n, i.e., 0 to 8,000.

9.13 (a) H_0: $(p_1 - p_2) = 0$, H_a: $(p_1 - p_2) > 0$ (*Note:* We conducted an upper one-tailed test because we wanted to determine whether the drug was effective, i.e., whether $p_1 > p_2$.) (b) The upper tail of the z distribution (c) $(\hat{p}_1 - \hat{p}_2) = .057$; $\widehat{\text{SE}}(\hat{p}_1 - \hat{p}_2) = .028$ (d) $z = 2.04$; the p-value is approximately .02. (e) Since the p-value for the test is less than, say $\alpha = .05$, there is sufficient evidence to indicate that the new drug is effective.

9.15 (a) H_0: $(p_1 - p_2) = 0$, H_a: $(p_1 - p_2) < 0$ (*Note:* We conducted a lower one-tailed test because we wanted to determine whether the proportion of "unable to determine" responses increased, i.e., whether $p_1 < p_2$, from 1970–72 to 1974–78.) (b) $(\hat{p}_1 - \hat{p}_2) = -.016$; $\widehat{\text{SE}}(\hat{p}_1 - \hat{p}_2) = .041$ (d) $z = -.39$; the p-value is larger than .25; since this p-value is larger than $\alpha = .05$, there is insufficient evidence to indicate that the proportion of "unable to determine" responses increased. (c) No relevance

9.17 Yes; $z = 2.286$; the p-value is approximately .01; this small p-value provides strong evidence (we would reject H_0 for an α as small as .01) to indicate that the drug is effective.

9.19 246

9.21 267

9.23 350

9.25 (a) 12.5916 (b) 11.3449 (c) 2.70554

9.27 (a) 7.81473 (b) 6.63490 (c) 7.77944

9.29 (a) 1 (b) Slightly larger than .10

9.31 (a) 3.025 (b) df = 2; larger than .10 (c) .220 (d) Yes (*Note:* This large p-value provides insufficient evidence to indicate a dependence between participation in organized sports and the participation in self-directed games.)

9.33 (a) We would expect the proportion of assaults at the non–air-conditioned prison to be higher during the summer months than during the winter and the proportion of assaults at the air-conditioned prison to be approximately the same during the summer months as during the winter. (b) H_0: The effect of time of year on whether an assault occurs is independent of prison type; H_a: The effect of time of year on whether an assault occurs is dependent on prison type. (c) $\chi^2 = 6.617$ with df = 2; the p-value for the test is .037; we would reject H_0 for $\alpha = .05$ and conclude that the effect of time of year on the frequency of assaults is dependent on prison type.

9.35 (a) H_0: $(p_1 - p_2) = 0$, H_a: $(p_1 - p_2) = 0$ (b) $z = 4.77$, $z^2 = 22.75$ (c) The approximate p-value corresponding to $z = 4.78$ is less than .001. This agrees (except for rounding errors) with the p-value for Exercise 9.34.

9.37 When the contingency table is 2×2 and the z test is two-tailed

9.39 .082 to .108; the students selected in the survey represent a random sample from the UBC student body.

9.41 (a) Let p_1 be the probability that a person with 15 years or less of education has been convicted and let p_2 be the corresponding probability for a person with 16 or more years of education. H_0: $(p_1 - p_2) = 0$, H_a: $(p_1 - p_2) \neq 0$, i.e., whether a person has been convicted depends on the person's education. (b) $z = 1.53$, $z^2 = 2.34$ (c) The approximate p-value corresponding to $z = 1.53$ is equal to .126; there is insufficient evidence to indicate a difference in p_1 and p_2 for $\alpha = .10$, i.e., that whether a person has been convicted depends on the person's education.

9.43 No. For a regression analysis, the dependent variable y must be quantitative.

9.45 (a) JAS score and the qualitative variable, survival status (b) $\chi^2 = 2.563$; df = 2; the p-value is .278; since the p-value is not less than .05, there is insufficient evidence to indicate that the qualitative variable, survival status, is dependent on JAS score. (c) Since the observed value of chi-square is less than $\chi^2_{.100} = 4.60517$, the p-value for the test is larger than .100. (d) No; a two-tailed z test is equivalent to a chi-square test only for 2×2 contingency tables.

Chapter 10

10.1 The comparison of mean responses for a quantitative variable y for different settings of one or more qualitative variables

10.3 It compares the variation between sample means with the variation within samples.

10.5 Total SS = SST + SSE

10.7 4; 45

10.9

Source	df	SS	MS
Treatments	$k - 1$	SST	MST
Error	$n - k$	SSE	MSE
Total	$n - 1$	Total SS	

10.11

Source	df	SS	MS
Treatments	5	2.1	.42
Error	18	.9	.05
Total	23	3.0	

10.13

Source	df	SS	MS
Treatments	3	31.5	10.5
Error	12	6.7	.56
Total	15	38.2	

10.15 $F = 7.14$ with 2 numerator and 15 denominator degrees of freedom

10.17 (a) 3.86 (b) 6.93 (c) 2.61

10.19 Between .05 and .10; for $\alpha = .05$, there is insufficient evidence to indicate differences in the population means.

10.21 Between .05 and .10; for $\alpha = .10$, reject H_0 and conclude that there is sufficient evidence to indicate differences in the population means.

10.23 (a)

Source	df	SS	MS
Treatments	2	64.31	32.16
Error	25	338.02	13.52
Total	27	402.33	

(b) $F = 32.16/13.52 = 2.38$; since $F = 2.38$ is less than $F_{.10} = 2.53$, the p-value is larger than .10. (c) 1.55 to 8.51 (d) 1.17 to 6.21

10.25 (a)

Source	df	SS	MS
Treatments	3	5,734.105263	1,911.37
Error	15	202.0	13.47
Total	18	5,936.105263	

(b) No; $F = 141.9$; $F = 141.9$ is larger than $F_{.005} = 2.54$; therefore, the p-value is less than .005 and less than $\alpha = .05$; there is strong evidence (considering the small p-value) to indicate a difference in the percentage germinating among the four age groups.

10.27 There is no evidence of a difference among the means for populations 3, 4, and 5; the mean for population 1 is larger than the means for populations 3, 4, and 5, and the mean for population 2 is larger than all of the others; the probability of incorrectly concluding that two means differ for at least one of the ten possible two-way comparisons is only .05.

10.29 11.58

10.31 It would not; the sample sizes must be equal.

10.33 A setting for a factor is called a factor level. For example, if a cake in Exercise 10.32 is baked at 350°F, 350 is a level for the factor "temperature."

10.35 An experiment involving two factors, one at 3 levels and one at 5 levels. A single replication of the experiment is one in which we observe y for each of the 15 different factor level combinations.

10.37 When the effect of one factor on a response variable y depends on the level of a second factor, we say that the factors interact.

10.39 (a)

Source	df	SS	MS	F
Factor A	2	4.38	2.19	1.99
Factor B	4	6.04	1.51	1.37
$A \times B$ interaction	8	25.12	3.14	2.85
Error	30	33.00	1.1	
Total	44	68.54		

(b) The F values are shown in column 5 of the ANOVA table. The F value for the $A \times B$ interaction, $F = 2.85$, is between $F_{.025} = 2.65$ and $F_{.01} = 3.17$ (8 and 30 df). Therefore, the p-value for the test is between .01 and .025. This small p-value provides sufficient evidence to indicate an $A \times B$ interaction.

The F value, $F = 1.99$, for main effect, factor A, is less than $F_{.10} = 2.49$ (2 and 30 df). Therefore, the p-value for the test is larger than .10. This large p-value provides no evidence of a main effect due to factor A.

The F value, $F = 1.37$, for main effect, factor B, is less than $F_{.10} = 2.14$ (4 and 30 df). Therefore, the p-value for the test is larger than .10. This large p-value provides no evidence of a main effect due to factor B.

(c) The key test is the test for interaction. Since there is evidence of interaction, we would want to continue our analysis by comparing the mean responses for various combinations of levels of A and B. This could be done using the HSD procedure (Section 10.6). The tests for main effects are not particularly relevant, now that we know that the factors interact.

10.41 (b) Yes; $F = 8.48$; the p-value is .0001; since the p-value is less than $\alpha = .05$, there is sufficient evidence to indicate that the two qualitative factors, design and foreman, interact. (c) *Both factors affect profit,* i.e., profit is dependent on the design–foreman combinations. (d) The tests show that there is evidence of a main effect due to B but not due to A; i.e., it suggests that B (design) affects profit but A (foreman) does not. This information is irrelevant and misleading because we have already shown that A and B interact. Therefore, *both* factors, A and B, affect profit y but their effect is dependent on the factor level combinations. (e) A_1B_3, although there is no evidence of a difference between A_1B_3 and A_3B_1, A_4B_2, A_1B_1, and A_2B_3.

10.43 Denote "training" as factor A and "ability" as factor B. (a) Since A is at two levels and B is at three levels, df for A is $(a - 1) = (2 - 1) = 1$, df for B is $(b - 1) = (3 - 1) = 2$, and df for the $A \times B$ interaction is $(a - 1)(b - 1) = (1)(2) = 2$; the df for Total is $(n - 1) = 60 - 1 = 59$ and, since the degrees of freedom for all sources must equal the df for Total, the df for Error is $59 - 1 - 2 - 2 = 54$. (c) The factors appear to interact because the difference in means, Active − Passive, for the high level of spatial ability appears to be much larger than for medium and low. (d) $F = 4.42$ and the p-value is .017; since the p-value is less than $\alpha = .05$, there is sufficient evidence to indicate interaction between training condition and spatial ability. (e) $F = 3.66$ and the p-value is .061; there is insufficient evidence, for $\alpha = .05$, to indicate that training affects terrain visualization ability; this disagrees with the results of our interaction test, which indicated that both training and spatial ability affect terrain visualization ability. Which is correct? Tests for main effects are irrelevant if factor interaction is present. (f) $q_{.05}$ for df = 54 and $k = 6$ is (from Table 7 of Appendix 2) approximately 4.18 and $W = 7.03$. The six means and the comparisons are shown graphically below.

Identification					
1	2	3	4	5	6
1.610	1.728	5.031	5.648	9.508	17.895

There is no evidence of a difference in population means corresponding to ID numbers 1, 2, 3, and 4, nor is there for 3, 4, and 5. The mean corresponding to ID 6 differs from all the others. (g) High spatial ability combined with the active training condition will produce the highest mean visualization score.

10.45 (a) The F values for foreman (A), shift (B), and $A \times B$ interaction are, respectively, 26.43, .18, and 56.23. (b) The p-value for the test for interaction is .0001; this small p-value provides strong evidence (for α as small as .0001) to indicate that there is an interaction between foreman A and shift B. There is no point in testing for main effects because we have already shown that the levels of foreman and shift both affect production. (c) $q_{.05}$ for df = 12 and $k = 6$ is (from Table 7 of Appendix 2) 4.75 and $W = 73.6$. The six means and the comparisons are shown graphically below:

Identification					
1	2	3	4	5	6
A_1A_3	A_2B_1	A_1B_2	A_1B_1	A_2B_2	A_2B_3
450.00	486.67	498.33	601.67	601.67	656.67

There is no evidence of differences among population means 1, 2, and 3, nor among 4, 5, and 6. Each of the means 4, 5, and 6 differs from means 1, 2, and 3.

(d) Use foreman A_1 on the day shift and A_2 on the 12 A.M. to 8 A.M. shift

Chapter 11

11.1 Both test for a shift in location for two populations.

11.3 The t statistic uses the actual numerical values of the measurements; nonparametric statistics use the ranks of the observations

11.5 (a) One-tailed; if we wanted to detect a shift in distribution 2 either to the right or the left of distribution 1, we would use a two-tailed test. (b) U_2 (see the box); small (c) .0760 (see Table 8 of Appendix 2); the probability of observing a value of U_2 less than or equal to 7 is .0760. (d) Reject H_0 and conclude that distribution 2 is shifted to the right of distribution 1.

11.7 Switch the subscripts on the samples.

11.9 (a) $T_1 = 34$, $T_2 = 21$, $U_1 = 0$, $U_2 = 24$; the test statistic is $U_1 = 0$. (b) The p-value for the test is $2(.0048) = .0096$. (c) We would reject H_0 and conclude that the distributions are shifted for $\alpha = .10$ (or for any value of α larger than .0096; this provides sufficient evidence to indicate that the distribution of wing stroke velocities for populations 1 and 2 differ in location. It appears that the velocities for population 1 tend to be larger than those for population 2. (d) The p-value, part (b), based on Table 8 of Appendix 2, is correct. The value given on the Minitab printout is close to but not equal to this value. (Minitab uses an approximation procedure to calculate its p-value. This explains the slight discrepancy between the p-values in Exercises 11.8 and 11.9.) (e) Using the U test, we do not have to worry about assumptions concerning the properties of the sampled populations. (f) The tests reach the same conclusion; the p-values are comparable; the p-value for the t test was between .01 and .02.

11.11 (a) $T_1 = 21$, $T_2 = 15$, $U_1 = 0$, $U_2 = 15$; the test statistic is $U_1 = 0$. (b) The p-value for the test is .0179. (c) We would reject H_0 and conclude that distribution 1 is shifted to the right of distribution 2 for any value of α equal to .0179 or larger; it indicates that the numbers of violations for rural areas tend to be larger than the numbers for urban areas. (d) The p-value, part (b), based on Table 8 of Appendix 2, is correct. The value given on the Minitab printout is close to but not equal to this value. (Minitab uses an approximation procedure to calculate its p-value. This explains the slight discrepancy between the p-values in Exercises 11.10 and 11.11.)

11.13 (a) H_0: The two population relative frequency distributions are identical; H_a: Distribution B (no antitoxin) is shifted to the right of distribution A (antitoxin). (b) $U_B = 11.5$; $z = -3.49$ (c) The p-value for the test is less than .001; the probability that z is less than -3.49 (this is a lower one-tailed test) is less than .001. (d) Reject H_0 and conclude that the antitoxin was effective.

11.15 (a) $U_2 = 27$ (b) Two-tailed test; $2(.0446) = .0892$ (c) $z = -1.74$; approximate p-value $= 2(.0414) = .0818$. (d) The p-value for this two-tailed test is less than $\alpha = .10$ but not .05. Therefore, there is sufficient evidence to indicate that distribution 2 is shifted either to the right or to the left of distribution 1 for $\alpha = .10$ but not for $\alpha = .05$.

11.17 (a) One-tailed (b) T^+; small (c) 17 or less (d) The p-value is .025. (e) There is sufficient evidence to indicate that distribution 2 is shifted to the right of distribution 1 for $\alpha = .05$.

11.19 (a) H_0: The two population relative frequency distributions are identical; H_a: Distribution 1 is shifted either to the right or to the left of distribution 2. (b) The p-value for the test is .575. This large p-value indicates that there is no evidence to indicate a shift of distribution 1 either to the right or to the left of distribution 2, i.e., no evidence of a bias in the tester readings. (c) The p-value for the t test was .3015; both p-values are large and reach the same test conclusion. The p-value for the t test will tend to be smaller than for the Wilcoxon signed rank T test when the assumptions for the t test are satisfied. This is because the t statistic uses more information than a T (i.e., it uses the measurements, rather than their ranks, and is more likely to detect a difference if one exists).

11.21 (a) Observations were matched on location. (b) H_0: The two population relative frequency distributions are identical; H_a: Distribution 1 is shifted either to the right or to the left of distribution 2. (c) The p-value is .059; therefore, there is sufficient evidence to indicate a bias in the air-ground temperature readings for $\alpha = .10$ but not for $\alpha = .05$.

11.23 (a) H_0: The k population relative frequency distributions are identical; H_a: At least two of the population distributions differ in location. (b) $H = 17.13$; the p-value is equal to .001. (c) There is sufficient evidence to indicate that the distributions of leaf length differ in location.

11.25 $F = 57.83$ and the p-value for the test is .000 compared with "less than .005" for the H test. The p-value for the ANOVA F test tends to be smaller than for the H test when the ANOVA assumptions are satisfied. This is because the F test uses more information (the actual values of the measurements) than the H test.

11.27 (a) $H = 12.48$; the value on the printout is 12.488. (b) $H = 12.48$ is larger than $\chi^2_{.005} = 10.5966$. Therefore the p-value for the test is less than .005. This agrees with the p-value shown on the printout.

11.29 See the box in Section 11.4.

11.31 $H = 10.3$ is between $\chi^2_{.025} = 9.34840$ and $\chi^2_{.01} = 11.3449$; therefore, the approximate p-value is between .01 and .025.

11.33 The rank of the y variable tends to increase as the rank of the x variable increases (or vice versa).

11.35 There is little or no relationship between the rank of y and the rank of x.

11.37 The rank of x and the rank of y in each pair are matched in reverse, i.e., the smallest x is matched with the largest y; the next smallest x is matched with the next largest y, etc.

11.39 .497

11.41 $-.441$

11.43 (a) $r_s = -.10398$; the value of r_s suggests a very weak negative rank correlation between the (x, y) values in the data points. (b) The pairs of (x, y) values associated with all fossils of this species in the Antarctic peninsula; the p-value for the test is .7750; this large p-value indicates that there is no evidence to indicate a rank correlation between fossil diameter and height.

11.45 $r_s = 1$; the difference in all rank pairs is equal to 0.

11.47 (a) The smaller of U_1 or U_2; $T_1 = 126$, $T_2 = 45$; the test statistic is $U_1 = 0$; the p-value for the test is .0000. The probability of observing a U value as small as 0, assuming no difference in the population distributions, is only .0000; this small p-value provides strong evidence to indicate a difference in scanning rates between deaf and hearing children. (Minitab uses an approximation procedure to calculate its p-value. This explains the slight discrepancy between the p-values in Exercises 11.46 and 11.47.)

11.49 Table 9, Appendix 2, shows that the probability that T is less than or equal to 2 (for a two-tailed test) is between .05 and .10; the t value for the paired difference test is 2.571 and the p-value is close to .05. Both p-values are less than $\alpha = .10$ and provide, for $\alpha = .10$, sufficient evidence to indicate a difference in the density for the two cake mixes; the p-value for the t test will tend to be smaller than for the Wilcoxon signed ranks test when the assumptions for the t test are satisfied. This is because the t statistic uses more information than a T (i.e., it uses the measurements, rather than their ranks, and is more likely to detect a difference if one exists).

index

Additive rule of probability, 126
Alternative hypothesis, 168
Analysis of variance (ANOVA), 386
  assumptions, 393
  comparing population means, 389
  F test, 394
  factorial experiments, 406
  table for independent random samples design, 389, 397
  table for *r* replications of a two-factor factorial experiment, 412

β_0, 284
β_1, 284
Bar graph, 29
Biased estimator, 165
Binomial experiment, 136, 337
  characteristics, 137
Binomial probability distribution, 136, 139
  mean, 139
  normal approximation, 155
  standard deviation, 139
Binomial probability table, 144
Binomial random variable, 136
Bivariate data, 15, 280
Box plot, 107

Categorical data, 9
Category frequency, 27
Category relative frequency, 27
Center line, 101
Central limit theorem, 157
Chi-square statistic, 359
Class, 35
Class frequency, 35
Class interval, 67
Class relative frequency, 35
Coefficient of correlation, Pearson, 289–291
Coefficient of determination, 289, 294, 296
Complement, 128
Confidence coefficient, 166, 199
Confidence interval, 166, 189, 200, 218
  difference between two population means, large paired samples, 261
  difference between two population means, large sample, 239
  difference between two population means, small paired samples, 261
  difference between two population means, small sample, 242
  difference between two population proportions, 349
  half-width of, 311
  individual means, 398
  mean of dependent variable, 310
  population mean, random sample, 189
    known standard deviation (large sample), 192
    unknown standard deviation (small sample), 192
  population proportion, 343
  slope of a straight line, 306
Contingency table, 356
  analysis, 356, 363
Continuous random variable, 134
Control chart, 101

Data, 9
  bivariate, 15, 280
  categorical, 9
  multivariate, 15
  quantitative, 9
  set, 9
  univariate, 15
Data point, 281
Decile, 105
Degrees of freedom, 184, 304, 359
Design of an experiment, 21, 230
  factorial, 406
  independent random samples, 231, 393
  matched-pairs, 232
  paired-difference, 233
  sample survey, 230
Discrete random variable, 132
Distribution
  mound-shaped, 40
  relative frequency, 34
  skewed, 40
  skewness of, 105
  symmetric, 40
Dot diagram, 84

Element, 7
Empirical Rule, 79
Equation of a line, 284
Error of estimation, 117
Estimated standard error, 184
Estimator, 164
  biased, 165
  interval, 166
  of mean of dependent variable, 310

Estimator (*continued*)
  point, 164
  pooled, of population variance, 242
  slope of least squares line, 309
  unbiased, 164
  of variance in linear regression analysis, 304
Expected value, 300
Experment
  binomial, 136
  factorial, 406, 410
  multinomial, 336
Experimental design, 21
  factorial, 406
  independent random samples, 393
Experimental unit, 8

F statistic, 394, 397
Factor, 406
Factor interaction, test for, 414
Factor level, 406
Factor main effects, test for, 415
Factorial experiments, 406, 410
  complete, 410
  multifactor, 406
  single-factor, 406
  single replication of, 410
First-order linear model, 321
Frame, 131
Frequency
  category, 27
  category relative, 27
  class, 35
  class relative, 35
  relative, 28

General linear model, 320–321

Hypotheses, tests of (*see* Tests of hypotheses)

Independent outcomes, 127
Independent random samples design, 231
Inference (*see also* Confidence interval: Tests of hypotheses), 17
  statistical, 17
Inner fence, 109
Interactions, factor, 409
Interquartile range (IQR), 107
Interval estimator, 166

Kruskal–Wallis *H* test, 449

Least squares
  estimator of variance, 304
  line, 289, 302
  method of, 289
Level, 335, 406
Line of means, 301
Linear regression analysis (*see* Regression analysis, linear)
Lower confidence limit, 166
Lower control limit, 102
Lower quartile (Q_1), 104

Main effects, 415
Mann–Whitney *U* test, 430
  compared to Student's *t* test, 434
  independent random samples design, 431
  large sample, 435
Margin of error, 166
Matched-pairs design, 232, 259, 267
Mean, 74, 117
  population, 117
  sample, 75, 117
Mean square for error (MSE), 302
  for independent random samples design, 392
Mean square for treatments, 392
Measure of central tendency, 74
Measure of relative standing, 103
Measure of variation (spread), 78
Median, 76, 104
Method of least squares, 288–289
Middle quartile (Q_2), 104
Model
  first-order linear, 321
  general linear, 321
  second-order linear, 321
  third-order linear, 321
Multinomial experiment, 336
  characteristics, 336
  relationship to binomial experiment, 337
Multiple comparisons
  Tukey's HSD procedure, 402
Multiple regression analysis (*see* Regression analysis, multiple)
Multiplicative rule of probability, 127
Multivariate data, 15
Mutually exclusive outcomes, 125

Nonparametric statistical test, 428
Normal approximation to binomial, 155
Normal probability distribution, 148
Normal random variable, 150
Null hypothesis, 168
Numerical descriptive measure
  for a population, 117
  for a sample, 116
Numerical descriptive method, 73

Observation, 7
One-tailed statistical test, 206
Outer fence, 109
Outlier, 107

P-value, 169
Paired-difference experiment, 233
Parameter, 117
Parametric test, 430
Pearson coefficient of correlation, 289–291
Percentile, 103
Pie chart, 29
Point estimator, 164
Poll, 21
Pooled estimator, 242
Population, 11
Population mean, 176
Population proportion, 339
Prediction equation, 289
Prediction interval, 300, 307
  half-width of, 311
  for particular value of dependent variable, 310
Predictor variable, 386
Probability, 123
  additive rule, 126
  multiplicative rule, 127
  relative frequency concept of, 125
  sample, 129
  theory of, 125
Probability distribution, 132
  binomial, 136, 139
  chi-square, 359
  continuous random variable, 134
  discrete random variable, 133
  normal, 148
  standard normal, 150

Process, 99
Proportion
    inferences, 338
    population, 339
    sample, 338–339

Qualitative variable, 9
Quantitative data, 9

Random number table, 130
Random sample, 129
Random variable, 132
    binomial, 136
    continuous, 134
    discrete, 132
    discrete, 132
    normal, 150
Range, 78
Regression analysis, linear, 299
    assumptions, 300
    computer outputs, 302
    estimator of variance, 302, 304
    finding the prediction equation, 302
    testing the prediction equation, 307
Regression analysis, multiple, 299
    computer outputs, 322
    general linear model, 321
Rejection region, 168
Relative frequency distribution, 34
    population, 39
    sample, 39
    skewness of, 105
Response variable, 386, 406
Risk, 170

Sample, 11–12
Sample mean, 75
Sample proportion, 339
    mean, 339
    standard error, 339
Sample size, 129
    for approximately normal distribution of sample mean, 159
    for estimating difference between two population means, dependent random samples, 271
    for estimating difference between two population means, matched-pairs, 271
    for estimating population mean, 219
Sample size (*continued*)
    for estimating population proportion, 354
Sample survey, 21
    design, 230
Sampling distribution, 156
    of the chi-square statistic, 359
    of the difference between two sample means, 235
    of the difference between two sample proportions, 348
    of the sample mean, 157, 176
    of the sample proportion, 339
    standard deviation, 156
    Student's *t*, 184
    theorems, 157
Sampling error, 117
Sampling unit, 8
Scattergram, 281
Second-order linear model, 321
Simple lienar regression analysis (*see* Regression analysis, linear)
Single replication, 422
Slope, 285
Slope–intercept form, 286
Spearman rank correlation coefficient, 455
Standard deviation, 79, 117
    approximating, 96
    calculating, 84
    population, 117
    sample, 79, 117
Standard error (SE), 156
Standard normal distribution, 150
Statistic, 116
Statistical control, 101
Statistical inference, 17
Statistical test, 167–168
Stem-and-leaf display, 64
Student's *t* statistic (*see* *t* statistic)
Studentized range, 403
Sum of squares
    of deviations ($SS_{\hat{y}}$), 295
        additivity of, 389
    for error (SSE), 289, 302, 389
    of residuals, 302
    for treatments (SST), 389

t distribution, 184
    comparison with standard normal z distribution, 185
t statistic, 184, 241, 246, 264
    degrees of freedom, 184, 241, 304
Test statistic, 168
Tests of hypotheses, 168, 205
    dependence between two qualitative variables, 360
    difference between two population means, large independent samples, 245
    difference between two population means, large paired samples, 264
    difference between two population means, small independent samples, 246
    difference between two population means, paired samples, 264
    difference between two population proportion, 350
    independent random samples, Kruskal–Wallis *H* test for more than two populations, 449
    independent random samples design, 397
    independent random samples design, Mann–Whitney *U* test, 431
    matched-pairs design, Wilcoxon signed ranks test, 443
    population mean, large sample, 208
    population mean, small sample, 209
    population proportion, 344
    slope of a straight line, 306
    Spearman's rank correlation test, 457
    treatment means, 416
Theory of probability, 125
Third-order linear model, 321
Treatment, 231
Tukey's HSD procedure, 402
Two-tailed statistical test, 207
Type I error, 170
Type II error, 170

Unbiased estimator, 164
Univariate data, 15
Universe, 11
Upper confidence limit, 166
Upper control limit, 102
Upper quartile (Q_3), 104

Variable, 8
- continuous random, 134
- discrete random, 132
- normal random, 150
- predictor, 386
- qualitative, 9
- quantitative, 9
- random, 132
- response, 386

Variance, 78–79, 117
- calculating, 84
- population, 117, 242
- sample, 117

Wilcoxon rank sum test, 430
Wilcoxon signed ranks test, 441
- matched-pairs design, 443

y-intercept, 284

z statistic, 149–150, 177, 236, 245, 264
- limitations, 183